AF788994

IET RADAR SERIES 34

Radar Micro-Doppler Signatures

Other volumes in this series:

Volume 1	**Optimised radar processors** A. Farina (Editor)
Volume 3	**Weibull radar clutter** M. Sekine and Y. Mao
Volume 4	**Advanced radar techniques and systems** G. Galati (Editor)
Volume 7	**Ultra-wideband radar measurements: analysis and processing** L. Yu. Astanin and A.A. Kostylev
Volume 8	**Aviation weather surveillance systems: advanced radar and surface sensors for flight safety and air traffic management** P.R. Mahapatra
Volume 10	**Radar techniques using array antennas** W. Wirth
Volume 11	**Air and spaceborne radar systems: an introduction** P. Lacomme (Editor)
Volume 13	**Introduction to RF stealth** D. Lynch
Volume 14	**Applications of space-time adaptive processing** R. Klemm (Editor)
Volume 15	**Ground penetrating radar, 2nd edition** D. Daniels
Volume 16	**Target detection by marine radar** J. Briggs
Volume 17	**Strapdown inertial navigation technology, 2nd edition** D. Titterton and J. Weston
Volume 18	**Introduction to radar target recognition** P. Tait
Volume 19	**Radar imaging and holography** A. Pasmurov and S. Zinovjev
Volume 20	**Sea clutter: scattering, the K distribution and radar performance** K. Ward, R. Tough and S. Watts
Volume 21	**Principles of space-time adaptive processing, 3rd edition** R. Klemm
Volume 22	**Waveform design and diversity for advanced radar systems** F. Gini, A. De Maio and L.K. Patton
Volume 23	**Tracking filter engineering: the Gauss-Newton and polynomial filters** N. Morrison
Volume 25	**Sea clutter: scattering, the K distribution and radar performance, 2nd edition** K. Ward, R. Tough and S. Watts
Volume 26	**Radar techniques using array antennas, 2nd edition** W. Wirth
Volume 33	**Radar automatic target recognition (ATR) and non-cooperative target recognition** D. Blacknell and H. Griffiths (Editors)
Volume 101	**Introduction to airborne radar, 2nd edition** G.W. Stimson

Radar Micro-Doppler Signatures

Processing and Applications

Edited by
Victor C. Chen, David Tahmoush
and William J. Miceli

The Institution of Engineering and Technology

Published by The Institution of Engineering and Technology, London, United Kingdom

The Institution of Engineering and Technology is registered as a Charity in England & Wales (no. 211014) and Scotland (no. SC038698).

© 2014 The Institution of Engineering and Technology

First published 2014

This publication is copyright under the Berne Convention and the Universal Copyright Convention. All rights reserved. Apart from any fair dealing for the purposes of research or private study, or criticism or review, as permitted under the Copyright, Designs and Patents Act 1988, this publication may be reproduced, stored or transmitted, in any form or by any means, only with the prior permission in writing of the publishers, or in the case of reprographic reproduction in accordance with the terms of licences issued by the Copyright Licensing Agency. Enquiries concerning reproduction outside those terms should be sent to the publisher at the undermentioned address:

The Institution of Engineering and Technology
Michael Faraday House
Six Hills Way, Stevenage
Herts, SG1 2AY, United Kingdom

www.theiet.org

While the author and publisher believe that the information and guidance given in this work are correct, all parties must rely upon their own skill and judgement when making use of them. Neither the author nor publisher assumes any liability to anyone for any loss or damage caused by any error or omission in the work, whether such an error or omission is the result of negligence or any other cause. Any and all such liability is disclaimed.

The moral rights of the author to be identified as author of this work have been asserted by him in accordance with the Copyright, Designs and Patents Act 1988.

British Library Cataloguing in Publication Data
A catalogue record for this product is available from the British Library

ISBN 978-1-84919-716-8 (hardback)
ISBN 978-1-84919-717-5 (PDF)

Typeset in India by MPS Limited
Printed in the UK by CPI Group (UK) Ltd, Croydon

Contents

Preface

The radar micro-Doppler signature is a distinctive characteristic of a target that provides evidence for the identity of the target with movement. Thus, the micro-Doppler signature has been widely applied in military and civilian uses, such as detection, tracking, and discrimination of dismounts, vehicles, or other targets of interest, identifying human movements, finding human subjects behind walls, detecting trapped victims, and many others.

Since 2011, we have organized a number of tutorials and workshops pertaining to the potential benefits of micro-Doppler phenomenology and processing. We have noted that participants tend to ask how to adapt general micro-Doppler signal analysis and processing to specific applications. This book will respond to such enquiries by selecting specific applications and dedicating a chapter contributed by internationally recognized authors. The book concentrates on the processing and application of radar micro-Doppler signatures in real-world situations, providing readers with a working knowledge on various applications of radar micro-Doppler signatures. We believe topics included will provide students, academics, and research professionals with a polyvalent reference book.

In the book, we first review the current progress, challenges, and perspectives on radar micro-Doppler research and, then, introduce the phenomenology of radar micro-Doppler signatures. The following chapters include progresses in combined micro-range and micro-Doppler signature analysis, human movement analysis using a super-high resolution radar, micro-Doppler signatures in through-wall radar, identification of human movements, micro-Doppler signature of helicopter rotor blades, micro-Doppler signature of small boats in sea clutter, bistatic/multistatic micro-Doppler signatures, decomposition of micro-Doppler signatures, micro-Doppler signatures in sonar, and micro-Doppler signatures of wind turbines.

We also provide supplementary material (MATLAB source codes, radar data, photos, and videos) at the URL *http://www.theiet.org/resources/books/rsna/rsdspa.cfm*. Source codes in Chapter 1 can be used to generate simulated radar signals and their micro-Doppler signatures for walking humans, helicopter rotor blades, and heavy top precession introduced in the chapter. Source codes in Chapter 2 are used to test some algorithms provided in this chapter. Source codes in Chapter 4 are used for micro-Doppler analysis of real radar data. These source codes are given on an as-is basis, and no warranties are claimed. The contributors of the source codes will not be held liable for any damage caused.

We would like to express our sincere thanks to contributors of chapters. We especially thank Dr. Luc Vignaud of the ONERA, French Aerospace Laboratory for kindly providing his super-high resolution X-band HYCAM radar data, images, and videos as described in Chapter 4. We are also grateful to the staff of IET Publishing for their interest and support in the publication of the book.

Chapter 1

Micro-Doppler Signatures – Review, Challenges, and Perspectives

Victor C. Chen, David Tahmoush, and William Miceli

1.1 Introduction

The micro-Doppler effect was originally introduced in coherent laser radar systems. Laser radar transmits light wave to an object, and receives the reflected wave to measure the object's range, velocity, and other properties. Coherent systems can preserve phase information of reflected light waves and can sense phase changes. As a consequence, in laser radar systems, even tiny vibration or rotation in an object can cause significant phase change and an observable Doppler frequency shift.

The micro-Doppler effect can also be observed in microwave radar systems. Compared to laser systems, microwave radar systems operate at much lower frequencies. However, if the object's velocity is high enough, the Doppler effect is still observable in microwave radar systems.

In many cases, an object or any structural component of the object may have oscillatory motion, called the micro motion. Micro motion is a broader usage of the "micro." In addition to the bulk translation of an object, any secondary motion of the object or its structural component can be called micro motion. The source of micro motion can be a rotating propeller of a fixed-wing aircraft, rotating rotor blades of a helicopter, a rotating antenna, flapping wings of birds, the swinging arms and legs of a walking person, or other causes. Human motion is an important topic in the study of micro-Doppler. Due to the high articulation and flexibility of the human body, human motion is a complex and interesting micro motion.

Micro motion can induce Doppler frequency modulation around the carrier frequency of the transmitted radar signals. If the micro motion is a pure periodic vibration or rotation, it generates side-band Doppler frequency shifts about the carrier frequency. The Doppler modulation frequencies and intensities are determined by the carrier frequency, the vibration or rotation rate, and the angle between the direction of vibration and the radar line of sight (LOS).

The frequency modulation, i.e., the micro-Doppler frequencies, enables us to determine the kinematic properties of the object of interest. How to represent the frequency modulation, how to extract the kinematic information of the object, and how to use the micro-Doppler effect for classification, recognition, and identification are still active research topics.

The micro-Doppler signature is a distinctive characteristic of the observed micro-Doppler effect in an object. The "signature" is commonly used to refer to the characteristic expression of an object or a process. When examining the micro-Doppler effect, the distinctive micro-Doppler characteristic, i.e., the micro-Doppler signature of an object, allows the recognition of an object's identity through its movement.

To extract micro-Doppler signature of an object by radar, a simple continuous wave (CW) radar is good enough. However, wide-band coherent Doppler radars, which use wide frequency bandwidth to gain high range resolution and measure both the range and Doppler information, are more useful and desirable in the radar micro-Doppler signature research.

In this chapter, we review current progress in the extraction, processing, analysis, and applications of radar micro-Doppler signatures, and discuss the challenges and perspectives in radar micro-Doppler signature research.

1.2 Review of Micro-Doppler Effect in Radar

When an object or a structural component in the object has oscillatory motion, such as vibration or rotation, the micro-Doppler signature of the object contains characteristics of the intricate frequency modulations induced by the micro motion. Micro-Doppler signatures are used to determine kinematic properties of the object, extract features of the object's movement, and identify the object of interest.

1.2.1 Micro-Doppler Signatures of Rigid Body Motion

An object can be a rigid body or a nonrigid body. If the distances between distinguishable points in an object remain constant throughout the object's motion, the object is called a rigid body. For rigid body motion, any rotation around a point can be described by the rotation axis and angle. Three rotation angles (φ, θ, ψ), called the Euler angles, represent rotations about three axes, where φ is rotating about the z-axis, θ is rotating about the y-axis, and ψ is rotating about the x-axis. A sequence of sequentially rotating about x-, y-, and z-axis is a commonly used sequence, called the *roll*, *pitch*, and *yaw* sequence. To completely describe a rigid body motion, its position, linear velocity and acceleration, angular velocity and acceleration, and even linear and angular momentum of each point in the rigid body must be defined.

If the description of a rigid body motion is available, the backscattering from the rigid body can be computed by combining radar returned signals from all scattering centers of the rigid body.

We summarize a few interesting micro-Doppler signatures of typical rigid body motion for reference as the following:

1.2.1.1 Rotating Tire

Motion of rotating tires and wheels on a vehicle is an important characteristic of the vehicle. By estimating motion kinematics, it is possible to discriminate and characterize movements and activities of the vehicle.

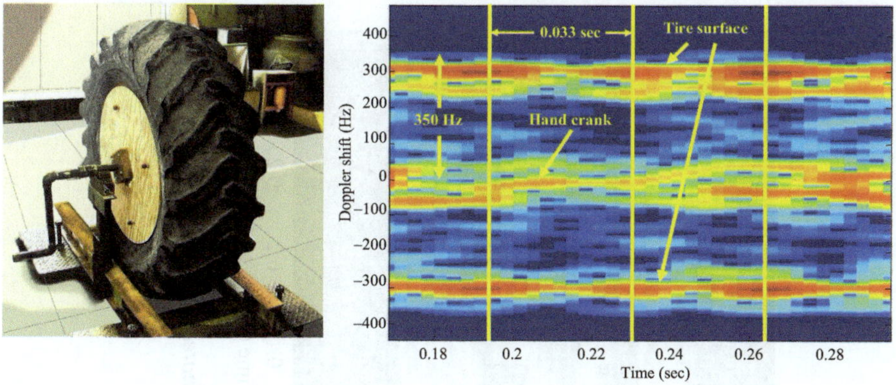

Figure 1.1 Micro-Doppler signature of a rotating tire

A big tire with 20 treads and a diameter of 1.32 m was used for studying the micro-Doppler effect in radar [1]. The tire was rotated by a hand crank at 1.5 turns/sec. The time interval between two successive treads was 0.033 sec and the angular velocity of each tread was 9.4 rad/sec or its tangential velocity was 6.19 m/sec. For the X-band radar used in the data collection, the calculated Doppler frequency shift of each tread was 396 Hz. An interesting example of the micro-Doppler signature of a rotating tire extracted from the real radar data is shown in Figure 1.1, where maximal Doppler frequency shift due to the tread is less than 396 Hz because of the radar LOS angle. The time interval between two strong Doppler returns of two adjacent treads is about 0.033 sec, which is the same as calculated. Also, the micro-Doppler signature of the hand crank is distinguishable in the data.

1.2.1.2 A Circling Vehicle

An SUV circling with a radius of 3.5 m and at a speed of 4 m/sec was used for studying the micro-Doppler effect detected by an X-band radar, as illustrated in Figure 1.2. The coherent integration time of the data is 2.56 sec. Because of the rotating tires and wheels, the range-Doppler image of the vehicle shows micro-Doppler features around the range cells where the tires and wheels are located.

From the vehicle motion parameters provided and the diameter of the tire, we can estimate the micro-Doppler signature pattern of the rotating tire and wheel. During the 2.56 sec coherent integration time, the total distance moved by the vehicle at a speed of 4 m/sec is 10.24 m. Because the diameter of the tire is 0.66 m, the calculated distance traveled by the tire in one revolution is 2.07 m. Therefore, during 2.56 sec time interval, the tire rotates five times. Thus, the rotation rate of the tire becomes 1.93 rev/sec, or 0.52 sec per revolution. Then, the rotating velocity of the surface of the tire becomes 4.06 m/sec, which corresponds to a 260 Hz Doppler frequency shift for this X-band radar. There are also five spokes on the wheel, which are observable. The rotating velocity at the center of the spokes is 1.9 m/sec, so the corresponding Doppler frequency is 121 Hz at X-band.

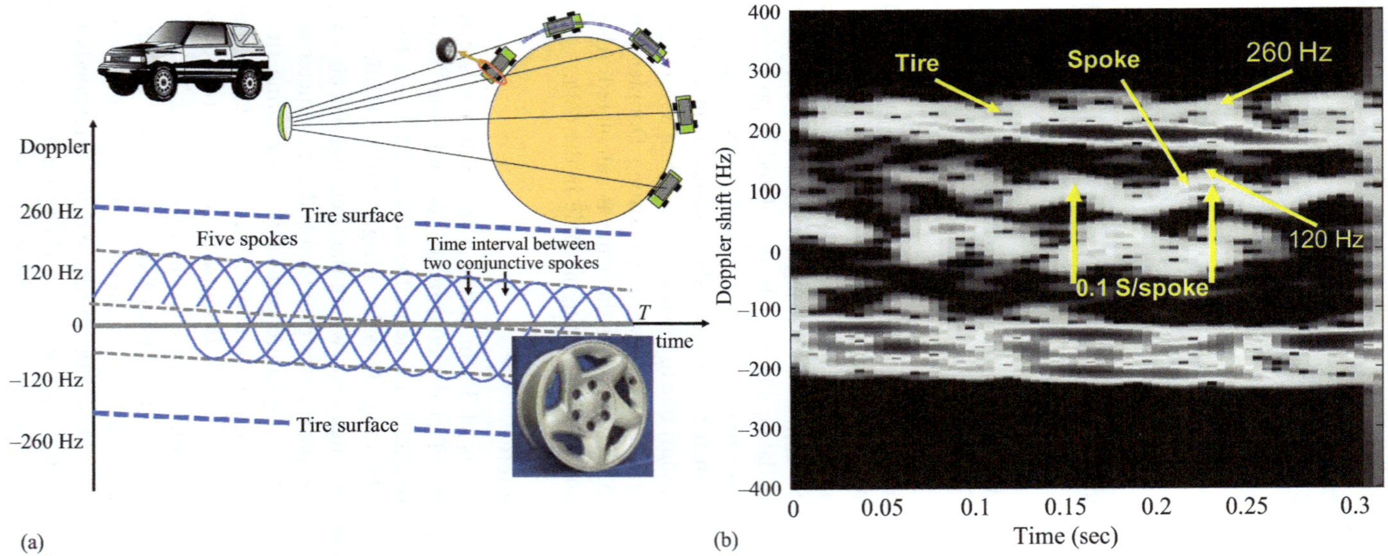

Figure 1.2 Micro-Doppler signature of rotating wheels of a vehicle driving in a circle (a) Estimated and (b) measured micro-Doppler signatures

From the above estimated Doppler shifts of the tire and the wheel, the micro-Doppler signature of the rotating tires and wheels of the circling vehicle is estimated and depicted in Figure 1.2(a) and the micro-Doppler signature of the tire and wheel extracted from the real radar data is shown in Figure 1.2(b), where the micro-Doppler signature of a spoke is also shown. Micro-Doppler features of rotating tires and wheels were also investigated in [2, 3].

1.2.1.3 Helicopter Rotor Blades

Radar scattering from helicopter rotor blades or aircraft propeller blades has been studied in [4, 5], where by modeling the blades or propellers as a set of metal skew-plates, radar scattered signals were modeled and simulated.

To simulate electromagnetic (EM) scattering from rotating blades of a helicopter, for simplicity, the rotor blade can be modeled as a rigid, homogeneous, and linear rod rotating about a fixed axis with a constant rotational rate. Additional complications to the model could be flapping, lagging, and feathering of the blades.

If a rotor with N blades rotates at a constant angular velocity Ω, through an angle $2\pi/N$, the backscattered signals from the rotor will repeat itself from its original position. Thus, the EM scattered field is a periodic function of time with a period of $T_c = \frac{2\pi}{N_c\Omega}$. Micro-Doppler signatures of a rotor with even number of blades and that with odd number of blades are different, as shown in Figure 1.3. From the micro-Doppler signature, the number of blades $\left(N = \frac{2\pi}{T_c\Omega}\right)$, the length of the blade $\left(L = \frac{c}{f}\frac{f_{D\max}}{2\Omega}\right)$, and the rotation rate of the rotor $\left(\Omega = \frac{2\pi}{T_c N}\right)$ can be estimated. For a helicopter, whether the number of blades is even or odd is discernible from whether the blade flashes are balanced in positive and negative Doppler or not. The rotor blade tip velocity can be measured from the maximum Doppler of the blade flash. The blade tip can also be observed, allowing the number of blades to be counted.

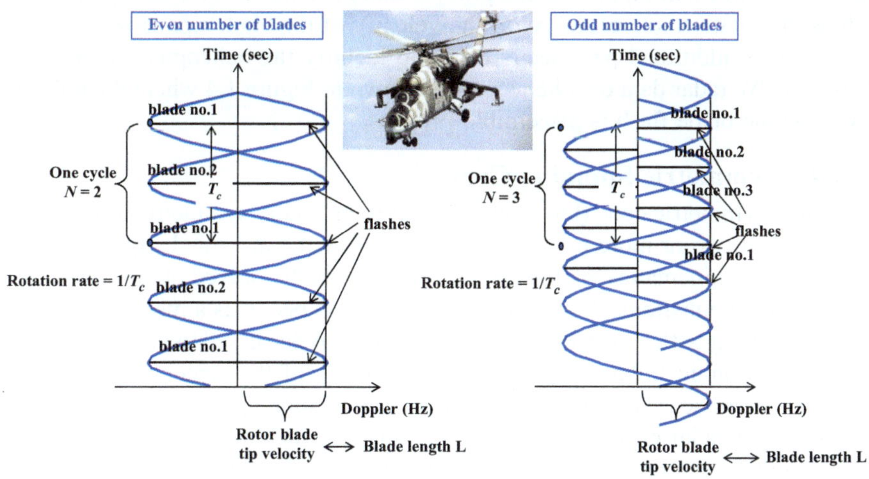

Figure 1.3 Micro-Doppler signature of rotor blades of a helicopter

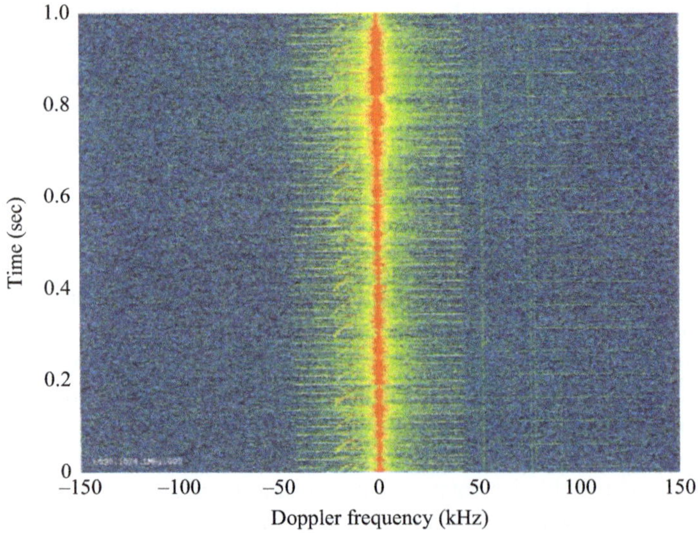

Figure 1.4 Micro-Doppler signature of tail and rotor blades of a helicopter

The rotation rate of the rotor can be estimated based on the observed cycle time of the rotors, and then the blade length can be calculated from the rotor blade tip velocity and rotation rate.

A method called *L*/*N* quotient proposed in [6] has been used as a criteria for helicopter classification, where the blade length *L* and the number of blades *N* are estimated from the spectrum of the scattered signals. However, different types of helicopter can have the same *L*/*N* value, so the *L*/*N* quotient method may not be a unique criterion for helicopter classification. The micro-Doppler signature of the rotor blades shown in Figure 1.3 can provide more detailed rotor blades parameters, such as the rotation rate of the rotor, the velocity of the blade tip, as well as the *L* and *N*. These additional parameters can help identify the helicopter's type [7, 8]. Measured CW radar data on a helicopter is shown in Figure 1.4 where the tail rotor as well as the blade flash is discernible [9].

1.2.1.4 Symmetric Spinning Cone

A spinning symmetric cone (top or gyroscope) is an axially symmetrical rigid body, where at least two or more principal moments of inertia are equal. By using Euler equations, the kinematics of the spinning cone can be solved. If the gravity is included in the equations, the torque-induced motion becomes a more complicated precession along with a nutation. For the spinning top discussed in [10], the angular velocities, dynamic Euler angles, and the trajectory of the center of mass of the spinning top can be easily calculated from the top's Euler equations. Using an exact EM model for radar cross section (RCS) prediction, including geometrical optics, physical optics, the physical theory of diffraction, and method of moments solutions, radar backscattered signals from a spinning symmetric top can be simulated.

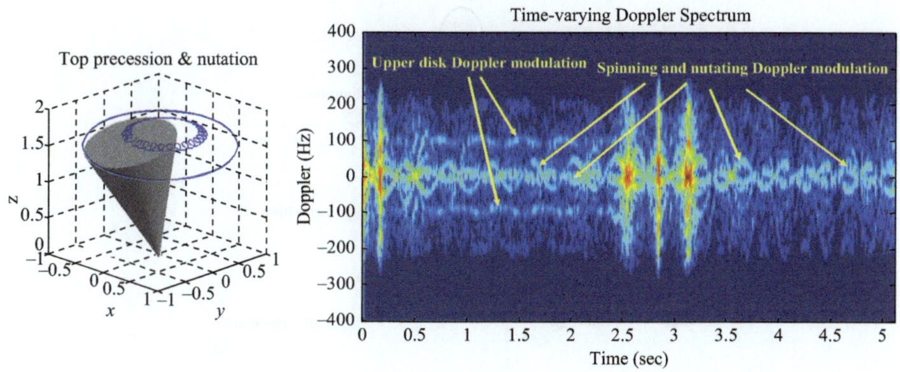

Figure 1.5 Micro-Doppler signature of a symmetric top with precession and nutation

The simulated and measured micro-Doppler signature of the top is shown in Figure 1.5. From the signature, we can clearly see the cycle of Doppler modulation of a top due to the precession [10]. During the entire observation time of 5.3 sec, the 12 Doppler modulation cycles of the spin and nutation, and the Doppler modulation induced by the top's upper disk can be seen in Figure 1.5.

1.2.2 Micro-Doppler Signatures of Nonrigid Body Motion

A human body's motion is a nonrigid and articulated motion. The locomotive motion of limbs in human body can be characterized as periodic motion. Human gait is an example of a complex human activity that can be decomposed into periodic motions in a gait cycle.

A popular human walking model was adapted from the biomechanical research and built from experimental data based on a wide range of normalized velocities [11, 12]. This model provides walking motion with time-dependent body trajectories. The motion is described by 12 trajectories, 3 translational and 14 rotational, 5 of which are duplicated to describe both sides of the body. These translations and rotations describe one cycle of walking motion and are dependent on a chosen walking speed. With these trajectories, the walking model is used to calculate the location of a series of naturally selected 17 reference points on the human body (Figure 1.6): head, neck, base of spine, shoulders, elbows, hands, hips, knees, ankles, and toes.

A more useful motion model is obtained from the motion-capture information. A human motion model with arms and legs movement can be created based on 3-D human motion data captured by CMU Motion Research Laboratory [13]. Another useful method is to extract data from animation routines used in certain 3-D computer programs from 3-D modeling software.

Radar scattering from a moving nonrigid body part can be computed by combining RCS of individual body parts. Similar to the case of rigid body, the parts can be modeled either by isotropic point scatterers with point-scatterer model distributed over the body parts or by a simple shaped part.

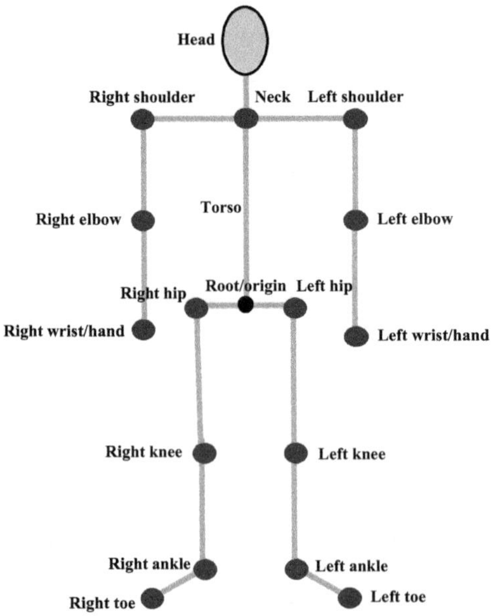

Figure 1.6 Seventeen joint points human model

In calculating radar scattering from human articulated body parts, a pre-defined 3-D model of human body is used, which consists of a number of rigid body parts specified by articulation points. Parts can be modeled with simple geometric shapes, such as cylinders or ellipsoids. With this approach the scattered RCS equation of each part is used to calculate relative amplitude of radar returns. Cylinders and ellipsoids can also be modeled by more accurate triangular facets.

An early study on radar micro-Doppler signature of a walking person was conducted in 1998, which is illustrated in Figure 1.7 [14]. Figure 1.7(a) shows a man walking toward the building at a speed of about 1.8 m/sec. The range profile has 64 range cells, with 1 024 pulses collected at a pulse repetition frequency (PRF) of 800 Hz. Figure 1.7(c) shows a radar range-Doppler image of the walking man using 64 range cells and 128 pulses, where the hot spot in the image indicates the body of the walking man, and smeared lines running across the Doppler direction around the body of the walking man can also be noticed.

The micro-Doppler signature of human gait in Figure 1.7(b) shows the Doppler shift of the body and micro-Doppler shifts of the swinging arms and legs. The Doppler shift of one arm is higher and the other is lower than the body Doppler frequency shift. We can see that the body's Doppler shift is almost constant with a slightly saw-tooth shape but the arm's and leg's micro-Doppler shifts are time-varying periodic curves. From the available time information, the swinging rate of the arm can be estimated to about 1.2 cycle/sec.

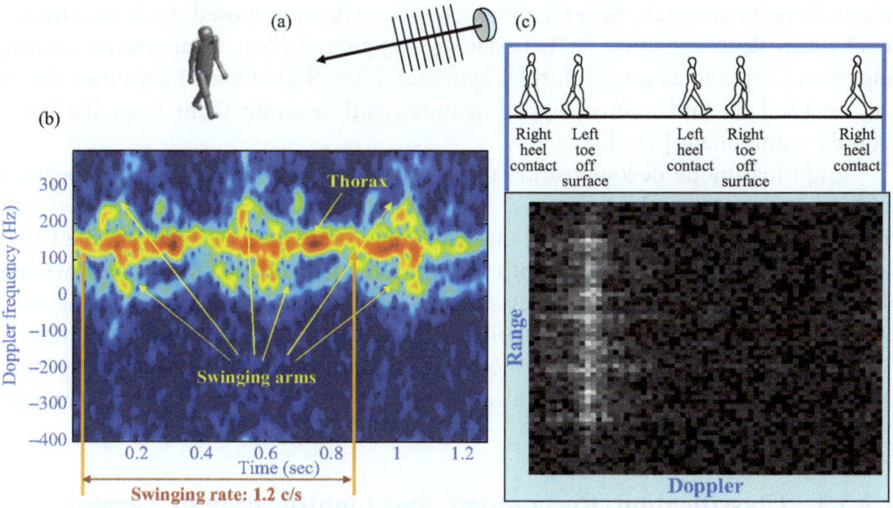

Figure 1.7 *(a) A man walks at a speed of about 1.8 m/sec; (b) time-frequency micro-Doppler signature of the walking man; and (c) radar range-Doppler image of a walking man where the hot spot in the image indicates the body of the walking man*

1.2.3 Review of Current Micro-Doppler Signature Research

In recent years, there has been a lot of progress in the extraction, processing, analysis, and applications of radar micro-Doppler signatures.

1.2.3.1 Estimation of Motion Parameters from Micro-Doppler Signatures

Various methods have been used to extract rigid body motion information from radar micro-Doppler signatures [15–18], especially for the estimation of kinematic parameters from vibration, rotation, or precession. An example of spinning and precession object is the top motion, which is a typical classical rigid body motion and has been studied for a long time [19, 20]. Spinning and precession cone-type object in free space is another example, which attracts many researchers' attention [21–24]. It has been shown that since the inertial parameters of the object are closely related to the state of its micro motion, the inertial ratio of the rigid cone can serve as an important merit index of the object for target discrimination [25, 26].

1.2.3.2 Separation of Micro-Doppler Components from the Main Doppler Component for Improving Imaging

Although micro-Doppler features provide additional information for target classification, recognition, and identification, they also can contaminate the signal when phase is used for imaging. Therefore, the separation of micro-Doppler components from the main Doppler component became another important issue in micro-Doppler research. Several approaches have been proposed, such as complex local mean decomposition [27], L-statistics approach [28], compressive sensing approach [29], and adaptive chirplet approach [30]. Radon and Hough transforms can be used to find micro-Doppler features and separate them from the main Doppler component [31, 32].

Empirical mode decomposition (EMD) is another method that decomposes a multi-component signal into mono-component constituents by a progressive sifting process that yields the bases called intrinsic mode functions (IMFs) [33]. The EMD adaptively decomposes a signal into a limited number of zero-mean, narrowband IMFs. The instantaneous frequency of each IMF is calculated by using the normalized Hilbert transform, called Hilbert–Huang transform (HHT). The combination of the Hilbert spectrum then becomes the complete time-varying frequency spectrum. The EMD method can also be used to separate rotating parts from the main part of an object [34].

1.2.3.3 Classification, Recognition, and Identification of Targets Based on Micro-Doppler Signatures

Target classification, recognition, and identification are extremely important research topics in radar systems. Classification resolves the class of a target, e.g., a ground vehicle or an aircraft. Recognition determines whether the ground vehicle is a trunk, a school bus, or a tank. Identification answers the type or model of the target, e.g., a T72 tank, or an M60 tank. Commonly used classification methods include Bayesian net, principal component analysis (PCA), or support vector machine (SVM). Features are characteristics extracted from a pattern, which can be a distinctive measurement, a transformation, or a structural component.

Micro-Doppler signatures can be seen as a statistical pattern. Thus, how to extract useful features from micro-Doppler signatures is a key issue in classification, recognition, and identification. There are many research efforts on micro-Doppler signature based target classification, recognition, and identification, especially for vehicle, human, and animal motion [35–37].

Li *et al.* [38–40] proposed approaches for classification of vehicles based on their micro-Doppler signatures. A Naïve Bayes nearest neighbor approach for classification of ground movers was proposed in [41].

1.2.3.4 Helicopter Identification from Micro-Doppler Signatures of Rotor Blades

Helicopter identification has been an attractive topic for radar researches. To identify a helicopter type, besides its shape and size, the number of blades, the length of the blade, and the rotation rate of the rotor are important features. These parameters can be estimated from the micro-Doppler signature of the helicopter.

Micro-Doppler signatures of helicopter rotor blades extracted by monostatic, bistatic, and multistatic radars have been studied [42–46]. They can be used to either estimate rotating parameters of the blades for identification or remove their effect on radar returned signals for imaging.

Other than studies on micro-Doppler signatures of conventional helicopter rotor blades, a study on multicopters, small unmanned helicopters, and small unmanned aerial vehicles (UAVs) has also been reported [47]. Especially, distinctive micro-Doppler features of different types of rotary-wing micro and mini UAVs are investigated and compared with theoretical models.

Another interesting study on radar micro-Doppler signatures is related to the use of passive radars. The possibility of extracting micro-Doppler signatures of helicopter rotor blades from a passive bistatic radar system with a GNSS transmitter was demonstrated [48].

1.2.3.5 Micro-Doppler Signatures of Human Motion

The extraction and analysis of various human body movements is motivated by visual surveillance, athletic performance analysis, and biometrics. Human gait has been studied for a long time in biomedical engineering, sports medicine, physiotherapy, medical diagnosis, and rehabilitation. Observation of human gait micro-Doppler signatures using radar was done in late 1990s [14].

Because the RCS of human body is small (about 0.5 m^2), radar returns from humans are relatively weak. Thus radars used for human motion study must have a relatively high power transmitter or should work at relatively short distances. In the real world, radar detection of humans is usually performed in a complex background clutter environment. Especially when humans move relatively slowly, the intensity of radar returns from clutter exceeds that from humans. How to detect weak human motion signals in clutter became an important research topic in human motion analysis.

Higher frequency can generate larger Doppler shifts even if the motion is slow. Thus, the use of higher frequency band, such as K-band and W-band, for micro-Doppler signatures of human motion study was investigated. In [49], a 77 GHz radar was used to observe micro-Doppler signatures of human gait to recognize multiple people.

In [50], an ultra-wide band (UWB) impulse radar was used for human gait research. The UWB impulse radar provides both high resolution range profiles and high resolution Doppler spectrogram, which helps to extract detailed micro-Doppler signatures like swinging arms. The unique characteristics of the very detailed signatures are used to recognize human activities, such as marching, walking, one-arm swinging, or two-arm swinging.

In [51], a combination of micro-Doppler signatures with micro-range features was proposed. The use of micro-range and micro-Doppler signatures can decompose Doppler and range features into human body parts for improving classification, recognition, and identification.

It was also found that through comprehensive measurement issues, radar micro-Doppler features can be enhanced. In [52], a measurement approach was proposed to enhance radar micro-Doppler signatures.

Another issue in human motion study is how to estimate the human gait parameters. Since the optimal estimation method uses nonlinear and non-Gaussian state-space models, it is not easy to derive an analytic solution. In [53] a particle filter method was proposed for tracking micro-Doppler features of human gait. The particle filtering method can numerically solve the estimation problems, and it is now routinely used in computer vision, robotics, and navigation.

Human gait studies are now more interested in multiple people and more complex human movements. How to distinguish multiple persons and different activities is still an open issue. An X-band radar imaging system with very high range resolution was used to study two persons cross-walking and cross-running, and a cyclist [54–56]. It was demonstrated that the high resolution ISAR images of moving objects combined with radar micro-Doppler signatures can improve features of object's movement. In [57], an efficient classifier was proposed for distinguishing micro-Doppler signatures of pedestrians, skaters, and cyclists.

The decomposition of human micro-Doppler signatures into human body parts may lead to the identification of human actions, emotions, and even the gender through micro-Doppler signatures. A micro-Doppler system and an efficient classifier were demonstrated to identify different individuals and even gender [58].

Interferometric radar for the detection and measurement of the signatures of walking human may have advantages in the case of slow moving objects. A 30 GHz interferometric radar was proposed with simulation to isolate human arm's and leg's micro-Doppler signatures for improving classification of slow moving humans [59].

Multistatic micro-Doppler signature formed from multi-angle observations in a radar network can improve the performance of oblique-angle classification. This issue was explored in [60] by using mutual information to find the degree of importance of features for the target classification issue [61].

Ultrasound can also be applied to extract micro-Doppler signatures of moving objects. [62] explored the utility of ultrasonic sensors to distinguish between people and animals.

1.2.3.6 Micro-Doppler Signatures of Quadrupedal Animal Motion

Quadrupedal animal motion can be easily distinguished from bipedal human motion in video, and the micro-Doppler signatures are quite different from that of a biped. In [63], radar micro-Doppler signatures of a walking horse carrying a rider and a trotting horse carrying a rider were analyzed and compared with human motion. In [64], X-band multiple frequency CW radar was built for studying micro-Doppler signatures including a dog. Quadrupedal animal motion can also be studied by computer animation approach as described in [65].

1.2.3.7 Micro-Doppler Signatures of Flying Bird Wing Flapping

Bird locomotion is a flapping (elevation and depression of its upper arm), twisting (rotation of the wing), and sweeping (forward or backward retraction of its shoulder) motion [20]. Based on aerodynamics, a kinematic model of bird wing locomotion was proposed in [66].

Radar micro-Doppler signatures of flying birds are generated by their locomotion, especially their elevating and depressing upper arms. Almost half-century ago, the Doppler spread of the bird wing flapping was utilized for identification of flying birds [67–69]. The time-varying Doppler spectrum of flying birds clearly indicates the flapping wings. For a flying Canada goose, the Doppler shifts from its wings can be significantly higher than that from its body. The bandwidth of the time-varying Doppler spectrum is related to the size of the bird, which could be used to discriminate birds.

In [70], different flapping styles of birds (passerine-like flapping style and swift-like flapping style) were observed. Their quite different patterns of time-varying Doppler spectrum, i.e., micro-Doppler signatures, are shown. The passerine-like flapping style has repeated clusters of larger fluctuations.

Some recent studies on micro-Doppler signatures of flying birds are also available [71–74]. In [71], a 24 GHz CW radar was used to detect bird and bats. Doppler signatures are used to discriminate different flying objects. The process of distinguishing flying birds from UAVs was also investigated [72]. A method that combined the micro-Doppler signatures with range-Doppler images for classification of single bird and bird flock was proposed in [74].

1.2.3.8 Through-the-Wall Micro-Doppler Signatures

Because of the relatively lower frequency used in through-wall radars, micro-Doppler frequency shift can be very low. This makes difficult detection of objects behind the wall. However, with advanced signal decomposition and processing techniques, radar can still sense human body motion, heartbeat, and even breathing, which can be used for detecting and monitoring human activities after an earthquake or an explosion. The use of radar micro-Doppler signatures to detect and identify the targets behind walls was proposed in [75, 76].

The impact of viewing micro-Doppler signatures through a wall was studied [77]. The presence of a wall does not change the pattern of the micro-Doppler signature. The wall only reflects or attenuates the radar signals, depending on the wall properties. In [78], radar micro-Doppler signatures of moving targets in urban sensing and through-the-wall are investigated, especially on the Cramer–Rao bound of micro-Doppler estimation. In [79], the wall effect on micro-Doppler shifts and through-the-wall recognition from radar micro-Doppler signatures were studied in detail. In [80], ultrahigh frequency (UHF) noise radar was used for through-the-wall imaging study, and a combined micro-Doppler signatures and EMD analysis was proposed for through-the-wall human detection.

1.2.3.9 Micro-Doppler Signatures of Vital Signs

Observing, measuring, and monitoring vital signs, such as heartbeat rate, pulse rate, or breathing rate, are very important for health care and for finding victims buried under earthquake rubble. Doppler radar returns modulated by vital signs can be used for these purposes. Specific modulations in reflected radar signals from humans, including heartbeats, thorax motion by breathing, and even vibration in larynx, can be detected by radar [81–89].

Micro-Doppler signatures of vital signs have been used to detect and estimate these vital signs. In [90], a 24 GHz CW radar was used for noninvasive estimation of respiratory and heartbeat rate from micro-Doppler features of the radar returned signals. In [91], UWB stepped frequency CW radar with MTI (moving target indicator) function was used to detect heartbeat from micro-Doppler characteristics of human body and observe Doppler signatures of multiple targets at different ranges. A millimeter-wave radar captured the micro-Doppler signatures of human respiration and heartbeat rates at distances up to 100 m [92].

In [93], a Doppler radar was used to recognize human breathing and torso bending from micro-Doppler signatures and with EMD analysis.

1.2.3.10 Micro-Doppler Signatures in Ultrasound Radar

Micro-Doppler signatures can also be observed in ultrasound frequency band because of much lower wave propagation speed of sound than light [62, 94–97]. An ultrasound radar operating at 80 kHz was built to collect data from human gait with various actions [94]. Classification of personnel targets was performed based on micro-Doppler signatures. In [95], a frequency-agile noncoherent ultrasound radar was also developed for the purpose of collecting micro-Doppler signatures of various human movements. A simple ultrasound radar was also built for acquiring gait signatures of humans and four-legged animals in indoor and outdoor environments [96]. Ultrasound micro-Doppler signatures were proposed to learn human activities using nearest neighbor classification [97].

1.2.3.11 Embedded Software and FPGAs for Micro-Doppler Signature Studies

Radar micro-Doppler signature studies include everything from signal processing algorithms to radar hardware and system integration. In [98], a UWB radar sensor for human health monitoring was demonstrated to localize patients and monitor their gait and vital signs using extracted micro-Doppler signatures with embedded software.

Building software defined radars using field-programmable gate arrays (FPGAs) is another progress in studying radar micro-Doppler signatures. The complex signal processing function in radar systems was usually based on digital signal processors (DSPs). But now signal demodulation and processing functions have been implemented in the FPGA without using DSPs [99–103]. A radar receiver based on the FPGA was proposed in. It processes the received radar signals in both the time and frequency domains, including FIR filter and digital down-converter (DDC) modules. The FPGA currently offers the best performance with the advantages of small size, light weight, and low power.

1.3 Challenges in Radar Micro-Doppler Signature Research

In the last section, we have read that radar micro-Doppler signatures have been applied to extract kinematic features of moving objects and identify the object of interest. However, to correctly interpret the extracted micro-Doppler signatures and establish a mapping between signatures and parts of the object is still a challenge. The success of solving the mapping between signatures and body parts can definitely improve the performance of the classification, recognition, and identification such that it is possible to further identify the intention and behavior from measured motion.

1.3.1 Decomposition of Micro-Doppler Features

To find the mapping between micro-Doppler signatures and body parts of an object, the first step is to decompose micro-Doppler signatures into component features that associate with body parts of the object. For example, for a micro-Doppler signature of human gait, we must decompose it into component signatures that associate with torso, arms, legs, and feet. Thus, the classifier may identify human actions, emotions, and even the gender information.

It is a challenging task to find an effective method for decomposing micro-Doppler signatures into components that can be associated with physical parts of the object. As a dynamic programming, the Viterbi algorithm has been used in the joint time-frequency domain and it has achieved some level of success as demonstrated in [104, 105]. However, how to effectively decompose micro-Doppler signatures into mono-components and how to measure the embedded kinematic/structural information from mono-component signatures is still an open issue.

1.3.2 Detection of Anomalous Human Behavior

As the number of threats grows and diversifies, an anomaly detection system becomes a required element of system security. Knowing what normal behavior usually looks like is the key to detecting possible threats. Anomaly detection is used to identify abnormalities by learning what normal behavior looks like. How to interpret human movements, track these movements, select appropriate features, and detect events of interest is still an open issue.

1.3.3 Feature Extraction and Target Identification Based on Micro-Doppler Signatures

Extraction of useful features for identifying targets of interest from radar micro-Doppler signatures is a challenge. For a simple target, such as a helicopter rotor, the identifying parameters can be extracted from the micro-Doppler signature. However, for a more complicated human gait, the extraction of these parameters for each human body part is challenging.

1.4 Perspectives of Micro-Doppler Signature's Research

Due to the short history of research efforts on radar micro-Doppler signatures, many aspects of research topics are still open and should be further explored. These topics include multistatic micro-Doppler signatures, micro-Doppler signature-based target classification, aural micro-Doppler, through-the-wall micro-Doppler, and polarimetric micro-Doppler.

1.4.1 Multistatic Micro-Doppler Signatures

Multistatic radar is considered a combination of multiple bistatic radars that observes targets from different aspects. Thus, the informative data acquired from targets is increased because of the multiple viewing angles, including the micro-Doppler. The combined micro-Doppler signature depends on the topology of the radar system, and the location and the moving direction of the target [35, 106]. The information encoded in the micro-Doppler signature may not be linearly increased with the number of viewing channels in the system because of the possible correlations between viewing channels. By combining information captured from multiple channels, the location, moving direction, and velocity of the target can be measured. With the increased information, the radar performance on target recognition is expected to be improved.

1.4.2 Micro-Doppler Signature-Based Target Classification

An operational radar system that uses Doppler spectrum for target classification has been investigated in [107, 108]. Humans, vehicles, helicopters, and ships were successfully classified with multiple Fisher linear discriminators on their Doppler spectra. Target classification based on Doppler spectrum in operational radar systems shows the feasibility of using micro-Doppler signatures for target classification.

Other micro-Doppler signature-based methods for target classification have been proposed [35, 36]. For identifying human activities, we must exploit the differences in micro-Doppler signatures to develop a better human activity classifier, because different movements have different micro-Doppler signatures. Features used in human activity classifier can be the torso signature, the maximal Doppler shift of the signature, the offset of the signature from the principal Doppler shift, the maximum Doppler variation of the torso line, the oscillation frequency or period of the motion, the kinematic parameters of limbs, and other available features.

1.4.3 Aural Micro-Doppler Signals for Target Classification

An audio depiction of micro-Doppler embedded signal, called an aural signal, may help a human listener to distinguish between different movements of a target of interest (human walking, running, or jumping), or distinguish between different targets (human and animal).

The advantage of auditory classification systems is that the human auditory classification process is particularly robust in the presence of noise. Aural signal classification is already used in sonar signal classification [109, 110]. The potential application of aural classification to micro-Doppler signatures is also possible by directly converting the baseband micro-Doppler signals into an audio signal for training listeners.

1.4.4 Through-the-Wall Micro-Doppler Signatures

The presence of a wall does not change the pattern of micro-Doppler signature. Therefore, radar micro-Doppler signatures can be used to detect the presence of human beings and their movements behind a wall. Because of the relatively lower frequency band used in through-wall radars, the micro-Doppler shift can be very low and may be buried in clutter. Effective signal processing and feature extraction methods must be further investigated.

1.4.5 Polarimetric Micro-Doppler Analysis

Fully polarimetric radar data can potentially improve the separability of different parts of human motion. One of the techniques used to determine whether someone is carrying something in their arms is looking at their arm motion [111]. The arm is often bent at the elbow, providing a surface similar to a dihedral that creates a double bounce of the radar signal. This is distinct from the more planar surfaces of the body and allows us to separate the signals from the arm (and knee) motion from the rest of the body. The double bounce can be measured in polarimetric radar data by measuring the phase difference between HH and VV [112]. This is one possible way to improve the extraction of relevant features for activity recognition.

References

[1] Chen, V.C., Miceli, W.J., and Himed, B., "Micro-Doppler analysis in ISAR – review and perspectives," *IEEE 2009 International Radar Conference*, Oct. 12–16, Bordeaux, France, 2009.

[2] Ghaleb, A., Vignaud, L., and Nicolas, J.M., "Micro-Doppler analysis of wheels and pedestrians in ISAR imaging," *IET Signal Processing*, vol. 2, no. 3, pp. 301–311, 2008.

[3] Ghaleb, A., *Analysis of Micro-Doppler in Imaging Radar*, PhD Thesis, Telecom Paris/Tech, France, 2009.

[4] Yang, S.Y., Yeh, S.M., Bor, S.S., Huang, S.R., and Hwang, C.C., "Electromagnetic backscattering from aircraft propeller blades," *IEEE Transaction on Magnetics*, vol. 33, no. 2, pp. 1432–1435, 1977.

[5] Martin, J. and Mulgrew, B., "Analysis of the theoretical radar return signal from aircraft propeller blades," *IEEE 1990 International Radar Conference*, pp. 569–572, 1990.

[6] Rotander, C. and von Sydow, H., "Classification of helicopters by the L/N quotient," Radar 97, *IEEE 1997 Radar Conference*, pp. 629–633, 1997.

[7] Cilliers, A. and Nel, W., "Helicopter parameter extraction using joint time-frequency and tomographic techniques," *Proceedings of IEEE 2008 International Radar Conference*, pp. 598–603, 2008.

[8] Yoon, S., Kim, B., and Kim, Y., "Helicopter classification using time-frequency analysis," *Electronics Letters*, vol. 36, no. 22, pp. 1871–1872, 2000.

[9] Wellman R.J. and Silvious, J.L., "Doppler signature measurements of an Mi-24 Hind-D Helicopter at 92 GHz," ARL-TR-1637, Army Research Laboratory, Adelphi, Maryland, 1998.

[10] Chen, V.C., Lin, C.T., and Pala, W.P., "Time-varying Doppler analysis of electromagnetic backscattering from rotating object," *IEEE 2006 Radar Conference Record*, April 24–27, Verona, NY, pp. 807–812, 2006.
[11] Boulic, R., Thalmann, M.N., and Thalmann, D., "A global human walking model with real-time kinematic personification," *The Visual Computer*, vol. 6, no. 6, pp. 344–358, 1990.
[12] van Dorp, P. and Groen, F.C.A., "Human walking estimation with radar," *IEE Proceedings Radar, Sonar and Navigation*, vol. 150, pp. 356–365, 2003.
[13] Website of the Motion Research Laboratory, Carnegie Mellon University: http://mocap.cs.cmu.edu.
[14] Chen, V.C., "Analysis of radar micro-Doppler signature with time-frequency transform," *Proceedings of the IEEE Workshop on Statistical Signal and Array Processing (SSAP)*, Pocono, PA, pp. 463–466, 2000.
[15] Clemente, C. and Soraghan, J.J., "Vibrating target micro-Doppler signature in bistatic SAR with a fixed receiver," *IEEE Transaction on GRS*, vol. 50, no. 8, pp. 3219–3227, 2012.
[16] Zhang, W., Tong, C., Zhang, Q., Zhang, Y., and Zhang, X., "Extraction of vibrating features with dual-channel fixed-receiver bistatic SAR," *IEEE Geoscience and Remote Sensing Letters*, vol. 9, no. 3, pp. 507–511, 2012.
[17] Chen, G., Yu, H., and Yang, X., "Micro-Doppler analysis and parameter estimation of the rotating linear rigid target," *Proceedings of IEEE International Conference on Computer Science and Automation Engineering*, vol. 1, pp. 94–97, 2012.
[18] Wang, Q., Xing, M., Lu, G., and Bao, Z., "High-resolution three-dimensional radar imaging for rapidly spinning targets," *IEEE Transaction on GRS*, vol. 46, no. 1, pp. 22–30, 2008.
[19] Goldstein, H., *Classical Mechanics*, 2nd ed., Reading, MA: Addison-Wesley, 1980.
[20] Chen, V.C., *The Micro-Doppler Effect in Radar*, Artech House, Boston/London, 2011.
[21] Lei, P., Sun, J., Wang, J., and Hong, W., "Micromotion parameter estimation of free rigid targets based on radar micro-Doppler," *IEEE Transaction on GRS*, vol. 50, no. 10, pp. 3776–3786, 2012.
[22] Yan, H., Fu, X., Lei, X., Li, P., and Gao, M., "Parametric estimation of micro-Doppler on spatial precession cone," 2011 *IEEE CIE International Conference on Radar*, vol. 1, pp. 613–616, 2011.
[23] Gao, H., Xie, L., Wen, S., and Kuang, Y., "Micro-Doppler signature extraction from ballistic target with micro-motions," IEEE Transaction on AES, vol. 46, no. 4, pp. 1968–1982, 2010.
[24] Zhao, Y., Shen, X., and Liang, J., "A novel micro-motion model for ballistic missile targets identification," 2011 *IEEE CIE International Conference on Radar*, vol. 1, pp. 604–608, 2011.
[25] Wang, T., Wang, X., Chang, Y., Liu, J., Xiao, S., "Estimation of precession parameters and generation of ISAR images of ballistic missile targets," *IEEE Transaction on AES*, vol. 46, no. 4, pp. 1983–1995, 2010.

[26] Lei, P., Wang, J., and Sun, J., "Analysis of radar micro-Doppler signatures from rigid targets in space based on inertial parameters," *IET Radar, Sonar, Navigation*, vol. 5, no. 2, pp. 93–102, 2011.

[27] Yuan, B., Chen, Z., and Xu, S., "Micro-Doppler analysis and separation based on complex local mean decomposition for aircraft with fast-rotating parts in ISAR imaging," *IEEE Transactions on Geoscience and Remote Sensing*, vol. PP, no. 99, p. 1, 2013. (IEEE Early Access Article)

[28] Stankovic, L., Thayaparan, T., Dakovic, M., and Popovic-Bugarin, V., "Micro-Doppler removal in the radar imaging analysis," *IEEE Transaction on AES*, vol. 49, no. 2, pp. 1234–1250, 2013.

[29] Stankovic, L., Orovic, I. Stankovic, S., and Amin, M., "Compressive sensing based separation of nonstationary and stationary signals overlapping in time-frequency," *IEEE Transactions on Signal Processing*, vol. 61, no. 18, pp. 4562–4572, 2013.

[30] Li, J. and Ling, H., "Application of adaptive chirplet representation for ISAR feature extraction from targets with rotating parts," *Proceedings of IEE Radar, Sonar, Navigation*, vol. 150, no. 4, pp. 284–291, 2003.

[31] Stankovic, L., Djurovic, I., and Thayaparan, T., "Separation of target rigid body and micro-Doppler effects in ISAR imaging," *IEEE Transaction on AES*, vol. 42, no. 4, pp. 1496–1506, 2006.

[32] Zhang, Q., Yeo, T.S., Tan, H.S., and Luo, Y., "Imaging of a moving target with rotating parts based on the Hough transform," *IEEE Transaction on GRS*, vol. 46, no. 1, pp. 291–299, 2008.

[33] Huang, N.E., *et al.*, "The empirical mode decomposition and the Hilbert spectrum for nonlinear and non-stationary time series analysis," *Proceedings Royal Society London, Series A*, vol. 454, pp. 903–995, 1998.

[34] Bai, X., Xing, M., Zhou, F., Lu, G., and Bao, Z., "Imaging of micromotion targets with rotating parts based on empirical mode decomposition," *IEEE Transaction on GRS*, vol. 46, no. 11, pp. 3514–3523, 2008.

[35] Smith, G.E., *Radar Target Micro-Doppler Signature Classification*, PhD Dissertation, Department of Electronic and Electrical Engineering, University College London, 2008.

[36] Anderson, M.G., *Design of multiple frequency continuous wave radar hardware and micro-Doppler based detection and classification algorithms*, PhD Dissertation, University of Texas at Austin, 2008.

[37] Bjoklund, S., Johansson, T., and Petersson, H., "Evaluation of a micro-Doppler classification method on mm-wave data," *Proceedings of 2012 IEEE Radar Conference*, pp. 934–939, 2012.

[38] Li, Y., Du, L., and Liu, H., "Moving vehicle classification based on micro-Doppler signature," *Proceedings of 2011 IEEE International Conference on Signal Processing, Communications and Computing*, pp. 1–4, 2011.

[39] Li, Y., Du, L., and Liu, H., "Noise robust classification of moving vehicles via micro-Doppler signatures," 2013 *IEEE Radar Conference*, Ottawa, Canada, May, 2013.

[40] Li, Y., Du, L., and Liu, H., "Hierarchical classification of moving vehicle based on empirical model decomposition of micro-Doppler signatures," *IEEE Transaction on Geoscience and Remote Sensing*, vol. 51, no. 5, pp. 3001–3013, 2013.

[41] Bar-Hillel, A., Bilik, I., and Hecht, R., "Naive Bayes nearest neighbor classification of ground moving targets," 2013 *IEEE Radar Conference*, Ottawa, Canada, 2013.

[42] Johnsen, T., Olsen, K.E., and Gundersen, R., "Hovering helicopter measured by bi-/multistatic CW radar," *Proceedings of IEEE 2003 Radar Conference*, pp. 165–170, May 2003.

[43] Olsen, K.E., Johnsen, T., Johnsrud, S., Tansem, I., and Sornes, P., "Multistatic and /or quasi monostatic radar measurements of propeller aircrafts," *Proceedings of 2007 IET International Radar Conference*, pp.1–6, Oct., 2007.

[44] Thayaparan, T., Abrol, S., Riseborough, E., Stankovic, L., Lamothe, D., and Duff, G., "Analysis of radar micro-Doppler signatures from experimental helicopter and human data," *IET Radar, Sonar, Navigation*, vol. 1, no. 4, pp. 289–299, 2007.

[45] Cilliers, A. and Nel, W., "Helicopter parameter extraction using joint time-frequency and tomographic techniques," *IEEE 2008 International Conference on Radar*, pp. 2–5, Adelaide, Australia, 2008.

[46] Molchanov, P., Totsky, A., Astola, J., Egiazarian, K., Leshchenko, S., and Rosa-Zurera, M., "Aerial target classification by micro-Doppler signatures and bicoherence-based features," *Proceedings of the 9th European Radar Conference*, pp. 214–217, Oct. 31–Nov. 2, Amsterdam, The Netherlands, 2012.

[47] de Witt, J.J.M., Harmanny, R.I.A., and Premel-Cabic, G., "Micro-Doppler analysis of small UAVs," *Proceedings of the 9th European Radar Conference*, pp. 210–213, Oct. 31–Nov. 2, Amsterdam, The Netherlands, 2012.

[48] Clemente, C. and Soraghan, J.J., "Passive bistatic radar for helicopters classification: a feasibility study," *Proceedings of 2012 IEEE Radar Conference*, pp. 946–949, 2012.

[49] Bjorklund, S., Petersson, H., Nezirovic, A., Guldogan, M.B., and Gustafsson, F., "Millimeter-wave radar micro-Doppler signatures of human motion," *Proceedings of 2011 International Radar Symposium* (IRS), pp. 167–174, 2011.

[50] Wang, Y. and Fathy, A.E., "Micro-Doppler signatures for intelligent human gait recognition using a UWB impulse radar," *IEEE 2011 International Symposium on Antennas and Propagation* (APSURSI), pp. 2103–2106, 2011.

[51] Fogle, O.R. and Rigling, B.D., "Micro-range/micro-Doppler decomposition of human radar signatures," *IEEE Transaction on AES*, vol. 48, no. 4, pp. 3058–3072, 2012.

[52] Sona, A., Giorgi, G., and Ricci, R., "A measurement approach based on micro-Doppler map for human motion analysis and detection," *IEEE 2012 International Instrumentation and Measurement Technology Conference*, pp. 354–359, 2012.

[53] Gulgogan, M.B., Gustafsson, F., Orguner, U., Bjorklund, S., Petersson, H., and Nezirovic, A., "Human gait parameter estimation based on micro-Doppler signatures using particle filters," *IEEE 2011 International Conference on Acoustics, Speech and Signal Processing* (ICASSP), pp. 5940–5943, 2011.

[54] Ghaleb, A., Vignaud, L., Nicolas, J.M., "Micro-Doppler analysis of wheels and pedestrians in ISAR imaging," *IET Signal Processing*, vol. 2, no. 3, pp. 301–311, 2008.

[55] Vignaud, L., Ghaleb, A., Le Kernec, J., and Nicolas, J.M., "Radar high resolution range & micro-Doppler analysis of human motions," *Proceedings of IEEE 2009 International Radar Conference*, pp. 1–6, Oct. 12–16, 2009.

[56] Ghaleb, A., Vignaud, L., Nicolas, J.M., "A refine micro-Doppler analysis of pedestrians in ISAR imaging," *Proceedings of 2008 7th European Conference on Synthetic Aperture Radar* (EUSAR), pp. 1–4, June 2–5, 2008.

[57] Garreau, G., Nicolaou, N., Andreou, C., D'Urbal, C., Stuarts, G., and Georgiou, J., "Computationally efficient classification of human transport mode using micro-Doppler signatures," 2011 45th Annual Conference on Information Sciences and Systems (CISS), pp. 1–4, 2011.

[58] Garreau, G., Andreou, C.M., Andreou, A.G., Georgiou, J., Dura-Bernal, S., Wennekers, T. *et al.*, "Gait-based person and gender recognition using micro-Doppler signatures," *IEEE Biomedical Circuits and Systems Conference* (BioCAS), pp. 444–447, 2011.

[59] Nanzer, J.A., "Simulations of the millimeter-wave interferometric signature of walking humans," *IEEE 2012 Antennas and Propagation Society Symposium* (APSURSI), pp. 1–2, 2012.

[60] Tekeli, B., Gurbuz, S.Z., Yuksel, M., Gurbuz, A.C., and Guldogan, M.B., "Classification of human micro-Doppler in radar network", 2013 *IEEE Radar Conference*, Ottawa, Canada, 2013.

[61] Tekeli, B., Gurbuz, S.Z., and Yuksel, M., "Mutual information of features extracted from human micro-Doppler," 2013 *21st Signal Processing and Communications Applications Conference* (SIU), pp. 1–4, 2013.

[62] Damarla, T., Bradley, M., Mehmood, A., and Sabatier, J.M., "Classification of Animals and People Ultrasonic Signatures," *IEEE Sensor Journal*, vol. 13, no. 5, pp. 1464–1472, 2013.

[63] Tahmoush, D., Silvious, J., and Clark, J., "An UGS radar with micro-Doppler capabilities for wide area persistent surveillance," *Proceedings of the SPIE, Radar Sensor Technology XIV*, vol. 7669, pp. 766904-11, 2010.

[64] Anderson M.G. and Rogers, R.L., "Micro-Doppler analysis of multiple frequency continuous wave radar signatures," *Proceedings of SPIE, Radar Sensor Technology XI*, vol. 6547, pp. 65470A-10, 2007.

[65] Ram, S.S. and Ling, H., "Micro-Doppler signature simulation of computer animated human and animal motion," *IEEE 2008 Antennas and Propagation Society International Symposium*, AP-S, pp. 1–4, 2008.

[66] Ramakrishnananda, B. and Wong, K.C., "Animated Bird Flight Using Aerodynamics," *The Visual Computer*, vol. 15, pp. 494–508, 1999.

[67] Vaughn, C.R., "Birds and insects as radar targets: a review," *Proceedings of the IEEE*, vol. 73, no. 2, pp. 205–227, 1965.

[68] Green, J.L. and Balsley, B., "Identification of flying birds using a Doppler radar," Proceedings Conference Biological Aspects Bird/Aircraft Collision Problem, Clemson University, pp. 491–508, 1974.

[69] Spruyt, J.A. and van Dorp, P., *Detection of Birds by Radar*, TNO Report, FEL-95-A244, TNO Physics and Electronics Laboratory, 1996.

[70] Zaugg, S., *et al.*, "Automatic identification of bird targets with radar via patterns produced by wing flapping," *Journal of the Royal Society Interface*, Sep. 6; vol. 5, no. 26, pp. 1041–1053. doi: 10.1098, 2008.

[71] Ozcan, A.H., Baykut, S., Sahinkaya, D.S.A., and Yalcin, I.K., "Micro-Doppler effect analysis of single bird and bird flock for linear FMCW radar," *20th Signal Processing and Communications Applications Conference* (SIU), 2012.

[72] Alabaster, C.M. and Hughes, E.J., "Is it a bird or is it a plane?" *Proceedings IEEE 6th International Conference on Waveform Diversity & Design*, Lihue, HI, January 22–27, 2012.

[73] Torvik, B., Olsen, K.E., and Griffiths, H., "K-band radar signature analysis of a flying mallard duck", 2013 *14th International Radar Symposium* (IRS), vol. 2, pp. 584–591, Dresden, Germany, June 19–21, 2013.

[74] Ozcan, A.H., Baykut, S., Sahinkaya, D.S.A., and Yalcin, I.K., "Micro-Doppler effect analysis of single bird and bird flock for linear FMCW radar," *The 20th Signal Processing and Communications Application Conference* (SIU), pp. 1–4, 2012.

[75] Chen, V.C., *et al.*, "Radar micro-Doppler signatures for characterization of human motion," Chapter 15, in *Through-the-Wall Radar Imaging*, M. Amin, ed., CRC Press, 2010.

[76] Ram, S.S., Christianson, C., Kim, Y., and Ling, H., "Simulation and analysis of human micro-Doppler in through-Wall Environments," *IEEE Transaction on GRS*, vol. 48, no. 4, pp. 2015–2023, 2010.

[77] Liu, X., Leung, H., and Lampropoulos, G.A., "Effects of non-uniform motion in through-the-wall SAR imaging," *IEEE Transactions on Antenna and Propagation*, vol. 57, no. 11, 2009, pp. 3539–3548.

[78] Setlur, P., *Statistical Algorithms an Bounds for Moving Targets in Urban Sensing and Through-the-Wall Applications*, PhD Thesis, Dept. of Electrical Engineering, Villanova University, Villanova, PA, May 2010.

[79] Liu, X.X., *A Study Of Through-The-Wall Radar And Its Application To Urban Sensing*, PhD Thesis, Dept. of Electrical and Computer Engineering, University of Calgary, Calgary, Alberta, CA, 2009.

[80] Chen, P.H., Shastry, M.C., Lai, C.P, and Narayanan, R.M., "A portable real-time digital noise radar system for through-the-wall imaging," *IEEE Transaction on GRS*, vol. 50, no. 10, Part 2, pp. 4123–4134, 2012.

[81] Gu, C., Li, C., Lin, J., Long, J., Huangfu, J., and Ran, L., "Instrument-based noncontact Doppler radar vital sign detection system using heterodyne

digital quadrature demodulation architecture," *IEEE Transaction on Instrumentation and Measurement*, vol. 59, no. 6, pp. 1580–1588, June 2010.

[82] Aho, J., Salmi, J., and Koivunen, V., "Adaptive processing and realistic signal propagation modeling for multiantenna vital sign radar," 2013 *IEEE Radar Conference*, Ottawa, Canada, 2013.

[83] Li, J., Liu, L., Zeng, Z., and Liu, F., "Advanced signal processing for vital sign extraction with applications in UWB radar detection of trapped victims in complex environments," *IEEE Journal of Selected Topics in Applied Earth Observations and Remote Sensing*, vol. 7, no. 3, pp. 1–9, 2014. (IEEE Early Access Article)

[84] Mikhelson, I.V., Bakhtiari, S., Elmer, T.W., and Sahakian, A.V., "Remote sensing of heart rate and patterns of respiration on a stationary subject using 94-GHz millimeter-wave interferometry," *IEEE Transaction on Biomedical Engineering*, vol. 58, no. 6, pp. 1671–1677, 2011.

[85] Fletcher, R. and Jing Han, J., "Low-cost differential front-end for Doppler radar vital sign monitoring," *Digest of IEEE MTT-S International Microwave Symposium*, pp.1325–1328, 2009.

[86] Bugaev, A.S., *et al.*, "Radar methods of detection of human breathing and heartbeat," *Journal of Communications Technology and Electronics*, vol. 51, no. 10, pp. 1154–1168, 2006.

[87] Lubecke, V.M., *et al.*, "Through-the-wall radar life detection and monitoring," in *Proceedings of the 2007 IEEE Microwave Theory and Techniques Symposium*, Honolulu, HI, 2007, pp. 769–772.

[88] Chia, M.Y.W., *et al.*, "Through-wall UWB radar operating within FCC's mask for sensing heart beat and breathing rate," *Proceedings of the 2005 European Radar Conference*, Paris, France, 2005, pp. 267–270.

[89] Bugaev, A.S., *et al.*, "Through wall sensing of human breathing and heart beating by monochromatic radar," *Proceedings of the 10th International Conference on Ground Penetrating Radar, Delft, the Netherlands*, 2004, pp. 291–294.

[90] Salmi, J., Luukkonen, O., and Koivunen, V., "Continuous wave radar based vital sign estimation: modeling and experiments," *IEEE 2012 Radar Conference*, pp. 564–569, 2012.

[91] Shirodkar, S., Barua, P., Anuradha, D., and Kullor, R., "Heart-beat detection and ranging through a wall using ultra wide band radar," *International Conference on Communications and Signal Processing* (ICCSP), pp. 579–583, 2011.

[92] Moulton, M.C., Bischoff, M.L., Benton, C., and Petkie, D.T., "Micro-Doppler radar signatures of human activity," *SPIE*, vol. 7837, pp. 78370L-1–78370L-7, 2010.

[93] Narayanan, R.M., "Earthquake survivor detection using life signals from radar micro-Doppler," *Proceedings of the 1st International Conference on Wireless Technologies for Humanitarian Relief*, pp. 259–264, 2011.

[94] Balleri, A., Chetty, K., and Woodbridge, K., "Classification of personnel targets by acoustic micro-Doppler signatures," *IET Radar, Sonar and Navigation*, vol. 5, no. 9, pp. 943–951, 2011.

[95] Balleri, A., Woodbridge, K., and Chetty, K., "Frequency-agile non-coherent ultrasound radar for collection of micro-Doppler signatures," *IEEE Radar Conference*, pp. 45–48, 2011.

[96] Zhang Z., Pouliquen, P., Waxman, A., and Andreou, A.G., "Acoustic micro-Doppler gait signatures of humans and animals," *The 41st Annual Conference on Information Sciences and Systems*, pp. 627–630, 2007.

[97] Dura-Bernal, S., Garreau, G., Andreou, C., Andreou, A., Georgiou, J., Wennekers, T. *et al.*, "Human action categorization using ultrasound micro-Doppler signatures," *Proceedings of the 2nd International Conference on Human Behavior Understanding*, pp. 18–28, 2011.

[98] Chu, T., Roderick, J., Chang, S., Mercer, T., Du, C., and Hashemi, H., "A short-range UWB impulse-radio CMOS sensor for human feature detection," *Digest of IEEE 2011 International Solid-State Circuits Conference*, pp. 294–296, 2011.

[99] Park, J., Johnson, J.T., Majurec, N., Frankford, M., Culpeper, E., Reynolds, J. *et al.*, "Software defined radar studies of human motion signatures," *Proceedings of IEEE 2012 Radar Conference*, pp. 596–601, May 7–11, 2012.

[100] Wang, Y., Liu Q., and Fathy, A.E., "CW and pulse-Doppler radar processing based on FPGA for human sensing applications," *IEEE Trans. on Geoscience and Remote Sensing*, vol. 51, no. 5, pp. 3097–3107, 2013.

[101] Liu, Q., Wang, Y., and Fathy, A., "Towards Low Cost, High Speed Data Sampling Module for Multifunctional Real-Time UWB Radar," *IEEE Transaction on AES*, vol. 49, no. 2, pp. 1301–1316, 2013.

[102] Godana, B., Barroso, A., and Leus, G., "Estimating human movement parameters using a software radio-based radar," *International Journal on Advances in Systems and Measurements*, vol. 4, no. 1 & 2, pp. 20–31, 2011.

[103] Yang, Y. and Fathy, A.E., "Development and implementation of a real-time sea-through-wall radar system based on FPGA," *IEEE Trans. on Geoscience and Remote Sensing*, vol. 47, no. 5, pp. 1270–1280, 2009.

[104] Thayaparan, T., Stankovic, L., and Djurovic, I., "Micro-Doppler-based target detection and feature extraction in indoor and outdoor environments," *Journal of the Franklin Institute*, vol. 345, no. 6, pp. 700–722, Sep. 2008.

[105] Li, P., Wang, D.C., and Wang, L., "Separation of micro-Doppler signals based on time frequency filter and Viterbi algorithm," in *Signal, Image and Video Processing*, Springer, Sep. 2011.

[106] Smith, G.E., Woodbridge, K., Baker, C.J., and Griffiths, H., "Multistatic micro-Doppler radar signatures of personnel targets," *IET Signal Processing*, vol. 4, no. 3, pp. 224–233, 2010.

[107] Stove, A.G. and Sykes, S.R., "A Doppler-based automatic target classifier for a battlefield surveillance radar," 2002 *International Radar Conference*, Edinburgh, U.K., pp. 419–423, 2002.

[108] Stove, A.G. and Sykes, S.R., "A Doppler-based target classifier using linear discriminants and principal components," *Proceedings of the 2003 International Radar Conference*, Adelaide, Australia, pp. 171–176, Sep. 2003.

[109] Hines, P.C. and Ward, C.M., "Classification of marine mammal vocalizations using an automatic aural classifier," *Journal of the Acoustical Society of America*, vol. 127, no. 1970, 2010, DOI: 10.1121/1.3385036, 2010.

[110] Allen, N., *et al.*, "Study on the human ability to aurally discriminate between target echoes and environmental clutter in recordings of incoherent broadband sonar," *Journal of the Acoustical Society of America*, vol. 119, no. 3395, 2006.

[111] Tahmoush, D. and Silvious, J., "Human polarimetric micro-Doppler," *Proceedings of SPIE*, vol. 8021, Orlando, April 2011.

[112] Freeman, A., "Fitting a two-component scattering model to polarimetric SAR data from forests," *IEEE Transactions on Geoscience and Remote Sensing*, vol. 45, no. 8, August 2007.

Chapter 2

Phenomenology of Radar Micro-Doppler Signatures

Victor C. Chen

2.1 Introduction

When the radar transmits an electromagnetic signal to an object, the signal interacts with the object and reflects back to the radar. If the object is moving, the carrier frequency of the returned signal will have Doppler frequency shift. The Doppler frequency is determined by the radial velocity of the object and the radar transmitted wavelength [1]. The roundtrip Doppler shift is given by $f_D = -f\frac{2v}{c}$, where c is the speed of wave propagation, f is the radar transmitted carrier frequency, and v is the velocity of the object along the line of sight of the radar. The velocity is defined to be positive when the object is moving away from the radar and, thus, the Doppler frequency shift is negative. If the radar is also moving, the radial velocity v is determined by the relative velocity between the radar and the object.

If the object or any structure on the object has micro motions, such as rotation or vibration, in addition to its translational motion, the micro motion will induce frequency modulation on the returned signal and generate side-bands about the Doppler shifted carrier frequency due to translation. This is called the micro-Doppler effect [2]. Micro-Doppler effect was originally observed in the coherent laser radar systems and is also observable in coherent microwave radars [2–4].

Radar returned signals from an object that has rotating or vibrating structures, such as propellers of a fixed-wing aircraft, rotor blades of a helicopter, blade assemblies of a jet aircraft, a human gaiting with swinging arms and legs, or a bird's flying wings, will contain micro-Doppler signatures related to these structures. For example, micro-Doppler signature of a gas turbine engine in a tank is different from that of a diesel engine in a bus. Thus, micro-Doppler signatures can be used to identify specific types of vehicles and determine the movement and speed of their engines.

To exploit these unique time-varying micro-Doppler characteristics, a high-resolution and time-frequency analysis method must be used. In this chapter, we introduce mathematics of the micro-Doppler effect in radar, derive basic formulas of micro-Doppler shifts due to vibration, rotation, tumbling and conning, and extract micro-Doppler signatures using time-frequency analysis.

2.2 Micro-Doppler Effect Induced by Micro Motion

The micro-Doppler effect induced by micro motion of an object or structures on the object can be formulated in mathematics. The formulas can be derived by applying micro motions to the conventional Doppler effect [2].

Motion includes rigid body motion (such as helicopter rotor blades, wind turbines, or spinning/precession tops) and nonrigid body motion (such as human movements, quadrupedal motion, or bird's flipping). Nonrigid body is a deformable body, i.e., a force acts on the body that will lead to the body changing its shape. For a free-form deformation, each particle in the body moves from an initial position P to a new position $P_t = f(t, P)$ at time t, which is a function of t and P. Stress and strain may occur everywhere in the body. To compute the deformation of a complex body, more complicated methods are required. However, for simplicity, any nonrigid body motion can be modeled by jointly connected rigid body parts [5, 6]. For example, in robotics, the robot arm is considered to be a flexible part of a mechanical system. The connection of two rigid segments of the arm is defined by the kinematical constraint on the joint that restricts the relative motions of the two individual rigid segments. When studying radar scattering from a nonrigid body motion, the nonrigid body (such as a walking human or a flying bird) is modeled as several jointly connected rigid body segments. The motion of each segment is treated as a rigid body motion.

To describe the motion of a rigid body, two coordinate systems: the global or space-fixed system (X, Y, Z) and the local or body-fixed system (x, y, z), which is rigidly fixed in the body, are commonly used (Figure 2.1). The range vector $\boldsymbol{R}$ is from the origin of the space-fixed system to the origin of the body-fixed system. Let the origin of the body-fixed system be the center of mass (CM) of the body. Then, the orientation of the axes of the body-fixed system relative to the axes of the space-fixed system is given by three independent angles. Therefore, the rigid body

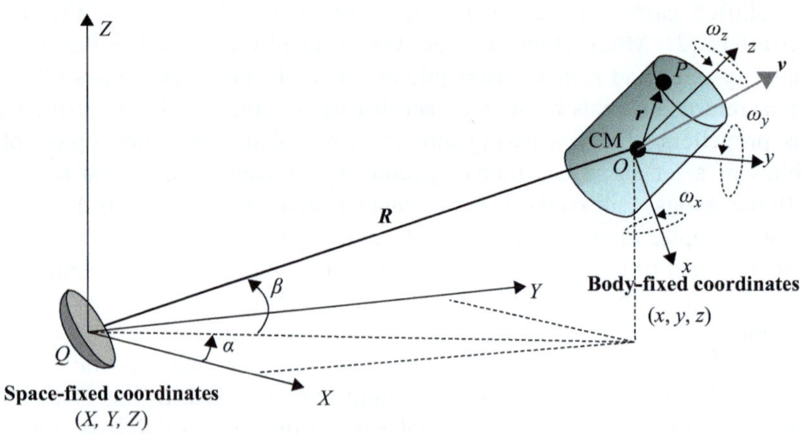

Figure 2.1 Two coordinate systems: the space-fixed system (X, Y, Z) and the body-fixed system (x, y, z) used to describe motion of an object

becomes a mechanical system with six degrees of freedom. Let $\boldsymbol{r}$ denote the position of an arbitrary particle P in the body-fixed system. Then, its position in the space-fixed system is given by $\boldsymbol{R}+\boldsymbol{r}$, and its velocity is

$$\frac{d}{dt}(\boldsymbol{R}+\boldsymbol{r}) = \boldsymbol{v} + \boldsymbol{\Omega} \times \boldsymbol{r} \tag{2.1}$$

where $\boldsymbol{v}$ is the translation velocity of the CM of the rigid body and $\boldsymbol{\Omega} = (\omega_x, \omega_y, \omega_z)^T$ is the angular velocity vector of the body rotation, where ω_x, ω_y, and ω_z are the angular rotation velocities about the body-fixed x-, y-, and z-axis, respectively. The direction of $\boldsymbol{\Omega}$ is along the axis of rotation. Thus, a rigid body motion consists of the translational motion and the micro motion (such as rotation and/or vibration).

2.2.1 Euler Angles and Rotation Matrices

To represent the orientation of an object, Euler angles and rotation matrices are commonly used. In a rigid body, rotation about an axis can be described by the axis and the rotating angle using an angular velocity vector $\boldsymbol{\Omega}$. The direction of the vector is along the axis of the rotation. Rotation about an axis can also be described by rotations about three axes of a coordinate system. Euler's rotation theorem tells us that any two independent orthonormal coordinates are related by a sequence of rotations about coordinate axes [7]. The rotation angles (φ, θ, ψ) are called the Euler angles, where φ, θ, and ψ are defined as counterclockwise rotations about the z-, y-, and x-axis, respectively. Euler angles are commonly used to represent three successive rotations in a given rotation sequence.

A commonly used rotation sequence is the x-y-z sequence, called the *roll-pitch-yaw* convention. To describe an object heading in the x-axis, with its left side toward the y-axis, and the upper side to the z-axis, three angles rotating about the x-, y-, and z-axis are used, as illustrated in Figure 2.2. Pitch is defined by the rotation of θ between $-\pi/2$ and $\pi/2$ about the y-axis; roll is defined by the rotation of ψ between $-\pi$ and π about the x-axis; and yaw or heading is defined by the rotation of φ between $-\pi$ and π about the z-axis.

For the *roll-pitch-yaw* convention, the rotation is in the *roll-pitch-yaw* (ψ-θ-φ) sequence. The first step is rolling about the x-axis, $x = [1\ 0\ 0]^T$, by an angle ψ. The second step is pitching about the new y-axis by an angle θ. The third step is yawing

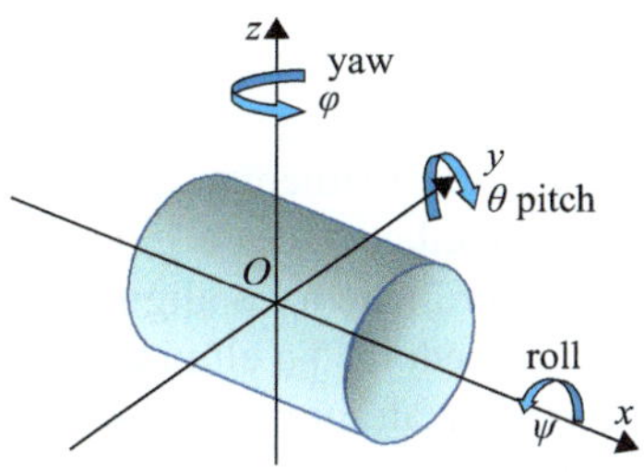

Figure 2.2 The roll-pitch-yaw convention used to describe an object's rotation

about the new z-axis by an angle φ. The rotation matrix of the *roll-pitch-yaw* sequence is described by

$$\Re_{X-Y-Z} = \Re_X(\psi)\Re_X(\theta)\Re_X(\varphi) = \begin{bmatrix} r_{11} & r_{12} & r_{13} \\ r_{21} & r_{22} & r_{23} \\ r_{31} & r_{32} & r_{33} \end{bmatrix} \tag{2.2}$$

where the components of the rotation matrix are

$$\begin{aligned} r_{11} &= \cos\theta\cos\varphi \\ r_{12} &= \sin\psi\sin\theta\cos\varphi + \cos\psi\sin\varphi \\ r_{13} &= -\cos\psi\sin\theta\cos\varphi + \sin\psi\sin\varphi \\ r_{21} &= -\cos\theta\sin\varphi \\ r_{22} &= -\sin\psi\sin\theta\sin\varphi + \cos\psi\cos\varphi \\ r_{23} &= \cos\psi\sin\theta\sin\varphi + \sin\psi\cos\varphi \\ r_{31} &= \sin\theta \\ r_{32} &= -\sin\psi\cos\theta \\ r_{33} &= \cos\psi\cos\theta \end{aligned} \tag{2.3}$$

Another commonly used rotation sequence is the z-x-z sequence, called the x-convention. The first step is rotating about the z-axis by an angle φ. The second step is rotating about the new x-axis by an angle θ. The third step is rotating about the new z-axis by an angle ψ. The rotation matrix of the z-x-z sequence is

$$\Re_{Z-X-Z} = \Re_Z(\varphi)\Re_X(\theta)\Re_Z(\psi)\begin{bmatrix} r_{11} & r_{12} & r_{13} \\ r_{21} & r_{22} & r_{23} \\ r_{31} & r_{32} & r_{33} \end{bmatrix} \tag{2.4}$$

where the components of the rotation matrix are

$$\begin{aligned} r_{11} &= -\sin\varphi\cos\theta\sin\psi + \cos\varphi\cos\psi \\ r_{21} &= -\cos\varphi\cos\theta\sin\psi - \sin\varphi\cos\psi \\ r_{31} &= \sin\theta\sin\psi \\ r_{12} &= \sin\varphi\cos\theta\cos\psi + \cos\varphi\sin\psi \\ r_{22} &= \cos\varphi\cos\theta\cos\psi - \sin\varphi\sin\psi \\ r_{32} &= -\sin\theta\cos\psi \\ r_{13} &= \sin\varphi\sin\theta \\ r_{23} &= \cos\varphi\sin\theta \\ r_{33} &= \cos\theta \end{aligned} \tag{2.5}$$

Euler angle rotation matrices are mathematically simple to handle and easy to understand. However, the Euler method faces a gimbal lock problem [8]. Gimbal lock happens when two of the coordinate axes align with each other. It results in a temporary loss of one degree of freedom. To eliminate the gimbal lock problem, an alternative method that uses the quaternion algebra can be applied. Quaternion algebra has been commonly used in computer graphics to represent orientations in 3-D space [9].

2.2.2 *Mathematics of Micro-Doppler Effect*

The mathematics of the micro-Doppler effect can be derived by introducing micro motion to the conventional Doppler analysis [2].

As shown in Figure 2.3, the radar is stationary and located at the origin Q of the radar-fixed coordinate system (X, Y, Z). The object has translation and rotation with respect to the radar coordinates and is described in a local coordinate system (x, y, z) attached to the object. To observe the object's rotations, a reference coordinate system (X', Y', Z') is introduced, which shares the same origin with the object local coordinates and has the same translation as the object but no rotation with respect to the radar coordinates. The origin O of the reference coordinates is assumed to be at a distance R_0 from the radar.

Suppose the object is moving with a velocity $\boldsymbol{v}$ with respect to the radar and, at the same time, with an angular rotation velocity vector $\boldsymbol{\Omega}$, which can be either represented in the object local coordinates or in the reference coordinates by $\boldsymbol{\Omega} = (\omega_x, \omega_y, \omega_z)^T = (\omega_{X'}, \omega_{Y'}, \omega_{Z'})^T$. For simplicity, an object is represented by a set of point scatterers. Thus, the process that a point scatterer P at time $t = 0$ moves to the point P' at time t can be represented by two steps, as shown in Figure 2.3: (1) translation from P to P'' with a velocity $\boldsymbol{v}$, i.e., $\overline{OO_t} = \boldsymbol{v}t$, and (2) rotation from P'' to P' with an angular velocity $\boldsymbol{\Omega}$. If we observe the movement in the reference coordinate system, the point scatterer P is located at $\boldsymbol{r}_0 = (X'_0, Y'_0, Z'_0)^T$, and the rotation from P'' to P' is described by a rotation matrix $\Re_t$. Then, at time t the location of P' is at

$$\boldsymbol{r} = \overline{O_tP'} = \Re_t\overline{O_tP''} = \Re_t\boldsymbol{r}_0 \tag{2.6}$$

The range vector from the radar at Q to the scatterer at P' can be derived as

$$\overline{QP'} = \overline{QO} + \overline{OO_t} + \overline{O_tP'} = \boldsymbol{R}_0 + \boldsymbol{v}t + \Re_t\boldsymbol{r}_0 \tag{2.7}$$

and the scalar range becomes

$$r(t) = \|\boldsymbol{R}_0 + \boldsymbol{v}t + \Re_t\boldsymbol{r}_0\| \tag{2.8}$$

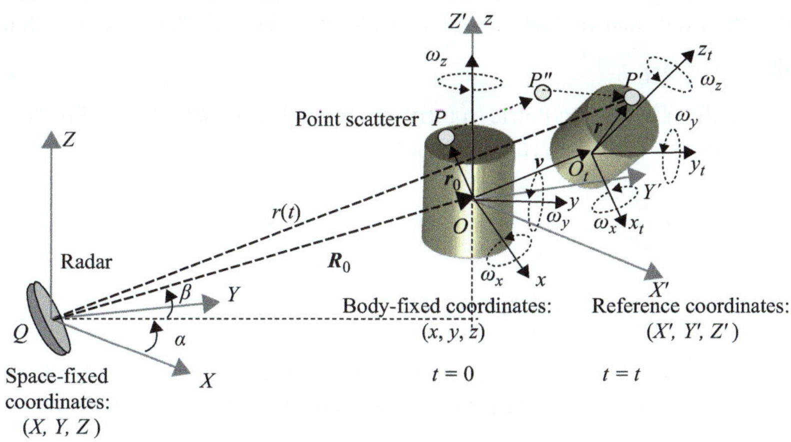

Figure 2.3 Geometry of the radar and an object with translation and rotation

To derive the Doppler frequency shift caused by the movement, we assume the radar transmits a sinusoidal waveform with a carrier frequency f. Then, the signal returned from the point scatterer P is a function of $r(t)$:

$$s(t) = \rho(x, y, z) \exp\left\{j2\pi f \frac{2r(t)}{c}\right\} = \rho(x, y, z) \exp\{j\Phi[r(t)]\} \tag{2.9}$$

where $\rho(x, y, z)$ is the reflectivity function of the point scatterer P described in the object local coordinates (x, y, z), and the phase of the base-band signal is

$$\Phi(r) = 2\pi f \frac{2r(t)}{c} \tag{2.10}$$

By taking the time derivative of the phase, the Doppler frequency shift induced by the object's motion can be derived:

$$\begin{aligned} f_D &= \frac{1}{2\pi}\frac{d\Phi(t)}{dt} = \frac{2f}{c}\frac{d}{dt}r(t) \\ &= \frac{2f}{c}\frac{1}{2r(t)}\frac{d}{dt}\left[(\boldsymbol{R}_0 + \boldsymbol{v}t + \Re_t\boldsymbol{r}_0)^T(\boldsymbol{R}_0 + \boldsymbol{v}t + \Re_t\boldsymbol{r}_0)\right] \\ &= \frac{2f}{c}\left[\boldsymbol{v} + \frac{d}{dt}(\Re_t\boldsymbol{r}_0)\right]^T \boldsymbol{n} \end{aligned} \tag{2.11}$$

where $\boldsymbol{n} = \frac{\boldsymbol{R}_0 + \boldsymbol{v}t + \Re_t\boldsymbol{r}_0}{\|\boldsymbol{R}_0 + \boldsymbol{v}t + \Re_t\boldsymbol{r}_0\|}$ is the unit vector of $\overline{QP'}$.

For further deriving the Doppler shift induced by the rotation, a useful relationship $\boldsymbol{u} \times \boldsymbol{r} = \hat{u}\,\boldsymbol{r}$ is used, which can be found in [2] and is useful in the analysis of the special orthogonal matrix group or $SO(3)$ group, also called the 3-D rotation matrix [10]. Thus, in the reference coordinate system, the angular rotation velocity vector can be described by $\boldsymbol{\Omega} = (\omega_{X'}, \omega_{Y'}, \omega_{Z'})^T$, and the object will rotate along the unit vector $\frac{\boldsymbol{\Omega}}{\|\boldsymbol{\Omega}\|}$ with a scalar angular velocity $\|\boldsymbol{\Omega}\| = \Omega$. Assuming a high pulse repetition frequency (PRF) and a relatively low angular velocity, during the time interval t, the rotational motion can be considered to be infinitesimal, and thus

$$\Re_t = \exp\{\hat{\omega}t\} \tag{2.12}$$

where $\hat{\omega}$ is the skew symmetric matrix associated with $\boldsymbol{\Omega}$ [2, 4]. Finally, the Doppler frequency shift in (2.11) becomes

$$\begin{aligned} f_D &= \frac{2f}{c}\left[\boldsymbol{v} + \frac{d}{dt}(e^{\hat{\omega}t}\,\boldsymbol{r}_0)\right]^T \boldsymbol{n} = \frac{2f}{c}(\boldsymbol{v} + \hat{\omega}\,e^{\hat{\omega}t}\,\boldsymbol{r}_0)^T\boldsymbol{n} \\ &= \frac{2f}{c}(\boldsymbol{v} + \hat{\omega}\boldsymbol{r})^T\boldsymbol{n} = \frac{2f}{c}(\boldsymbol{v} + \boldsymbol{\Omega} \times \boldsymbol{r})^T\boldsymbol{n} \end{aligned} \tag{2.13}$$

If $\|\boldsymbol{R}_0\| \gg \|\boldsymbol{v}t + \Re_t\,\boldsymbol{r}\|$, $\boldsymbol{n}$ can be approximated as $\boldsymbol{n} = \frac{\boldsymbol{R}_0}{\|\boldsymbol{R}_0\|}$, which is the direction of the radar LOS. The Doppler frequency shift is then approximately

$$f_D = \frac{2f}{c}[\boldsymbol{v} + \boldsymbol{\Omega} \times \boldsymbol{r}] \cdot \boldsymbol{n} \tag{2.14}$$

where the first term is the Doppler shift due to the object's translation:

$$f_{Trans} = \frac{2f}{c} \boldsymbol{v} \cdot \boldsymbol{n} \tag{2.15}$$

and the second term is the micro-Doppler due to the rotation:

$$f_{mD} = \frac{2f}{c} [\boldsymbol{\Omega} \times \boldsymbol{r}] \cdot \boldsymbol{n} \tag{2.16}$$

If the rotation is time-varying, the angular rotation velocity becomes a function of time, which can be expressed by a polynomial function:

$$\boldsymbol{\Omega}(t) = \boldsymbol{\Omega}_0 + \boldsymbol{\Omega}_1 t + \boldsymbol{\Omega}_2 t^2 + \cdots \tag{2.17}$$

If no more than second-order terms are considered, the micro-Doppler shift can be expressed by

$$f_{mD} = \frac{2f}{c} [\boldsymbol{\Omega}(t) \times \boldsymbol{r}] \cdot \boldsymbol{n} = \frac{2f}{c} \left[\boldsymbol{\Omega}_0 \cdot (\boldsymbol{r} \times \boldsymbol{n}) + \boldsymbol{\Omega}_1 \cdot (\boldsymbol{r} \times \boldsymbol{n}) t + \boldsymbol{\Omega}_2 \cdot (\boldsymbol{r} \times \boldsymbol{n}) t^2 \right] \tag{2.18}$$

where vector operation $(\boldsymbol{a} \times \boldsymbol{b}) \cdot \boldsymbol{c} = \boldsymbol{a} \cdot (\boldsymbol{b} \times \boldsymbol{c})$ has been applied.

Micro-Doppler shifts for four special micro motions – vibration, rotation, tumbling, and coning – are analyzed below.

2.2.2.1 Vibration Induced Micro-Doppler Shift

Figure 2.4 is the geometry of a vibrating point scatterer. The radar is located at the origin of the radar coordinate system (*X*, *Y*, *Z*) and a point scatterer *P* is vibrating about a center point *O*, which is situated at a distance R_0 from the radar and is the origin of the reference coordinate system (X', Y', Z'). If the azimuth and elevation angles of the point *O* with respect to the radar are α and β, respectively, the location of the point *O* is at

$$(R_0 \cos\beta \cos\alpha,\ R_0 \cos\beta \sin\alpha,\ R_0 \sin\beta) \tag{2.19}$$

in the space-fixed radar coordinates (*X*, *Y*, *Z*). The unit vector of the radar LOS is

$$\boldsymbol{n} = [\cos\alpha \cos\beta,\ \sin\alpha \cos\beta,\ \sin\beta]^T \tag{2.20}$$

Assume the point scatterer *P* is vibrating at a frequency f_v with a displacement of D_v and the azimuth and elevation angles of the vibrating direction are α_P and β_P, respectively, defined in the reference coordinates (X', Y', Z'). Thus, as shown in Figure 2.4, the vector from the radar to the scatterer *P* is $\boldsymbol{R}_t = \boldsymbol{R}_0 + \boldsymbol{D}_t$ and, thus, the range from the radar to the scatterer *P* is

$$\begin{aligned} R_t = |\boldsymbol{R}_t| = \big[& (R_0 \cos\beta \cos\alpha + D_t \cos\beta_P \cos\alpha_P)^2 \\ & + (R_0 \cos\beta \sin\alpha + D_t \cos\beta_P \sin\alpha_P)^2 \\ & + (R_0 \sin\beta + D_t \sin\beta_P)^2 \big]^{1/2} \end{aligned} \tag{2.21}$$

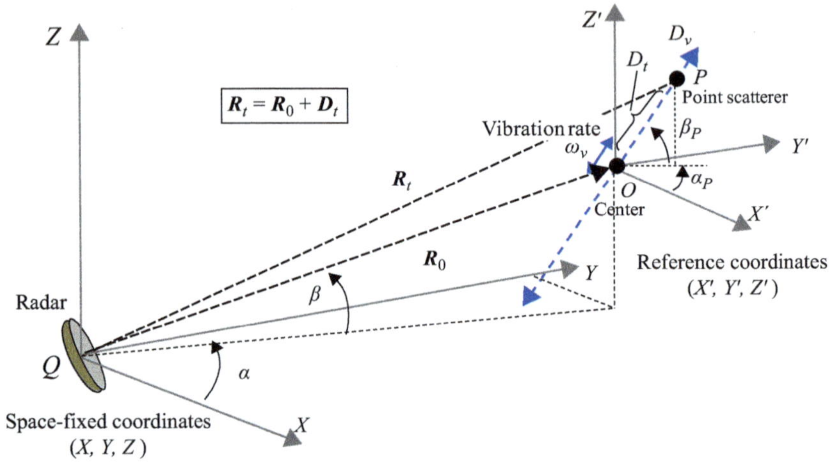

Figure 2.4 Geometry of the radar and a vibrating point scatterer

If $R_0 \gg D_t$, the range is approximately

$$\begin{aligned} R_t &= \left\{R_0^2 + D_t^2 + 2R_0 D_t[\cos\beta\cos\beta_p\cos(\alpha - \alpha_p) + \sin\beta\sin\beta_p]\right\}^{1/2} \\ &\approx R_0 + D_t[\cos\beta\cos\beta_p\cos(\alpha - \alpha_p) + \sin\beta\sin\beta_p] \end{aligned} \tag{2.22}$$

where $D_t = D_v \sin\omega_v t$ and $\omega_v = 2\pi f_v$.

If the azimuth angle α and the elevation angle β_P of the scatterer P are all zero, and $R_0 \gg D_t$, then we have

$$R_t = \left(R_0^2 + D_t^2 + 2R_0 D_t \cos\beta\cos\alpha_P\right)^{1/2} \cong R_0 + D_t\cos\beta\cos\alpha_P \tag{2.23}$$

The range of the scatterer is

$$R(t) = R_t = R_0 + D_v \sin\omega_v t\cos\beta\cos\alpha_P \tag{2.24}$$

and the received signal becomes

$$s_R(t) = \rho\exp\left\{j\left[2\pi f_0 t + 4\pi\frac{R(t)}{\lambda}\right]\right\} = \rho\exp\{j[2\pi f_0 t + \Phi(t)]\} \tag{2.25}$$

where ρ is the reflectivity of the point scatterer, f_0 is the carrier frequency of the transmitted signal, λ is the wavelength, and $\Phi(t) = 4\pi R(t)/\lambda$ is the phase function.

Substituting (2.24) into (2.25) and denoting $B = (4\pi/\lambda)D_v\cos\beta\cos\alpha_P$, the received signal can be rewritten as

$$\begin{aligned} s_R(t) &= \rho\exp\left\{j\frac{4\pi}{\lambda}R_0\right\}\exp\{j2\pi f_0 t + B\sin\omega_v t\} \\ &= \rho\exp\left(j\frac{4\pi}{\lambda}R_0\right)\sum_{k=-\infty}^{\infty} J_k(B)\exp[j(2\pi f_0 + k\omega_v)t] \end{aligned} \tag{2.26}$$

where

$$J_k(B) = \frac{1}{2\pi}\int_{-\pi}^{\pi} \exp\{j(B\sin u - ku)\,du \tag{2.27}$$

is the Bessel function of the first kind of order.

From (2.26), we can see that the frequency spectrum of the received signal consists of pairs of spectral lines around the center frequency f_0 and with spacing $\omega_v/(2\pi)$ between adjacent lines.

Because of vibration, if the point scatterer P is initially located at $[X_0, Y_0, Z_0]^T$ in the reference coordinates (X', Y', Z') at time $t=0$, at time t it will move to

$$\begin{bmatrix} X_1 \\ Y_1 \\ Z_1 \end{bmatrix} = D_v \sin(2\pi f_v t)\begin{bmatrix} \cos\alpha_P\cos\beta_P \\ \sin\alpha_P\cos\beta_P \\ \sin\beta_P \end{bmatrix} + \begin{bmatrix} X_0 \\ Y_0 \\ Z_0 \end{bmatrix} \tag{2.28}$$

Thus, the velocity of the vibrating scatterer P is

$$\boldsymbol{v} = D_v 2\pi f_v \cos(2\pi f_v t)[\cos\alpha_P\cos\beta_P,\ \sin\alpha_P\cos\beta_P,\ \sin\beta_P]^T \tag{2.29}$$

The micro-Doppler shift induced by the vibration is

$$\begin{aligned} f_{mD} &= \frac{2f}{c}(\boldsymbol{v}^T\cdot\boldsymbol{n}) \\ &= \frac{4\pi f f_v D_v}{c}[\cos(\alpha-\alpha_P)\cos\beta\cos\beta_P + \sin\beta\sin\beta_P]\cos(2\pi f_v t) \end{aligned} \tag{2.30}$$

2.2.2.2 Rotation Induced Micro-Doppler Shift

The geometry of the radar and a rotating object is illustrated in Figure 2.5. The radar coordinate system is (X, Y, Z), the target local coordinate system is (x, y, z), and the reference coordinate system is (X', Y', Z'). Assume that the azimuth and elevation angles of the object in the radar coordinates (X, Y, Z) are α and β, respectively.

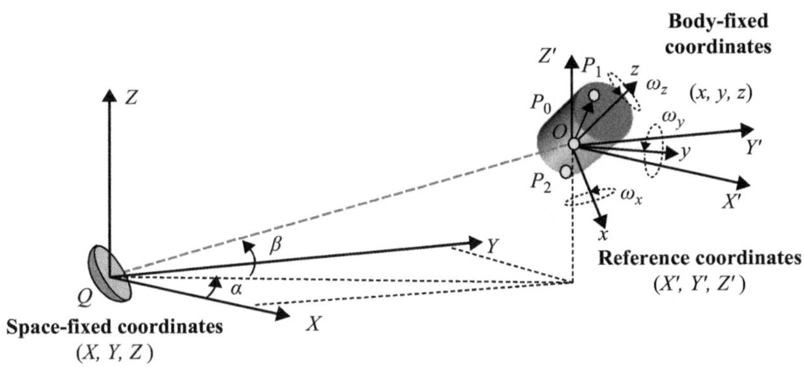

Figure 2.5 Geometry of the radar and a rotating object

For a rotating object, any point on it will move to a new position in the reference coordinates (X', Y', Z') and is calculated by multiplying its initial position vector with a rotation matrix. It can be a commonly used z-x-z rotating sequence by first rotating an angle φ_0 about the z-axis, then rotating an angle θ_0 about the x-axis, and, finally, rotating an angle ψ_0 about the z-axis again. The corresponding initial rotation matrix is defined by

$$\Re_{Init} = \Re_Z(\varphi_0) \cdot \Re_X(\theta_0) \cdot \Re_Z(\psi_0) \tag{2.31}$$

Viewed in the object local coordinate system, when an object rotates about its axes x, y, and z with angular velocity $\boldsymbol{\omega} = (\omega_x, \omega_y, \omega_z)^T$, a point scatterer P at $\boldsymbol{r_P} = [x_P, y_P, z_P]^T$ in the local coordinates (x, y, z) will move to a new location in the reference coordinates by $\Re_{Init} \cdot \boldsymbol{r_P}$ and the unit vector of the rotation is

$$\Re_{Init} \cdot \frac{\boldsymbol{\Omega}}{\|\boldsymbol{\Omega}\|} = \boldsymbol{\omega}' = (\omega'_x, \omega'_y, \omega'_z)^T \tag{2.32}$$

with the scalar angular velocity $\|\boldsymbol{\Omega}\| = \Omega$. Thus, according to Rodrigues formula [10], at time t the rotation matrix is

$$\Re_t = I + \hat{\omega}' \sin \Omega t + \hat{\omega}'^2 (1 - \cos \Omega t) \tag{2.33}$$

where $\hat{\omega}'$ is the skew symmetric matrix

$$\hat{\omega}' = \begin{bmatrix} 0 & -\omega'_z & \omega'_y \\ \omega'_z & 0 & -\omega'_x \\ -\omega'_y & \omega'_x & 0 \end{bmatrix} \tag{2.34}$$

Therefore, viewed in the reference coordinate system (X', Y', Z'), at time t the scatterer P will move from its initial location to a new location $\boldsymbol{r} = \Re_t \cdot \Re_{Init} \cdot \boldsymbol{r_P}$. Thus, the micro-Doppler frequency shift induced by the rotation is approximately

$$\begin{aligned} f_{mD} &= \frac{2f}{c}[\Omega \boldsymbol{\omega}' \times \boldsymbol{r}]_{radial} = \frac{2f}{c}(\Omega \hat{\omega}' \boldsymbol{r})^T \cdot \boldsymbol{n} = \frac{2f}{c}[\Omega \hat{\omega}' \Re_t \cdot \Re_{Init} \cdot \boldsymbol{r_P}]^T \cdot \boldsymbol{n} \\ &= \frac{2f\Omega}{c}\{[\hat{\omega}'^2 \sin \Omega t - \hat{\omega}'^3 \cos \Omega t + \hat{\omega}'(\boldsymbol{I} + \hat{\omega}'^2)]\Re_{Init} \cdot \boldsymbol{r_P}\}^T \cdot \boldsymbol{n} \end{aligned} \tag{2.35}$$

If the skew symmetric matrix $\hat{\omega}'$ associated with $\boldsymbol{\omega}'$ is defined by a unit vector, then $\hat{\omega}'^3 = -\hat{\omega}'$ and the rotation induced micro-Doppler frequency becomes

$$f_{mD} = \frac{2f\Omega}{c}[\hat{\omega}'(\hat{\omega}' \sin \Omega t + I \cos \Omega t)\Re_{Init} \cdot \boldsymbol{r_P}]_{radial} \tag{2.36}$$

2.2.2.3 Tumbling Induced Micro-Doppler Shift

A tumbling object has both translation and rotation, as illustrated in Figure 2.6(a). Assume the object has an initial translation velocity v along the x-axis and acceleration of $g = 9.8$ m/sec^2 due to gravity. At the same time, the object is also rotating along the y-axis with $(0, \omega_y, 0)$. The azimuth and elevation angles of the object with respect to the radar are α and β, respectively. Then, the unit vector of the radar LOS is expressed by $\boldsymbol{n} = [\cos\alpha\cos\beta, \sin\alpha\cos\beta, \sin\beta]^T$.

The reference coordinate system (X', Y', Z') has the same initial velocity and acceleration as the object but without rotation with respect to the space-fixed radar coordinates. The location of any point scatterer P of the object described in the reference coordinates can be represented by the point location $\boldsymbol{r}_0 = [x, y, z]^T$ described in the body-fixed coordinates multiplied by the initial rotation matrix $\Re_{Init}(\varphi, \theta, \psi)$, i.e., $\Re_{Init} \cdot \boldsymbol{r}_0$.

Suppose the object rotates along the y-axis with a velocity ω_y (rad/sec). An illustration of the trajectories of the object's center O and a point scatterer P are plotted in Figure 2.6(b), where the dot-line is the center's trajectory and the

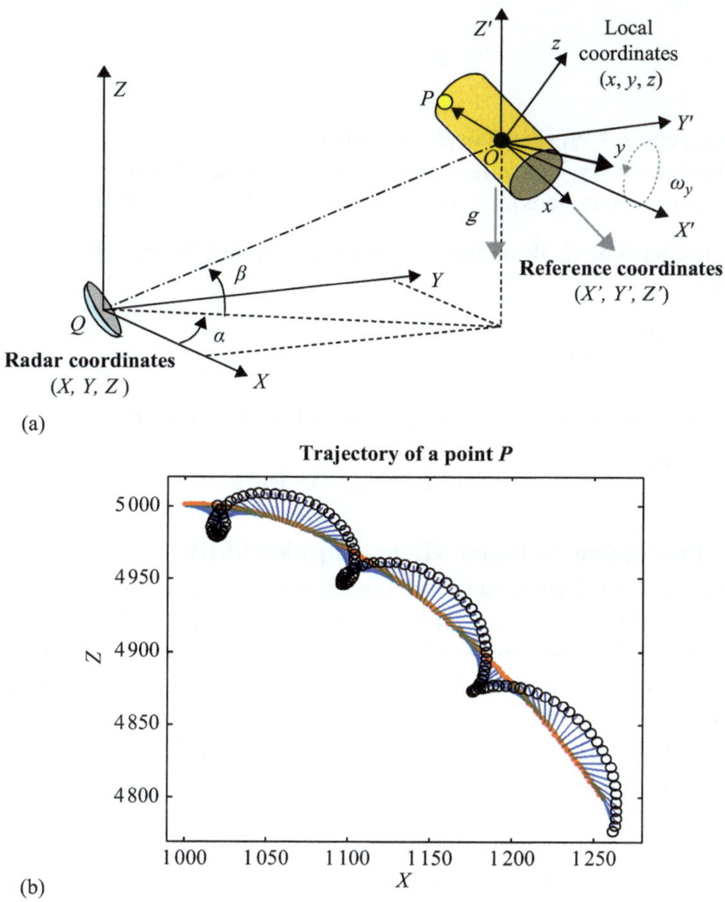

Figure 2.6 (a) Geometry of radar and the tumbling object and (b) trajectories of the object's center O and a point scatterer P

circled-line is the trajectory of the point P. During tumbling, the length of the segment $\overline{OP}$ stays constant. Viewing from the reference coordinates (X', Y', Z'), the angular velocity of the object is $\mathbf{\Omega}' = \omega_y\,[r_{12}, r_{22}, r_{32}]^T$ and the location of the point scatterer P will move to

$$\boldsymbol{r} = \Re_{Init} \begin{bmatrix} \cos\omega t & 0 & \sin\omega t \\ 0 & 1 & 0 \\ -\sin\omega t & 0 & \cos\omega t \end{bmatrix} \boldsymbol{r}_0 \tag{2.37}$$

At time t, the velocity of the object becomes $\boldsymbol{v} = [vr_{11}, vr_{21}, vr_{31} - gt]^T$.

The Doppler shift induced by the tumbling is

$$\begin{aligned} f_D &= \frac{2f}{c}[\boldsymbol{v} + \mathbf{\Omega}' \times \boldsymbol{r}]_{radial} \\ &= \frac{2f}{c}(-\sin\beta gt + t_1) + \frac{2f\omega_y}{c}[(xt_2 + zt_3)\cos\omega_y t + (zt_2 - xt_3)\sin\omega_y t] \end{aligned} \tag{2.38}$$

where

$$\begin{aligned} t_1 &= v[(r_{11}\cos\alpha + r_{21}\sin\alpha)\cos\beta + r_{31}\sin\beta] \\ t_2 &= [(r_{22}r_{31} - r_{32}r_{21})\cos\alpha + (r_{32}r_{11} - r_{12}r_{31})\sin\alpha]\cos\beta + (r_{12}r_{21} - r_{22}r_{11})\sin\beta \\ t_3 &= [(r_{22}r_{33} - r_{32}r_{23})\cos\alpha + (r_{32}r_{13} - r_{12}r_{33})\sin\alpha]\cos\beta + (r_{12}r_{23} - r_{22}r_{13})\sin\beta \end{aligned}$$

The first term of (2.38) is the Doppler shift induced by the initial velocity and the gravity:

$$f_{D_{Trans}} = \frac{2f}{c}(-\sin\beta \cdot gt + t_1) \tag{2.39}$$

The second term is micro-Doppler induced by the rotation:

$$f_{mD} = \frac{2f\omega_y}{c}[(xt_2 + zt_3)\cos\omega_y t + (zt_2 - xt_3)\sin\omega_y t] \tag{2.40}$$

2.2.2.4 Precession Induced Micro-Doppler Shift

Precession is a typical combination of coning, spinning, and nutation of an object. Coning is a rotation about an axis that intersects with an axis of the body-fixed local coordinates. A whipping top usually undergoes coning motion while its body is spinning with an angle φ around its axis of symmetry. It has a fixed tip point and the axis of symmetry rotates about another axis, called the precession axis, that intersects with the symmetric axis, as shown in Figure 2.7. The angle between the symmetric axis and the precession axis is the precession angle ψ.

If the axis of symmetry does not remain at a constant angle to the axis of coning, it will oscillate up and down, called nutation, between two limits. The up and down angle is called the nutation angle $\theta(t)$.

Without considering spinning and nutation, assume an object has only pure coning motion about an axis $\overline{SN}$, which intersects with the Z'-axis at the point $S(X' = 0, Y' = 0, Z' = Z_0)$ of the reference coordinates, as shown in Figure 2.8(a).

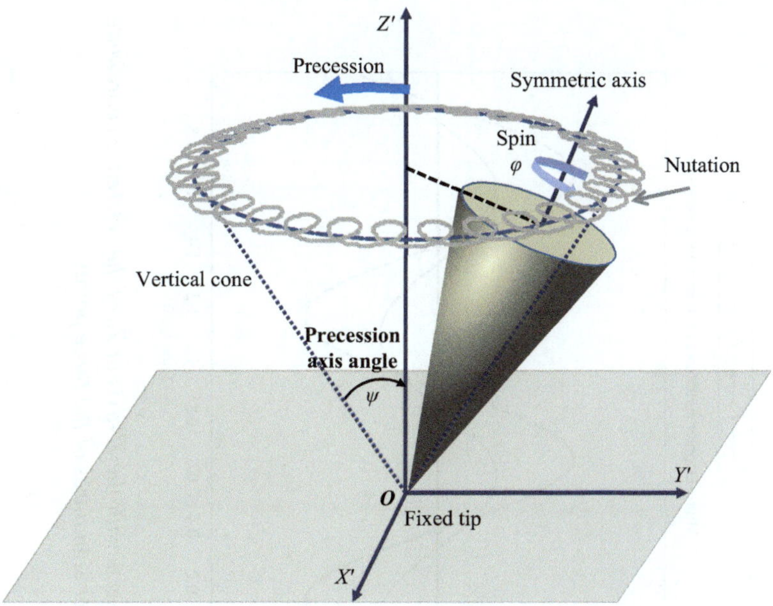

Figure 2.7 Precession of a spinning symmetric top

Assume the azimuth and elevation angles of the object center with respect to the radar are α and β, respectively, and the azimuth and elevation angles of the coning axis $\overline{SN}$ with respect to the reference coordinates (X', Y', Z') are α_N and β_N, respectively. Thus, the unit vector of the radar LOS is $\boldsymbol{n} = [\cos\alpha\cos\beta, \sin\alpha\cos\beta, \sin\beta]^T$, and the unit vector of the coning axis in the reference coordinates (X', Y', Z') is

$$\boldsymbol{e} = [\cos\alpha_N \cos\beta_N, \sin\alpha_N \cos\beta_N, \sin\beta_N]^T \tag{2.41}$$

Assume the initial location of a point P is $\boldsymbol{r} = [x, y, z]^T$ in the object local coordinates. Then, the location of the point P in the reference coordinates (X', Y', Z') can be calculated through its local coordinates by subtracting the coordinates of the intersect point S and multiplying the initial rotation matrix $\Re_{Init}$ determined by the initial Euler angles (φ, θ, ψ). Viewing from the reference coordinates (X', Y', Z'), the location of the point P is at $\Re_{Init}\,[x-0, y-0, z-Z_0]^T$. Suppose the object has a coning motion with an angular velocity of ω (rad/sec). According to the Rodrigues's formula, at time t the rotation matrix in the reference coordinates (X', Y', Z') is

$$\Re_t = I + \hat{e}\sin\omega t + \hat{e}^2(1-\cos\omega t) \tag{2.42}$$

where, in terms of (2.32), (2.34), and (2.41), the skew symmetric matrix is given by

$$\hat{e} = \begin{bmatrix} 0 & -\sin\beta_N & \sin\alpha_N\cos\beta_N \\ \sin\beta_N & 0 & -\cos\alpha_N\cos\beta_N \\ -\sin\alpha_N\cos\beta_N & \cos\alpha_N\cos\beta_N & 0 \end{bmatrix} \tag{2.43}$$

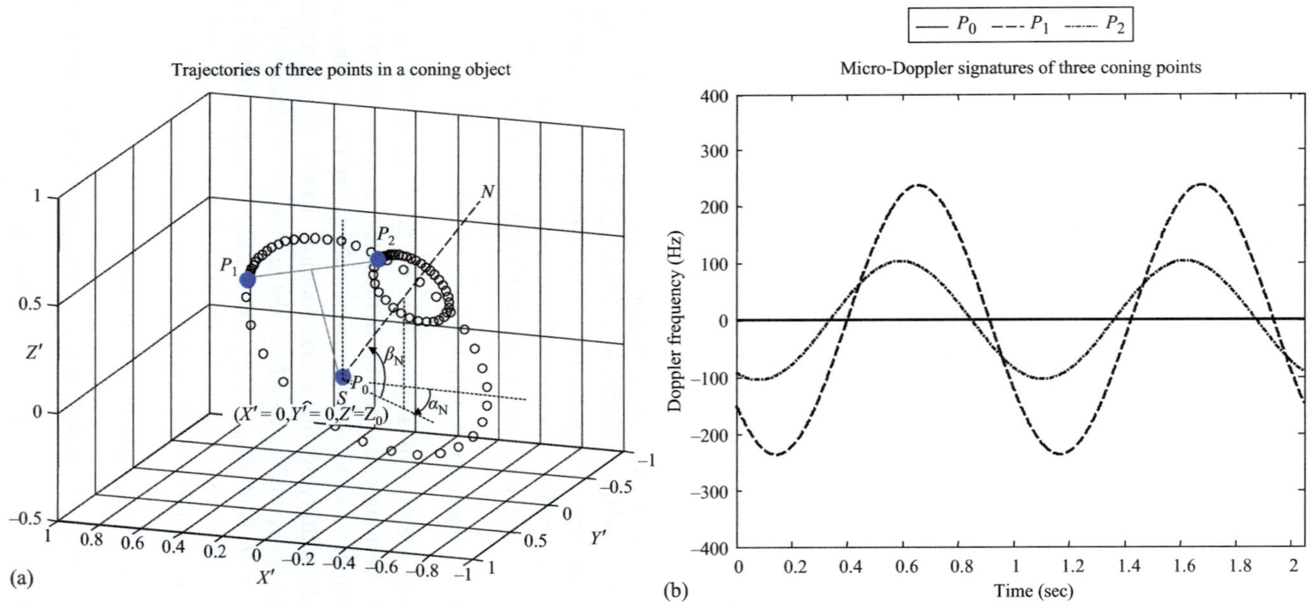

Figure 2.8 An object with only pure coning motion about an axis that intersects with the Z'-axis at Z_0 of the reference coordinates. (a) Trajectories of three points: P_0, P_1, and P_2 (b) micro-Doppler signatures of the three points

Therefore, at time t the point P will move to

$$\boldsymbol{r}(t) = \Re_t \cdot \Re_{Init} \cdot [x - 0, y - 0, z - Z_0]^T \tag{2.44}$$

If the point S is not too far from the object center of mass, we can approximate the radar LOS as the radial direction of the point S with respect to the radar. Thus, the micro-Doppler modulation induced by the coning motion is approximately

$$\begin{aligned} f_{mD}\big|_{coning} &= \frac{2f}{c}\left[\frac{d}{dt}\boldsymbol{r}(t)\right]_{radial} = \frac{2f}{c}\left\{\left[\frac{d}{dt}\Re_t\right]\Re_{Init}\cdot[x, y, z - Z_0]^T\right\}^T \cdot \boldsymbol{n} \\ &= \frac{2f\omega}{c}\left\{(\hat{e}\cos\omega t + \hat{e}^2\sin\omega t)\cdot\Re_{Init}\cdot[x, y, z - Z_0]^T\right\}^T \cdot \boldsymbol{n} \end{aligned} \tag{2.45}$$

Figure 2.8(b) shows micro-Doppler signatures of three points in a coning object, where P_0 is at (0,0,0), P_1 is at (0.3,0,0.6), and P_2 is at (−0.3,0,0.6). The coning rate is 2 Hz with an initial yaw, pitch, and roll angle of 10°, 10°, and −20°, respectively. The coning axis has an azimuth angle of 160° and an elevation angle of 50°.

When a coning body is also spinning around its axis of symmetry, the precession matrix $\Re_{prec}$ is the product of the coning matrix $\Re_{coning}$ and the spinning matrix $\Re_{spin}$:

$$\Re_{prec} = \Re_{coning} \cdot \Re_{spin} \tag{2.46}$$

where the coning matrix is

$$\Re_{coning} = I + \hat{\omega}_{coning}\sin(\Omega_{coning}t) + \hat{\omega}_{coning}{}^2[1 - \cos(\Omega_{coning}t)] \tag{2.47}$$

and the spinning matrix is

$$\Re_{spin} = I + \hat{\omega}_{spin}\sin(\Omega_{spin}t) + \hat{\omega}_{spin}{}^2[1 - \cos(\Omega_{spin}t)] \tag{2.48}$$

where $\boldsymbol{\Omega}_{coning}$ and $\boldsymbol{\Omega}_{spin}$ is the coning and spinning angular velocity, respectively. The skew symmetric matrix of the coning motion is defined by

$$\hat{\omega}_{coning} = \begin{bmatrix} 0 & -\omega_{coning_z} & \omega_{coning_y} \\ \omega_{coning_z} & 0 & -\omega_{coning_x} \\ -\omega_{coning_y} & \omega_{coning_x} & 0 \end{bmatrix} \tag{2.49}$$

and the skew symmetric matrix of the spinning motion is defined by

$$\hat{\omega}_{spin} = \begin{bmatrix} 0 & -\omega_{spin_z} & \omega_{spin_y} \\ \omega_{spin_z} & 0 & -\omega_{spin_x} \\ -\omega_{spin_y} & \omega_{spin_x} & 0 \end{bmatrix} \tag{2.50}$$

A point P at $\boldsymbol{r_P} = [x_P, y_P, z_P]^T$ represented in the object local coordinates (x, y, z) can be described in the reference coordinate system by $\Re_{Init} \cdot \boldsymbol{r_P}$. Viewing from the reference coordinate system (X', Y', Z'), at time t, the scatterer P will move from its initial location to a new location $\boldsymbol{r} = \Re_t \cdot \Re_{Init} \cdot \boldsymbol{r_P}$, where $\Re_t = \Re_{coning} \cdot \Re_{spin}$ is the product of the coning matrix and the spinning matrix.

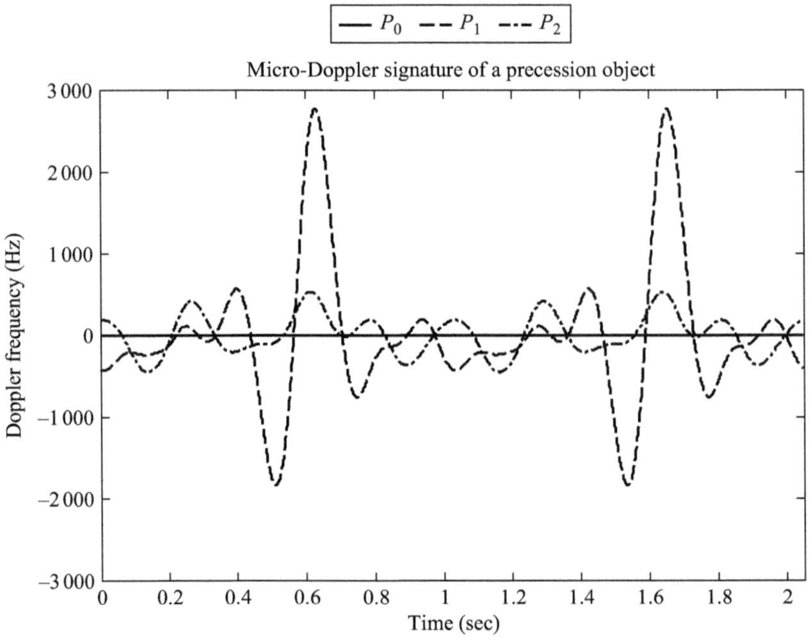

Figure 2.9 Micro-Doppler signatures of the three points in an object with coning and spinning. The coning motion is about an axis that intersects with the Z'-axis at Z_0 of the reference coordinates

The micro-Doppler modulation of the point scatterer P in the object body due to the precession is formulated as follows:

$$\begin{aligned} f_{mD}\big|_{coning+spin} &= \frac{2f}{c}\left[\frac{d}{dt}\boldsymbol{r}\right]_{radial} = \frac{2f}{c}\left[\frac{d}{dt}(\Re_{prec}\cdot\Re_{Init}\cdot\boldsymbol{r_P})\right]_{radial} \\ &= \frac{2f}{c}\left[\frac{d}{dt}(\Re_{coning}\cdot\Re_{spin})\cdot\Re_{Init}\cdot\boldsymbol{r_P}\right]_{radial} \\ &= \frac{2f}{c}\left[\left(\frac{d}{dt}\Re_{coning}\cdot\Re_{spin}+\Re_{coning}\cdot\frac{d}{dt}\Re_{spin}\right)\cdot\Re_{Init}\cdot\boldsymbol{r_P}\right]_{radial} \end{aligned} \tag{2.51}$$

Figure 2.9 shows micro-Doppler signatures of three points in a precession object, where P_0 is at (0,0,0), P_1 is at (0.3,0,0.6), and P_2 is at (−0.3,0,0.6). The coning rate is 2 Hz and the spinning rate is 3.9 Hz with an initial yaw, pitch, and roll angle of 10°, 10°, and −20°, respectively. The coning axis has an azimuth angle of 160° and an elevation angle of 50°.

2.3 How to Analyze Time-Varying Micro-Doppler Shifts

Fourier analysis can only show the frequency distribution of a signal during the entire analyzing time period. To analyze time-varying frequency distribution of a signal, we need a method to represent frequency distribution as a function of time.

Because the frequency is determined by the time derivative of the phase function, the phase function can be used to calculate the instantaneous frequency of a signal. However, the instantaneous frequency is only suitable for mono-component or single-tone signals, but not for those containing multiple components. To deal with multiple components, an approach that decomposes a multi-component signal into multiple addible mono-component signals, such as the empirical mode decomposition [11], should be used. Then, the complete time-frequency distribution of the multi-component signal is derived by computing the instantaneous frequency for each of the mono-component signals and adding these instantaneous frequencies of mono-component signals together.

Various time-frequency analyses have been developed by the signal processing community. These can be divided into two classes: linear time-frequency analysis and bilinear analysis.

The well-known linear time-frequency analysis of a signal is the Gabor transform [12]. It is also known as the short-time Fourier transform (STFT). Gabor suggested that the time and frequency characteristics of a signal $s(t)$ can be simultaneously observed by using the expansion

$$s(t) = \sum_{m=-\infty}^{\infty} \sum_{n=-\infty}^{\infty} a_{mn} G(g, n, m) \tag{2.52}$$

where $G(g, n, m)$, called the Gabor function, is expressed in terms of a Gaussian window $g(t)$ by

$$G(g, n, m) = g(t - m\Delta T)\, e^{jn\Delta Ft} \tag{2.53}$$

ΔT and ΔF are the time and frequency lattice intervals, respectively, and the Gaussian window is defined by

$$g(t) = \frac{1}{\pi^{1/4}\sqrt{\sigma}} \exp\left\{ -\frac{t^2}{2\sigma^2} \right\} \tag{2.54}$$

Gabor claimed that the basis functions $G(\bullet)$ used in this time-frequency decomposition have the minimum area in the joint time-frequency plane.

2.3.1 Joint Time-Frequency Analysis of Micro-Doppler Signature

2.3.1.1 Linear Analysis – Spectrogram

Spectrogram is a widely used method to display time-varying spectral density of a time-varying signal. It is a spectro-temporal representation and provides the actual change of frequency contents of a signal over time. The spectrogram is calculated by using the STFT and is represented by the squared magnitude of the STFT without keeping phase information of the signal:

$$Spectrogram(t,f) = |STFT(t,f)|^2 \tag{2.55}$$

The STFT is basically a moving window Fourier transform. By examining the frequency content of the signal as the time window is moved, a 2-D time-frequency distribution called the spectrogram is generated. The spectrogram contains information on the frequency content of the signal at different time instances. The drawback of the STFT is its resolution limit imposed by the length of the window function. A shorter time window results in better time resolution, but leads to worse frequency resolution, and vice versa. To overcome the resolution limit of the STFT, various bilinear time-frequency representations have been proposed.

2.3.1.2 Bilinear Analysis – Wigner–Ville Distribution

To better analyze the time-varying micro-Doppler frequency characteristics and visualize the localized joint time and frequency information, a signal must be analyzed by using a high-resolution time-frequency transform to characterize the spectral and temporal behavior of the signal. Bilinear transforms, such as the Wigner–Ville distribution (WVD), are high-resolution time-frequency transforms. The WVD of a signal $s(t)$ is defined by the Fourier transform of the time-dependent auto-correlation function:

$$WVD(t,f) = \int s\left(t + \frac{t'}{2}\right) s^*\left(t - \frac{t'}{2}\right) \exp\{-j2\pi f t'\} dt' \tag{2.56}$$

where $s(t + \frac{t'}{2}) s^*(t - \frac{t'}{2})$ can be interpreted as a time-dependent auto-correlation function. The bilinear WVD has better joint time-frequency resolution than any linear transform, such as the STFT. However, it suffers from the problem of cross-term interference, i.e., the WVD of the sum of two signals is not the sum of their individual WVDs. If a signal contains more than one component in the joint time-frequency domain, its WVD will contain cross-terms that occur halfway between each pair of auto-terms. The magnitude of these oscillatory cross-terms can be twice as large as the auto-terms. To reduce the cross-term interference, filtered WVDs have been used to preserve the useful properties of the time-frequency transform with a slightly reduced time-frequency resolution and largely reduced cross-term interference. The WVD with a linear low-pass filter belongs to the Cohen's class [13, 14].

The general form of Cohen's class is defined by

$$C(t, f) = \iint s\left(u + \frac{\tau}{2}\right) s^*\left(u - \frac{\tau}{2}\right) \phi(t - u, \tau) \exp\{-j2\pi f \tau\} du\, d\tau \tag{2.57}$$

The Fourier transform of the low-pass filter $\phi(t, \tau)$, denoted as $\Phi(\theta, \tau)$, is called the kernel function. If $\Phi(\theta, \tau) = 1$, then $\phi(t, \tau) = \delta(t)$ and the Cohen class reduces to the WVD. The Cohen's class with different kernel functions, such as the pseudo Wigner, the smoothed pseudo Wigner–Ville (SPWV), the Choi–Williams distribution, and the cone kernel distribution, can be used to largely reduce the cross-term interference in the WVD.

2.3.2 Doppler Aliasing in Micro-Doppler Signatures

Coherent Doppler radars retain the phase of the transmitted signals and track the phase changes in the received signals. Phase change rate determines the Doppler frequency shift of the received signals. If the phase change is more than $\pm\pi$, the estimated Doppler frequency becomes ambiguous, called the Doppler aliasing [2]. It is caused by discrete time sampling of a continuous time signal.

Sampling process can be represented by the multiplication of the continuous time signal $s(t)$ with a sequence of delta functions $\delta(t)$. Taking the Fourier transform of the time sampled signal, the discrete time sampled signal is transformed to the frequency domain represented by a discrete form of the Fourier transform:

$$s(t) \times \sum_n \delta(t - n\Delta t) \Rightarrow S(f) \otimes \sum_m \delta(f - m/\Delta t) \tag{2.58}$$

where Δt is the time-sampling interval, the convolution operator $\otimes$ in the frequency domain makes the signal spectrum $S(f)$ to be replicated at a period of $1/\Delta t$. If the frequency bandwidth of the signal spectrum is greater than the Nyquist frequency $1/(2\Delta t)$, or the sampling rate is lower than the half bandwidth of the signal, this replication will cause signal spectrum overlap and produce ambiguity, called aliasing phenomenon.

The aliasing causes the true frequency values to be offset by multiples of $(1/\Delta t)$ until they fall into the Nyquist cointerval. Assume a radar received signal has maximum bandwidth of 1 000 Hz and, thus, the Nyquist frequency should be $2\times 1\,000$ Hz. Figure 2.10 illustrates conventional spectrum and time-varying

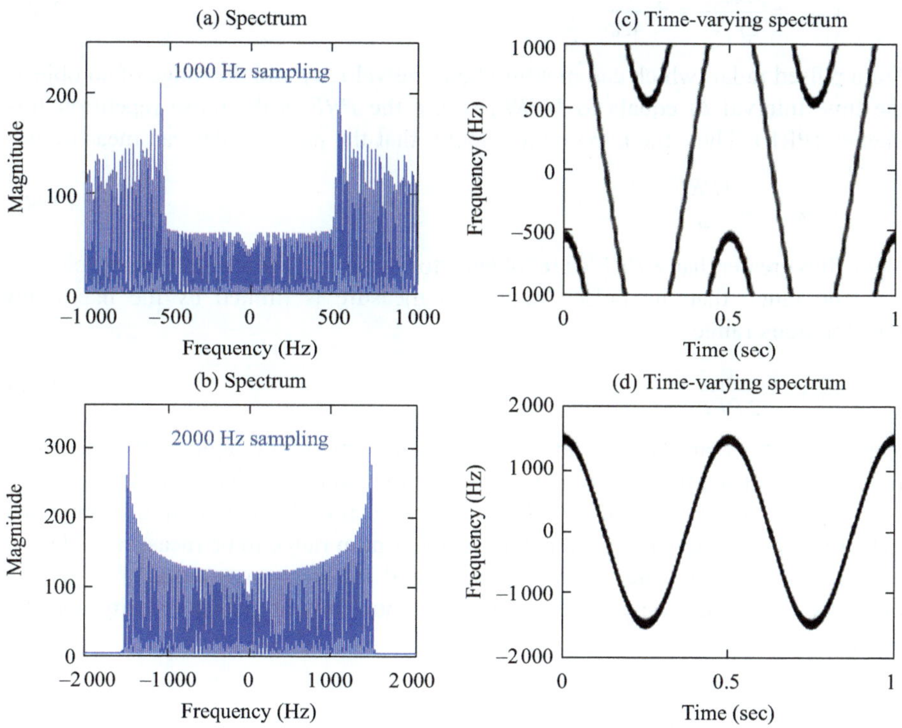

Figure 2.10 The aliasing phenomenon

spectrum of the signal under two different sampling rates. Figure 2.10(a) is the conventional spectrum under 1 000 Hz sampling rate. Because the signal bandwidth is also 1 000 Hz, a frequency value of +1 500 Hz is aliased to +1 500 Hz –2×1 000 Hz = –500 Hz, and –1 500 Hz is aliased to –1 500 Hz + 2×1 000 Hz = +500 Hz. In this case, the true frequency values are offset by one multiple of ($1/\Delta t$) to fall into the Nyquist interval. The aliasing phenomenon can be seen in the folded time-varying spectrum in Figure 2.10(c).

To resolve the aliasing ambiguity, either increasing the sampling rate or any technique that interpolates missing data points can be used. Figure 2.10(b) is the spectrum of the same signal sampled with twice sampling rate (2 000 Hz). The time-varying spectrum with twice sampling rates is shown in Figure 2.10(d), where the full time-varying spectrum is restored.

2.3.3 PRF Selection Determined by Unambiguous Velocity and Range

Generally, the required unambiguous radial velocity must be at least the radial velocity with which the object moves. Thus, the unambiguous velocity that a radar can measure is determined by the transmitted signal wavelength $\lambda = c/f$ and the time interval Δt between two sampling points:

$$v_{max} = \frac{f_{Dmax}}{2f/c} = \pm\frac{\lambda}{4\Delta t} \tag{2.59}$$

For a pulsed radar, which can measure both the velocity and the range of an object, the time interval Δt equals to $1/PRF$, where the PRF is the pulse repetition frequency (PRF). Thus, the maximum velocity that the pulsed radar can measure is

$$v_{max} = \pm\frac{\lambda PRF}{4} \tag{2.60}$$

Velocities greater than $\lambda\, PRF/4$ are folded into $\pm\lambda\, PRF/4$, called the Nyquist velocity.

The range that the pulsed radar can measure is limited by the maximum unambiguous range:

$$r_{max} = \frac{c}{2PRF} \tag{2.61}$$

Ranges greater than the r_{max} are folded into the first range region. Without additional information, the correct range information cannot be determined.

From (2.60) and (2.61), the PRF is proportional to the maximum unambiguous velocity, but inversely proportional to the maximum range to be measured. Thus, a compromise between the maximum range and the maximum velocity is always desirable. However, the product of the maximum unambiguous velocity and the unambiguous range:

$$v_{max}r_{max} = \pm\frac{c\lambda}{8} \tag{2.62}$$

is not directly related by the PRF. It is only determined by the wavelength λ. For a given frequency wavelength, the product of the maximum unambiguous velocity and the unambiguous range is fixed. Increasing the maximum unambiguous velocity will decrease the maximum unambiguous range, and vice versa.

2.3.4 Illustration of Extracted Micro-Doppler Signature

Figure 2.11 illustrates the micro-Doppler signature extracted from a rotating object. This is a rotating air-launched cruise missile (ALCM) provided by the simulation software in [15, 16], which can be used to generate radar returned signals from it. The length of the cruise missile is 6.4 m and its wingspan is about 3.4 m. A burst of 1 μsec chirp pulse radar operating in X-band is assumed to simulate the EM backscattering field. The radar transmits 8 192 pulses with the pulse repetition interval of 67 μsec during a period of 0.55 sec to cover the total target's rotation angle of 360°, as shown in Figure 2.11(a).

Micro-Doppler features of the rotating ALCM is extracted from the radar range profiles by summing range profiles over range cells where the object is located, as illustrated in Figure 2.11(b). The micro-Doppler signature can be observed in the frequency domain and with much more clear view in the joint time-frequency domain [17]. Figure 2.11(c) shows the joint time-frequency micro-Doppler signature of the rotating ALCM. Recall that the missile rotating rate is about 1.8 cycle/sec because it takes 0.55 sec to complete a rotation of 360°. For the ALCM model, (1) the missile head tip, (2) head joint, (3) wing joint, (4) turbine engine intake, (5) tail fin and tail plane, and (6) tail tip and engine exhaust are located at about −2.5 m, −1.8 m, 0.2 m, 2.5 m, 3.5 m, and 4.2 m, respectively, off the pivot point at 0. If these parts are considered to be the dominant scatterers, the maximum Doppler shifts induced by rotations of these parts would appear at those positions when their angular velocities are nearly in parallel with the radar LOS. That is, when the missile is at 90° or 270° aspect to the radar, or at the elapsed time of 0.14 sec or 0.41 sec, respectively, as shown in Figure 2.11 (c). When the missile is at 90° aspect, the induced Doppler shifts are −1 917 Hz, −1 380 Hz, 153 Hz, 1 917 Hz, 2 684 Hz, and 3 217 Hz, respectively. For these dominant scatterers, the traces of induced Doppler shifts by rotations are clearly shown in Figure 2.11(c). At the aspects from 180° to 90° and from 180° to 270°, the induced Doppler shifts of the ALCM model have the same magnitudes but with opposite signs. The same motion kinematics is also seen at the missile aspects from 0° to 90° and from 360° to 270°. These Doppler shifts gradually reduce to zero when their moving directions are perpendicular to the radar LOS, i.e., when the missiles are approximately at 180° and 360° aspects. Notice that the induced Doppler shifts are further dispersed for two scatterers near the missile tail where two turbines are located. Thus, the missile kinematic motion can be well characterized by its micro-Doppler signature, which can be used to identify distinctive target features.

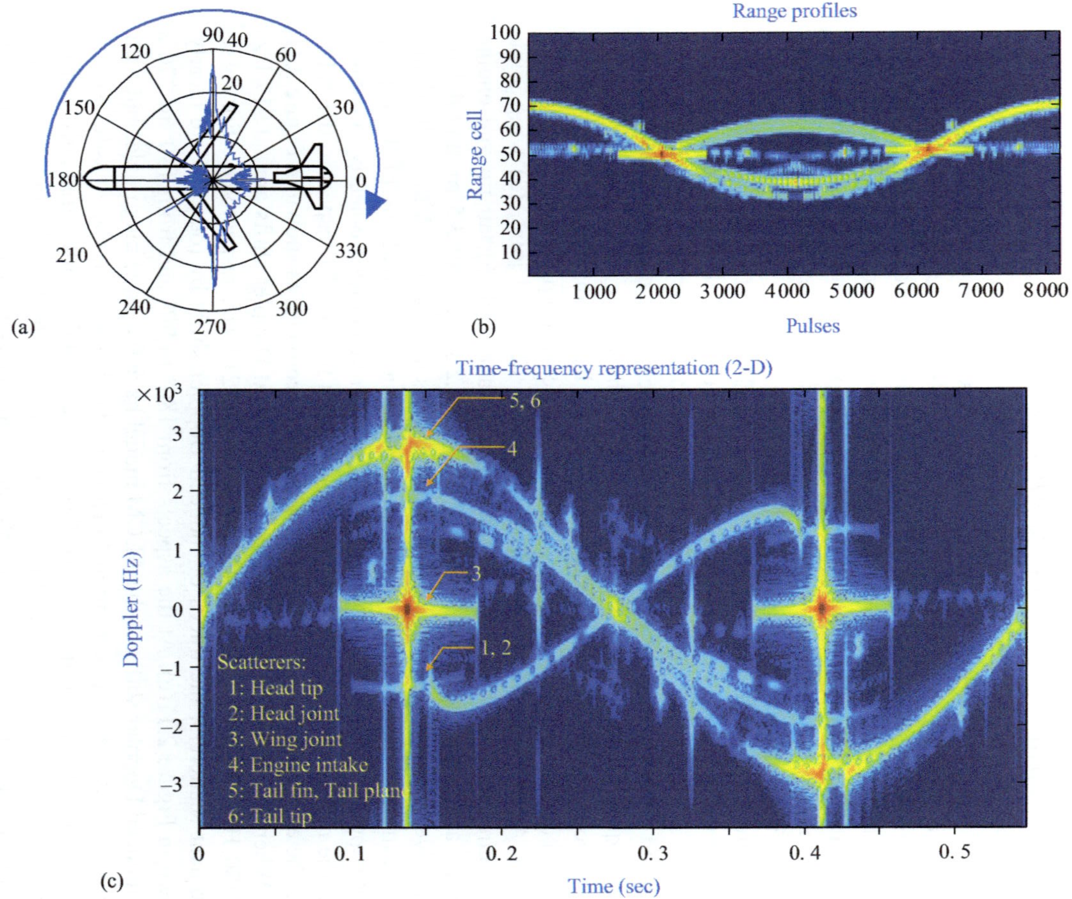

Figure 2.11 Micro-Doppler signature of a simulated rotating ALCM

References

[1] Gill, T.P., *The Doppler Effect – the theory and applications*, Academic Press, NY, 1965.

[2] Chen, V.C., *The Micro-Doppler Effect in Radar*, Artech House, Boston/London, 2011.

[3] Chen, V.C., "Analysis of radar micro-Doppler signature with time-frequency transform", *Proceedings of the 10th IEEE Workshop on Statistical Signal and Array Processing*, pp. 463–466, August, 2000.

[4] Chen, V.C., *et al.*, "Micro-Doppler Effect in Radar—phenomenon, model, and simulation Study", *IEEE Transaction on AES*, vol. 42, no. 1, 2006, pp. 2–21.

[5] Magnus, K., *Dynamics of multibody systems*, Springer Verlag, Berlin, 1978.

[6] Shabana, A.A., *Dynamics of multibody systems*, (3rd ed.), Cambridge University Press, Cambridge, UK, 2005.

[7] Goldstein, H., Poole, C., and Safko, J., *Classical Mechanics*, (3rd ed.), Addison-Wesley, Boston, NY, 2002.

[8] Kuipers, J.B., *Quaternions and Rotation Sequences*, Princeton University Press, Princeton, NJ, 1999.

[9] Mukundand, R., "Quaternions: from classical mechanics to computer graphics, and beyond," *Proceedings of 7th Asian Technology Conference in Mathematics* (ATCM), 2002, pp. 97–106.

[10] Murray, R.M., Li, Z., and Sastry, S.S., *A Mathematical Introduction to Robotic Manipulation*, CRC Press, NY, 1994.

[11] Huang, N.E., *et al.*, "The empirical mode decomposition and the Hilbert spectrum for nonlinear and non-stationary time series analysis," *Proceedings Royal Society London, series A*, vol. 454, 1998, pp. 903–995.

[12] Gabor, D., "Theory of communication," *Journal IEE (London)*, vol. 93, Part III, no. 26, 1946, pp. 429–457.

[13] Cohen, L., *Time-Frequency Analysis*, Prentice Hall, Englewood Cliffs, NJ, 1995.

[14] Qian, S., and Chen, D., *Introduction to Joint Time-Frequency Analysis—Methods and Applications*, Prentice Hall, Englewood Cliffs, NJ, 1996.

[15] Shirman, Y.D., *Computer Simulation of Aerial Target Radar Scattering, Recognition, Detection, and Tracking*, Artech House, Boston, 2002.

[16] Gorshkov, S.A., *et al.*, *Radar Target Backscattering Simulation—Software and User's Manual*, Artech House, Boston, 2002.

[17] Chen, V.C., Lin, C.-T., and Pala, W.P., "Time-varying Doppler analysis of electromagnetic backscattering from rotating object," *The IEEE Radar Conference Record*, April 24–27, Verona, NY, 2006, pp. 807–812.

Chapter 3

Analysis of Human Signatures using High-Range Resolution Micro-Doppler Radar

O. Ryan Fogle and Brian D. Rigling

3.1 Introduction

As seen in previous chapters, Doppler radars with long coherent processing intervals (CPIs) offer means for observing target state changes that are minute relative to the overall target signature. The small state changes, or micro-motion (e.g., vibrations, rotations), cause frequency modulations about the bulk target Doppler frequency response [1]. These detected frequency shifts form the micro-Doppler signature. For a pulse-Doppler radar operating with sufficiently high range resolution, the target micro-motion may cause observable displacements from the target "macro-range" response. Following the same naming convention, we term these responses as micro-range signatures. The micro-range signatures provide an additional and perhaps more intuitive domain in which to study and exploit target signatures.

For example, automated analysis of the intermixed patterns of time-varying range or Doppler signatures tends to be very difficult. In these scenarios, the radar takes three-dimensional physical data and projects it onto a single dimension range. As such, the radar observes moving scattering centers overlapping in trajectories and scintillating as the target maneuvers. The lack of persistent characteristics makes discriminating and tracking individual scatterers difficult. However, by simultaneously utilizing both range and Doppler observations, many of the scatterer crossings may be disambiguated since the scatterers are uniquely resolved in range-Doppler. The elimination of such scatterer interactions facilitates extraction of useful target features. Straightforward extraction of point scatterers in terms of their range-Doppler trajectories and scattering coefficients may potentially be exploited for target characterization, discrimination, and tracking.

Effective exploitation of target signature features hinges on a fundamental understanding of target-radar scattering dynamics and phenomenology. Past studies of micro-Doppler target behavior have included wind turbines [2], helicopters [1, 3–6], airplanes [7, 8], ballistic warheads [9–11], animals [12–16], humans [14, 17–23], and many others [21, 24–26]. These studies have enabled identification of critical target information that may be used for detection (false alarm reduction), classification, and discrimination. Features of interest have included the number of helicopter blades, gait lengths, micro-Doppler bandwidth, rotational velocities, etc. Of the targets previously studied, humans (or dismounts) have recently been of increasing interest, with ongoing investments in measurement, understanding, and exploitation of radar-human phenomenology.

Radar dismount detection and characterization could be useful for border protection, military surveillance, airport security and numerous other defense, law enforcement, and homeland security purposes. However, radio frequency (RF) dismount sensing offers some unique challenges that must be addressed. First, humans have comparatively small radar cross-sections (RCS). When observed from long standoff distances, the small RCS results in low intensity returns. Second, dismounts are often observed in high-clutter areas, particularly in applications of air-to-ground surveillance where the ground reflects much of the radar energy. Consequently, low signal-to-clutter ratios may be the norm in many instances. Third, humans tend to move relatively slowly, oftentimes placing the human response in the clutter filter notch and further reducing signal power. Also, dismounts are usually only distinct from one another via micro-motion features with small range-Doppler displacements, thus motivating use of micro-Doppler–micro-range features, which demand radars operating with ultrafine range and Doppler resolutions.

In the remainder of this chapter, we will highlight some of our contributions to the micro-Doppler–micro-range body of knowledge. Particularly, section 3.2 will briefly discuss the micro-Doppler and micro-range signature phenomenon. Section 3.3 will introduce a micro-Doppler–micro-range decomposition algorithm. Section 3.4 will summarize the progress made to date.

3.2 Micro-Range Micro-Doppler Human Radar Signature Analysis

In this section, an intuitive explanation of micro-range–micro-Doppler signatures is provided. For a pulse-Doppler radar, a finite duration signal is emitted from the transmitter. The signal propagates to the target where a portion of the signal energy may be reflected by the target's constituent scatterers and observed by the radar. To develop a theory for the recovery of these physical features, we turn to Joseph Keller's 1962 paper on the Geometric Theory of Diffraction [27], which offered some simplifying insights on electromagnetic scattering. Of particular interest, Keller illustrated that, assuming a sufficiently high operating frequency, a complex target's scattering behavior may be succinctly modeled as a summation of scattering responses from simple scattering mechanisms. Hence, a physically relevant model for a measured radar return, $S(\omega, k)$, may be constructed in canonical form as [28, 29]

$$S(\omega,k) = \sum_{m=1}^{N_S} S_{\Gamma_m}(\omega,k;\underline{\theta}_m(k))\, e^{-j\frac{4\pi\omega}{c}r_m(k)} + \nu(\omega,k) \tag{3.1}$$

where N_S is the model order, $S_{\Gamma_m}(\cdot)$ represents the scattering behavior model of type Γ_m, which is a function of the sampled frequencies ω and the pulse number

k and has parameters $\underline{\theta}_m(k)$. The mth scatterer's range is $r_m(k)$ during pulse k, while c denotes the speed of light and $\nu(\omega, k)$ is additive white Gaussian noise. The observations are an aggregation of radar signal returns for pulses $k = 1, \ldots, N_K$ and frequencies ω_n for $n = 1, \ldots, N_F$.

Many primitive scattering types (e.g., top hat, dihedral, sphere) have been studied with application to various radar modes [30–35]. In many instances where target features are physically small relative to the radar resolution, restricting (3.1) to a summation of point-type scatterers may be sufficient for modeling all target dynamics of interest and removes from consideration many free variables that may not be accurately or unambiguously estimated. As such, in many circumstances, the scattering behavior model may be simplified to $S_{\Gamma_m}(\omega, k; \underline{\theta}_m(k)) = A_m(k)$, such that [29]

$$S(\omega, k) = \sum_{m=1}^{N_S} A_m(k)\, e^{-j\frac{4\pi\omega}{c} r_m(k)} + \nu(\omega, k) \tag{3.2}$$

where $A_m(k)$ is the mth feature's scattering coefficient at pulse k. Thus, a target may be viewed as a number of point scatterers propagating through space, though we note that other researchers [28] have employed more complicated models made up of cylinders and ellipsoids. Each scatterer reflects a phase-delayed (proportional to the scatterer-radar range) and frequency-shifted (proportional to the scatterer-radar velocity) version of the transmitted signal. A portion of this reflected signal, along with energy reflected from other target scatterers and clutter, is received by the radar. When a desired number of pulses have been observed, the recorded phase histories may be matched filtered via pulse-Doppler compression, resulting in peaks proportional to the scatterer's range and Doppler. This signature will evolve over time based on the individual scatterer trajectories relative to the radar line-of-sight, which may even change in number and be dynamic in terms of their scattering behavior, depending on the overall target shape and movement.

To illustrate the micro-motion signature concept, consider measured data of a human walking. The data sets employed in this chapter contain radar returns of individuals walking various paths of the layout shown in Figure 3.1. The data was obtained using a radar operating with a center frequency of 10 GHz and a bandwidth of 4 GHz. The resultant resolution is approximately 3.75 cm. The pulse repetition frequency (PRF) was 1 kHz. The data was collected in a controlled environment with relatively little Doppler clutter via a stationary background and stationary radar mounting.

The target signatures may be decomposed along a number of independent axes. Consider first a range-Doppler-time portrayal, which permits some intuitive observations. A range-Doppler-time cube is formed first by range compression of each pulse, computed via inverse Fourier transform ($F_1^{-1}\{\cdot\}$) with respect to ω in (3.2), to form range profiles and then by calculating a short-time Fourier transform ($STFT\{\cdot\}$) [36] across pulses for each range bin. It should be noted that the STFT has a trade-off between spectral resolution and temporal resolution. When larger windows are used, better Doppler frequency resolution may be obtained, while the

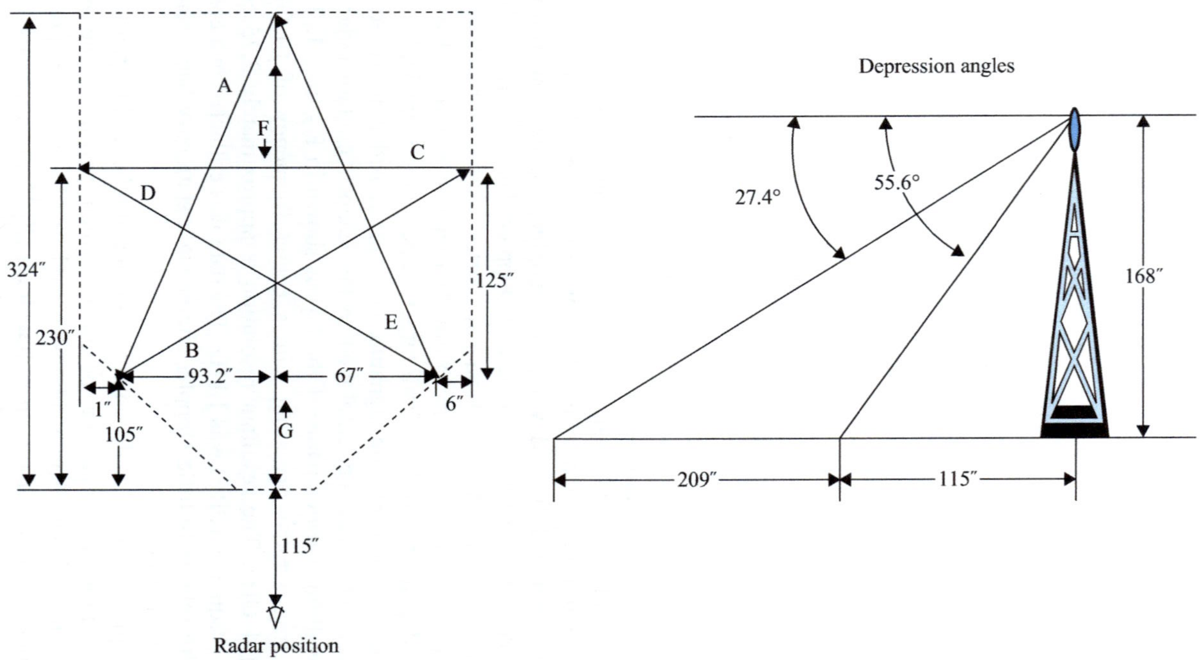

Figure 3.1 The paths walked by measured human subjects are labeled A through G (see [37])

moment of observation becomes less precise and vice versa. The application of a time-frequency transform to the range-time data creates a three-dimensional (range-Doppler-time) data cube denoted by $I_0(\underline{x}, n_{cpi})$, where $\underline{x} = [f, r]$ denotes range-Doppler coordinates and n_{cpi} indexes the current position of sliding window in the STFT.

Figure 3.2 draws a visual connection with simultaneously recorded video frames of the human walking and a coincident range-Doppler image. Specifically, this depiction allows range-Doppler responses to be intuitively associated with components of the human body. As the limbs move to their greatest displacement relative to the torso, their micro-Doppler approaches the target's gross Doppler shift, as seen in the right column of Figure 3.2. When the limbs are closest in range to the torso, they reach their maximum velocities or greatest micro-Doppler offset as seen in the left column of Figure 3.2. The radar is also known to be in front of the human subject and looking down with a depression angle between 27.4° and 55.6°; thus, range further distinguishes scatterer responses. The near pair corresponds to the arms while the farther pair is the legs. The left and right limbs are ambiguous in the RF domain, but the video allows positive micro-Doppler shifts to be associated with forward-swinging limbs and vice versa. Furthermore, Figure 3.2 shows that

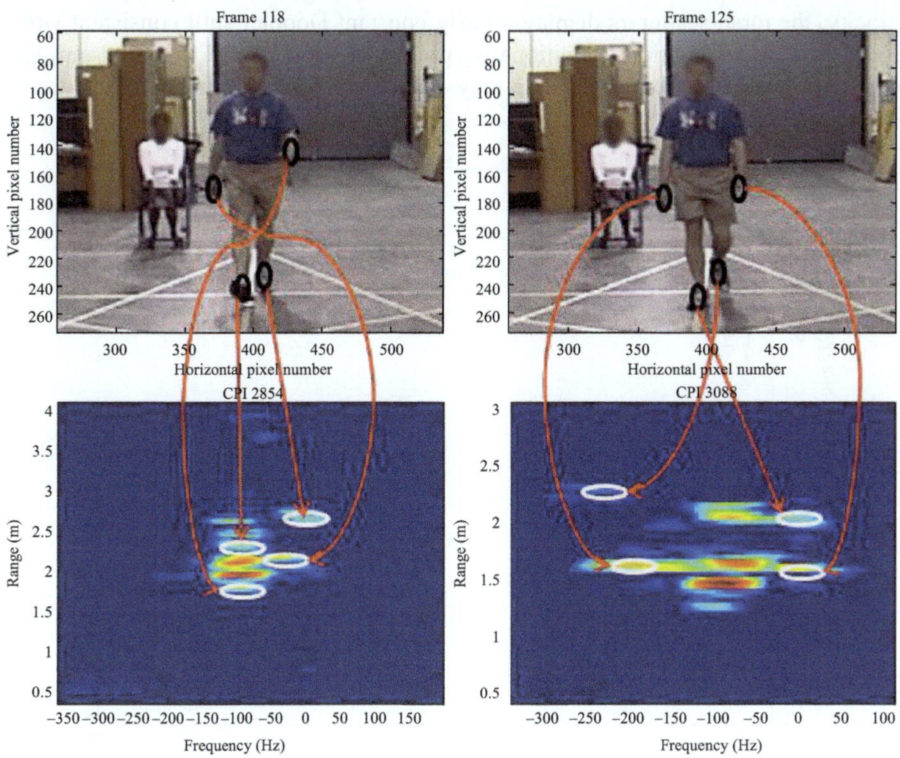

Figure 3.2 A portion of a human gait cycle in both the electro-optical and range-Doppler domains

the dismount signature appears to be composed of a finite number of simple scatterers as seen by the resolved range-Doppler blobs.

The nonrigid dynamic human has many moving parts and complex contours resulting in the potential for sophisticated scattering behavior. However, at a 10 GHz carrier frequency, past investigations have shown that 6–8 point scatterers (4 torso scatterers and 2–4 limbs scatterers) are oftentimes sufficient to model normal human radar returns [29]. Indeed, in examining a sampling of 1 200 pulses, we estimate the location of the 15 largest scatterers in each. The residual energy was then averaged over the pulses for each model order and is plotted relative to the total energy in each range profile in Figure 3.3. To select a knee in the curve, model orders of 8 and greater should model more than 95% of the dismount data energy. And, as illustrated via Figure 3.2, a model consisting of a parsimonious number of scatterer trajectories can be fairly intuitive to interpret in micro-range–micro-Doppler imagery.

For a more simplistic analysis, we may alternately eliminate the range or Doppler dimensions. For example, the walking human micro-Doppler signature is depicted in the time-frequency plot of Figure 3.4, which shows a projection of the range-Doppler-time signature along the range axis. Here, one would correctly expect the arms and legs to induce larger frequency modulations than the torso scatterers. Indeed, as shown for this subject walking at a nominally constant velocity, the torso scatterers display a fairly constant Doppler shift consistent with the target's macro-motion. Additionally, given the proportional size and similar composition of limbs and torso, one may expect greater scattering intensity from

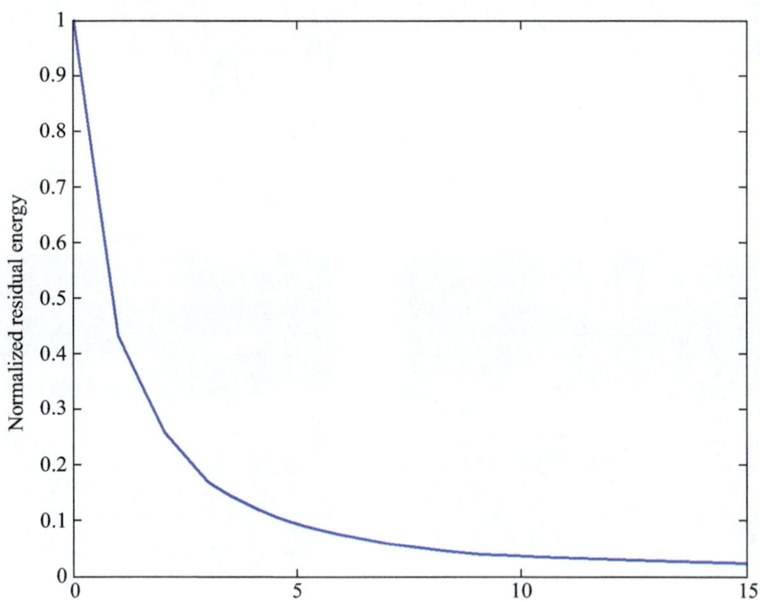

Figure 3.3 Fraction of the dismount response modeled on average for different orders of (3.2)

the torso. One is left to differentiate arms and legs based on their micro-Doppler behavior given the lack of range information to provide a spatial separation.

In contrast, the same signature is now portrayed in range-time in Figure 3.5, having projected along the Doppler axis. Using the insight that the target should be

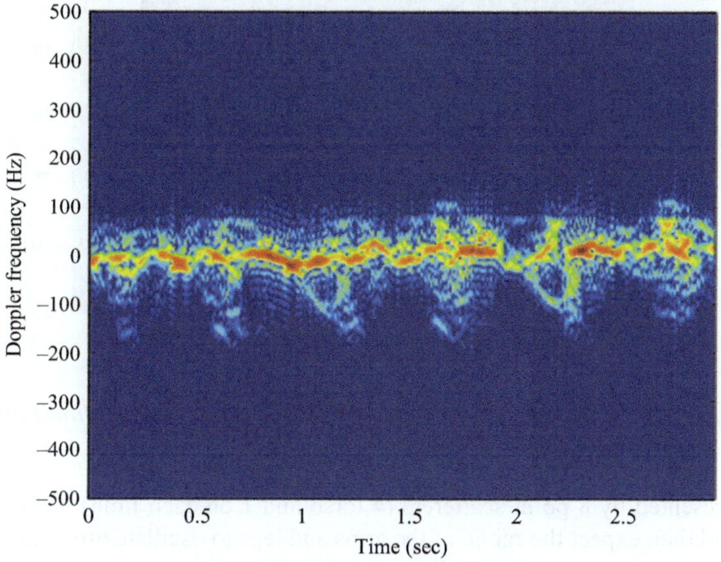

Figure 3.4 Dismount Doppler signature

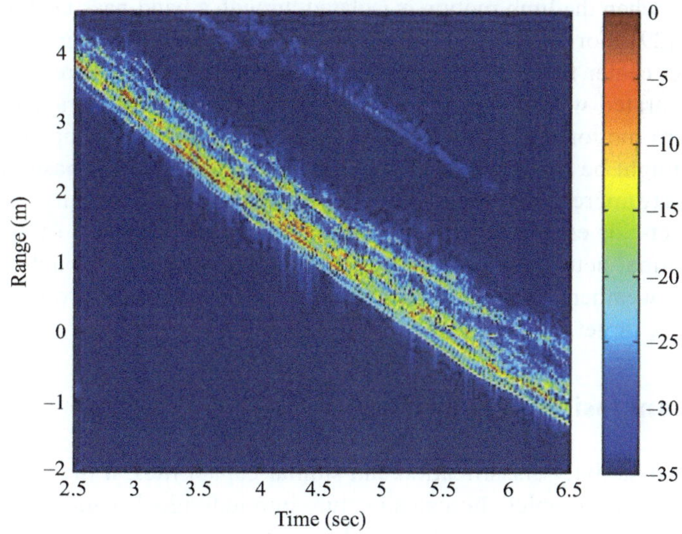

Figure 3.5 Range-time plot of measured data illustrates interaction of macro- and micro-motion features

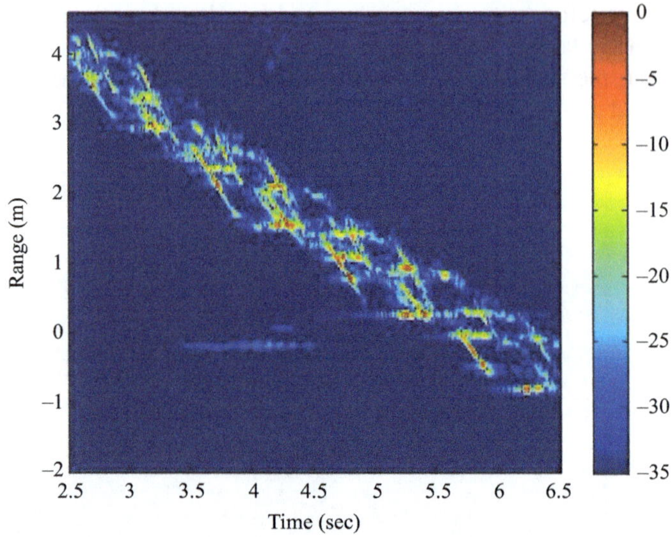

Figure 3.6 The isolated micro-motion data obtained from the μD filterbank shows the arm and leg responses

well represented by 8 point scatterers (4 torso and 1 on each limb) or even fewer, one would then expect the range of the arms and legs to oscillate around the targets macro-range motion and to be displaced in range from one another, given the down-looking orientation of the radar in Figure 3.1. Furthermore, there should be two limb pairs, each pair having an in-phase arm and leg. These concepts are readily seen when the limb motion is isolated through a band-pass Doppler filter in Figure 3.6 [29]. For comparison, applying a similar band-pass Doppler filter, but with a lower center frequency, isolates the torso scatterers in Figure 3.7.

As illustrated with this example, a far greater level of understanding of the human micro-motion signature can be obtained through micro-range exploitation than what might be extracted in micro-Doppler alone. In which case, the human signature may more easily be exploited to classify human activity, detect concealed objects, or estimate the human poses. Such abilities may offer means to detect abnormal activities from large standoff distances and through fog and in other poor weather conditions that prohibit many sensors from adequately observing the target.

3.3 Decomposition Algorithm

To aid human activity classification and similar capabilities, it may be beneficial to decompose the complete human signature into individual scatterer trajectories. These individual trajectories will provide information on the motion of specific body components, perhaps allowing for intuitive inference on human pose, motion, and activity. However, manual decomposition of the human signature may be tedious and time consuming. Therefore, automated decomposition algorithms are needed. Such a decomposition algorithm has been previously developed [20]. The first step in the decomposition process attempts to extract the scattering features from the range-Doppler images using nonlinear least squares (NLS). The data points are then kinematically associated via expectation maximization (EM). The process will be briefly reviewed, while a more detailed description may be found in [20].

3.3.1 Range-Doppler Feature Extraction

The feature extraction algorithm estimates the largest M peaks in each range-Doppler frame. In cases when a signal model is understood, an NLS optimization [38] can estimate the signal parameters. This may be viewed as iteratively finding the M scatterer parameter estimates that minimize the squared-difference between the measured signal and the expected model response:

$$\hat{\boldsymbol{\Theta}}_i(n_{cpi}) = \arg\min_{\Theta(n_{cpi})} \sum_{\underline{x}} \left| I_{i-1}(\underline{x}, n_{cpi}) - TF_2\{F_1^{-1}\{\hat{S}(\omega, k; \Theta(n_{cpi}))\}\} \right|^2 \quad (3.3)$$

where

$$I_i(\underline{x}, n_{cpi}) = I_{i-1}(\underline{x}, n_{cpi}) - TF_2\{F_1^{-1}\{\hat{S}(\omega, k; \hat{\boldsymbol{\Theta}}_i(n_{cpi}))\}\} \quad (3.4)$$

and $\hat{S}(\omega, k; \hat{\boldsymbol{\Theta}}_i(n_{cpi}))$ is the modeled scattering response defined by the parameter set $\hat{\boldsymbol{\Theta}}_i(n_{cpi})$. The minimizer of (3.3) is found by gradient descent with initial range-Doppler position vector $\hat{\underline{x}}_i = \lfloor \hat{f}_i, \hat{r}_i \rfloor$, which denotes the location of the peak

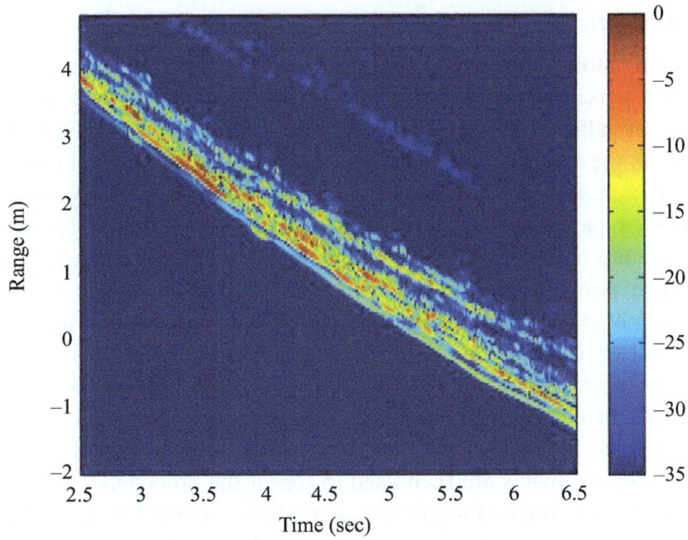

Figure 3.7 The isolated macro-motion data obtained from the μD filterbank shows the principal head and torso responses

intensity of $I_i(\underline{x}, n_{cpi})$. Following the CLEAN approach [39], after estimating each scatterer's parameters, that scatterer is subtracted from the frame via (3.4), and (3.3) is reinitialized with the location of the next strongest peak. We repeat this process until M peaks have been removed. This process is computationally intense. However, it may be accelerated by accepting the initial range-Doppler location of the scatterer as the estimated location, thus foregoing numerical optimization at the expense of accuracy. A mask is placed over the neighborhood of each subsequent peak to prevent artifacts from the scatterer extraction process from distracting the algorithm, though the size of this mask must be carefully chosen.

3.3.2 Feature Association

The range-Doppler feature extraction method provides an ensemble of measurement tuples (micro-range, micro-Doppler frequency, and scattering coefficient). In general, the coefficients of the scatterers will stochastically vary. Since the scattering measurements are innately ordered by amplitude, oftentimes the ith measurement tuple, corresponds to different scattering centers as a function of n_{cpi}, thus requiring an additional step of association similar to multiple target tracking. The algorithm in [20] uses EM to kinematically sort the range-Doppler scattering centers. To aid the EM algorithm, we first separate scatterers displaying macro- and micro-motion, and we then further parse the micro-motion scatterers to eliminate crossing range-time trajectories. The remainder of this section is organized as follows. We describe the EM process applied to the macro-motion scatterers and the two sets of micro-motion scatterers. We then detail the processes for separating macro and micro-motion scatterers and for parsing the micro trajectories.

3.3.3 Expectation–Maximization

Following application of the NLS algorithm and the parsing processes described subsequently, a parsed parameter set, $\boldsymbol{\Theta}$, contains elements we seek to assign to a finite number of salient features. This organization could be achieved straightforwardly via the range and/or Doppler estimates, if the two-dimensional range-Doppler trajectories of each scatterer feature were known. Likewise, if the correct associations were known, parameterized curves could be fit through the range data points to estimate the feature trajectories. Unfortunately, neither the correct curve parameters nor correct data point associations are known. In such cases, the EM algorithm is useful. The EM algorithm is an iterative two-step process alternating between an expectation step (E-step) and a maximization step (M-step). In this particular problem, the E-step assigns labels to the parameter vector estimates based upon their proximity to estimated feature trajectories. Whereas, the M-step uses the previous associations defined in the E-step to update estimates of the trajectories. The new scatterer trajectory estimates are then used to repeat the procedure.

During the E-step, the EM algorithm attempts to find the "best" or "expected" labeling of the NLS-estimated parameter vectors. Each parameter vector, $\underline{\theta}_i(k)$, is labeled with a value $\hat{q}_i(k)$ from the set $\mathbf{L} = \{0, 1, 2, \ldots, N_T\}$ where 0 represents no association, 1 corresponds to the first scatterer, 2 corresponds to the second scatterer, and so on until the last scatterer N_T. In order to determine the "optimal" labelings, a common association scheme used in radar tracking [40] known as gated global nearest neighbors (GNN) will be applied. An advantage to using gated GNN is that the gating avoids potential outliers and allows for the deviation-sensitive L_2 norm to be used in the M-step. In gated GNN, the objective is to find the associations from the set of valid associations for each pulse, $\mathbf{Q}(k)$, that result in minimum total normalized distance for each pulse. The GNN optimization is given by

$$\underline{q}(k) = \arg \min_{q(k) \in Q(k)} \sum_{i=1}^{N_S} c_1 \left(\check{r}(k; \hat{q}_i(k)) - \hat{r}_i(k) \right)^2 + c_2 \left(\check{f}(k; \hat{q}_i(k)) - \hat{f}_i(k) \right)^2 \tag{3.5}$$

where the label assignment vector is $\underline{q}(k) = \lfloor q_1(k), q_2(k), \ldots, q_{N_S}(k) \rfloor^T$ for the kth pulse. The estimated range trajectory for the lth scatterer is $\check{r}(k; l)$ with corresponding Doppler trajectory $\check{f}(k, l)$. The weights c_1 and c_2 are chosen to provide appropriate scaling to the range and Doppler measurements. The rules for determining valid associations to populate $\mathbf{Q}(k)$ may be summarized as follows [41]:

- The normalized distance between the position estimate, $\hat{r}_i(k) \in \underline{\theta}_i(k)$, and the fit function, $\check{r}(k; l)$, must be less than or equal to the gate, χ_g.
- A scatterer with only one range estimate meeting the gate criteria automatically has that estimate associated to it, assuming no other scatterer is also singly validated with the same estimate (in which case, the point is assigned to the closest scatter).
- Each range estimate may only be associated to one scatterer.
- A feature with no scatterer trajectories within the gate has no estimate assigned and receives label 0.

The set of valid associations, $\mathbf{Q}(k)$, is thus a subset of all association permutations. Furthermore, the valid association set may change from pulse-to-pulse based upon the trajectory estimates. Since the validity of all possible permutations are examined, the computational complexity of the algorithm grows rapidly with the number of scatterers and pulses, making the gating rule a valuable tool for eliminating invalid associations early. The classic GNN association has good performance with low computational expense. For more sophisticated association techniques, see [40, 41].

The M-step attempts to find parameterized scatterer trajectory estimates, $\check{r}(k; l)$, based on the associated feature estimates $\hat{r}_i(k)$ with labels $\hat{q}_i(k) = l$. In order to achieve this goal, a basis for the trajectory functions must first be chosen. There are multiple basis options, each with different advantages and disadvantages. For instance, an nth-degree polynomial, with basis functions given by $\phi_p(k) = k^p$ for $p = 0 \ldots n$, may be adequate and computationally inexpensive for fitting translational motion. On the other hand, radial basis functions, given by $\phi_p(k) = e^{-\beta|k - x_p|}$ for $p = 0 \ldots n$, may offer the flexibility needed to fit multiple nonuniform scatterer range oscillations. Meanwhile, sinusoids, given by $\phi_p(k) = \sin(2\pi\omega_p k)$ for $p = 1 \ldots n$, may parsimoniously model strictly oscillatory behavior.

By weighting the basis functions $\phi_p(k)$ with coefficients $\hat{c}(p,l)$, and summing over p, a trajectory estimate is formed (i.e., $\check{r}(k;l) = \sum_p \phi_p(k)\hat{c}(p,l)$). Thus after a basis has been chosen, the task of finding optimal scatterer trajectory estimates is reduced to calculating the maximum-likelihood basis parameters. The optimization for this step is

$$\{\hat{c}(p,l)\} = \arg\min_{c(p,l)} \sum_{k=0}^{N_{K-1}} \sum_{i=1}^{N_T} \left(\left[\sum_{p=1}^{n} \phi_p(k)\hat{c}(p,\hat{q}_i(k)) \right] - \hat{r}_i(k) \right)^2 \tag{3.6}$$

which may be found for $l = 1 \ldots N_T$ and $p = 1 \ldots n$ via linear least squares. In the sum over i in (3.6), unassociated feature ranges are skipped. These parameters will be used for the next (E-step) iteration. The algorithm continues until either a user-defined maximum number of iterations is exceeded or the change in parameters from one iteration to the next is less than a predefined threshold.

The EM algorithm, in general, has many potential convergence points. Depending upon the initialization, some of the resulting estimates model the data well, but many may result in poor trajectory estimates. It is therefore imperative that the first associations and trajectory estimates are thoughtfully chosen. However, the initialization scheme complexity may be reduced by separating the different trajectories into smaller subsets. We propose using post-extraction Doppler filters to separate macro-Doppler from the micro-Doppler scatterers. In certain cases, the micro-Doppler features may be further separated by incorporating typical target knowledge. These steps will be discussed next.

3.3.4 Macro-/Micro-Doppler Separation

Separating the macro-Doppler and micro-Doppler features greatly reduces the computational complexity of the EM algorithm. Oftentimes, the macro-Doppler signature will be particularly simplistic containing no overlapping scatterer range trajectories, whereas the majority of the signature crossings will be present in the micro-Doppler signatures. For human-radar signatures, we have found that macro- and micro-Doppler features can be cleanly separated by designating respective pass bands after feature extraction.

Given an estimate for the target's gross macro-Doppler trajectory, $\check{f}(n_{cpi})$, the separation process is straightforward. After extraction, the data points may be categorized into feature subsets defined as macro-Doppler:

$$\mathbf{\Theta}_M = \left\{ \hat{f}_i(n_{cpi}) \in \hat{\mathbf{\Theta}}_i(n_{cpi}) \middle| \left(\hat{f}_i(n_{cpi}) - \check{f}(n_{cpi}) \right)^2 \le \beta, \forall i, n_{cpi} \right\} \tag{3.7}$$

or as micro-Doppler:

$$\mathbf{\Theta}_\mu = \left\{ \hat{f}_i(n_{cpi}) \in \hat{\mathbf{\Theta}}_i(n_{cpi}) \middle| \left(\hat{f}_i(n_{cpi}) - \check{f}(n_{cpi}) \right)^2 > \beta, \forall i, n_{cpi} \right\} \tag{3.8}$$

where $\check{f}(n_{cpi})$ is the target's gross Doppler estimate. The bandwidth, $\sqrt{\beta}$, is set based upon radar parameters and known target dynamics. The bandwidth should be set large enough to incorporate small target speed deviations but small enough not to include larger target oscillations.

The macro-/micro-Doppler separation process should simplify the EM process for the resultant subsets. Since the macro-Doppler signature should portray the target's gross motion, the related range signature of the macro-Doppler signature should contain fewer scatterers (or at least no more scatterers) and few, if any, crossings in range. On the other hand, target signatures corresponding to the micro-motion will contain the majority of range crossings and gaps (partially introduced by the filtering) but will contain less scatterers than the entire target signature. At this point, the EM algorithm can be applied to the macro- and micro-motion data sets. For more complex signatures, however, it may be beneficial to further segregate the micro-Doppler features [20].

3.3.5 Decomposition Example

The proposed feature extraction method is applied to the previously described data set of Figure 3.5. The estimated range and frequency parameters are plotted in Figure 3.8. It is difficult to see discernible scattering paths, though the decoupling process will aid this immensely. First, following section 3.4, the macro- and micro-motion data sets are isolated as seen in Figures 3.9 and 3.10, which are consistent with Figures 3.6 and 3.7, respectively. At this point, the scatterer trajectories are initialized, and the EM algorithm is applied. The result of the decomposition algorithm can be seen in Figure 3.11, which shows the associated limb and torso features in a single plot consistent with Figure 3.8. The figure illustrates that the dismount data is composed of 4 torso scatterers and 4 limbs scatterers. Additionally, the decomposed target signature now has extracted features, or namely,

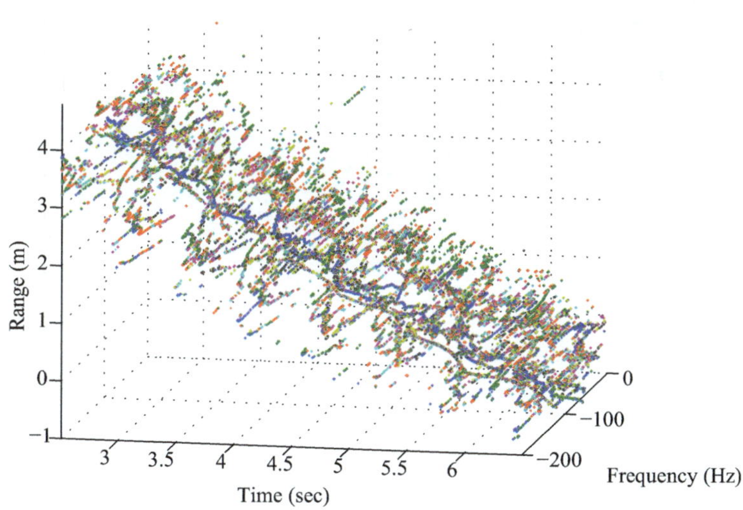

Figure 3.8 Doppler frequency, range, and time plot of the extracted points where the colors correspond to relative amplitudes

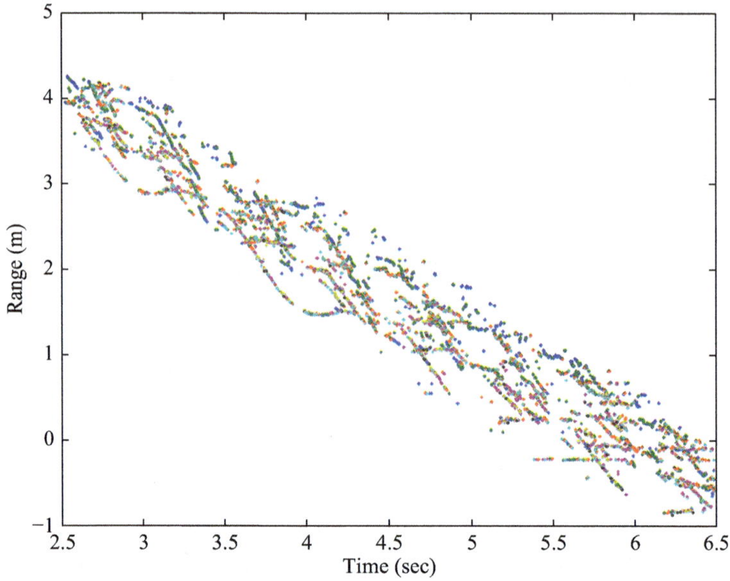

Figure 3.9 Post-extraction micro-motion limb data shows quasi-periodic behavior about a linear trajectory

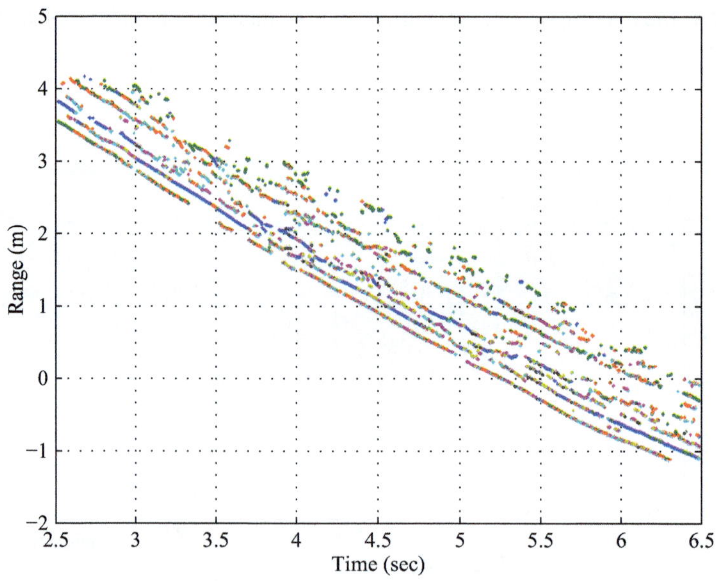

Figure 3.10 Range-time plot of macro-motion features displays a roughly linear trajectory

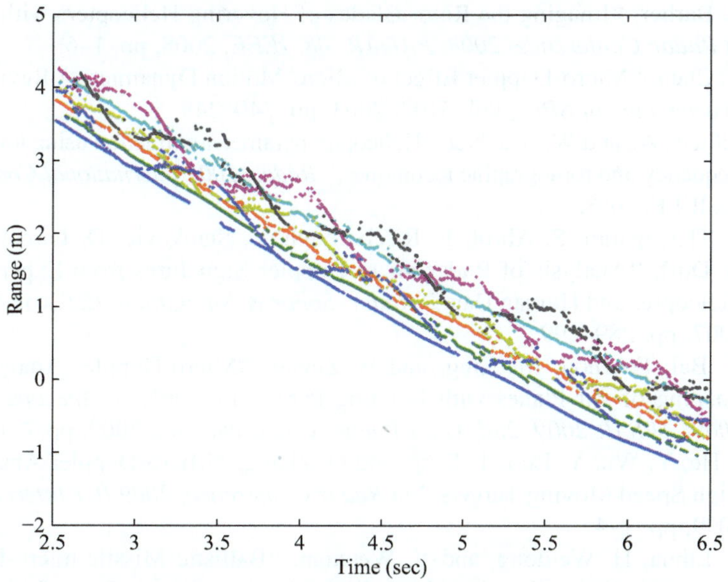

Figure 3.11 Scattering signature for an individual walking path A, toward the radar at an angle of 74° off to the left of the radar's line-of-sight

scattering coefficients, ranges, and frequencies organized by physical body component of origin. This organization allows the scattering and RCS from particular body regions to be statistical characterized.

3.4 Conclusion

In this chapter, the micro-range and micro-Doppler phenomenon has been discussed. The simultaneous utilization of both the micro-range and micro-Doppler domains provides an extra dimension in which to resolve and isolate individual scatterer paths corresponding to different body elements. Extracting information regarding specific body parts may be exploited for improved target detection, classification, and discrimination. An algorithm for automated micro-range–micro-Doppler target signature decomposition has been discussed. These concepts and algorithms were illustrated on measured human signatures.

References

[1] V. Chen and R. Lipps, "Time Frequency Signatures of Micro-Doppler Phenomenon for Feature Extraction," in *Proceedings of SPIE*, vol. 4056, 2000, pp. 220–226.

[2] B. Gallardo-Hernando, F. Perez-Martinez, and F. Aguado-Encabo, "Statistical Characterization of Wind Turbine Clutter in C-band Radars," in *Radar, 2008 International Conference on*, 2008, pp. 360–364.

[3] B. Barber, "Imaging the Rotor Blades of Hovering Helicopters with SAR," in *Radar Conference, 2008. RADAR '08. IEEE*, 2008, pp. 1–6.
[4] V. Chen, "Micro-Doppler Effect of Micro-Motion Dynamics: A Review," in *Proceedings of SPIE*, vol. 5102, 2003, pp. 240–249.
[5] Cilliers, A., and W. A. J. Nel, "Helicopter parameter extraction using joint time-frequency and tomographic techniques," *Radar, 2008 International Conference on*. IEEE, 2008.
[6] T. Thayaparan, S. Abrol, E. Riseborough, L. Stankovic, D. Lamothe, and G. Duff, "Analysis of Radar micro-Doppler Signatures from Experimental Helicopter and Human Data," *Radar, Sonar & Navigation, IET*, vol. 1, no. 4, 2007, pp. 289–299.
[7] X. Bai, F. Zhou, M. Xing, and B. Zheng, "Micro-Doppler Analysis and Imaging of Air-Planes with Rotating Parts," in *Synthetic Aperture Radar, 2009. APSAR 2009. 2nd Asian-Pacific Conference on*, 2009, pp. 610–613.
[8] J. He, Y. Wu, Y. Luo, T. Feng, and Q. Zhang, "Micro-Doppler Analysis of High Speed Moving Targets," in *Radar Conference, 2009 IET International*, 2009, pp. 1–4.
[9] L. Lihua, H. Weidong, and Y. Wenxian, "Ballistic Missile micro-Doppler Parameter Estimation based on Multi-Section Signals," in *Radar Conference, 2009 IET International*, 2009, pp. 1–4.
[10] H. Sun and Z. Liu, "Micro-Doppler Feature Extraction for Ballistic Missile Warhead," in *Information and Automation, 2008. ICIA 2008. International Conference on*, 2008, pp. 1333–1336.
[11] Z. Sun, B. Li, and Y. Lu, "Research on micro-Motion and micro-Doppler of Ballistic Targets," in *Radar Conference, 2009 IET International*, 2009, pp. 1–4.
[12] C. Baker, M. Vespe, and G. Jones, "Target Classification by Echo Locating Animals," in *Waveform Diversity and Design Conference, 2007. International*, 2007, pp. 348–352.
[13] L. Kong, T. Su, G. Cui, and J. Yang, "Life Detection Algorithm for Stepped-Frequency CW Radar," in *Radar Conference, 2009 IET International*, 2009, pp. 1–4.
[14] Z. Zhang, P. Pouliquen, A. Waxman, and A. Andreou, "Acoustic Micro-Doppler Gait Signatures of Humans and Animals," in *Information Sciences and Systems, 2007. CISS '07. 41st Annual Conference on*, 2007, pp. 627–630.
[15] F. Zhu, Y. Feng, Y. Luo, and Z. Qun, "A Novel Method of Identifying Moving Status of Bird based on Waveform Entropy Variance," in *Communications, Circuits and Systems, 2008. ICCCAS 2008. International Conference on*, 2008.
[16] F. Zhu, Y. Luo, Q. Zhang, Y. Feng, and Y. Bai, "ISAR Imaging for Avian Species Identification with Frequency-Stepped Chirp Signals," in *Geoscience and Remote Sensing Letters, IEEE*, 2010, pp. 151–155.
[17] A. Alejos, M. Sanchez, D. Iglesias, and I. Cuinas, "Real-Time Method for Human Presence Detection by using Micro-Doppler Signatures Information

at 24 GHz," in *Antennas and Propagation Society International Symposium, 2009. APSURSI '09. IEEE*, 2009, pp. 1–4.

[18] T. Dogaru and L. Nguyen, "FDTD Models of Electromagnetic Scattering by the Human Body," *IEEE Antennas and Propagation Society International*, 2006, pp. 1995–1998.

[19] T. Dogaru, C. Le, and G. Kirose, "Time-Frequency Analysis of a Moving Human Doppler Signature," *Army Research Laboratory, Technical Report*, 2009.

[20] O. Fogle and B. Rigling, "Micro-Range/Micro-Doppler Feature Extraction and Association for High-Resolution Radar," in *IEEE Transactions on Aerospace and Electronic Systems*, 2012.

[21] A. Ghaleb, L. Vignaud, and J. Nicolas, "Micro-Doppler Analysis of Wheels and Pedestrians in ISAR Imaging," in *Signal Processing, IET*, 2008, pp. 301–311.

[22] N. Yamada, Y. Tanaka, and K. Nishikawa, "Radar Cross Section for Pedestrian in 76 GHz Band," *R&D Review of Toyota CRDL*, vol. 39, no. 4, 2005, pp. 46–51.

[23] Z. Zhang and A. Andreou, "Human Identification Experiments using Acoustic micro-Doppler Signatures," in *Micro-Nanoelectronics, Technology and Applications, 2008. EAMTA 2008. Argentine School of*, 2008, pp. 81–86.

[24] H. Gao, L. Xie, and S. Wen, "Extraction of Micro-Doppler Signature of Cone-Shaped Target with Precession," in *Radar Conference, 2009 IET International*, 2009, pp. 1–4.

[25] L. Kangle, J. Weidong, L. Yongxiang, and L. Xiang, "Feature Extraction of Cone with Precession based on micro-Doppler," in *Radar Conference, 2009 IET International*, 2009, pp. 1–5.

[26] X. Xu, C. Ma, L. Zhang, and Y. Wang, "Modeling of Wideband Radar Micro-Doppler Signatures for a Coning Object," in *Radar Conference, 2009 IET International*, 2009, pp. 1–5.

[27] J. Keller, "Geometrical Theory of Diffraction," *Journal of the Optical Society of America*, vol. 52, no. 2, 1962, pp. 116–130.

[28] P. van Dorp and F. Groen, "Human Walking Estimation with Radar," *Radar, Sonar and Navigation, IEE Proceedings*, vol. 150, no. 5, 2003, October, pp. 356–365.

[29] O. Fogle and B. Rigling, "Parametric Model of High-Resolution Radio-Frequency Dismount Data," in *Aerospace and Electronics Conference, 2008. NAECON 2008. IEEE National*, 2008, pp. 349–352.

[30] M. Gerry, L. Potter, I. Gupta, and A. Van Der Merwe, "A Parametric Model for Synthetic Aperture Radar Measurements," *IEEE Transactions on Antennas and Propagation*, vol. 47, no. 7, 1999, pp. 1179–1188.

[31] J. Jackson, B. Rigling, and R. Moses, "Parametric Scattering Models for Bistatic Synthetic Aperture Radar," in *Radar Conference, 2008. RADAR '08. IEEE*, 2008, pp. 1–5.

[32] J. Jackson, B. Rigling, and R. Moses, "Canonical Scattering Feature Models for 3D and Bistatic SAR," in *IEEE Transactions on Aerospace and Electronic Systems*, 2008.

[33] L. Potter, D. Chiang, R. Carriere, and M. Gerry, "A GTD-based Parametric Model for Radar Scattering," *IEEE Transactions on Antennas and Propagation*, vol. 43, no. 1010, 1995, pp. 1058–1067.

[34] L. Potter and R. Moses, "Attributed Scattering Centers for SAR ATR," *IEEE Transactions on Image Processing*, vol. 6, no. 1, 1997, pp. 79–91.

[35] B. Rigling and R. Moses, "GTD-Based Scattering Models for Bistatic SAR," in *Algorithms for Synthetic Aperture Radar Imagery XI, Proceedings of SPIE*, 2004, p. 5427.

[36] L. Cohen, "Time-Frequency Distributions – A Review," *Proceedings of the IEEE*, vol. 77, no. 7, 1989, July, pp. 941–981.

[37] J. M. Blackaby, B. D. Rigling, and P. M. Hanna, "Simultaneous radar frequency/electro-optical tracking and characterisation of dismounts," in *Signal Processing*, IET, vol. 5, no. 1, 2011, pp. 49–57.

[38] Stoica, Peter, and Randolph L. Moses. *Spectral analysis of signals*. Upper Saddle River, NJ: Pearson/Prentice Hall, 2005.

[39] J. A. Hogbom, "Aperture Synthesis with a Non-Regular Distribution of Interferometer Baselines," in *Astronomy and Astrophysics Supplement*, 1974.

[40] Y. Bar-Shalom and T.E. Fortmann, *Tracking and data association*. Academic Press. 1988.

[41] S. S. Blackman, *Multiple-Target Tracking with Radar Application*. Artech House, 1986.

Chapter 4

Range and Micro-Doppler Analysis of Human Motion Using High Resolution Experimental HYCAM Radar

Antoine Ghaleb and Luc Vignaud

4.1 Introduction

In radar imaging it is well known that relative motion or deformations of parts of illuminated objects induce additional features in the Doppler frequency spectra. These features are called micro-Doppler effect and appear as sidebands around the central Doppler frequency. They can provide valuable information about the structure of the moving parts and may be used for identification purposes [1–4].

Most of related papers deal with time/micro-Doppler analysis [5, 6]. Here we propose to emphasize the analysis in a three-dimensional imaging space, using both the micro-Doppler and a high range resolution analysis over time. As a matter of fact, a fine range separation enables us to better discriminate the various effects caused by the time varying reflectors.

We concentrate our study on the human motion. We see how micro-Doppler signatures can be used to extract information on pedestrian gait.

We further process data recorded by the experimental HYCAM ultrawide band radar and build temporal series of range-Doppler images, which correspond to radar "movies" of humans in various motions. Meanwhile, we develop a simulator tool that enables us to link experimental data and models in order to extract physical features.

4.1.1 Micro-Doppler Effect in Radar Imaging

In traditional radar imaging, rigid-body motion is usually assumed without considering nonstationary targets [7–13]. But the relative motion, or the deformation of parts of illuminated objects, induces additional features in the Doppler frequency spectra. These features are called micro-Doppler effect and appear as sidebands around the central Doppler frequency [5, 6, 14–16]. Micro-Doppler spectrum and its evolution through time are then characteristic of the target in motion. They can provide valuable information about the structure of the moving parts and may be used for recognition purposes [17–21].

However, most of the recent works on the subject are still focused on characterizing the one-dimensional time-frequency evolution of micro-Doppler.

We show that adding high range resolution enables us to better analyze the signatures of time varying targets. The idea of combining time varying Doppler and high range resolution had already been introduced in [13] through the concept of "Kinematic radar imaging" (or radar "movies").

This implies of course the development of a dedicated experimental radar that can provide a super-high range resolution while preserving the micro-Doppler flow. The experimental HYCAM radar (Hyper CAMera [34, 35], developed by ONERA) has made this feasible through the adaptation of a special architecture and waveform design.

4.1.2 Application to Human Motion Analysis

Human motion is quite unique. Previous studies showed that radar signature can reveal information on the human's behavior [22–26]. When a human moves, different parts of his body (torso, arms, legs) have a particular motion that produces characteristic Doppler signatures. That is the reason why most of the methods are based on the spectral analysis of the received signal [22–27]. Considering the distribution and the evolution of the velocities over time, Doppler spectrum can be seen as a unique signature. As the evolution of the Doppler spectrum is periodic, a time Doppler variation analysis enables the feature extraction of the pedestrian's gait [22–24, 28–30].

In most of the papers relative to the subject, detection and recognition of pedestrians are only based on the motion rhythm analysis. Fourier transform and more recently time-frequency transforms (TFT) are used to analyze the Doppler and identify the different parts of the human body playing part during the motion [31–33]. Time-Doppler spectrograms enable us to retrieve some motion features such as velocity or gait cycle frequency.

However, such analysis of the human's motion may show some limitations. Analyzing the time-Doppler variation requires to observe the phenomenon during a long time, at least long enough to retrieve the gait cycle frequency. But other motion features could vary during this period, particularly the pace, the velocity, or the direction of the motion. In such cases, the analysis would become more delicate. Moreover, the presence of several characters simultaneously in the radar field of sight could involve interferences. Doppler frequencies would thus be mixed and the received signal would not be interpretable anymore.

Introducing the range dimension [1–4, 13] in the analysis shortens the time needed to observe the pedestrian, in order to detect him and distinguish the different parts of his body at a given time. A trained observer could indeed detect pedestrian characteristic features from a single image. A micro-Doppler analysis combined with a high range resolution enables a more accurate analysis of a moving pedestrian radar signature with a better separation of the different parts of his body. It also permits to localize and analyze simultaneously the gait of several peoples, as well as the signature of many other possible human activities (cycling, jumping, boxing, etc.).

The experimental HYCAM radar was used to capture data on pedestrians from the ONERA laboratory roof [1–4, 34, 35].

In addition to the experiments, we have developed a simulator tool that permits to link experimental data and models in order to extract physical features.

To better understand and validate experimental data, we have also worked with synthetic data provided by a motion capture system. These data have enabled us to compute simulated radar backscattering of every parts of the human body in order to perform spectrograms and range-Doppler images of pedestrians.

Using these data, we aim at emphasizing the complex interaction between a radar signal and a pedestrian (walking or running), by analyzing the phenomenon in three dimensions (range, Doppler, and time). We use both simulated and experimental data and take into consideration the angle between the pedestrian's direction and the radar's line of sight (LOS). This analysis demonstrates how that kind of tool can be useful to improve our knowledge in human recognition and gait analysis in radar imaging.

This chapter is organized into several sections. In the Section 4.2 we give some insight on the human motion analysis (especially the walking and running cases). In Section 4.3, we describe its implication on radar imaging and the signal processing tools used to extract relevant information on the motion. In Section 4.4 we simulate radar returns using motion capture data. The next two sections are experimental, Section 4.5 describes the HYCAM radar and the special waveform used, then Section 4.6 describes the experiment itself and the results on human motions. Section 4.7 has the conclusions.

4.2 Some Insights on Human Motion

The human gait is a complex mechanism where the body and limbs move in precise coordination to create human locomotion. After presenting the main features of walking and running, we will do a kinematic analysis of different parts of the body according to a human motion model.

4.2.1 Walking and Running Principle

A graphical representation of a typical walking cycle is shown in Figure 4.1(a). The gait cycle extends from heel strike to heel strike of one leg and includes the stance and swing phases of both legs. In the basic gait cycle the movements are divided into the times when the foot is on the ground (the stance phase that takes up 62% of the full gait cycle) and when the foot is off the ground (the swing phase that takes up the remaining 38% of one full gait cycle). The time period when both feet are on the ground will be referred to as double support. The gait cadence, alternatively referred to as the step rate, is defined as the number of steps taken per minute. The stride length is defined as the distance between two consecutive contacts of the same foot.

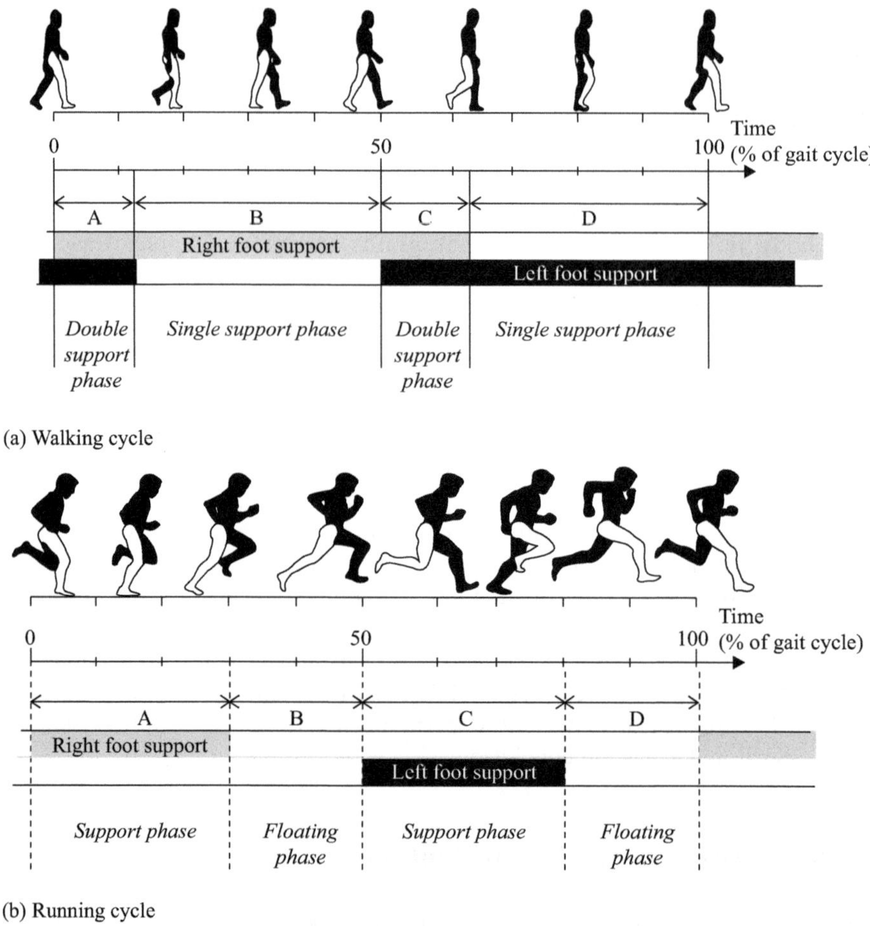

Figure 4.1 Gait cycle (drawing inspired by Adelaar [48])

Concerning the running cycle (Figure 4.1(b)), the main differences with the walking cycle are the following:

- The stance phase is shortened.
- The swing phase is lengthened.
- There is no time of double support.
- There is a nonsupport phase in which neither leg is weight bearing (double float).

4.2.2 Human Motion Model and Kinematic Analysis

Human motion has been the object of many studies. Muybridge (1830–1904) and Marey (1830–1904) invented and used chronophotography to decompose motion of human beings in time and space. Due to the large number of applications (animation, medicine, etc.), human motion is nowadays often studied. Most of the

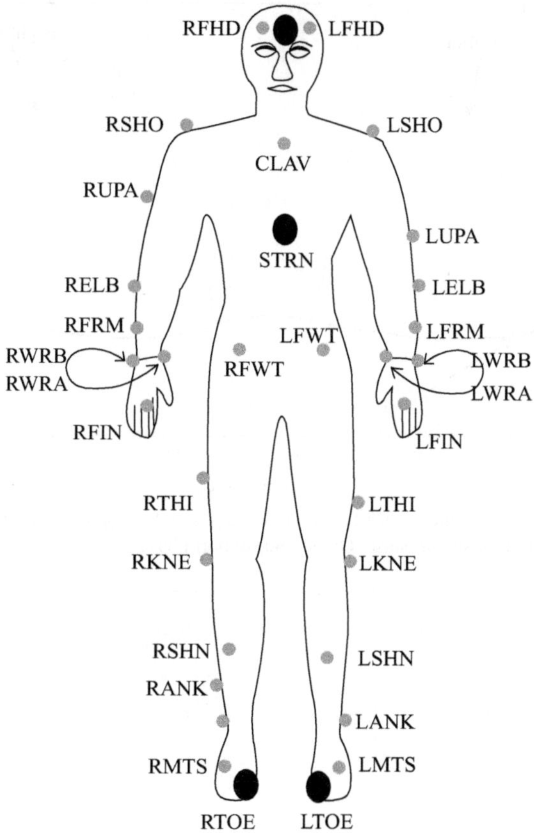

Figure 4.2 Human model

methods lie on biomechanical knowledge [36], mainly using the kinematic of rigid parts of the body [37]. Global motion model developed by Boulic [38, 39] based on empirical considerations can be used but is limited by the number of scenario possibilities, particularly when the motion becomes complex.

From the 1980s, the advent of motion capture system enabled significant progress in the human motion synthesis. Such system records the three-dimensional position over time for the different parts of the body through sensors distributed all over it. We have used some data that have been made publicly available for free on the Carnegie Mellon University website (mocap.cs.cmu.edu). They include the three-dimensional coordinates over time for 41 markers placed on the body as shown in Figure 4.2. Still, only few basic motions are available, but it turned out to be enough for our first radar simulations.

We began with a simple kinematic analysis in order to better understand the motion of the different parts of the human body, for walking and running. We draw the temporal evolution of the velocity, for walking (Figure 4.3(a)) and running (Figure 4.3(b)), of some representative points taken from the human motion model. These points correspond to torso and both feet (Figure 4.2). Because we wish to

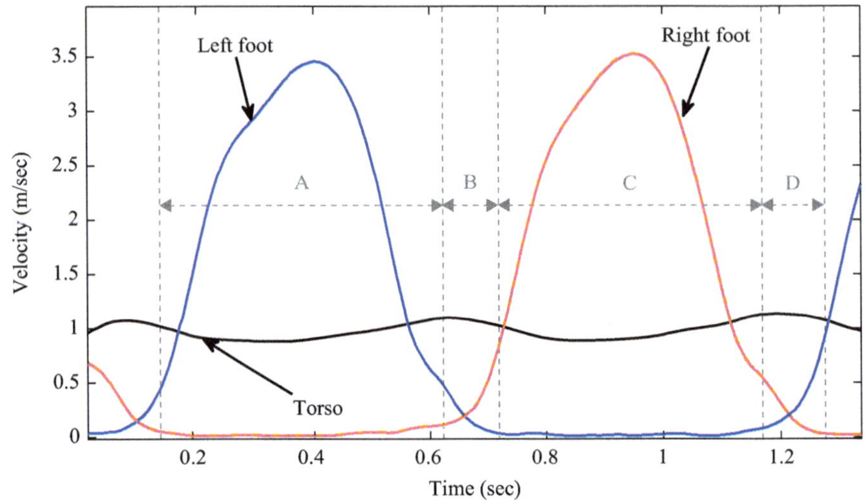

(a) Walking: A: Right foot stance phase/Left foot swing phase; B: Double support phase; C: Left foot stance phase/Right foot swing phase; D: Double support phase

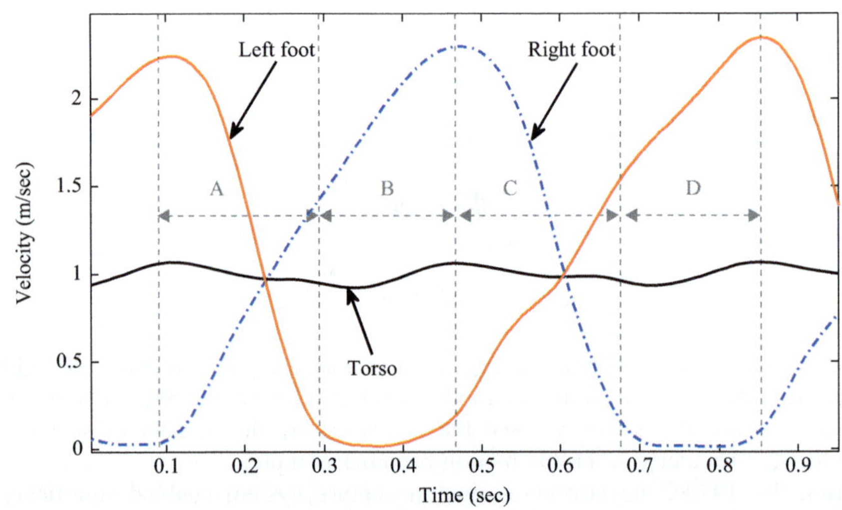

(b) Running: A: Floating phase; B: Left foot stance phase/Right foot swing phase; C: Floating phase; D: Right foot stance phase/Right foot swing phase

Figure 4.3 Kinematic analysis of feet and torso (walking and running)

compare the relative velocity of the different parts of the human body, we have normalized velocities by the mean pedestrian speed. Thus, the torso velocity is artificially set around 1 m/sec for both paces.

We observe that even the swinging motion of legs does not seem to fit a simple analytical model in three dimensions and in function of time. Human gait is indeed a complex motion of swinging arms and legs.

Next, we briefly describe the walking and the running gait cycle and try to explain the main differences between these two paces.

For each of the two paces, we can define motion cycle corresponding to a stride (two steps) defined by a time interval T_c and a length L_c. These two parameters are linked by the pedestrian's mean velocity V_0 as $L_c = V_0 T_c$. Let us note for both paces that hands make progress in distance steadily in a sinusoidal way, whereas feet alternatively stop during a part of the cycle.

For each foot, we can define stance phase when the foot is on the ground and a swing phase when the foot is off the ground.

We have observed that during the swing phase, the foot's velocity can reach 3.5 times $V_{0\,walk}$ in the walking case and 2.3 times $V_{0\,run}$ in the running case. The difference between these two values is due to the fact that during a complete cycle, the foot remains longer on the ground in the walking case than in the running case. The value reaches a higher value in the walking case because the foot needs to cover the same distance as the other parts of the body, and thus it has to compensate the distance uncovered during the stance phase.

There are also other major differences between walking and running. In the walking case, there is a time of double support where both feet are on the ground. Inversely for running, there is a nonsupport phase (floating phase) in which neither leg is weight bearing.

4.3 Signal Processing Tools for Radar Observation of Human Motion

4.3.1 Radar Asset for Human Observation

Radar can be very helpful for human detection. In spite of its low radar cross section (RCS), a pedestrian can be detected through the intermediary of its motion. When the radar remains still (ISAR like configuration), the target Doppler will naturally pop it outside of the clutter region that remains stable. Assuming the pedestrian is moving in a motionless environment, it is possible to distinguish a pedestrian through the Doppler processing. Indeed, the immobile elements are processed at zero Doppler and may be easily filtered out. Not only a pedestrian can be detected but his gait modifies its signature in a range-Doppler radar image. Indeed, the human motion induces a signature spreading both in distance and Doppler.

Let us go back to Figure 4.3 illustrating the velocity curves. At a given time, we see that different parts of the human body may have different velocities. There is a spreading of the different speeds around the mean velocity. In the Doppler frequency domain, it induces additional micro-Doppler frequencies around the central Doppler frequency.

From a spatial point of view, we can also notice that there is a spreading around the pedestrian's mean position. Simple geometrical considerations show that this spreading is more noticeable when the radar is not directly facing the pedestrian but when it is placed overhead. Changing the radar elevation enables to better highlight the different parts of the body. Elevating the radar enables to emphasize the human height, as shown in Figure 4.4. Thus we have positioned the radar such that the elevation is about 45°, which is a good trade-off to emphasize both head and foot range.

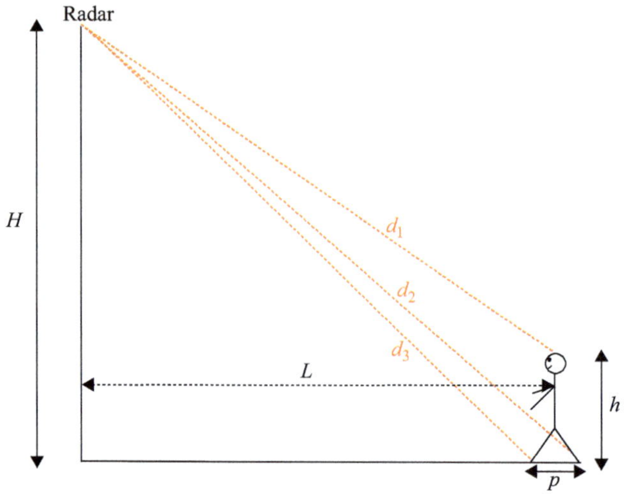

Figure 4.4 Geometry of an overhead radar observation

We then need a super-high range resolution radar to be able to take into account the micro-motions variation in range. We obviously need a resolution of few centimeters to be able to discriminate the different parts of the human body.

Moreover, the phenomenon is not stationary in both range and Doppler. It is thus interesting to analyze the phenomenon through time in addition to the range-Doppler plane. Hence the idea is to build range-Doppler "movies" through time.

These variations must be taken into account in the radar imaging processing as the range-Doppler-time parameters are coupled by nature. Nonstationary phenomenon may induce defocusing especially if the integration time is too long compared to the variation of the observed phenomenon.

4.3.2 Time-Doppler Analysis

Most of the papers dedicated to radar analysis of the human motion [22–27] and more recently [28, 29, 40–43] are based on the Doppler spectrum analysis over time. Fourier transform or time frequency approaches are used to analyze Doppler spectrums and identify the different parts of the body involved in the human motion.

Analyses are often displayed with a spectrogram-like view, which is the representation of the Doppler spectrum as a function of time: $I(f_d, t)$. A spectrogram can be processed by short-time Fourier transform (STFT), which is equivalent to processing a succession of Fourier transforms on short windows.

$$STFT\{h(t)\} = I(f_d, t) = \left| \int_{\tau} h(\tau) w(\tau - t)\, e^{-j2\pi f_d \tau}\, d\tau \right|^2 \tag{4.1}$$

where $h(t)$ denotes the analyzed signal, $w(t)$ the sliding window, t the time, and f_d the Doppler frequency that is linked to the relative velocity v with $f_d = \frac{2f_0 v}{c}$. The signal $h(t)$ is the backscattering function of the observed scene for the carrier frequency f_0.

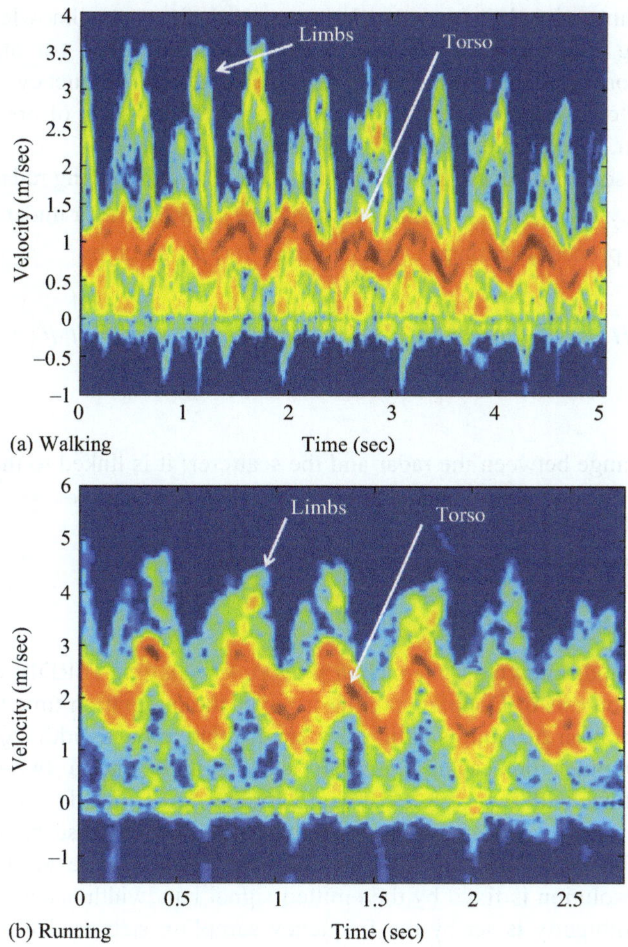

Figure 4.5 Comparison of walking and running using spectrograms of real data

Figure 4.5 is given as an example of such spectrograms. Data acquisition and image interpretation will be detailed further on. These representations enable us to distinguish walking from running, and to extract some motion features such as velocity, gait cycle frequency, and step length.

But this analysis can be improved using range diversity. Adding the range dimension to the micro-Doppler analysis will enhance the human motion radar analysis.

4.3.3 Range-Doppler Imaging

Range-Doppler imaging is a technique to compute the radar reflectivity map of a mobile object: elementary reflectors are localized by processing the scattering data that are collected by a monostatic coherent radar system, in combination with the relative motion of the object [7–9, 11, 13].

A two-dimensional image can be computed from the knowledge of the backscattering function on a domain in frequency and time (or angle) space. High resolution in range is obtained by a wide bandwidth frequency signal while Doppler (or "cross-range") resolution depends on the duration of observation (or on the angular aspect excursion) of the object.

The backscattering coefficient $H(f, t)$ collected by the imaging radar system for each frequency f, $f \in \left[f_0 - \frac{\Delta f}{2}; f_0 + \frac{\Delta f}{2}\right]$ and over slow time t (or under angles $\theta(t)$) enables us to process the range-Doppler image $I(x, f_d)$ given by

$$LRDI\{H(f,t)\} = I(x, f_d) = \left| \int_{\Delta f} \int_{\Delta t} H(f,t)\, e^{-\frac{4j\pi f x}{c}}\, e^{-2j\pi f_d t}\, dt\, df \right|^2 \tag{4.2}$$

where

- x is the range between the radar and the scatterer: it is linked to the round-trip delay of the signal $\frac{2x}{c}$, where c denotes the velocity of light.
- f_d is the Doppler frequency.
- f_0 is the carrier frequency of the emitted signal.
- Δf is the frequency bandwidth of the emitted signal.
- Δt is the Doppler integration time.

This method, known as *linear Range Doppler imaging* (LRDI), can be used under the hypothesis of Doppler stationarity over the integration time (of a narrow angular exploration $\Delta\theta$) and is valid for a small relative bandwidth ($\Delta f / f_0 \ll 1$).

Range-Doppler processing equation is simplified into a two-dimensional Fourier transform with two couples of dual variables ($x \leftrightarrow f$) and ($f_d \leftrightarrow t$). The first Fourier transform enables to process range profiles (or impulse response of the scattering) whereas the second achieves a Doppler analysis on every range cell.

Range resolution is fixed by the emitted signal bandwidth, such as $\delta x = \frac{c}{2\Delta f}$.

Range ambiguity is set by the frequency sampling step used for emission δf (or step between the frequencies of the emitted signal), such as $\Delta x = \frac{c}{2\delta f}$.

Thus the frequency step will be preferably chosen so that the image range ambiguity is large enough to include the target range total displacement.

Doppler ambiguity is set by the time $T_r = \frac{1}{PRF}$ between successive measurements of the emitted signal, such as $\Delta f_d = \frac{1}{T_r}$. This controls the available unambiguous Doppler spreading. Consequently the pulse repetition frequency (PRF) must be chosen according to the expected Doppler frequency extension of the observed phenomenon in order to avoid Doppler aliasing.

Doppler resolution δf_d is fixed by the chosen integration time $\Delta T = NT_r$ where N is the total number of sampled pulse $\delta f_d = \frac{1}{\Delta T}$.

4.3.4 *Range-Doppler "Movies"*

As stated before, the range-Doppler analysis supposes that the Doppler frequency remains constant during the integration time, which may not be the case when the target variations are fast. Then, there exists a trade-off between the Doppler resolution and the stationarity of the phenomenon.

We can then combine the two previous approaches by replacing the Doppler Fourier transform over time in (4.2) by a time-Doppler analysis such as (4.1):

$$I(x, f_d, t) = \left| \int_f \int_t H(f, \tau) w(\tau - t)\, e^{-\frac{4j\pi f x}{c}}\, e^{-2j\pi f_d \tau}\, d\tau\, df \right|^2 \tag{4.3}$$

In other words, it is equivalent to a time-Doppler analysis on each range cell. We call the three-dimensional image $I(x, f_d, t)$ range-Doppler movies or kinematical images [13].

4.4 Simulation of Human Motion Radar Observation

Our objective is to better understand the interaction between the transmitted signal and the different parts of the human body and analyze their respective contribution in radar imaging in range, Doppler and time.

A simulation tool is very desirable. On one hand, it helps us to understand the phenomenon from a determinist motion model; on the other hand, it should cross-check with the data obtained during the measurement campaign performed by the experimental high range resolution radar.

We choose the point model provided by a motion capture system (as described previously in Figure 4.2).

4.4.1 Simulation Configuration

We have restrained our model to simple motions: walking and running on a straight line at constant velocity. The objective is to faithfully reproduce some of the HYCAM experiment (cf. Figure 4.6) in order to compare images obtained with experimental and simulated data.

As illustrated in Figure 4.6, we can set up the pedestrian anywhere in the scene and put the radar at a fixed coordinates, according to the distance of observation and the angles θ and φ between the radar LOS and the pedestrian direction.

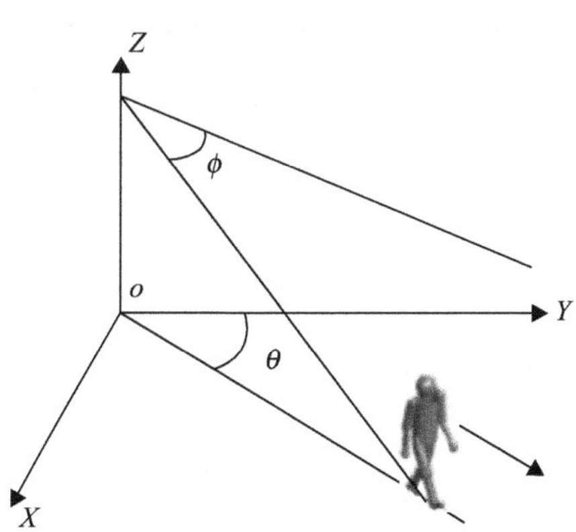

Figure 4.6 Angles between the pedestrian direction and the radar LOS

Let us define the angles between the pedestrian direction and the radar LOS:

- Azimuth θ is the angle between the pedestrian direction and the LOS projection on the ground plane (0, *X*, *Y*). When $\theta = 0°$, the pedestrian is moving in front of the radar.
- Elevation φ is the angle between the horizontal plan and the radar LOS. In the experiment, φ is comprised between 40° and 50°. In our simulations the elevation will be fixed (mostly at 0° or 45°).

As explained in section 3.1, the direction of observation (θ, φ) will have a strong influence on the interaction between the radar signal and the different parts of the body. It is also described in section 4.3.2.

4.4.2 Radar Signal Model

In the simulations, we assume that the emitted signal is entirely reflected by each point reflectors in the direction of the reception antenna. We also consider that all the reflectors are visible with a constant strength during the whole acquisition. Note that a refined model could be built taking into account the masking and the relative orientation of each point.

We also take the assumption that the motion is quasi-static in slow time, which means that we neglect the influence of motion within the fast time of an individual pulse measurement: the Doppler will only be considered in the "slow" time of pulses repetition. We are entitled to do so since the speeds involved are very slow.

The simulated backscattering function is the sum of the individual backscattering function of each point:

$$H(f,t) = \sum_{n=1}^{N} c_k\, e^{j\phi_k(f,t)} \tag{4.4}$$

with the phase $\phi_k(f,t) = \frac{4\pi f D_k(t)}{c}$, where D_k is the distance between the *k*th reflector and the radar.

This simulation supposes that all the frequency bandwidth can be emitted simultaneously and almost instantaneously. We will see that the experimental setup uses a special waveform called orthogonal frequency-division multiplexing (OFDM) to achieve the sampling over all frequencies at each time *t*.

This approach enables us to compute the backscattered coefficients at each time and for every transmitted frequency and then to process range-Doppler images by two-dimensional Fourier transform following (4.2) or range-Doppler movies through (4.3).

Analytic formulas may even be found, provided there exists some simple models for $D_k(t)$.

Parameters for the following simulations are given in Table 4.1. The pedestrian is moving with a constant velocity V_0 on a straight path.

4.4.3 Model-Based Time-Doppler Imaging

In the literature, most reports on human motion radar analysis are performed in corridors with the pedestrian facing the radar (or back to it). Here we also explore

Table 4.1 Motion simulation parameters

	Walking	Running
$V_0\left(\frac{\text{m}}{\text{sec}}\right)$	1.18	3.55
T_c(sec)	1.10	0.75
L_c(m)	1.30	2.66

how the relative orientation between the radar and the pedestrian can influence the radar signature.

The spectrograms displayed in the next paragraphs represent the time-velocity variation of three markers (torso and two feet). For comparative purposes, velocity is normalized by the mean velocity.

We can observe the evolution of the velocity during the gait cycle, in function of the relative orientation between the pedestrian's direction and the radar LOS θ and φ:

- When $\theta = 0$ the person is walking toward the radar LOS.
- When $\varphi = 90$ the radar is above the pedestrian's head.

4.4.3.1 Walking versus Running

Walking (Figure 4.7)

The sternum's velocity has slight variations. We can see that the parts move at different velocities during the gait cycle. Their contribution can be quite mixed up and it may be difficult to separate them accurately in real signals. Feet do show some characteristic patterns. One foot remains still while the other successively accelerates to reach it maximal velocity and decelerates to be still. It reaches a peak of velocity (which can be about $3V_0$ when the pedestrian is walking in front of the radar) in a very short time. These brutal variations have to be considered when imaging. Points will be not focused if the integration time is too long.

Running (Figure 4.8)

The running cycle shows major differences with the walking cycle. The stance phase is shortened, while the swing phase is lengthened. There is no time of double support but a nonsupport phase in which neither leg is weight bearing (double float).

4.4.3.2 Influence of the Radar Orientation

Doppler frequencies differ if the radar is not in the pedestrian's direction. Depending on the pedestrian's relative orientation, the Doppler frequencies are induced by different parts of the body.

Variation of θ

When the radar is in the motion pedestrian's plan (O, X, Y) (which means $\varphi = 0$) resulting Doppler is only induced by the translational velocity's component. Considering this component is preponderant in the motion, the Doppler spreading will be maximized in this configuration. When the pedestrian is not in front of the radar, the amplitude of the velocities decreases with θ because of the projection of the

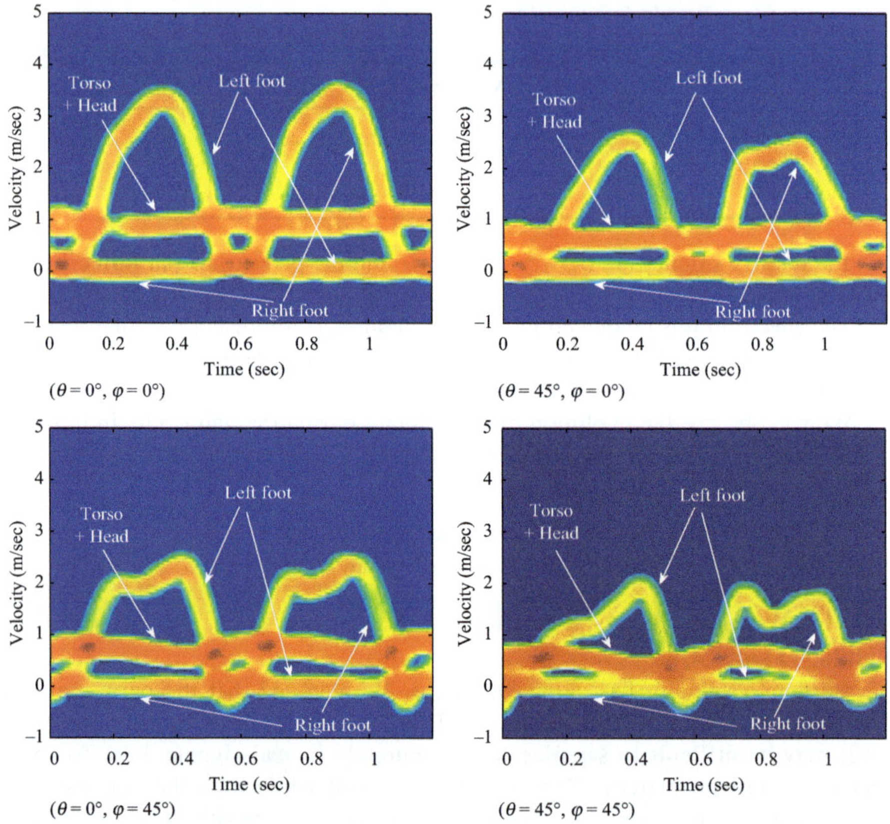

Figure 4.7 Model-based spectrograms (walking phase) with three markers (torso and two feet)

velocity vector. Spectrogram's form thus looks like velocity curves in Figure 4.3. The amplitude varies with $\cos(\theta)$.

When $\theta = 90$ (pedestrian moving perpendicular to the radar direction) only feet may have non-zero Doppler, even if it is not significant. It is due to the fact that their velocity vector does not remain on the direction plan.

Variation of φ

The higher the radar (the bigger is φ) the more the vertical velocity component is significant in the resulting Doppler. Torso vertical motion is consequently measurable especially for running. Even negative Doppler can appear with the limbs motion.

If the radar was just above the pedestrian (not represented here), the Doppler would only be induced by arm's and leg's vertical motion.

4.4.4 Model-Based Range-Doppler Movies

We have just seen that using spectrograms was not easy to fully separate the various parts of the moving body. Moreover, the motion parameters may vary over

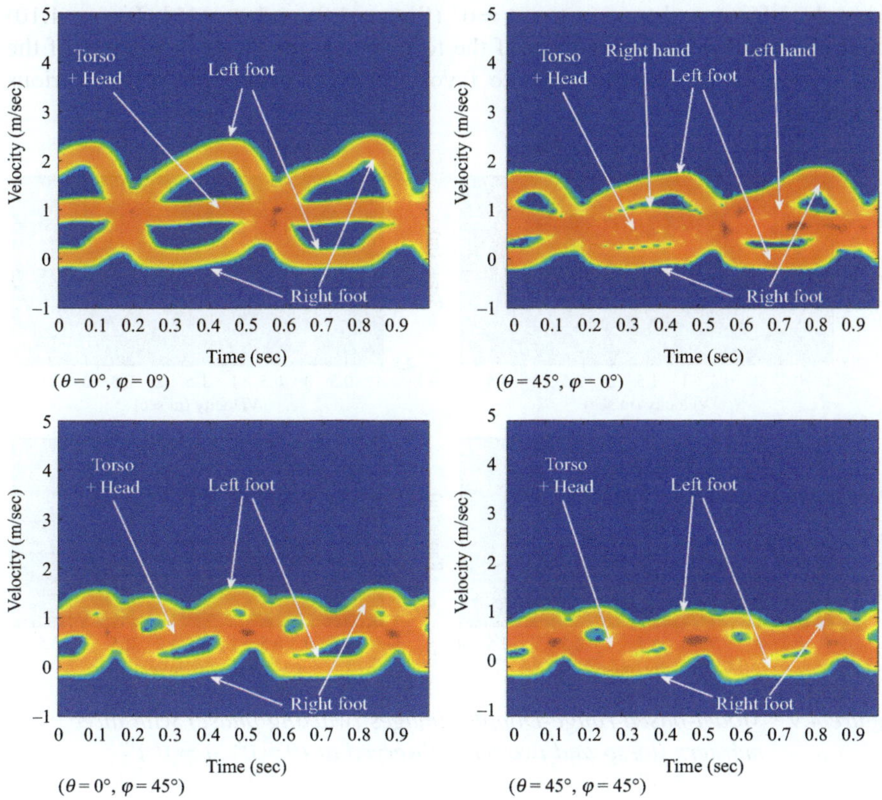

Figure 4.8 Model-based spectrograms (running phase) with three markers (torso and two feet)

time (pace, speed, direction, etc.) and even get mixed together if several pedestrians are in the same radar footprint. Then range analysis becomes mandatory for a refined analysis.

Model-based range-Doppler image series has been built using the three markers previously used (torso and two feet) and are presented in the following figures. In these simulations, the pedestrian is walking in the direction of the radar LOS ($\theta = 0$).

Walking

Here the range resolution is accurate enough to improve the Doppler analysis and thus the possibilities of identification by gait analysis.

We have decomposed the walking cycle in four images. In the first image, both feet have a velocity close to zero. The first foot (at the top) is already still while the second (at the bottom) is ready to stop. It is called the double support phase. In the second image the back foot accelerates to reach its maximum velocity in the third image and decelerates in the fourth image. During this phase, the second foot remains still and the body has a quite constant velocity.

The only difference between the $\phi = 0°$ (Figure 4.9) and $\varphi = 45°$ (Figure 4.10) lies in the relative range position of the torso when compared to the one of the legs: high grazing scenario tends to favor the range separation of the various body parts.

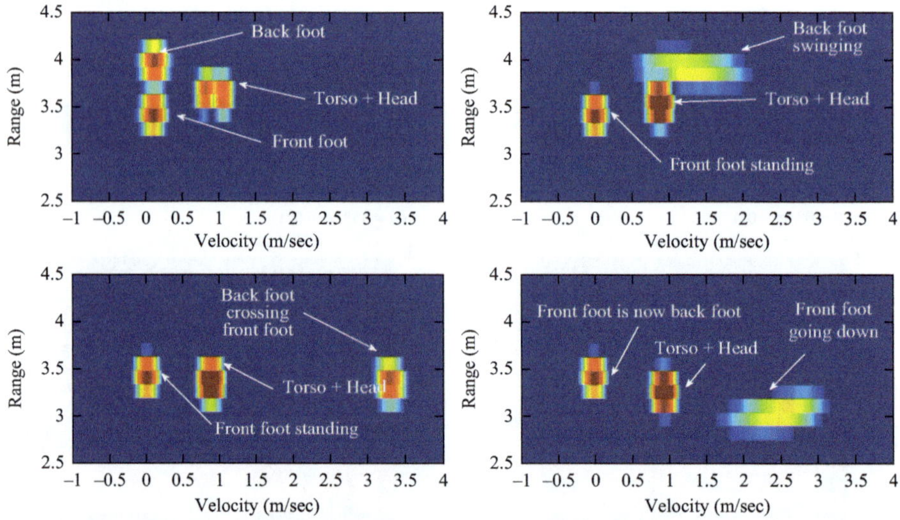

Figure 4.9 Model-based range-Doppler images (walking phase) with three markers (torso and two feet) observed at ($\theta = 0°$, $\varphi = 0°$)

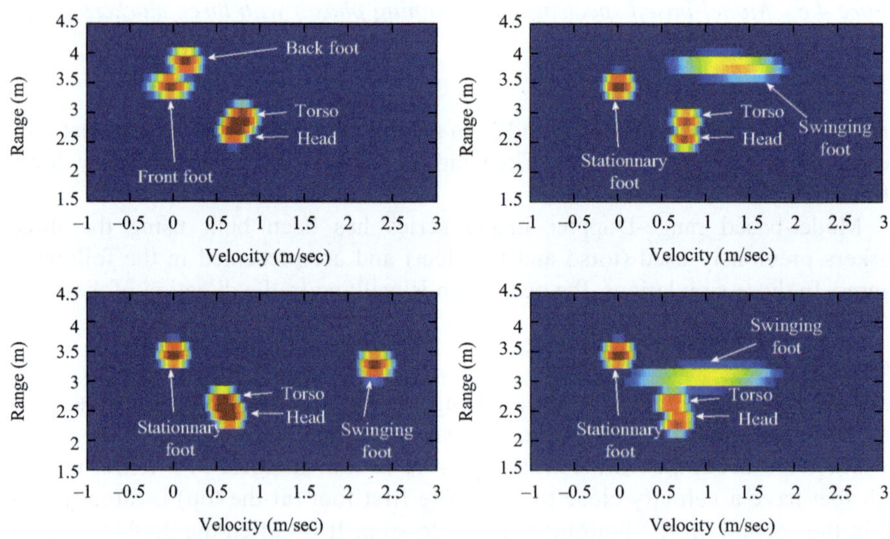

Figure 4.10 Model-based range-Doppler images (walking phase) with three markers (torso and two feet) observed at ($\theta = 0°$, $\varphi = 45°$)

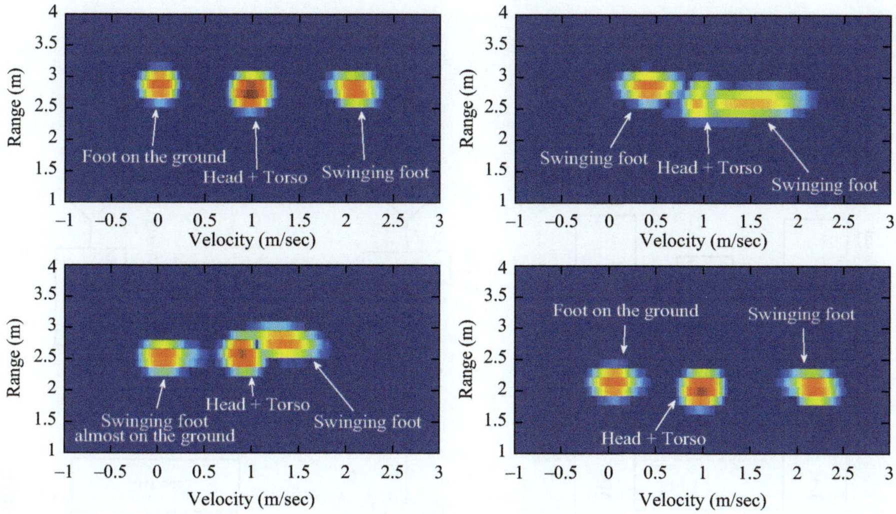

Figure 4.11 Model-based range-Doppler images (running phase) with three markers (torso and two feet) observed at ($\theta = 0°$, $\varphi = 0°$)

Running

We have decomposed the running cycle in four images (Figure 4.11). In the first image, the first foot is standing while the second is reaching its maximal velocity. When the second is decelerating, the first one leaves the ground and its velocity increases. Here there is no double support phase, but we can see a floating phase in the second image where both feet have a positive Doppler.

4.5 High Resolution Range-Doppler Radar HYCAM

4.5.1 General Concepts and Architecture

The experimental HYCAM prototype (for HYperfrequencyCAMera) is an original high resolution X band radar, built by ONERA, The French Aerospace Lab, in order to measure the signature of targets with a time varying behavior [34, 35]. In order to make a good instantaneous measurement of the target backscattering, it must be done with the shortest possible sampling, while preserving a very large bandwidth for the range resolution.

Thus, HYCAM was designed with the idea of putting the digital-analog EM conversion at the closest distance of the antennas. The heart of this radar is a digital-analog converter that makes use of a specially designed waveform [44, 45].

The other components are an analog frequency shifter, wideband antennas, and a wideband receiver. A detailed diagram is given in Figure 4.12.

This compact system was designed, built, and used by several doctoral students from 2003 to 2008 [3, 34, 35, 46]. It was the prototype version of a much bigger HYCAM demonstrator radar that was built since and installed in 2011 on the roof of ONERA laboratory.

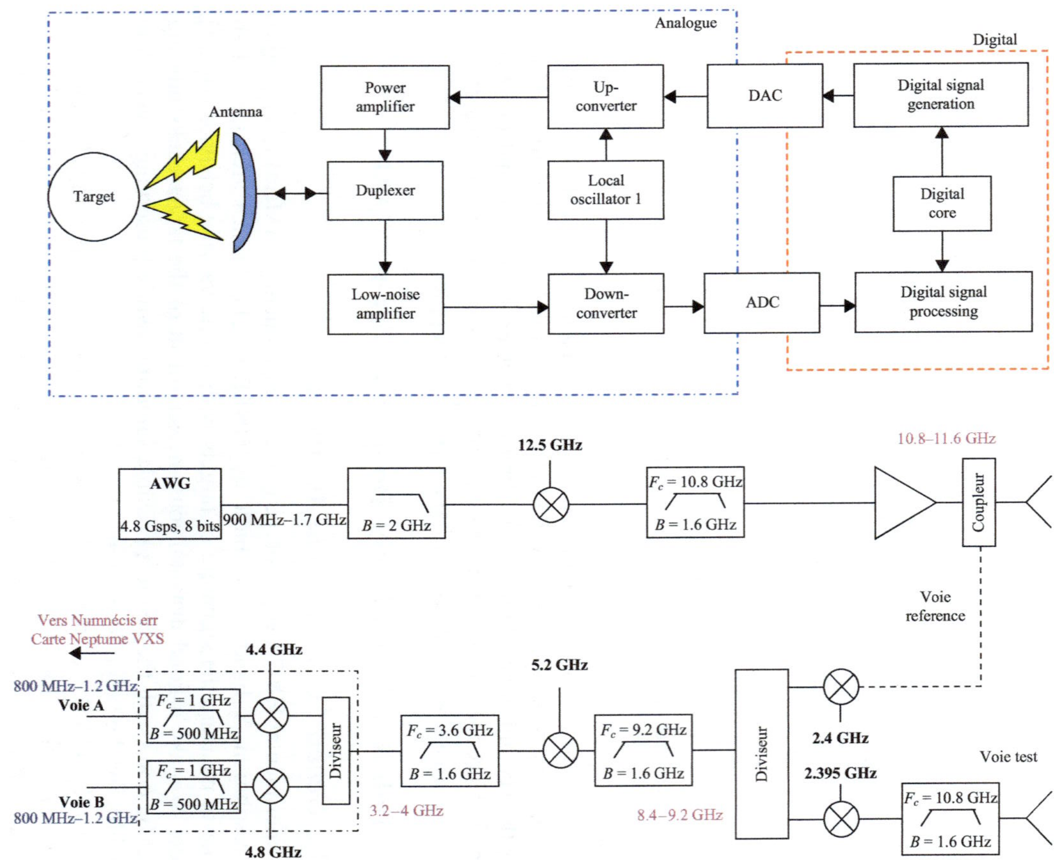

Figure 4.12 HYCAM prototype architecture

4.5.2 Waveform and Processing

The waveform is an OFDM frequency comb, where all the frequencies are emitted simultaneously within a very short time (below a micro-second) [44–47]. This enables us to measure the target scattering on a wide bandwidth in a very short time. The emitted signal (called "reference signal") is also recorded in order to be able to compensate for any possible errors in the emitting chain that may also change through time.

OFDM is a method of transmitting data simultaneously over multiple equally spaced carrier frequencies, using Fourier transform processing for modulation and demodulation. The method has been proposed for many types of radio systems, such as wireless local area networks and digital audio and video broadcasting but it is relatively new for the radar community. By dividing the bandwidth into many small orthogonal frequencies (efficiently achievable using the fast Fourier transform), the data can be transmitted across multiple narrowband channels having overlapping frequency spectra. The emitted frequencies are multiple of a frequency step δf: $f = (n_0 + n)\,\delta f$ with $n \in [1, N]$, and $\Delta f = N\delta f$ is the frequency bandwidth. The emitted frequencies are thus orthogonal on a period $T_0 = 1/\delta f$, which defines the minimal signal duration.

The received signal (or "test signal") is interleaved with the reference signal (shifted by $\delta f/2$) to form a single signal, which simplifies the wideband receiver architecture. This frequency interleaving is possible due to the orthogonal properties of this special waveform. In Figure 4.13, we depict in more detail the frequency interleaved front end.

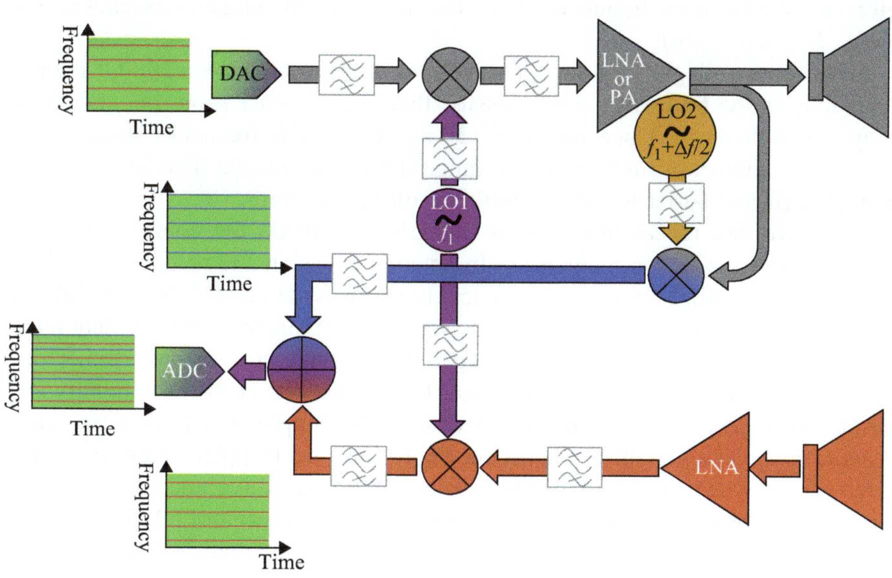

Figure 4.13 Frequency interleaved radar front end for two channels

The signal is generated in IF and a low pass filter removes the image signal. The IF signal is up-converted in RF by FLO. The up-converted signal goes through a band pass filter to remove the image frequency after up-conversion. The signal is then amplified either by a low noise amplifier (LNA) for short range applications or a power amplifier for applications requiring longer ranges. A directional coupler (DC) is plugged at the output of the amplifier. The coupled output of the DC is used as reference channel and the direct path output is plugged directly to the antenna.

The backscattered signal is received by the second antenna. This path is the test channel. An LNA is used to amplify the signal before down-conversion with the LO frequency used for up-conversion. The signal in the reference channel is attenuated and down-converted with an LO frequency equal to FLO + $\delta f/2$. With this operation the reference channel is shifted by $\delta f/2$ with respect to the test channel. Both signals are sent to a power combiner; thus, both signals merge together and the frequency spacing is now $\delta f/2$. The signal goes through an anti-aliasing band pass filter, is amplified, and finally digitized by an analog to digital converter (ADC).

This architecture presents the advantage of digitizing the reference and test signals on the same ADC, thus reducing by half the number of required ADC channels.

Sampling is done at the frequency comb interval by projecting the signal on the theoretical frequencies

$$S_{test}(n) = \langle s(t), e^{2j\pi ft} \rangle_T, \text{ where } \langle s_1(t), s_2(t) \rangle_T = \frac{1}{T}\int_T s_1(t) s_2^*(t)dt \quad \text{and} \quad n \in [1, N]$$

This processing will also enable us to separate the reference signal from the test signal. Calibration is then undertaken by simply dividing the test signal by the reference one for each frequency; thus, the instantaneous target backscattering is $S(n) = S_{test}(n)/S_{ref}(n)$.

The range ambiguity of such a waveform is given by $\Delta d = c/2\delta f$. Of course, depending on the frequency comb density, this can be rather short, but enough to study a specific surveillance zone (e.g., 30 m for a 5 MHz frequency step).

The instantaneous high resolution range (HRR) profile can then be formed by taking the (fast) Fourier transform on the resulting spectrum.

Now we can repeat this process through time to do the coherent Doppler processing ("slow" time integration). In practice emission and reception are done continuously, but for obvious technological issues, it is not possible to record all the samples received at a very high sampling frequency (typically 4 Go/sec data rate). Thus the signal is sequenced via a triggered sampling; in a similar manner one collects pulsed waveforms. The equivalent PRF is also the Doppler ambiguity and must be fixed greater than the maximum observable Doppler extension in the scene to avoid Doppler overlap. We must be aware that the Doppler extension is the difference between the extreme observable Doppler (positive and negative) and may not be related to the maximum absolute Doppler (maximum relative speed)!

Doppler resolution will depend not only on the integration time (i.e., the number of sampled "pulses") but also on the target motion to avoid range migration problems.

The result is a matrix $S(n, m)$ representing the target hologram sampled over frequencies f and (slow) time t (same as $H(f, t)$ in section 4.3). At the first order, the two-dimensional Fourier transform of this hologram will give the radar range-(micro-)Doppler image.

4.6 Experimental Setup and Results

4.6.1 Experimental Setup

The HYCAM prototype was used in an ONERA lab looking down a parking lot (see Figure 4.14), at the rather short distance of 25 m. The elevation angle is then close to 45° (see Figures 4.14 and 4.15).

The two emission/reception antennas are separated by 50 cm. Each antenna is a rectangle horn (with 13 cm by 11 cm aperture). The beam projection on ground is then a 10 m by 8 m zone enabling us to observe a slow target for a few seconds.

The main radar parameters are given in Table 4.2:

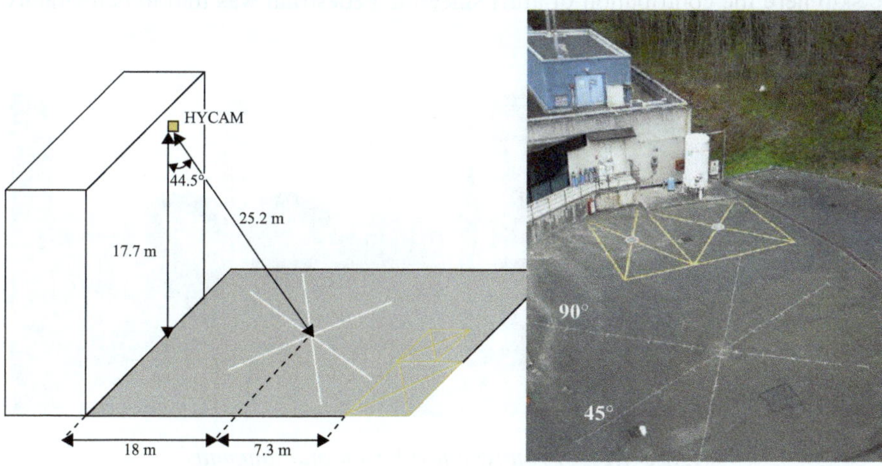

Figure 4.14 Experimental setup and observation scene from the radar

Table 4.2 HYCAM main parameters

Waveform	**OFDM**
Polarization	VV
Bandwidth	800 MHz
Central frequency	10.4 GHz
Frequency step	2 MHz
Number of frequencies	400
Reception bandwidth	2 GHz
PRF/Doppler sampling	Variable

The radar range resolution is then 18.5 cm, enabling us to decompose the observation scene into about 64 range cells. Doppler ambiguity and resolution will depend on the target motion and are fixed for each experiment. For instance, a 1 KHz PRF is enough to cover the + or –7 m/sec velocity rate, which could be used typically for walking pedestrian observation.

RCS calibration can be conducted on typical trihedral targets on ground.

The radar beam will obviously intercept a large amount of fixed object in the scene (mainly ground and buildings through multiple reflections), thus creating a large scattering contribution at 0 Hz Doppler for any range in the scene. This phenomenon is known as "zero Doppler Clutter" or ZDC. Since ZDC has a stable behavior through time, it is possible to filter its contribution by subtracting the signal mean in the temporal domain.

4.6.2 Results on Pedestrian Motions

One pedestrian walking (Figure 4.16)

We will focus on two phases of walking: stance phase and swing phase. Let's note we skip here the contribution of arms since the pedestrian was told to deliberately keep his arms along his body.

Figure 4.15 Prototype test bench and antennas

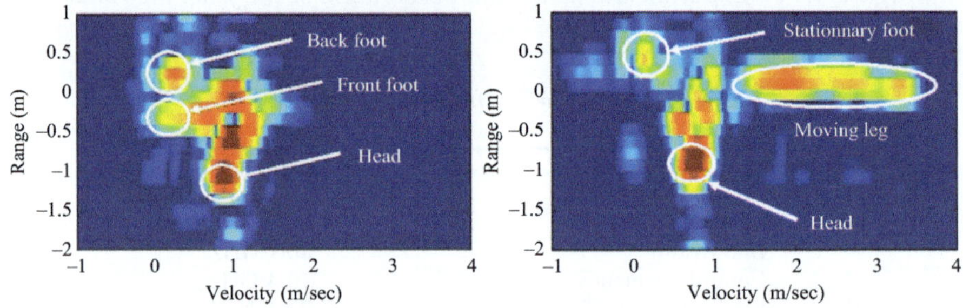

(a) Double support phase (b) Swing phase

Figure 4.16 Experimental range-Doppler image (pedestrian walking) ($\theta = 0°$, $\varphi = 45°$)

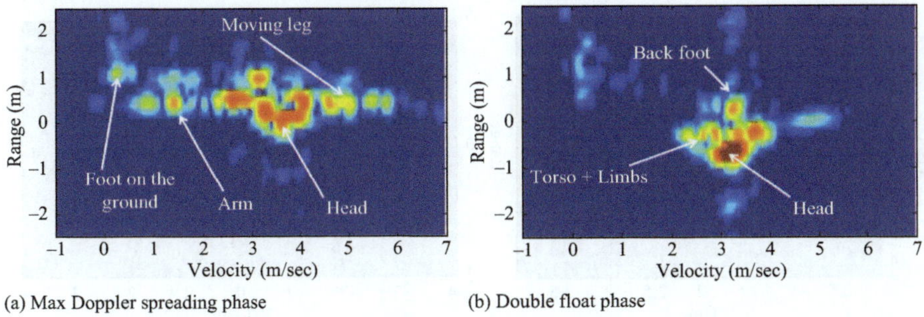

Figure 4.17 Experimental range-Doppler image (pedestrian running) ($\theta = 0°$, $\varphi = 45°$)

On each image, we can easily emphasize the head's contribution on the radar image, which is the closest reflecting point from the radar.

The range spreading is due to the radar observing from above. It enables us to measure the person height: here the projected range spreading is about 130 cm, which corresponds to the actual height of the pedestrian (185 cm).

We can get even more insight from these two phases:

- Double support phase: both feet are at a standstill, visible by their zero Doppler. The image is comparable to the model-based one presented in Figure 4.10 (first image). On this case it is even possible to assess the step length by measuring the number of range cell between the two reflectors (about 60 cm).
- Swing phase: one foot is moving while the other is swinging. This image also perfectly agrees with the model-based image shown in Figure 4.10 (fourth image). The Doppler spreading is linked to the velocity distribution along the leg and the range spreading still depends on radar height.

One pedestrian running (Figure 4.17)

The first image (a) shows the phase of maximal Doppler spreading, which is the moment when one foot is on the ground while the other is reaching its maximum velocity. The second one (b) shows the phase of double float when no foot touches the ground. Legs are crossing, which means that all the points have almost the same velocity. Contrary to the walking phases it is not obvious to assess the step length because both feet are not on the ground at the same moment.

Two people crossing each other (Figure 4.18)

Spectrogram can only analyze the Doppler's variation. They don't take into account the motion's range component. When several people move in the radar field of view, there are interferences between Doppler frequencies and it is not possible anymore to interpret the received signal (Figure 4.18(a)). Adding range resolution to the system enables a clear separation of the pedestrians, to localize them in range and to distinguish the evolution of the different parts of the body both in range and Doppler (Figure 4.18(b)).

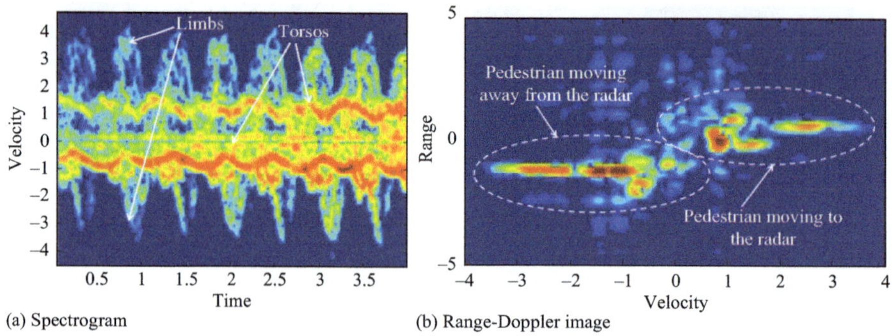

Figure 4.18 Experimental range-Doppler image (two pedestrians crossing)

4.7 Conclusions and Perspectives

We have presented a refined radar analysis of some pedestrian motion mostly focused on walking and running.

We have shown that coupling range resolution with time-Doppler analysis was very valuable for this task.

To better understand and analyze the range and Doppler's evolution of the different body parts, we have used a pedestrian's model provided by a motion capture system. It was then possible to obtain the three-dimensional coordinates of reflectors distributed on the pedestrian. This model has permitted to isolate the motion of each part. This model has then been injected in a range-Doppler imaging simulator. Knowing the kinematic of every part, it has been possible to better understand the composition of a range-Doppler image resulting from the interaction between a pedestrian and a microwave signal.

Images from simulated data have been compared with images from experimental data and results are particularly satisfying. They first show that HYCAM is a device capable to detect micro-Doppler induced by pedestrian's limbs. The high range resolution has enabled us to discriminate in range and Doppler the different parts of the body and even several people moving in a scene.

We have started to work on the influence of the various motion parameters, including the relative position of the observed pedestrian toward the radar.

The investigation field on human motion analysis is very wide and offers many perspectives. We wish to go deeper in our analysis to link real and simulated data exploring the huge experimental database that was acquired by the HYCAM prototype. We think this can be partly done by improving the reflector model (using for instance a continuous distribution of reflectors) and by deriving more analytic human motion model.

This study leads also to several technological issues: micro-Doppler in ultra-wide band radar is still a challenging task for radar and waveforms designers.

Future applications will need robust model inversion not only to extract and track the body parts from real data, but also to derive possible intents of the human under observation.

References

[1] A. Ghaleb, L. Vignaud, and J. M. Nicolas, "Micro-Doppler analysis of wheels and pedestrians in ISAR imaging", in *IET Signal Processing*, vol. 2, no. 3, September 2008, pp. 301–311.

[2] L. Vignaud, A. Ghaleb, and J. M. Nicolas, "Micro-Doppler analysis of pedestrians in ISAR imaging", in *IEEE Radar Conference*, Rome, May 2008.

[3] A. Ghaleb, "Analyse des micro-Doppler de cibles mobiles déformables en imagerie radar" (*Micro-Doppler analysis of time varying targets in radar imaging*), Ph.D. dissertation, University Télécom ParisTech, February 2009.

[4] L. Vignaud, A. Ghaleb, J. L. Kernec, and J. M. Nicolas, "Radar high resolution range and micro-Doppler analysis of human motions", in *IEEE Radar Conference*, Bordeaux, September 2009.

[5] V. C. Chen and H. Ling, *Time Frequency Transforms for Radar Imaging and Signal Analysis*, Artech House, 2002.

[6] V. C. Chen, "Micro-Doppler effect in radar: phenomenon, model, and simulation study", in *IEEE Transactions on Aerospace and Electronics Systems*, vol. 42, no.1, 2006, pp. 2–21.

[7] C. Chen and H. Andrews, "Target motion induced radar imaging", in *IEEE Transactions on Aerospace Electronics Systems*, vol. 16, 1980, pp. 2–14.

[8] J. L. Walker, "Range-Doppler imaging of rotating objects", in *IEEE Transactions on Aerospace Electronics Systems*, vol. 16, 1980, pp. 23–52.

[9] H. J. Li, S. H. Yang, T. Y. Liu, S. L. Yen, Y. Shen, and N. H. Farhat, "Microwave imaging of moving objects", in *International Journal of Imaging Systems and Technology*, vol. 2, 1990, pp. 219–230.

[10] R. Wehner, *High Resolution Radar*. Artech House, 1991.

[11] D. Mensa, *High Resolution Radar Cross-Section Imaging*, Artech House, 1991.

[12] W. Carrara, R. Goodman, and R. Majewski, *Spotlight Synthetic Radar: Signal Processing Algorithms*, Artech House, 1995.

[13] L. Vignaud, "Imagerie micro-ondes des scènes instationnaires" (*Microwave imaging on non-stationary scenes*), Ph.D. dissertation, Université Paris 6, 1996.

[14] V. C. Chen, "Analysis of micro-Doppler signatures with time-frequency transform", in *Proceedings of the 10th IEEE Workshop on Statistical Signal and Array Processing*, 2000, pp. 463–466.

[15] V. C. Chen, F. Li, S. Ho, and H. Wechsler, "Analysis of micro-Doppler signatures", in *IEE Proceedings on Radar, Sonar and Navigation*, vol. 150, 2003, pp. 271–276.

[16] V. C. Chen, "Doppler signatures of radar backscattering from objects with micro-motions", in *IET Signal Processing*, vol. 2, 2008, pp. 291–300.

[17] J. Li and H. Ling, "Micro-Doppler features extraction from ISAR data using adaptive chirplet representation", in *IEEE Radar Conference*, 2002.

[18] T. Sparr and B. Krane, "Micro-Doppler analysis of vibrating targets in SAR", in *IEE Proceedings on Radar, Sonar and Navigation*, vol. 150, 2003, pp. 277–283.

[19] T. Thayaparan, S. Abrol, and E. Riseborough, "Micro-Doppler radar signatures for intelligent target recognition", in *Technical Report DRDCOttawa TM*, vol. 170, 2004.

[20] T. Thayaparan, S. Abrol, E. Riseborough, L. Stankovic, D. Lamothe, and G. Duff, "Analysis of radar micro-Doppler signatures from experimental helicopter and human data", in *IEE Proceedings on Radar, Sonar and Navigation*, vol. 4, 2007, pp. 288–299.

[21] T. Thayaparan, L. Stankovic, and I. Djurovic, "Micro-Doppler-based target detection and feature extraction in indoor and outdoor environments", in *J. Franklin Institute*, 2008.

[22] A. M. Sabatini and V. Colla, "A method for sonar based recognition of walking people", in *Robotics and Autonomous Systems*, vol. 25, 1998, pp. 117–126.

[23] J. L. Geisheimer, W. Marshall, and E. Greneker, "A continuous-wave (CW) radar for gait analysis", in *IEEE Conference Signals, Systems and Computers*, vol. 1, 2001, pp. 834–838.

[24] J. L. Geisheimer, W. Marshall, and E. Greneker, "High-resolution Doppler model of the human gait", in *Proceedings of SPIE*, 2002, pp. 8–18.

[25] G. Greneker, "Very low cost stand-off suicide bomber detection system using human gait analysis to screen potential bomb carrying individuals", in *Proceedings of SPIE*, vol. 5788, 2005.

[26] M. Otero, "Application of a continuous wave radar for human gait recognition", in *Proceedings of SPIE, Signal Processing Sensor Fusion and Target Recognition*, vol. 5809, 2005, pp. 538–548.

[27] P. van Dorp and F. Groen, "Human walking estimation with radar", in *IEE Proceedings on Radar, Sonar and Navigation*, vol. 150, 2003, pp. 356–365.

[28] Y. Yang, W. Zhang, and C. Lu, "Classify human motions using micro-Doppler radar", in *Proceedings of the SPIE*, vol. 6944, 2008.

[29] P. van Dorp and F. Groen, "Feature-based human motion parameter estimation with radar", in *IET Radar, Sonar and Navigation*, vol. 2, 2008, pp. 135–145.

[30] G. E. Smith, K. Woodbridge, and C. J. Baker, "Micro-Doppler signature classification", in *IEEE Radar Conference*, 2006.

[31] S. S. Ram and H. Ling, "Analysis of micro-Doppler from human gait using reassigned joint time-frequency transform", in *Electronics Letters*, vol. 43, 2007.

[32] M. G. Anderson and R. L. Rogers, "Micro-Doppler analysis of multiple frequency continuous wave radar signatures", in *Proceedings of SPIE*, vol. 6547, 2007.

[33] R. G. Raj, V. C. Chen, and R. Lipps, "Analysis of radar human gait signatures", in *IET Signal Processing*, vol. 4, 2010, pp. 234–244.

[34] Y. Paichard, "Caméra hyperfréquence pour la mesure et l'analyse de la SER de cibles scintillantes" (*Hyperfrequency camera for scintillating targets RCS measurement and analysis*), Ph.D. dissertation, Universite Paris-Sud, 2007.

[35] J. L. Kernec, "Contribution des multi-tons au radar logiciel ultralarge bande" (*Multitone contribution to the ultra-wideband software defined radar*), Ph.D. dissertation, Universite Paris-VI, 2011.

[36] D. Winter, *Biomechanics and Motor Control of Human Movement*, Wiley, 1990.

[37] F. M. *et al.*, "Computer animation of human walking: a survey", in *The Journal of Visualization and Computer Animation*, vol. 10, no. 1, 1999, pp.39–54.

[38] R. Boulic, M. Thalmann, and D. Thalmann, "A global human walking model with real-time kinematic personification", in *Vision Computer*, vol. 6, 1990, pp. 344–358.

[39] R. Boulic, B. Ulciny, and D. Thalmann, "Versatile walk engine", in *Journal Of Game Development*, vol. 1(1), 2004, pp. 29–50.

[40] Z. Zhang and P.Pouliquen, "Acoustic micro-Doppler radar for human gait imaging", in *The Journal of the Acoustical Society of America*, vol. 121, 2007, pp. 110–113.

[41] S. S. Ram, Y. Li, A. Lin, and H. Ling, "Doppler-based detection and tracking of humans in indoor environments", in *J. Franklin Institute* (Special issue on Indoor Radar Imaging), 2008.

[42] G. E. Smith, K. Woodbridge, and C. J. Baker, "Multistatic-micro-Doppler signature of personnel", in *IEEE Radar Conference*, Rome, Italy, 2008.

[43] M. G. Anderson, *Design of multiple frequency continuous wave radar hardware and micro-Doppler based detection and classification algorithms*, Ph.D. dissertation, University of Texas at Austin, 2008.

[44] D. Poullin, "L'utilisation active des COFDM en radar multistatique", *REE*, 2006, pp. 36–42.

[45] D. Poullin, and M. Flecheux, "Recent progress in Passive Coherent Location (PCL) concepts and technique in France using DAB or FM broadcasters", 2008, pp. 1–5.

[46] J. Le Kernec, "UWB radar for people location and tracking", *Cork Institute of Technology*, 2006.

[47] J. Castelli, G. Bobillot, "Case study analysis of linear chirp and multi-tones radar signals through simulations and measurement with HYCAM research test-bench", 2009.

[48] R. Adelaar, "The practical biomechanics of running", *The American Journal of Sports Medicine*, vol. 14, 1986, pp. 497–500.

Chapter 5

Through-the-Wall Micro-Doppler Signatures

Dustin P. Fairchild, Ram M. Narayanan, Eric R. Beckel, Wai K. Luk, and Gerard A. Gaeta

5.1 Introduction

Radar technology has been utilized for through-the-wall operation in recent years. Through-the-wall radars have become an area of interest due to their usefulness in surveillance, search and rescue operations, and others. They can be used to map the layout of a building, detect and track human targets, or even in aiding firefighters and rescue workers by alerting them if someone is trapped inside a structure.

Due to the very nature of electromagnetic propagation through various types of media, through-the-wall sensing poses additional challenges that must be considered and overcome in the design of any through-the-wall radar system. Micro-Doppler radars are no exception to these challenges. Once the radar has been designed appropriately, the limitations imposed by the wall barrier can be minimized via appropriate waveform design and signal processing techniques. Consequently, a through-the-wall micro-Doppler radar system can be operated in much the same manner as a barrier-free radar system. Some of the considerations that must be taken into account include reflections from the front and back faces of the wall (and possibly resonant cavities within the wall), attenuation of electromagnetic waves through different materials, and the dispersive characteristics of the media.

It is well established that higher frequencies exhibit much larger losses than lower frequencies. For this reason through-the-wall radars typically operate at frequencies of 5 GHz or less. Wall losses also reduce the signal-to-noise ratio of the radar returns, thus limiting the maximum range of the radar. This also means that some walls may induce too great a loss to detect some targets due to their material properties or simply because of the thickness of the wall itself.

In the following sections, we address the specific challenges that through-the-wall radars endure, and present simulations and experimental data for through-the-wall micro-Doppler signatures. In section 5.2, the design considerations, including wall attenuation, frequency selection, and dispersion are discussed. In section 5.3, some commonly used time-frequency transforms are briefly discussed with a focus on how they can be used to analyze micro-Doppler signatures. Section 5.4 demonstrates some of the effects that a barrier such as a wall has on the return micro-Doppler signals and discusses how these differ from that of micro-Doppler signals that arise without any barrier. Sections 5.5 and 5.6 present some experimental data of micro-Doppler signatures for targets with and without a translational motion. Some models of common human activities and motions are also discussed and compared to the experimentally measured through-the-wall micro-Doppler signals.

5.2 Design Considerations for Through-the-Wall Radars

When considering micro-Doppler signatures through a barrier such as a wall, foliage, etc. one must understand the limitations that must be overcome. Through-the-wall radars must have the ability to detect targets through a lossy material and must operate in a high clutter environment. Observing the Doppler features of the reflected signal enables the clutter to be easily suppressed because, in general, traditional clutter elements such as walls, furniture, and other stationary objects will not exhibit a Doppler shift. In contrast, a human, despite staying still, may have some deliberate (e.g., waving arms) or involuntary (e.g., breathing) movements that can give rise to micro-Doppler signatures, thereby permitting the use of Doppler as both a discriminant from clutter and a means of recognizing the movements to infer possible intent.

The electromagnetic properties of building materials are critical in through-the-wall radar design and data analysis. Attenuation, dispersion, reflection, and refraction all are important considerations when dealing with through-the-wall micro-Doppler signatures. The building material may not pose a considerable challenge if it is wood, glass, or drywall where the attenuation is relatively small. On the other hand, materials such as concrete, cinder block, or brick may cause significant attenuation of the signal to the point where micro-Doppler cannot be detected. As we would expect, the material properties are frequency dependent; therefore, selecting an appropriate radar carrier frequency will alleviate some of the challenges that arise due to barrier effects.

5.2.1 Wall Attenuation

The attenuation of electromagnetic waves through materials depends on the properties of the material. The permittivity (ε), permeability (μ), and conductivity (σ) all vary depending on the type of material in question and on the frequency of the incident wave. If these material properties are known, then the attenuation constant can be approximated as

$$\alpha = \frac{\sigma}{2}\sqrt{\frac{\mu}{\varepsilon}} \tag{5.1}$$

Here we are assuming that the wall is made of a dielectric material because conducting walls act as opaque barriers for through-the-wall radars. Often, the loss tangent (tan δ) is reported instead of the attenuation constant, where $\delta = \frac{1}{\alpha}$. Because ε and σ are frequency dependent, the attenuation constant is also frequency dependent. Table 5.1 shows the attenuation as measured experimentally in [1] for some selected materials at 100 MHz, while Table 5.2 shows the attenuation measured experimentally in [2] for some common wall materials at 500 MHz. Because the loss

Table 5.1 Attenuation and relative permittivity of some selected materials at 100 MHz [1]

Material	Attenuation constant (α in dBm^{-1})	Relative permittivity (ε_r)
Air	0	1
Asphalt	2–15	2–4
Dry concrete	2–12	4–10
Wet concrete	10–25	10–20
Sandy dry soil	0.1–2	4–6
Sandy wet soil	1–5	15–30

Table 5.2 Attenuation of typical wall materials at 500 MHz [2]

Wall type	Two-way attenuation (dB)
Typical interior wall (two 1/2 inch gypsum panels filled with wood)	8
Reinforced concrete wall (6 inch deKhotinsky concrete with 1/2 inch rebar on 3 feet centers)	11
Stucco/concrete wall (1 inch 22AWG chicken-wire with 1/2 inch concrete)	19
Chain-link (cyclone) fence	11

increases as the frequency increases, through-the-wall radars are limited to lower frequencies. The loss that is encountered may not be significant for many materials; however, when considering a material like concrete or cinder block the loss can have a crippling effect on the received power by the radar. It is also important to remember that the losses that are reported in Table 5.1 are one-way loss factors. The loss experienced by the radar system is two-way and the loss factor must be doubled.

A more complete plot of the attenuation through various materials at frequencies ranging from 4 to 100 GHz is shown in Figure 5.1 [3]. Because through-the-wall radars are often desired to work in worst-case scenarios, the radar operating frequency is chosen to be less than 5 GHz so that the total attenuation due to two-way propagation through concrete block is less than 30 dB.

Additional measurements by [4] for 20 cm thick concrete and wooden walls at a frequency of 2.4 GHz show a transmission loss at the air–wall interface of 3.9 and 0.9 dB, respectively. The through-the-wall propagation loss was found to be 1.2 and 0.7 dB for 20 cm thick concrete and wooden walls, respectively. Many other resources can be found that discuss attenuation through wall materials. A sampling of some of these resources can be found in [5–7].

5.2.2 Wall Reflection

As discussed in the previous section, the material properties determine the loss due to propagation through lossy media, such as a wall. These same material properties are responsible for reflecting part of the incident wave back to the radar without penetrating the wall. The reflections are caused by the discontinuity in the intrinsic

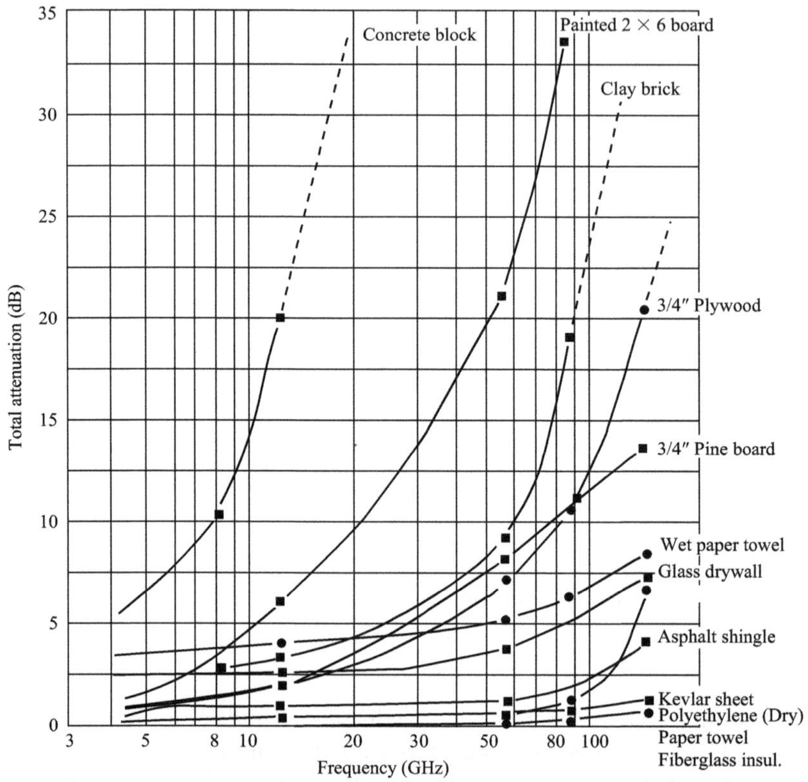

Figure 5.1 One-way attenuation of EM waves (in dB) through various materials [3]

impedance of the material through which the wave is traveling. The reflection coefficient Γ and the transmission coefficient T (assuming normal incidence) for a wave traveling from Medium 1 into Medium 2 are well known and are given in (5.2) and (5.3), respectively, where $\eta = \sqrt{\frac{j\omega\mu}{\sigma + j\omega\varepsilon}}$ is the intrinsic impedance of the material. Because transmission through conductive materials is not practical due to high losses, we can assume that the wall material is a good dielectric, for which the intrinsic impedance can be approximated as $\eta \approx \sqrt{\frac{\mu}{\varepsilon}}$.

$$\Gamma = \frac{\eta_2 - \eta_1}{\eta_1 + \eta_2} \tag{5.2}$$

$$T = \frac{2\eta_2}{\eta_1 + \eta_2} = 1 + \Gamma \tag{5.3}$$

The average reflected and transmitted power densities S^r_{avg} and S^t_{avg}, respectively, are given by

$$\left|S^r_{avg}\right| = |\Gamma|^2 S^i_{avg} \tag{5.4}$$

and

$$\left|S_{avg}^{t}\right| = (1 - |\Gamma|^2)S_{avg}^{i} \tag{5.5}$$

Because a large fraction of the power is reflected and never reaches the other side of the wall, reflections are another source of losses in through-the-wall radars.

5.2.3 Dispersion

Many Doppler radars utilize a CW wave that consists of a single frequency. For these radars, dispersion obviously does not play a role. However, for pulsed Doppler, multi-frequency, or ultra wideband (UWB) radars, dispersion can have some degrading effects on radar performance. Dispersion is caused because the individual dipoles created by an induced electric field take some small amount of time to align themselves in the direction of the electric field. In the presence of an oscillating electric field, the response time of the dipoles will depend on the frequency of the electric field oscillations. This will cause the different frequency components of a UWB signal to travel at different speeds. Because of this, dispersive media will have the effect of broadening pulses and distorting the signal. The broadening of the pulse will effectively decrease the signal bandwidth.

Most materials can be assumed to follow classical models such as the well-known Debye equation. This model can be used to calculate the complex permittivity as a function of frequency. Using the Debye equation, the complex permittivity of a material is

$$\epsilon_r^{\cdot}(\omega) = \epsilon_r^{\prime}(\omega) - j\epsilon_r^{\prime\prime}(\omega) = \epsilon_{r\infty}^{\prime} + \frac{\epsilon_{rs}^{\prime} - \epsilon_{r\infty}^{\prime}}{1 + j\omega\tau_e} \tag{5.6}$$

where $\epsilon_{rs}^{\prime}$ and $\epsilon_{r\infty}^{\prime}$ are the real parts of the complex permittivity at zero frequency and at a very large frequency (optical frequency), respectively, and τ_e is a new relaxation time constant that is related to the original relaxation time constant τ by

$$\tau_e = \tau \frac{\epsilon_{rs}^{\prime} + 2}{\epsilon_{r\infty}^{\prime} + 2} \tag{5.7}$$

For a more complete and thorough discussion of dispersion, including the Lorentz model that accounts for multiple resonances (Debye model only accounts for a single resonance), see [8, 9].

5.3 Time-Frequency Transforms

When analyzing micro-Doppler signals, the time-varying nature of the signal makes traditional analysis techniques such as the Fourier transform inadequate for getting a complete picture of how the target is moving relative to the radar. The Fourier transform assumes a linear time-invariant system. Because the Doppler frequency changes over time, we can easily see how the Fourier transform is not suitable for analyzing micro-Doppler signatures. While the Fourier transform can be useful in many applications to Doppler radar, other techniques must be explored if knowledge of how the velocity of the target changes over time is desired. Among the simplest of these is the short-time Fourier transform (STFT). Many other techniques have also been developed. Wavelet theory has recently been gaining in popularity, and other techniques such as the Wigner–Ville transform (WVT), Choi–Williams distribution (CWD), and the Hilbert–Huang transform (HHT) are also widely used in practice. For an in-depth discussion of these and other time-frequency transforms, see [10–15].

5.3.1 Short-Time Fourier Transform

The Fourier transform is the fundamental signal processing tool today. However, it has drawbacks when analyzing nonlinear and nonstationary signals. The Fourier transform projects a signal $x(t)$ onto a set of orthonormal bases that are complex sinusoids with infinite extent, as shown in (5.8). The Fourier transform assumes linearity and time-invariance and these basis functions are formulated in such a manner that time-frequency information cannot be extracted from the signal. The Fourier transform can only give information as to which frequencies were present within the duration of the signal, but cannot give information as to when they occurred (except that they occurred sometime within the duration of the signal).

$$X(\omega) = \int_{-\infty}^{+\infty} x(t)e^{-j\omega t}dt \tag{5.8}$$

In order to evaluate how a function changes over time, the simplest approach is to window the data and slide the window across the entire duration of the signal. This approach is the well-known STFT and is expressed in (5.9), where $w(t)$ is the window function. Here the signal is projected onto the basis functions: $w(t-\tau)e^{j\omega t}$.

$$STFT(t,\omega) = \int x(\tau)w(\tau - t)e^{-j\omega t}\, d\tau \tag{5.9}$$

The uncertainty principle as applied to radar states that $\Delta_t\Delta_\omega \geq 1/2$, where Δ_t is the time resolution and Δ_ω is the frequency resolution; thus, the smaller the time window, the poorer the frequency resolution. Because of the windowing operation, in order to gain information on how the signal varies with time, we must sacrifice some of our frequency resolution.

The window function and the width of the window both must be selected appropriately depending on the application. Hanning, Hamming, and rectangular windows are frequently used. Some are selected for their simplicity, others for their sidelobe suppression. A Gaussian window, as shown in (5.10), is sometimes chosen because it has the desirable property of achieving the lowest possible time-frequency product ($\Delta_t\Delta_\omega = 1/2$).

$$w(t) = \frac{1}{\pi^{1/4}\sqrt{\sigma}}e^{-\frac{t^2}{2\sigma^2}} \tag{5.10}$$

5.3.2 Continuous Wavelet Transform

In the STFT discussed above, the frequency resolution is a constant. Low frequency signals and high frequency signals alike will have the same frequency resolution. In many cases, this is not desirable. For example, assume that the frequency resolution of the STFT is 10 Hz and the signal to be analyzed has a frequency of 20 Hz for the first half of the signal duration and a frequency of 1 kHz for the second half of the signal duration. The frequency resolution is 50% of the 20 Hz component but is only 1% of the 1 kHz component. When analyzing such a signal, having equally spaced frequency resolution bins is not desirable.

The continuous wavelet transform (CWT) solves this problem by introducing multi-resolution capability. The mathematical expression for the CWT is shown in (5.11), where a is the scale parameter, b is the translation parameter, and ψ is the mother wavelet. The scale parameter is analogous to frequency and the translation parameter is analogous to time, resulting in a "scale-translation" or time-frequency representation of the signal. Common wavelet functions include the Haar, Daubechies, Symlet, Coiflet, Mexican Hat, Morlet, and Meyer wavelets. The wavelet selection is dependent on the application. In the cases of the Fourier transform and the STFT, the signal was projected onto the basis functions: $e^{j\omega t}$ or $w(t)e^{j\omega t}$. The CWT no longer uses complex exponentials for basis functions, but instead uses scaled and translated versions of the mother wavelet. By scaling the mother wavelet, multi-frequency resolution is attained [14, 15].

$$CWT(a, b) = \frac{1}{\sqrt{|a|}} \int_{-\infty}^{\infty} x(t)\psi^* \left(\frac{t-b}{a}\right) dt \tag{5.11}$$

5.3.3 The Hilbert–Huang Transform and Empirical Mode Decomposition

The HHT is an adaptive time-frequency technique that is well-suited for nonlinear and nonstationary time series. It combines a sifting process called empirical mode decomposition (EMD) to decompose a signal into its various components with the Hilbert transform to analyze the instantaneous frequency of each signal component. Because of the adaptive nature of EMD, there is no solid theoretical foundation on which this time-frequency transform is based upon. Instead, empirical testing has shown it to be reliable in decomposing signals into components which are physically meaningful. One advantage that EMD has is that it does not require a priori knowledge of the signal and is also not dependent on the selection of a kernel function like the wavelet transform and others.

EMD will decompose a signal into its intrinsic oscillatory modes (also called intrinsic mode functions (IMFs)), based on the time-scale of the oscillations. The faster oscillations in the signal will be present in the lower-indexed IMFs and the slower oscillations will be present in the higher-indexed IMFs. These oscillatory modes, called IMFs, are components of the original signal and each IMF is orthogonal to all of the other IMFs. The orthogonality of the IMFs allows for EMD to be used in a number of ways. Filtering of the signal can be performed by selectively adding together IMFs of the signal and omitting others. In addition to this, the mechanism that produces the signal can be analyzed by inspecting IMFs that have known properties [16].

The EMD algorithm consists mainly of a sifting process. In this process, the extrema of the signal $s(t)$ are identified. From the extrema, the envelope of the minima, $x_{min}(t)$, and the envelope of the maxima, $x_{max}(t)$, are formed by interpolation techniques. Next, the mean of these two envelopes is calculated as

$$m(t) = \frac{(x_{max}(t) + x_{min}(t))}{2} \tag{5.12}$$

and the mean envelope is subtracted from the signal to complete the first iteration of the sifting process. The resulting signal is denoted by $x_j^k(t)$ $(k=1, j=1)$, where k indicates the iteration and j indicates the index of the IMF. This sifting process is repeated until $\langle x_j^k(t)\rangle = 0$ (the mean envelope is equal to zero at all times) and the number of zero crossings of $x(t)$ differs from the number of extrema by no more than 1. Once this condition has been satisfied, we set $x_j(t) = I_j(t)$, and $I_j(t)$ is the j-th IMF $(I_1(t))$ of $x(t)$. Next, the residue is calculated as $x_j^r(t) = x(t) - I_1(t)$. The sifting process is then repeated, where the residue is $x_j^r(t) = x(t) - \sum_{i=1}^{i=j} I_i(t)$. This process is repeated until the residue is either monotonically increasing or monotonically decreasing (having only a single zero crossing). Figure 5.2 illustrates the process of decomposing a signal into its components using EMD.

The Hilbert transform can then be performed on the collection of IMFs that result from the EMD process. This is the second part of the HHT algorithm. The Hilbert transform is a method of calculating the instantaneous frequency of a real-valued signal and it is related to the number of zero crossings that occur in a given time [11, 17]. The Hilbert transform can be calculated by

$$y(t) = \mathcal{H}[x(t)] = \frac{1}{\pi} PV \int_{-\infty}^{\infty} \frac{x(\tau)}{t-\tau} d\tau \tag{5.13}$$

In (5.13), PV denotes the principal value of the singular integral. Once the Hilbert transform has been found, the analytic signal is

$$z(t) = x(t) + jy(t) = x(t) + j\mathcal{H}[x(t)] = a(t)e^{j\theta(t)} \tag{5.14}$$

The instantaneous amplitude $a(t)$ and instantaneous phase $\theta(t)$ are then found simply by

$$a(t) = \sqrt{x^2 + y^2} \tag{5.15}$$

and

$$\theta(t) = \arctan\frac{y}{x} \tag{5.16}$$

The instantaneous frequency can then be determined simply by

$$\omega(t) = \frac{d\theta}{dt} \tag{5.17}$$

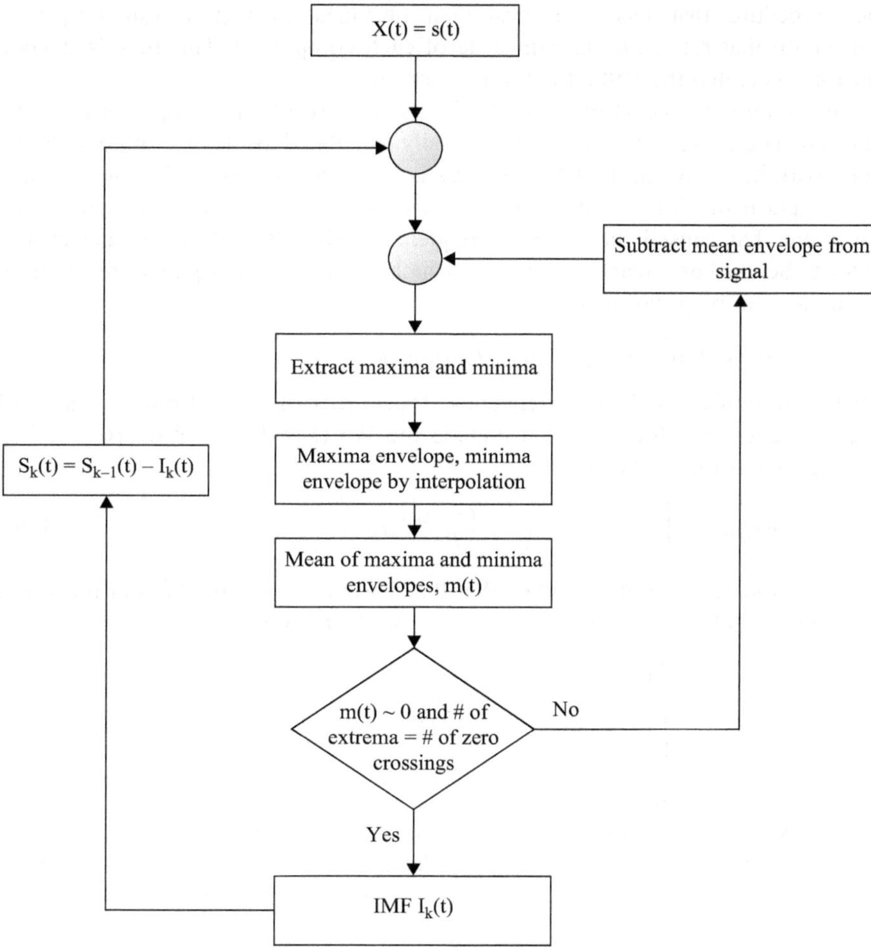

Figure 5.2 EMD algorithm flowchart

Past applications of the Hilbert transform have been limited to narrow band signals. This is the reasoning behind performing the EMD sifting operation first so that monocomponent signals are considered. The HHT algorithm calculates the instantaneous frequency and instantaneous amplitude of each IMF component separately. After HHT has been performed, if it is desired, the original signal can be represented as in (5.18) based on its instantaneous amplitude and instantaneous frequency components. Compare this to the Fourier representation in (5.19). In the Fourier case, both a_i and ω_i are constants. It is apparent that the IMF represents a generalized Fourier expansion. By permitting the amplitude and frequency to vary with time, HHT allows for nonlinear and nonstationary signals to be analyzed.

$$x(t) = \Re\left\{\sum_{i=1}^{n} a_i(t)\exp\left[j\int \omega_i dt\right]\right\} \tag{5.18}$$

$$x(t) = \Re\left[\sum_{i=1}^{n} a_i e^{j\omega_i(t)t}\right] \tag{5.19}$$

The procedure that has been described provides us with a time-frequency distribution that relies on the amplitude of each component. This time-frequency transform is called the Hilbert amplitude spectrum.

In addition to the Hilbert spectrum, the energy of the components can be analyzed. The energy of a signal can be easily calculated as the inner product of the signal with itself. When the EMD process is conducted on micro-Doppler signals, the collection of IMF energies provides us with a vector that is unique to the movement that caused the Doppler frequency shift. This allows for the energy vector to be used as a feature vector, on which classification or pattern recognition techniques can be performed.

5.3.4 Other Time-Frequency Transforms

Other commonly used time-frequency transforms include Cohen's class of time-frequency transforms, which include the WVD and the CWD [10, 13, 14]. The expression for the WVD is

$$WVD(t,\omega) = \int x\left(t+\frac{\tau}{2}\right)x^*\left(t-\frac{\tau}{2}\right)e^{-j\omega\tau}d\tau \tag{5.20}$$

By inspection of (5.20) it is evident that the WVD is simply the Fourier transform of the autocorrelation function. The expression for the CWD is

$$CWD(t,\omega) = \iint \frac{1}{\sqrt{4\pi\alpha(\tau)^2}} e^{-\frac{(t-u)^2}{4\alpha(\tau)^2}} x\left(u+\frac{\tau}{2}\right)x^*\left(u-\frac{\tau}{2}\right)e^{-j\omega t}dud\tau \tag{5.21}$$

Both the WVD and the CWD are special cases of Cohen's class of time-frequency transforms. The general form of Cohen's class is shown in (5.22), where ϕ is a two-dimensional kernel function. The kernel determines the distribution and also the properties of the distribution. If ϕ is chosen to be 1, then it reduces to the WVD.

$$C(t,\omega) = \frac{1}{4\pi^2}\iiint e^{-j\theta t - j\tau\omega + j\theta u}\phi(\theta,\tau)x\left(u+\frac{\tau}{2}\right)x^*\left(u-\frac{\tau}{2}\right)dud\tau d\theta \tag{5.22}$$

Because θ does not appear in the signal itself, this general form of Cohen's class can be simplified by defining a new kernel, r, which is the Fourier transform of the original kernel, ϕ.

$$r(t,\tau) = \frac{1}{2\pi}\int \phi(\theta,\tau)e^{-jt\theta}\,d\theta \tag{5.23}$$

Using this new kernel, the general form of Cohen's class becomes

$$\iint r(t-u,\tau)x\left(u+\frac{\tau}{2}\right)x^*\left(u-\frac{\tau}{2}\right)e^{-j\omega t}dud\tau \tag{5.24}$$

If r is chosen to be

$$r(t,\ \tau) = \frac{1}{4\pi\alpha(\tau)^2}e^{-\frac{t^2}{4\alpha(\tau)^2}} \tag{5.25}$$

then (5.24) is the expression for the CWD. As previously mentioned, the kernel function determines properties of the distribution. Because the WVD is simply the Fourier transform of the autocorrelation function, it suffers from having significant cross-term interference (two separate components in the signal present themselves as a new term in the distribution, which is the product of the two components). The kernel for the CWD is chosen to significantly reduce the cross terms.

5.4 Wall Effects on Micro-Doppler Signatures

As discussed in section 5.2, electromagnetic waves that travel through media will experience the effects of attenuation and dispersion. These effects are nonspecific to micro-Doppler signatures. Aside from these effects, additional phenomena occur when observing micro-Doppler. The first and the simplest to overcome is a constant phase offset that occurs because of the stationary wall. The second is that the signal-to-noise ratio (SNR) of the micro-Doppler signature significantly decreases due to attenuation and this will greatly reduce the maximum detectable range.

5.4.1 Constant Phase Offset

The familiar expressions for the Doppler frequency shift and phase shifts are shown in (5.26) and (5.27), respectively, where $v(t)$ is the time-varying velocity, $r(t)$ is the time-varying distance to the target, and λ is the wavelength of the carrier signal.

$$f_d(t) = \frac{2v(t)}{\lambda} \tag{5.26}$$

$$\varphi_d(t) = \frac{4\pi r(t)}{\lambda} \tag{5.27}$$

Each scatterer in the radar beam will introduce a component to the total received radar signal, some of which will have a micro-Doppler component and some of which will not. If, for simplicity, we assume a single frequency continuous wave Doppler radar system and that the received radar signal consists of only two components, one from a wall and another from an object that exhibits micro-Doppler motion, the received signal can be expressed as

$$r_x(t) = a_{wall}\cos[2\pi f_c t + \varphi_{wall}] + a_{obj}(t)\cos[2\pi f_c t + \varphi_{obj}(t)] \tag{5.28}$$

Because the range to the wall is constant, it does not produce a time-varying phase shift. Conversely, the object that exhibits micro-Doppler motion will have a range to the target, $r(t)$ that will vary with time and therefore the Doppler phase shift will vary with time. In addition to this, the amplitude of the wall reflection will not vary with time because its RCS remains constant. The amplitude of the object, however, will generally vary with time because its RCS will depend on its orientation relative to the radar. Typically the carrier signal is removed via frequency mixing and filtering, which will simplify (5.28) to

$$r_x(t) = a_{wall}\cos(\phi_{wall}) + a_{obj}(t)\cos(\phi_{obj}(t)) \tag{5.29}$$

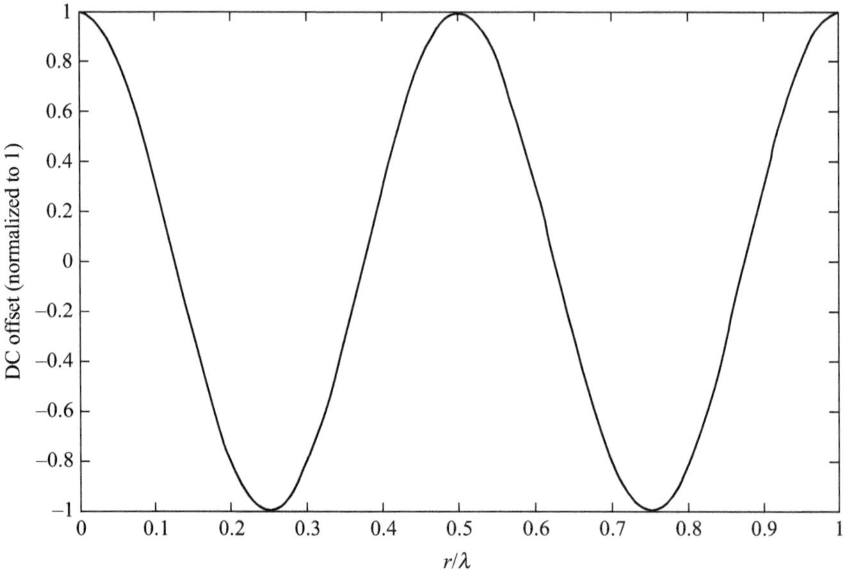

Figure 5.3 Theoretical DC offset versus range to the wall as a fraction of λ

The first term of (5.29) shows that there will be a DC component present in the micro-Doppler signature because of the wall. This DC component depends on the magnitude of the wall component and also on the distance to the wall as a fraction of the carrier wavelength. Because the phase can be represented by (5.27) and because the cosine function repeats every 2π, the DC offset will repeat when the distance from the radar to the wall is a multiple of $\lambda/2$. This is illustrated in Figure 5.3 and experimentally verified by measurements shown in Figure 5.4. Because the wavelength for the experiments was small (on the order of a couple of centimeters), it was very difficult to measure the exact distance to the wall, which led to some error in the measurements; however, these experiments show that the phenomenon of a constant offset due to a stationary barrier will have an effect on the radar micro-Doppler returns. Based on Figure 5.3, the DC offset should achieve negative values for some distances. This is not the case in Figure 5.4, which is due to antenna coupling and other additional components. In the development of the theoretical DC offset, only a single constant phase term was considered for simplicity. In general, this is not realistic. In practical measurements, multipath, antenna coupling, and clutter will all produce constant phase terms contributing to the DC offset. In Figure 5.4 the combined DC value of the other components caused the DC offset to remain positive for all but a single distance to the wall. It should also be noted that if a target exhibits micro-motions, but is otherwise remaining stationary, the distance to the target can be expressed as $R_0 + r(t)$. This constant distance will also produce a constant DC term, which will add together with the wall's DC component.

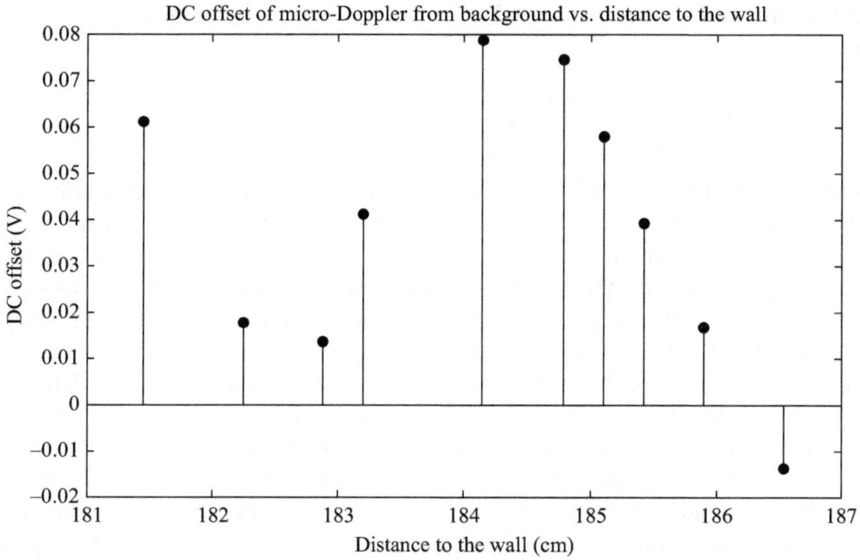

Figure 5.4 Experimental DC offset versus range to the wall as a fraction of λ

The DC component does not have a great effect on the micro-Doppler signature; however, when viewing the signal using a time-frequency representation, low frequency signals may be obscured by a very strong, constant DC offset. This can be remedied by simply subtracting the DC offset from the signal.

5.4.2 Signal-to-Noise Ratio and Maximum Detectable Range

As with all radar systems, micro-Doppler radars have a maximum detectable range that depends on SNR. Because of the attenuation that electromagnetic waves experience when traveling through wall materials, the SNR will decrease and thus decrease the maximum detectable range of a micro-Doppler radar. Equation (5.30) shows the radar range equation in a form that shows the maximum detection range at which a target with a given RCS can be detected with a given SNR.

$$R_{max} = \left[\frac{P_t G_t G_r \lambda^2 \sigma}{(4\pi)^3 SNR_{min}\, L} \right]^{\frac{1}{4}} \tag{5.30}$$

In (5.30), P_t is the transmit power, G_t is the gain of the transmitting antenna, G_r is the gain of the receiving antenna, λ is the wavelength, σ is the RCS, and SNR is the signal-to-noise ratio. Here all of the losses from wall attenuation are included in the term L ($L \geq 1$). This means that the maximum detectable range of the radar will decrease significantly as L increases. However, due to reflections, not all of the power is transmitted through the wall, so (5.30) must be modified according to the transmission power from (5.5). Because of two-way transmission, (5.30) becomes

$$R_{max} = \left[\frac{P_t G_t G_r \lambda^2 (1 - |\Gamma|^2)^2}{(4\pi)^3 SNR_{min}\, L} \right] \tag{5.31}$$

As an example, let us consider a two-way wall loss of 20 dB and a maximum range of 50 m when operating in an environment without a wall ($L = 1$). For now, neglect the effect of wall reflections and only consider wall losses. A 20 dB wall attenuation is a loss factor of 100 and will decrease our maximum range by a factor of $(1/100)^{1/4} = 1/\sqrt{10} \approx 0.32$. This means that our radar can only detect targets up to a range of $50 \times 0.32 = 15.8$ m. At higher frequencies for materials such as concrete, cinder block, or brick, the losses may be significant enough that it is impossible to detect micro-Doppler signals at any distance.

The distances from the radar to the wall and from the radar to the target are also important. Using our previous example, suppose that the wall is located at 20 m from the radar. This would mean that it is necessary for the target to be in front of the wall in order to detect the micro-Doppler. In this case the wall no longer has an effect and the wall losses are irrelevant.

Now, let us consider a target that is located 10 m behind a wall. The two-way wall loss is again 20 dB. In order to detect this object, the radar must be located closer than 5.8 m from the wall. This ensures that the target is within the maximum detectable range of 15.8 m. Figure 5.5 illustrates the effect of attenuation on the maximum range. It is clear that the maximum range is halved for every additional 12 dB of loss.

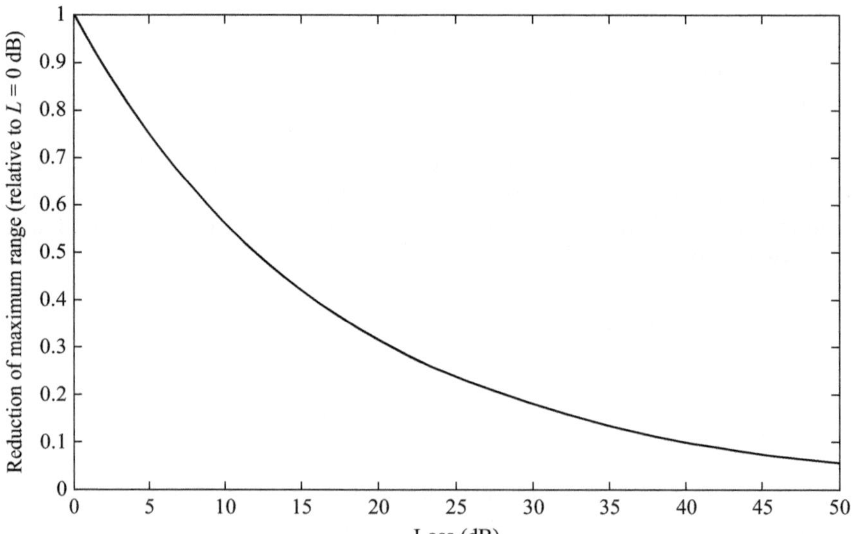

Figure 5.5 Maximum detectable range for a given loss (relative to L = 1)

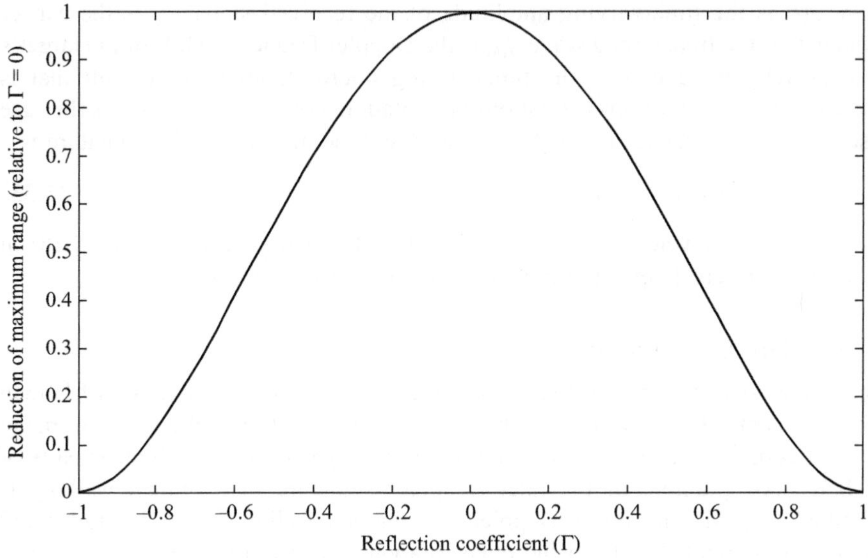

Figure 5.6 Maximum detectable range versus Γ

As another example, consider the first scenario once again, except we will now include a reflection coefficient of $|\Gamma| = 0.5$. The $(1 - |\Gamma|^2)^2$ term is equal to 0.5625, which corresponds to an additional 2.5 dB of loss. Figure 5.6 illustrates the effect of the wall reflections on maximum range. Larger values for the magnitude of Γ will increase the loss significantly and if we have a perfect conductor ($|\Gamma| = 1$) then, as should be expected, the target cannot be detected.

5.5 Micro-Doppler Signals of Targets with a Translational Velocity

Some common applications of micro-Doppler radar include identification of signatures from human gait, vehicles, and animals. If vehicles or animals are to be considered, in many cases a through-the-wall environment is not considered. However, in some cases it may be desirable to sense objects in an outdoor environment from within a building. In these cases, a through-the-wall environment is applicable.

Under normal circumstances, vehicle micro-Doppler will include some translational velocity, in which case the micro-Doppler will be centered around the constant Doppler frequency that arises from the translational motion. This is also the case for human gait, where the translational velocity of the human will shift the micro-Doppler from the legs, arms, etc. away from being centered at DC.

In the case of a signal with a constant translational velocity combined with a micro-Doppler frequency component, the radar received signal can be simulated as

$$x_r(t) = a(t)\cos(2\pi f_c t + 2\pi f_{d0} t + \varphi(t)) \tag{5.32}$$

where $a(t)$ is the time-varying amplitude of the received signal, f_c is the carrier frequency of the transmitted wave, f_{d0} is the Doppler frequency shift of the target's constant velocity, and φ is the time-varying micro-Doppler phase shift that is produced by micro-motions (vibrations and rotations) of the target. As was the case in section 5.4, the carrier is usually removed and the micro-Doppler signature is

$$x_r(t) = a(t)\cos(2\pi f_{d0}t + \varphi(t)) \tag{5.33}$$

Here the constant velocity produces a constant frequency according to (5.26) in addition to a micro-Doppler component with a time-varying phase.

5.5.1 Walking Humans

A common area of current research is in the analysis and identification of human gait movements. The motion of walking is a complex motion that consists of many different components, all of which are moving in synchronism with other parts of the body. An example of the micro-Doppler signature of a human walking is described in [18]. The micro-Doppler of a human walking toward a radar and moving from about 8 m from the radar to about 2 m from the radar is shown in Figure 5.7 [19]. This time-frequency representation shows a strong response from the torso and weaker responses due to the arms and legs. The torso is restricted to a small bandwidth, whereas the arms and legs exhibit larger Doppler shifts. The mean frequency of the walking signature is shifted in the positive direction, indicating a translational velocity toward the radar. In [20], the human walking movement is simulated. This simulation is shown in Figure 5.8 and clearly shows some of the distinct Doppler traces that arise from each moving part of the body (left toe, left ankle, torso, right ankle, right toe, etc.).

Figure 5.9 shows a human walking toward a noncoherent micro-Doppler radar operating in the S-Band. The person was located behind a 4 inch thick brick wall.

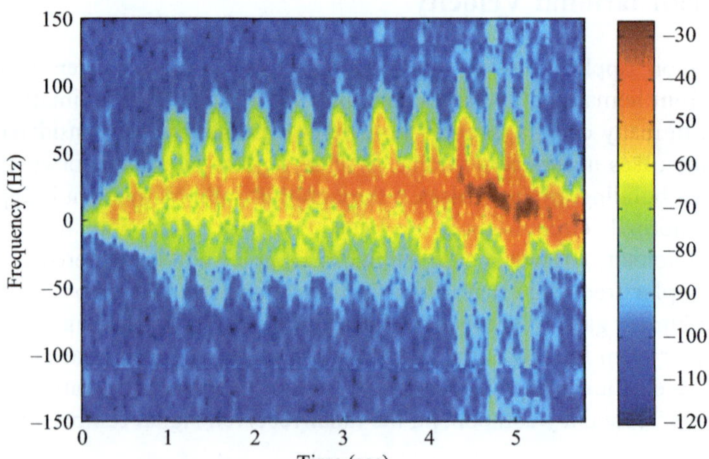

Figure 5.7 Spectrogram of a human walking toward a radar [19]

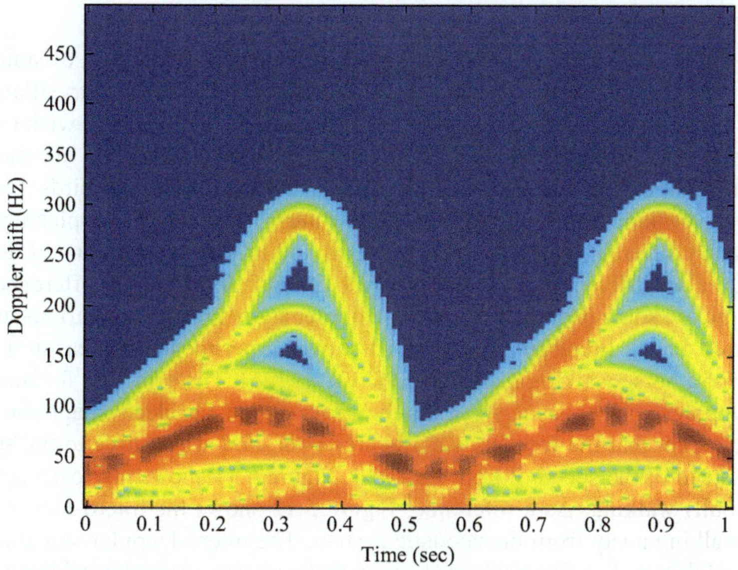

Figure 5.8 Spectrogram of a human walking toward a radar [18]

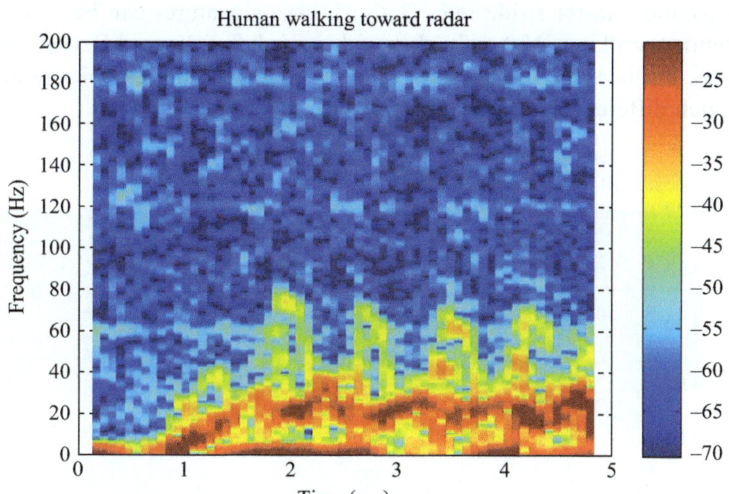

Figure 5.9 Spectrogram of a human walking using an S-band, noncoherent, micro-Doppler radar

Because the radar is noncoherent, only positive frequencies can be observed. Nonetheless, many of the same features can be clearly seen in the time-frequency representation of the walking signal and, in fact, the walking signature appears very similar to the walking signature of Figure 5.7 that was reported in [19].

For additional information with more detail on human gait micro-Doppler, see [18–24].

5.5.2 Animals

The ability to differentiate human signatures from the signatures of animals is another ongoing research topic. This is especially important in surveillance and security applications, where animals do not pose a threat and one may wish to ignore their signatures. Animals move in such a way that their movements are much different than that of humans. For example, consider dogs, horses, and birds. Dogs and horses both walk on four legs and birds have wings. These micro-Doppler signatures are all much different than human movements and in fact, the motions of dogs and horses are also different from each other (albeit much more subtle differences).

Below are some examples of micro-Doppler signatures due to various animals, as reported in the literature. Figure 5.10 shows experimentally measured micro-Doppler of a horse walking using acoustic waves [25]. The principle for measuring micro-Doppler with acoustic waves is the same as that for electromagnetic micro-Doppler except that the frequency shifts are lower, and are relative to the speed of sound instead of the speed of light for radar. Figure 5.11 shows micro-Doppler of a dog (pitbull) walking. As a reference, Figure 5.12 shows the micro-Doppler from human walking away from the acoustic system. The micro-Doppler signature from a horse exhibits a slow repetition frequency that corresponds to the stride rate of the horse. For dogs, we observe a much faster stride rate. Intuitively, this makes sense because a horse has very long legs and takes longer strides, whereas a dog has shorter legs and a faster stride rate. Both of these signatures can be distinguished from a human walking. Not only does a human being have different body part proportions, but because the animals walk on four legs, the Doppler components will be much different from that of a human [25].

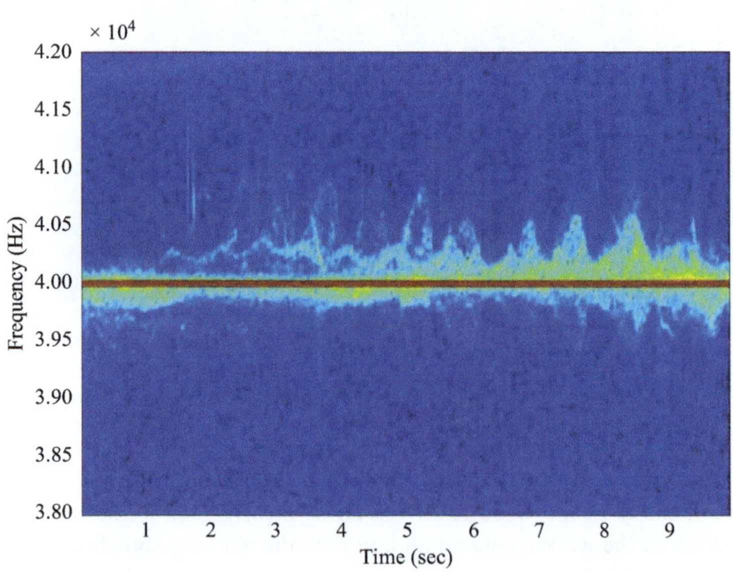

Figure 5.10 Spectrogram of a horse walking toward a radar [25]

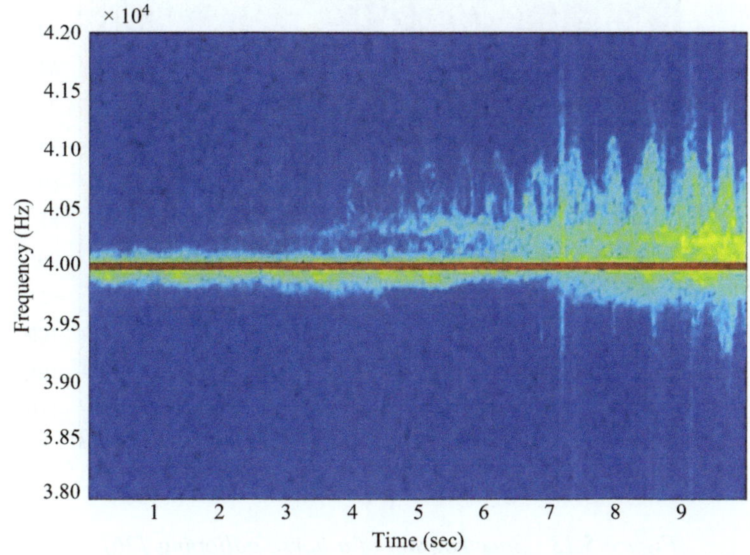

Figure 5.11 Spectrogram of a pitbull walking toward a radar [25]

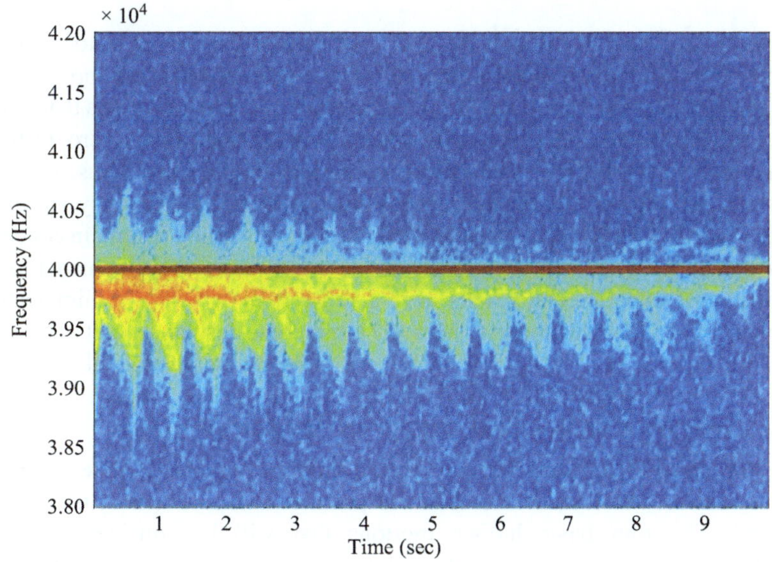

Figure 5.12 Spectrogram of a human walking away from a radar [25]

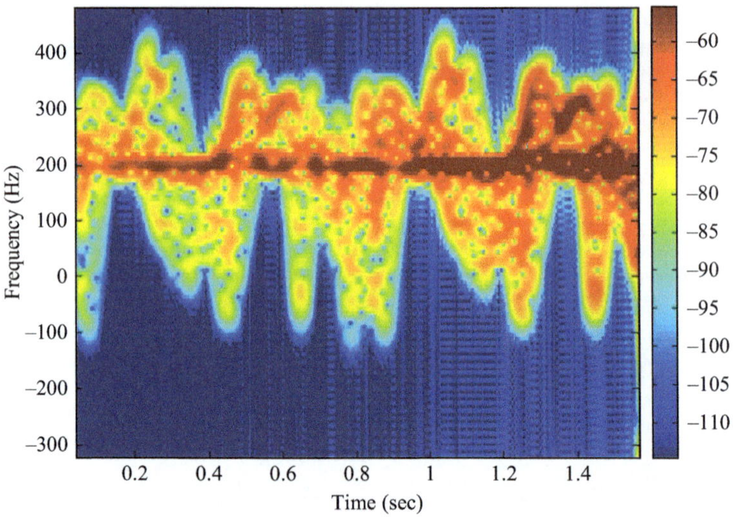

Figure 5.13 Spectrogram of a horse galloping [26]

The motion of a galloping horse has also been simulated in [26]. This simulation result is shown in Figure 5.13.

5.6 Micro-Doppler Signals of Stationary Targets

As is usually the case in through-the-wall applications, the target is located in an indoor environment and is therefore limited to the distances that it can move. For this reason, another class of micro-Doppler movements is considered: targets that have no translational velocity. In this situation, the mean Doppler frequency is zero. Detecting these motions is more difficult than detecting motions due to gait because the movements are much slower and produce much smaller Doppler frequency shifts. The ability to detect these motions is critical when the target is trying to remain concealed. Here, we develop some simple models to describe some basic human motions. The models and experimental results have also been shown for a millimeter-wave (mm-wave) micro-Doppler radar system [27], but these results are not shown here because of the inability of the mm-wave radar to penetrate wall materials.

5.6.1 Experimental Micro-Doppler of Human Motions

Before discussing the models for some basic human motions, some experimental results are shown to illustrate the different parts of each movement that must be considered. The four basic human motions that will be considered here are (1) breathing, (2) swinging arms, (3) picking up an object, and (4) transitioning from crouching to standing. The micro-Doppler signature of a pendulum is also considered. The micro-Doppler signals and the corresponding STFTs of the signals are shown in Figures 5.14–5.18. The micro-Doppler time signature is shown on the

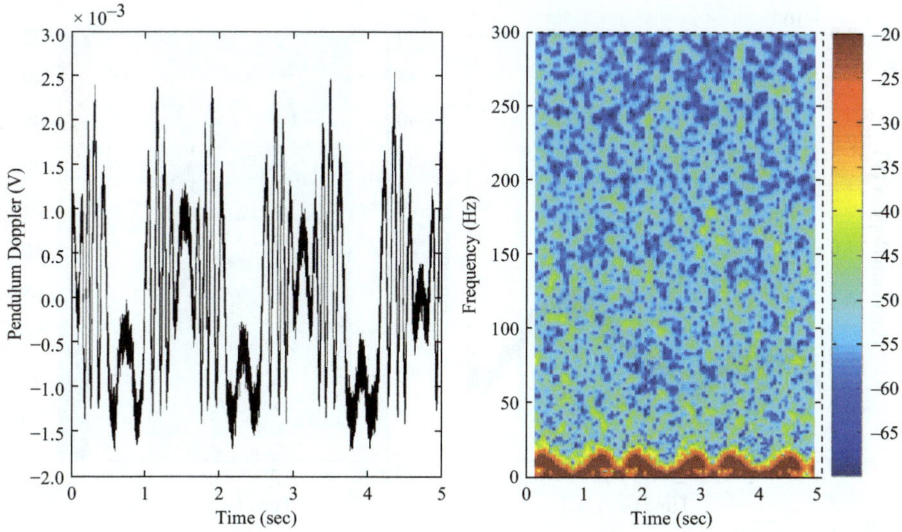

Figure 5.14 (Left) Experimentally measured micro-Doppler of a pendulum and (Right) STFT of the micro-Doppler signal

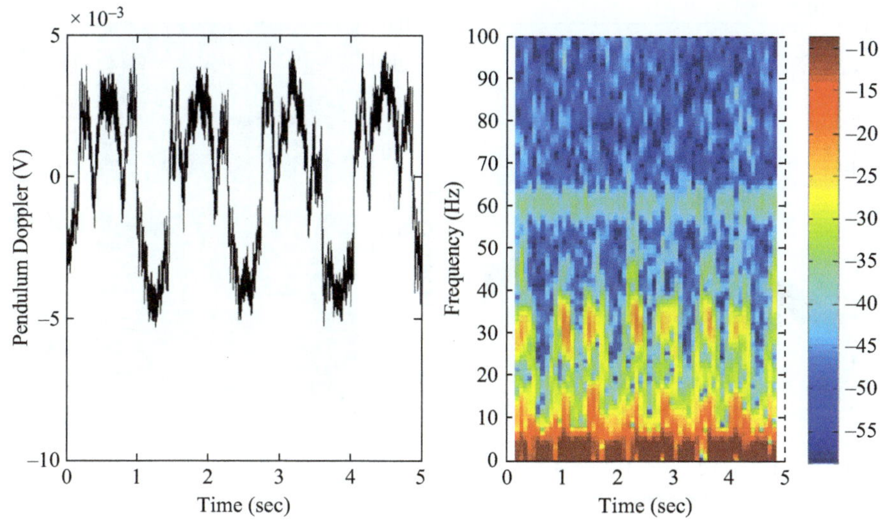

Figure 5.15 (Left) Experimentally measured micro-Doppler of a person swinging their arms and (Right) STFT of the micro-Doppler signal

left and the STFT is shown on the right. These experiments were performed with a noncoherent CW Doppler radar operating in the S-band. Because of the noncoherence of the radar, the direction of travel cannot be determined. Any motion that creates a negative Doppler frequency shift will be displayed as a positive frequency since the direction cannot be determined.

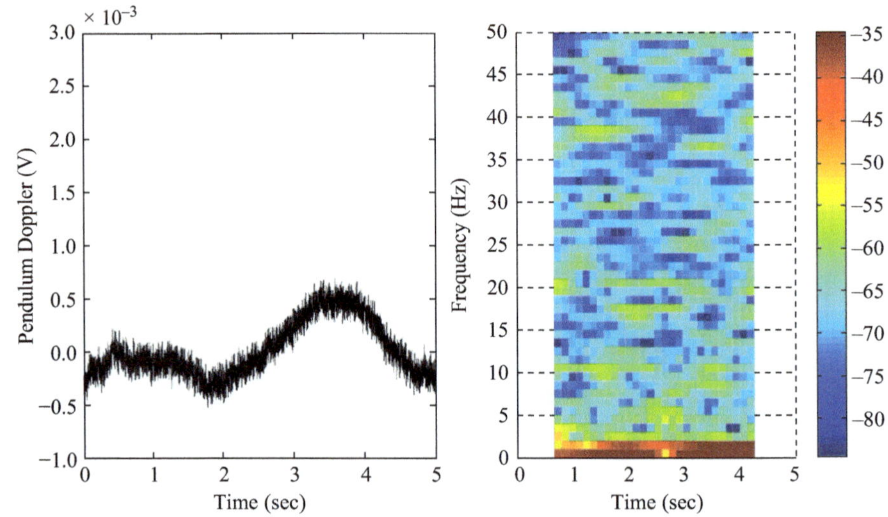

Figure 5.16 (Left) Experimentally measured micro-Doppler of a person breathing and (Right) STFT of the micro-Doppler signal

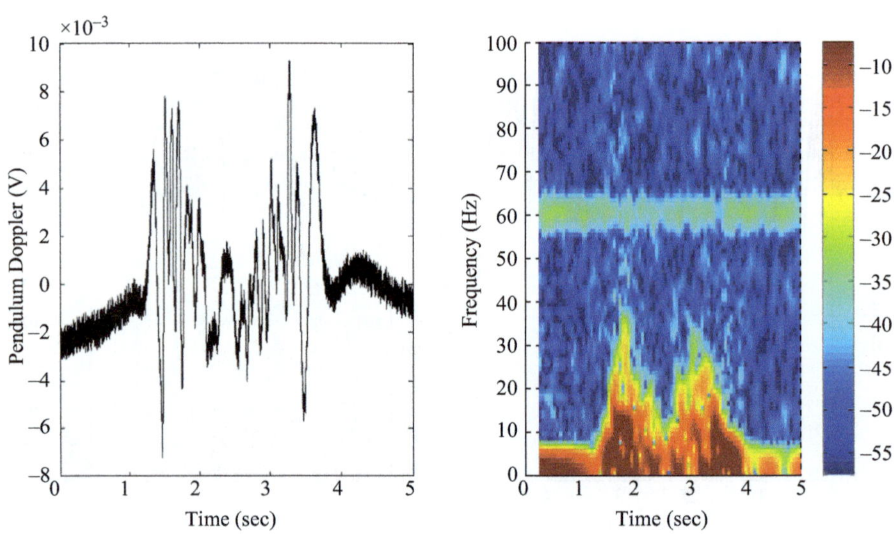

Figure 5.17 (Left) Experimentally measured micro-Doppler of a person picking up an object and (Right) STFT of the micro-Doppler signal

For the human motions in Figures 5.15–5.18, the radar was operating in an indoor environment with the human located at approximately 3 feet behind a 4 inch thick cinderblock wall. The radar was located approximately 9 feet in front of the wall.

Figure 5.14 shows the motion that is exhibited by a swinging pendulum. The pendulum is a long flat metal plate that swings at a pivot point. The motion is

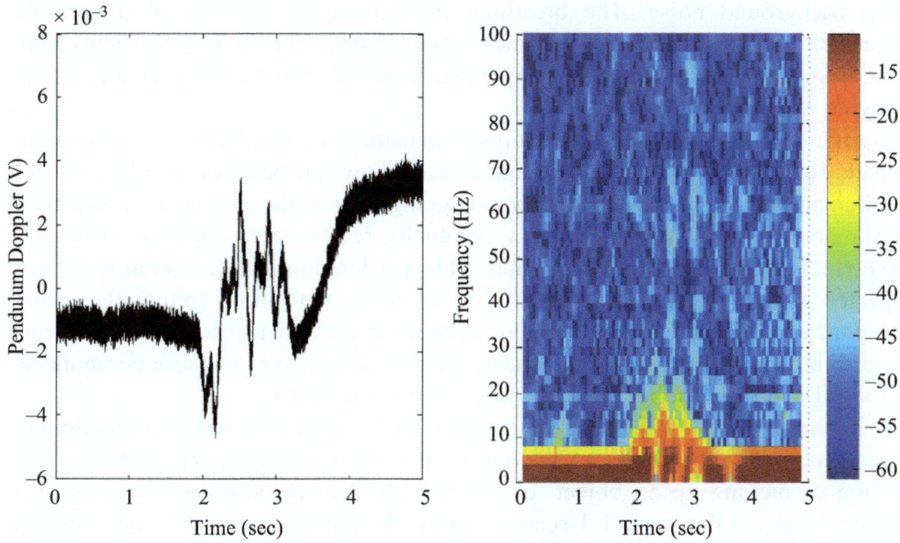

Figure 5.18 (Left) Experimentally measured micro-Doppler of a person transitioning from crouching to standing and (Right) STFT of the micro-Doppler signal

sinusoidal and because the pendulum is not forced, friction causes the pendulum to slow down gradually. This is clearly seen in the STFT plot where the maximum frequency of each successive swing reaches a lower frequency (hence a slower speed) than the previous swing.

Figure 5.15 shows the micro-Doppler signature due to a human swinging their arms. While this motion is similar to the motion of a pendulum, it is evident from the STFT plot that the arm swinging of a human is more complex and contains more components. The additional components are mainly due to the torso swaying to maintain balance and the fact that each arm must be considered as two separate components, namely the upper arm and the lower arm. The upper and lower arms must be considered as separate components because any bend in the elbow joint will cause the lower arm to travel at a faster angular velocity than the upper arm. In addition to these extra components, the human body cannot be considered as a point scatterer. Each point along the arm will contribute its own Doppler frequency shift and will be proportional to the location of that point on the arm. The components that are closer to the end of the arm will produce larger Doppler frequency shifts because this point on the arm is traveling faster than a point on the arm that is close to the shoulder joint (where the radial velocity is zero).

Figure 5.16 shows the micro-Doppler signature due to a human breathing. This motion is very slow and therefore produces only very low frequency micro-Doppler signals. In this case the STFT does not have a high enough resolution to accurately determine the time-frequency characteristics of the signal. However, it is still clear that this motion is distinct from other human motions and is still distinguishable from background noise. The breathing motion mainly consists of a periodic expansion and contraction of the chest cavity. Other nonideal components may arise from the human subject swaying either in an effort to maintain balance or the swaying may be inadvertent.

Figure 5.17 shows the micro-Doppler signature due to a human picking up an object. This motion is nonperiodic and clearly shows an acceleration phase where the human bends at the waist and begins going toward the ground followed by a deceleration phase where the human eventually reaches zero velocity when they have reached the object on the ground. This acceleration and deceleration is then repeated for when the person returns to an upright position. Intuitively, many components contribute to this motion: bending at the waist, reaching of the arms toward the object, bending of the knees, etc. However, many of these components are small and may not be obvious when viewing the STFT.

Figure 5.18 shows the micro-Doppler signature due to a human transitioning from a crouching position to a standing position. This motion is very similar to the motion of picking up an object, except that the human is already starting at a position close to the ground. Because of this, the human body only goes through one acceleration–deceleration cycle and this is clearly seen in the STFT of the crouching to standing micro-Doppler signal. As in the case of picking up an object, other body motions contribute to the signal, but may not have a large effect. Because the person is crouching, the bending of the knees may play a more significant role in the micro-Doppler signature of transitioning from crouching to standing than in picking up an object. However, this contribution is still small compared to the very large contribution of the torso.

Similar experimental data has been obtained for various human motions using a 750 MHz radar [28]. The lower carrier frequency means that lower Doppler frequencies are observed; however, similar characteristics are found in these data as was found in the data for the S-band radar above. The results obtained are shown in Figure 5.19. Both the raw micro-Doppler and the energy of each IMF from EMD processing are shown. The micro-Doppler signals of Figure 5.19 also reveal that the micro-Doppler signatures are similar for different stand-off distances.

Additional information on human motions, including the motions of breathing and heartbeat, can be found in [29–41]. These results show the ability of radars to detect vital signs for medical applications and these radars are specifically designed to operate in a through-the-wall environment.

5.6.2 Models of Simple Human Motions Micro-Doppler Signal of a Pendulum

The first motion that is considered is that of a person swinging their arms. This motion is similar to that of a pendulum; therefore, it will be used as a building block to create the swinging arm model. The pendulum is among the simplest of periodic motions that can be considered when discussing micro-Doppler. The motion of a

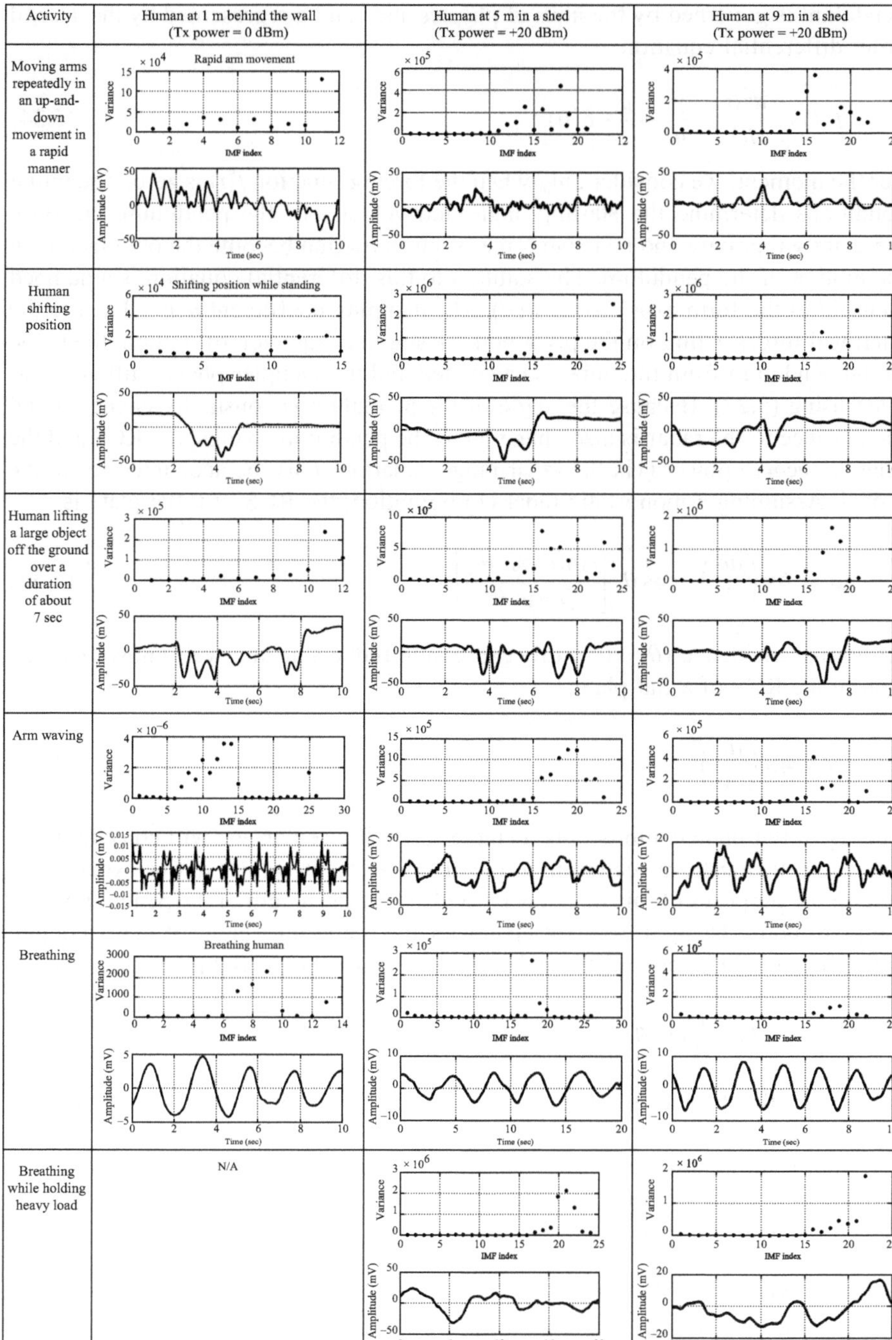

Figure 5.19 Micro-Doppler characteristics of various human activities [28]

pendulum is governed by the laws of physics and can be represented by the second order differential equation:

$$f(t) = \frac{d^2\theta}{dt^2} + \frac{\gamma}{ml} + \frac{g}{l}\sin(\theta) \tag{5.34}$$

For the moment, we consider only when the forcing function $f(t)$ is zero. The initial conditions determine the starting angle and velocity of the pendulum, but once these are set, gravity and resistive forces such as friction become the only factors in the motion of the pendulum. The solution to this differential equation is a damped sinusoid of the form $\exp(-\alpha t)\cos(\omega t)$. The damping coefficient α is related to the frictional forces acting on the pivot point. Using this equation of motion, the range of the pendulum from the radar is calculated and the Doppler phase shift is calculated using (5.27). Because the angle of the pendulum is constantly changing and the distance to the radar is also changing as the pendulum swings, the RCS and the relative receive power (via the radar range equation) must be accounted for in the model. Assuming a monostatic radar configuration, the RCS of a flat plate is

$$\sigma = 4\pi\left(\frac{ab}{\lambda}\right)^2 \cos\theta_i^2\left[\frac{\sin(\beta b\sin\theta_i)}{\beta b\sin\theta_i}\right]^2 \tag{5.35}$$

At normal incidence ($\theta_i = 0°$), this expression reduces to the well-known expression of the RCS of a flat plate:

$$\sigma = 4\pi\left(\frac{ab}{\lambda}\right)^2 \tag{5.36}$$

Because a flat plate exhibits only a single scattering center, we will consider radar returns occurring from scattering at the antenna height, which in our experiments and for the purposes of this model is located at half the length of the pendulum from the pivot point. Using a constant height for the scattering center of the pendulum and using simple geometry, the radial distance from the pendulum to the radar is

$$r(t) = R_0 + \frac{L}{2}\tan\theta_i \tag{5.37}$$

where R_0 is the distance to the pendulum at normal incidence and L is the length of the pendulum.

The resulting simulated Doppler signal is shown in Figure 5.20. Comparing this signal to the experimentally measured signal in Figure 5.14, we observe good agreement with real data.

5.6.2.1 Swinging Arms

The motion of a human swinging arm is a more complex motion than a simple pendulum due to the nonrigid nature of the human body, in particular, the elbow joint. The swinging arm causes the torso to sway in order for the person to maintain balance and the arm itself contains multiple components due to the joints at the

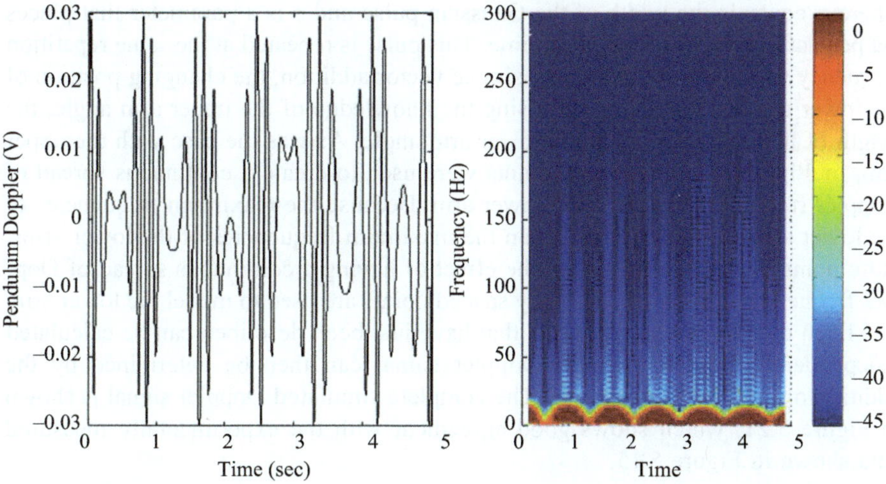

Figure 5.20 (Left) Simulated micro-Doppler of a pendulum and (Right) STFT of the simulated micro-Doppler signal of the pendulum

shoulder and elbow. As was the case with the pendulum, the angle of the upper arm relative to vertical is modeled as moving sinusoidally (with $\alpha = 0$). The torso is also modeled as swaying sinusoidally, but the torso movements are 180° out of phase with the arm angle. The lower arm angle relative to the upper arm is modeled as moving in a Gaussian shaped trajectory. Based on our observations, this is a reasonable estimation of the lower arm motion. Many of our test subjects kept their arms straight for the majority of the swinging motion and only bent their elbows near the forward most point of the swing. The Gaussian trajectory also allows for the elbow joint to maintain a positive angle relative to the upper arm, which is in accordance with the physical limits of the human body.

Here, the torso movement is simulated with a large, near constant RCS and is oscillating forward and backward with an amplitude of approximately 2 cm and a frequency of 0.7 Hz. The upper arm motion is modeled as a pendulum swinging at the same frequency as the torso oscillations, but with 180° difference in the phase. The length of the arm is modeled with a length of 0.8 m and its RCS is considered to be that of an ellipsoid. The maximum arm angle is set to be about 5°. Because our experimental measurements display a continuous spread of frequencies and not a single component for the upper arms, the micro-Doppler frequency shift is calculated at 10 different points along the arm. The 10 points are randomly located on the length of the arm. If the points were located at evenly spaced distances instead placed randomly, the signals would add in phase and produce a signal that is visually similar to an impulse train.

The angle that the lower arm makes relative to the upper arm is modeled as a Gaussian, given by

$$\theta(t) = \theta_0 \exp\left(-\frac{(t-c)^2}{2\sigma^2}\right) \tag{5.38}$$

where σ controls the width of the Gaussian pulse and c is a parameter that places the peak at the center of the pulse time. This pulse is repeated at the same repetition frequency as the upper arm. Using simple vector addition, the changing position of the lower arm can be calculated using the knowledge of the upper arm angle, the length of the upper arm, and the lower arm angle. As was the case with the upper arm, multiple randomly spaced points were used to create a continuous spread of Doppler frequency shifts for the lower arm. Because the maximum frequencies of the lower arms are much larger than the maximum frequencies of the lower arms, more points are needed to create the effect of having a continuous spread of Doppler frequencies. Here, 25 randomly spaced points are used to model the lower arm.

Each of the three components that have just been described can be calculated independently. The total micro-Doppler signal can then be determined by the addition of all three components. The complete simulated Doppler signal is shown in Figure 5.21, which shows good agreement with the experimentally measured data shown in Figure 5.15.

5.6.2.2 Breathing

The motion of a human breathing is fairly simple. There is only one main component that must be accounted for, namely, the chest expansion and contraction. However, this motion is highly irregular in humans due to the long length of time that it takes to complete a single breath. Because of the long duration required for a single cycle of the motion, the frequency signature varies significantly from breath to breath. Adding even more difficulty is the fact that individuals often breathe differently when they are conscious of how they are breathing as opposed to when they are not paying attention to how they breathe. Despite this fact, a breathing motion can be reasonably approximated as a smooth transition to an inhaled

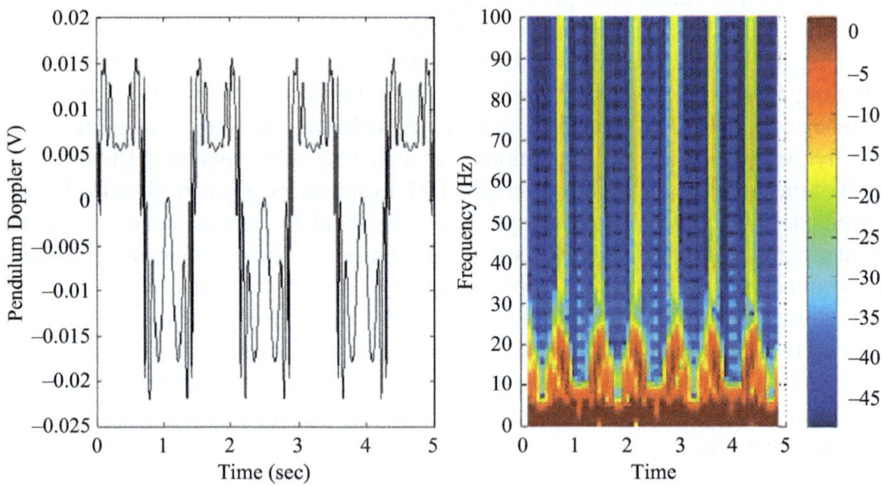

Figure 5.21 (Left) Simulated micro-Doppler of swinging arms and (Right) STFT of the simulated micro-Doppler signal

position, a short pause, a smooth transition to an exhaled position, a short pause, etc. To model this mathematically, a rectangular pulse train with a 50% duty cycle is constructed. This models the short pauses after inhalation and exhalation, but fails to accurately model a smooth transition between the two states. To achieve this, passing the rectangular pulses through a simple RC filter, shown in (5.39), smoothes the transition regions and the transition time can be adjusted by changing the RC time constant, $\tau = RC$. Performing both forward and reverse filtering allows for both transition regions to be "smoothed out" appropriately.

$$H(z) = \frac{1}{1 - \tau z} \tag{5.39}$$

The simulation of human breathing is shown in Figure 5.22. Because breathing is a much slower motion than swinging arms, 10 sec are shown instead of 5 sec. It is evident from this figure that breathing consists of extremely slow motions, making Doppler shifts extremely difficult to detect. In fact the STFT of the breathing motion is incapable of showing any detail in the time-frequency representation of the micro-Doppler signal. Compare the simulated breathing signal of Figure 5.22 to the experimentally measured signal of Figure 5.16. Micro-Doppler signals due to breathing are comprised of very low frequency components and the Doppler phase does not change rapidly enough for frequency variations to be observed. For this reason, a slowly varying Doppler phase shift can be thought of in a similar manner as was considered in section 5.4.1, where the wall produced a constant DC offset. A slowly varying phase can be thought of as a slowly varying DC offset. This phenomenon is observed in both the simulated and experimental breathing data.

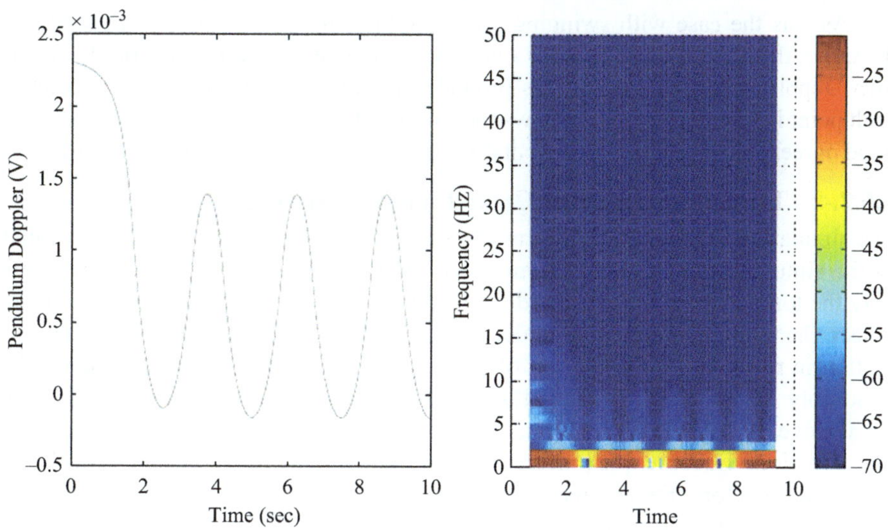

Figure 5.22 (Left) Simulated micro-Doppler of breathing and (Right) STFT of the simulated micro-Doppler signal

5.6.2.3 Picking Up an Object

The motion of a human picking up an object is a complex motion that involves the interaction of multiple body parts to complete the motion. The knees and waist bend in an effort to lower the person closer to the ground and the arms reach down and slightly outward to grasp the object. These motions are then repeated in reverse to return the human to an upright position. For this simulation, only the torso motion is considered. The torso has the largest RCS of the body components and obscure the other components for this reason and also because it causes larger Doppler frequency shifts than the other body parts. Also, the arms can be neglected because their motion mainly consists of reaching downward toward the object (and not radially toward the radar). Bending of the knees, while consisting mainly of a radial velocity toward the radar, is a short and slow motion that is obscured by the torso and is neglected for this reason.

To model this motion, the torso is considered to bend only at the waist. It starts in an upright position (0°) and rotates so that the torso is almost parallel to the floor (80°). This motion is then repeated in reverse. To achieve this, the angle between the torso and vertical is considered to be Gaussian (a reasonable estimate based on [42]) so that the angle can be written as

$$\theta(t) = ae^{-\frac{(t-t_0)^2}{2\sigma^2}} \tag{5.40}$$

where $a = \theta_{max}$ determines the maximum angle of the torso, t_0 is the time that the trajectory achieves its maximum angle, and σ determines how quickly the torso transitions from $\theta = 0°$ to $\theta = \theta_{max}$. Using θ, the length of the torso, and some basic trigonometry and calculus, the velocity of the torso can be calculated. The Gaussian function allows for the desired acceleration–deceleration trajectory where the velocity is zero at the beginning, middle, and end of the motion.

As was the case with swinging arms, the human body is not a point scatterer; therefore, the torso is considered as a continuous line of scatterers, each producing its own Doppler frequency shift. This produces a continuous spread of frequencies at each point in time. Figure 5.23 shows the simulated micro-Doppler signal which is in good agreement with the experimentally measured signal shown in Figure 5.17.

5.6.2.4 Transitioning from Crouching to Standing

The simulation of the motion for transitioning from crouching to standing is similar to the simulation for picking up an object. For the reasons previously discussed, only the torso motion is considered for the simulation. In the case of picking up an object, the torso accelerated, then decelerated to zero velocity before repeating this motion in reverse. For the motion of transitioning from crouching to standing, the torso only experiences the second acceleration–deceleration phase of the motion. To achieve this, the angle between the torso and vertical is modeled as a sigmoid-like function. Because [42] describes how human motions typically display a Gaussian trajectory, the angular velocity is modeled as a Gaussian. The result is an initial torso angle of 0° and a smooth transition to $\theta_{max} = 90°$ at the end, given by

$$\theta(t) = \frac{\theta_0}{2}\left[1 - \text{erf}\left(\frac{t-t_0}{\sigma\sqrt{2}}\right)\right] \tag{5.41}$$

where the error function erf(x) is given by

$$\text{erf}(t) = \frac{2}{\pi}\int_0^t e^{-t^2}\,dt \tag{5.42}$$

This trajectory (shifted in time) provides the desired acceleration–deceleration cycle that is exhibited in Figure 5.24. Comparison of this simulation to the experimentally measured signal in Figure 5.18 reveals good agreement.

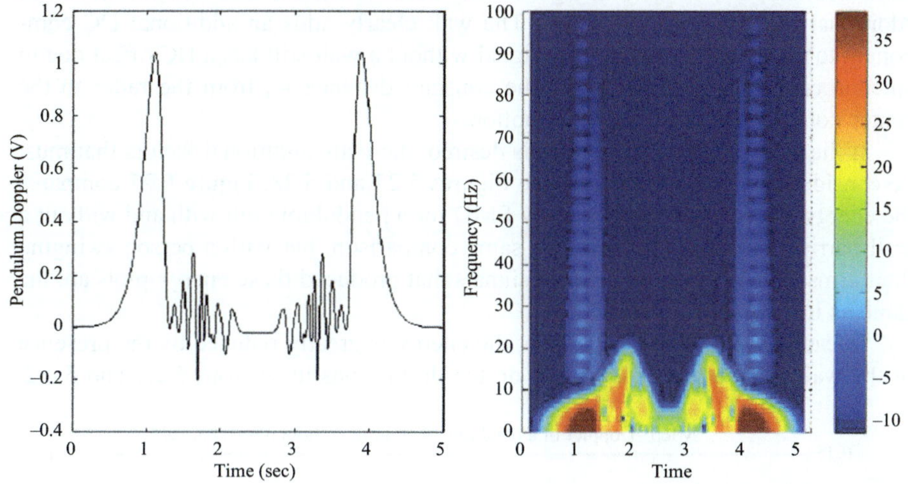

Figure 5.23 (Left) Simulated micro-Doppler of picking up an object and (Right) STFT of the simulated micro-Doppler signal

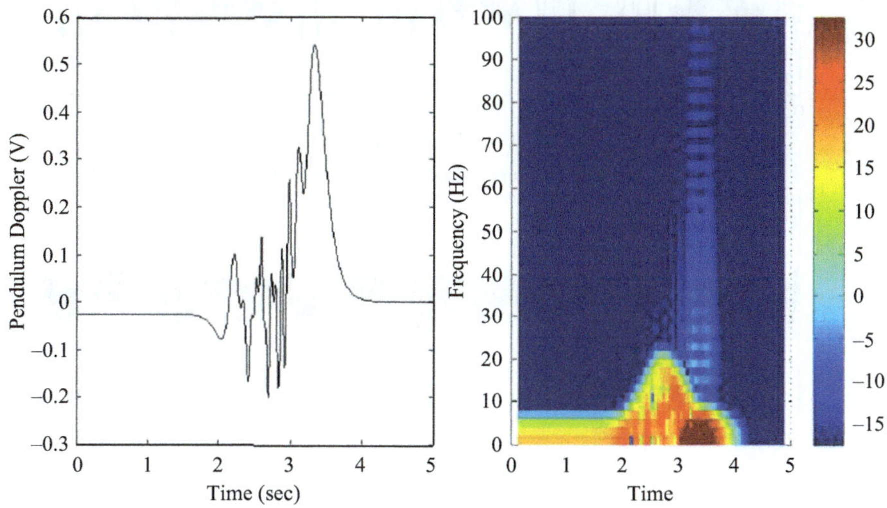

Figure 5.24 (Left) Simulated micro-Doppler of transitioning from crouching to standing and (Right) STFT of the simulated micro-Doppler signal

5.6.3 Comparison of Through-the-Wall Micro-Doppler Signatures versus Non-Through-the-Wall Micro-Doppler Signatures

In order to visualize how a wall affects micro-Doppler signatures, two examples are briefly discussed now. The first example compares the micro-Doppler due to a pendulum for both through-the-wall and barrier-free situations. This is shown in Figure 5.25. The second example is of a human target that is swinging their arms. The human comparison for swinging arms is shown in Figure 5.26. As expected, for both examples, the SNR is significantly reduced and the wall has introduced an additional DC offset to the signal. The wall clearly adds an additional DC component to the signal; however, the signal without a wall still has a DC offset due to antenna coupling, multipath, and the constant distance R_0 from the radar to the center point of the target's swing motion.

If the classification of motions is desired, there are additional factors that must be considered. To illustrate this, see Figures 5.27 and 5.28. Figure 5.27 compares the energy of each IMF produced by EMD for a pendulum both with and without a wall barrier. Figure 5.28 shows the same comparison, but with a person swinging their arms. The raw micro-Doppler signals that produced these energy plots are the same as in Figures 5.25 and 5.26.

In each energy plot comparison, the energy is greatly reduced by the presence of the wall. This is expected based on the discussions in sections 5.2.1 and 5.4.2.

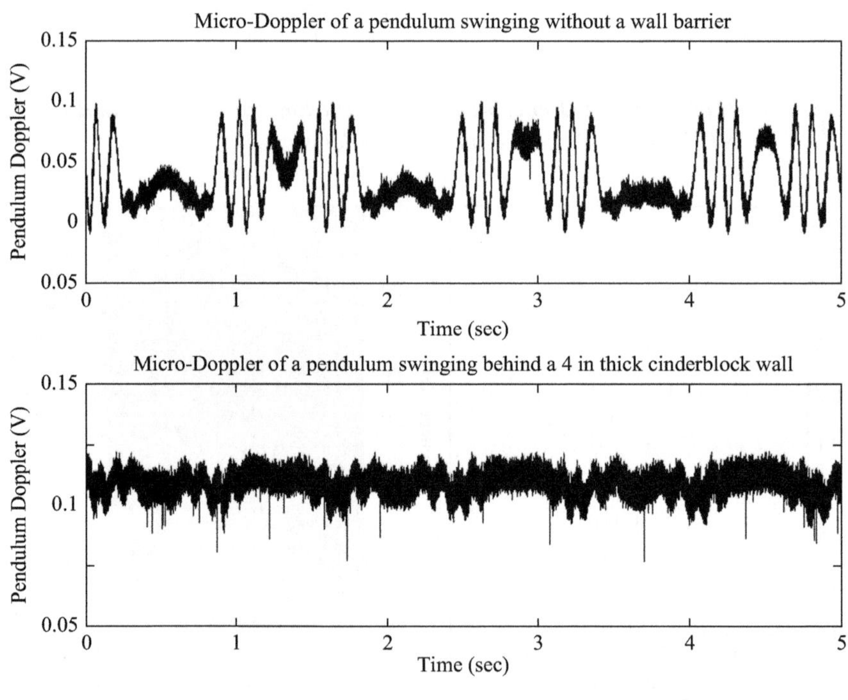

Figure 5.25 Comparison of the micro-Doppler signal of a pendulum (Top) without a wall barrier and (Bottom) with a wall barrier

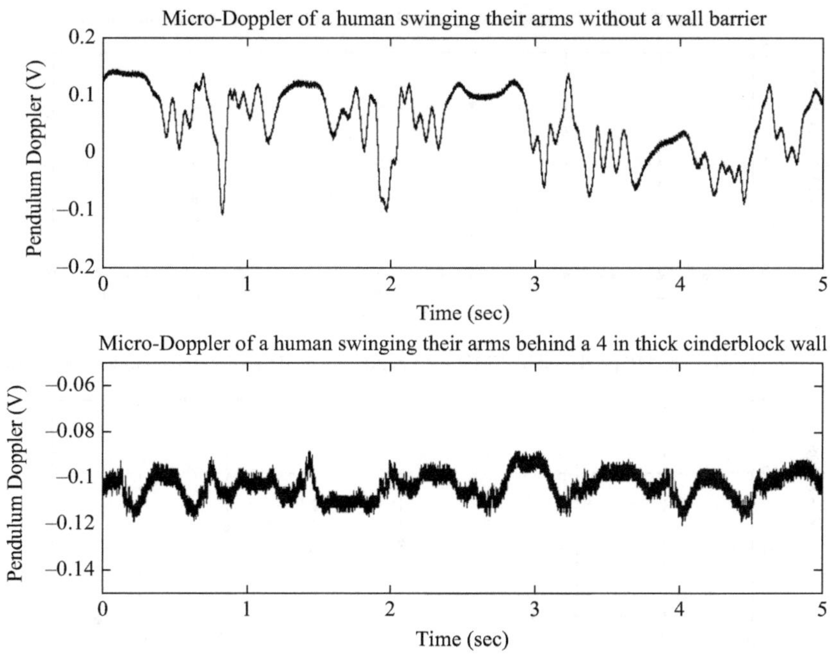

Figure 5.26 Comparison of the micro-Doppler signal of a pendulum (Top) without a wall barrier and (Bottom) with a wall barrier

If the energy of each IMF of a signal is to be used as a feature vector for classification, the amplitudes of each are critical to the accuracy of the classifier. Training a classifier under a scenario without a wall barrier may not produce accurate results if the test data are taken from a through-the-wall scenario. For this reason, the classifier must be trained and tested under the appropriate conditions and it is important to understand its limitations.

5.6.4 Classification of Micro-Doppler Signatures

The ability to identify human motions using micro-Doppler signatures has become an area of interest recently. Activity classification has potential applications for surveillance, search and rescue operations, and patient monitoring in hospitals. Knowledge of a human's movement can aid in improving the safety of security or surveillance personnel or it can potentially locate a person trapped inside a building. The identification of movements can be useful; however, the larger goal may be to infer the intent of the target, which is a much more difficult problem. The classification of human movements can be performed by utilizing a Doppler radar system. The micro-Doppler frequency will vary depending on the unique velocity profile of a radar target over time. Specific movements will have unique micro-Doppler signatures which can then be classified [19].

The feasibility of using EMD and a support vector machine (SVM) classifier to identify human movements has been demonstrated through the use of a 750 MHz

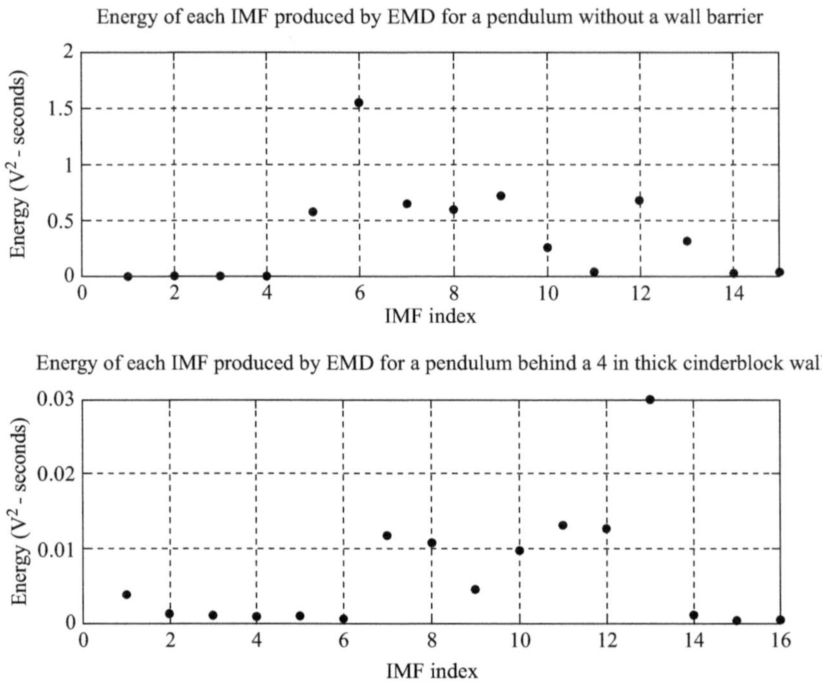

Figure 5.27 Comparison of the energy of each IMF produced by EMD for a pendulum (Top) without a wall barrier and (Bottom) with a wall barrier

micro-Doppler radar [28, 43]. This radar was designed to detect human targets concealed behind a wall [44]. The radar operates through direct transmission of a 760 MHz CW wave. EMD and classification are performed off line using Matlab. The classification is performed through the use of the LIBSVM software package [45, 46]. The same procedure has also been implemented for use with an S-band micro-Doppler radar and also a mm-wave micro-Doppler radar [27]. The higher frequency of operation allows these radars to pick out more subtle variations in the human movements because the Doppler frequency shift is directly proportional to the transmitted frequency; however, the larger carrier frequency also introduces a higher loss for through-the-wall applications. The S-band radar is capable of detecting micro-motions through a 4 inch thick cinderblock wall and accurately classifying those motions. The mm-wave radar is incapable of detecting through-wall signatures because of the high loss of the wall material, but it is capable of detecting micro-motions through light foliage.

The S-band radar is designed for close range, through-the-wall situations, whereas the mm-wave radar has been designed for longer range applications or applications where the target is concealed within light foliage. As we should expect, the mm-wave radar is not capable of detecting through-the-wall micro-Doppler signatures due to the extremely high losses that the wave will experience.

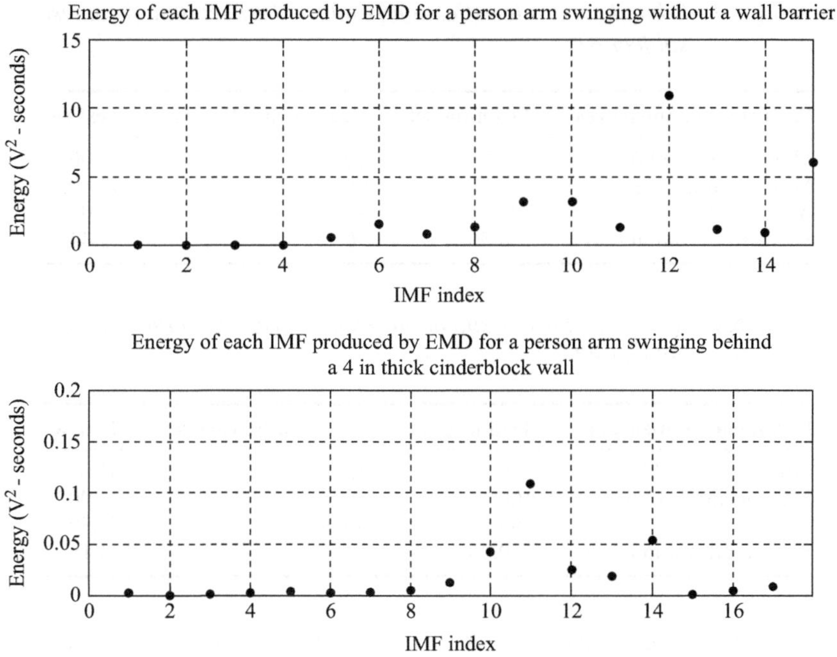

Figure 5.28 Comparison of the energy of each IMF produced by EMD for a person swinging arms (Top) without a wall barrier and (Bottom) with a wall barrier

Experiments have been conducted to test the accuracy of the micro-Doppler motion classification procedure. These experiments represent data taken for four different test subjects. For testing purposes, the data are rotated among these test subjects so that three of them are always used for training and cross-validation and the fourth subject is used for testing. This will allow for an accurate assessment of how the classifier responds to data that are both "known" and "unknown."

The S-band micro-Doppler has been tested for the situation of a person moving behind a 4 inch thick cinderblock wall. The person is located at 3 feet behind the wall and the radar is located at about 8 feet in front of the wall. The classification of motions for this radar considers five different motions: (1) background noise, (2) breathing, (3) swinging arms, (4) picking up an object, and (5) transitioning from a crouching position to a standing position. The training set consists of data from three of the four test subjects and was used both for training the classifier and in the cross-validation. The human test subjects represented wide variability in size and shape. The cross-validation set also consists of data from three of the four test subjects, but these data were used only for cross-validation and not used for training the classifier. The test set consists of data from one of the four test subjects and these data were used for neither training the classifier nor for cross-validation. Results are shown in Tables 5.3–5.6. The average accuracy is 76.25% when

Table 5.3 S-band radar classification results – Test set data from test subject #1

Performance measures	Training set	Cross-validation set	Test set
Min. accuracy (%)	98.33	80.00	72.00
Max. accuracy (%)	100	96.67	92.00
Avg. accuracy (%)	99.75	87.33	85.00
Standard deviation (%)	0.56	6.25	5.44

Table 5.4 S-band radar classification results – Test set data from test subject #2

Performance measures	Training set	Cross-validation set	Test set
Min. accuracy (%)	99.17	83.33	64.00
Max. accuracy (%)	100	93.33	76.00
Avg. accuracy (%)	99.92	90.00	71.80
Standard deviation (%)	0.26	3.14	4.16

Table 5.5 S-band radar classification results – Test set data from test subject #3

Performance measures	Training set	Cross-validation set	Test set
Min accuracy (%)	100	86.67	66.00
Max accuracy (%)	100	96.67	82.00
Avg. accuracy (%)	100	92.33	74.40
Standard deviation (%)	0.00	3.16	4.97

Table 5.6 S-band radar classification results – Test set data from test subject #4

Performance measures	Training set	Cross-validation set	Test set
Min accuracy (%)	99.17	76.67	66.00
Max accuracy (%)	100	93.33	80.00
Avg. accuracy (%)	99.92	89.00	73.80
Standard deviation (%)	0.26	4.73	7.76

combining the results of all four of the test subjects. As should be expected, the accuracy of the classification is dependent on the person performing the motions.

The mm-wave micro-Doppler radar has been tested for the situation of a person moving at a distance of 100 feet in an outdoor environment without any barriers. (The radar itself is capable of picking up Doppler signals at a range of over 300 feet; however, the classifier will become less accurate if the signal power decreases to where the SNR is very low.) At a range of 100 feet, this radar is sensitive enough to be able to classify all of the motions of interest (the same five motions as in the S-band classifier). Same training and test set procedure was used for classification with radar as was used for classification with the S-band radar. Results are shown in Tables 5.7–5.10. The average accuracy is 79.7% when combining the results of all four of the test subjects.

Table 5.7 MM-wave radar classification results – Test set data from test subject #1

Performance measures	Training set	Cross-validation set	Test set
Min accuracy (%)	98.33	86.67	62.00
Max accuracy (%)	99.17	100	76.00
Avg. accuracy (%)	98.75	94.33	68.00
Standard deviation (%)	0.44	4.46	4.22

Table 5.8 MM-wave radar classification results – Test set data from test subject #2

Performance measures	Training set	Cross-validation set	Test set
Min accuracy (%)	91.67	83.33	78.00
Max accuracy (%)	98.33	96.67	92.00
Avg. accuracy (%)	96.17	88.33	85.60
Standard deviation (%)	2.19	4.23	5.15

Table 5.9 MM-wave radar classification results – Test set data from test subject #3

Performance measures	Training set	Cross-validation set	Test set
Min accuracy (%)	97.50	86.67	72.00
Max accuracy (%)	100	96.67	86.00
Avg. accuracy (%)	98.67	91.00	77.00
Standard deviation (%)	0.70	3.87	5.44

Table 5.10 MM-wave radar classification results – Test set data from test Subject #4

Performance measures	Training set	Cross-validation set	Test set
Min accuracy (%)	91.67	83.33	82.00
Max accuracy (%)	98.33	96.67	94.00
Avg. accuracy (%)	96.08	88.33	88.20
Standard deviation (%)	1.97	4.51	3.94

The classifier above utilizes EMD to produce feature vectors for classification and the target remains stationary when performing the motions. Other work has been performed in this area, which concentrates on identifying human gait. One example is in [19], where the motions of running, walking, walking while holding a stick, crawling, boxing while moving forward, boxing while standing in place, and sitting still were considered for micro-Doppler classification. Here, features were selected based off of STFT characteristics. The features that were selected for classification were (1) torso Doppler frequency, (2) the total bandwidth of the Doppler signal, (3) the offset of the total Doppler, (4) the bandwidth without micro-Doppler, (5) the normalized standard deviation of the Doppler signal strength, and (6) the period of the limb motion. The data was classified using an SVM. The classification of these motions was reported to have an accuracy as high as 92.8%.

Another example of human activity classification can be found in [47]. The technique is different from the above techniques in that it transmits an ultrawide-band signal and uses principal component analysis as the feature extraction method. The overall classification accuracy for the classification of eight different human activities is 85.3%. The eight activities are walking, running, rotating, punching, jumping, transitioning between standing and sitting, crawling, and standing still. Even though this technique does not utilize the micro-Doppler phenomenon, it is always of interest to investigate other methods.

References

[1] D. J. Daniels, "Surface-penetrating radar," *Electronics Communication Engineering Journal*, vol. 8, no. 4, pp. 165–182, August 1996.

[2] D. G. Falconer, K. N. Steadman, and D. G. Watters, "Through-the-wall differential radar," in *Proceedings of SPIE Conference on Command, Control, Communications, and Intelligence Systems for Law Enforcement*, vol. 2938, (Boston, MA), pp. 147–151, November 1996.

[3] N. C. Currie, D. D. Ferris, R. W. McMillan, and M. C. Wicks, "New law enforcement applications of millimeter-wave radar," in *Proceedings of SPIE Conference on Radar Sensor Technology II*, vol. 3066, (Orlando, FL), pp. 2–10, April 1997.

[4] D. P. Nguyen, N. Petrochilos, A. Host-Madsen, V. Lubecke, and O. Boric-Lubecke, "A versatile through-the-wall Doppler radar using BSS algorithms," in *Proceedings of the 2007 IEEE Antennas and Propagation Society International Symposium*, (Honolulu, HI), pp. 273–276, June 2007.

[5] A. Muqaibel, A. Safaai-Jazi, A. Bayram, A. M. Attiya, and S. M. Riad, "Ultrawideband through-the-wall propagation," *IEE Proceedings on Microwaves, Antennas and Propagation*, vol. 152, no. 6, pp. 581–588, December 2005.

[6] A. Safaai-Jazi, S. M. Riad, A. Muqaibel, and A. Bayram, "Ultra-wideband propagation measurements and channel modeling," DARPA NETEX Program, Technical Report on "Through-the-wall propagation and material

characterization," November 2002. Available online at http://www.darpa.mil/ato/solicit/netex/docs/models.pdf.

[7] R. Wilson, "Propagation losses through common building materials: 2.4 GHz vs. 5 GHz," Technical Report, University of Southern California, August 2002.

[8] M. G. Amin, *Through-the-wall Radar Imaging*. CRC Press, 2011.

[9] K. Oughstun and N. Cartwright, "On the Lorentz-Lorenz formula and the Lorentz model of dielectric dispersion," *Optics Express*, vol. 11, no. 13, pp. 1541–1546, June 2003.

[10] L. Cohen, *Time-Frequency Analysis*. Prentice Hall PTR, 1995.

[11] N. E. Huang and S. P. Shen, *Hilbert–Huang Transform and Its Applications*. World Scientific Publishing Co., 2005.

[12] N. E. Huang, Z. Shen, S. R. Long, M. C. Wu, H. H. Shih, Q. Zheng, N. C. Yen, C. C. Tung, and H. H. Liu, "The empirical mode decomposition and the Hilbert spectrum for nonlinear and nonstationary time series analysis," *Proceeding of the Royal Society A*, vol. 454, no. 1971, pp. 903–995, March 1998.

[13] V. Chen and H. Ling, *Time-Frequency Transforms for Radar Imaging and Signal Analysis*. Artech House, 2002.

[14] F. Hlawatsch and F. Auger, *Time-Frequency Analysis: Concepts and Methods*. Wiley, 2005.

[15] G. Strang and T. Nguyen, *Wavelets and Filter Banks*. Wellesley-Cambridge Press, 1996.

[16] Q. H. Du and S. N. Yang, "Application of the EMD method in the vibration analysis of ball bearings," *Mechanical Systems and Signal Processing*, vol. 21, no. 6, pp. 2634–2644, August 2007.

[17] R. M. Narayanan, M. C. Shastry, P. H. Chen, and M. Levi, "Through-the-wall detection of stationary human targets using Doppler radar," *Progress in Electromagnetics Research B*, vol. 20, pp. 147–166, 2010.

[18] R. G. Raj, V. C. Chen, and R. Lipps, "Analysis of radar dismount signatures via non-parametric and parametric methods," in *Proceedings of the 2009 IEEE Radar Conference*, (Pasadena, CA), pp. 1–6, May 2009.

[19] Y. Kim and H. Ling, "Human activity classification based on micro-Doppler signatures using a support vector machine," *IEEE Transactions on Geoscience and Remote Sensing*, vol. 47, no. 5, pp. 1328–1337, May 2009.

[20] R. G. Raj, V. C. Chen, and R. Lipps, "Analysis of radar human gait signatures," *IET Signal Processing*, vol. 4, no. 3, pp. 234–244, June 2010.

[21] A. Lin and H. Ling, "Through-wall measurements of a Doppler and direction-of-arrival (DDOA) radar for tracking indoor movers," in *Proceedings of the 2005 IEEE Antennas and Propagation Society International Symposium*, vol. 3B, (Washington, DC), pp. 322–325, July 2005.

[22] R. Boulic, N. M. Thalmann, and D. Thalmann, "A global human walking model with real-time kinematic personification," *The Visual Computer*, vol. 6, no. 6, pp. 344–358, 1990.

[23] C. P. Lai, R. M. Narayanan, Q. Ruan, and A. Davydov, "Hilbert-Huang transform analysis of human activities using through-wall noise and noise-like radar," *IET Radar, Sonar Navigation*, vol. 2, no. 4, pp. 244–255, August 2008.

[24] V. Chen, *The Micro-Doppler Effect in Radar*. Artech House, 2011.

[25] Z. Zhang, P. Pouliquen, A. Waxman, and A. G. Andreou, "Acoustic micro-Doppler gait signatures of humans and animals," in *Proceedings of the 41st Annual Conference on Information Sciences and Systems (CISS'07)*, (Baltimore, MD), pp. 627–630, March 2007.

[26] S. S. Ram and H. Ling, "Micro-Doppler signature simulation of computer animated human and animal motions," in *Proceedings of the 2008 IEEE Antennas and Propagation Society International Symposium,* (San Diego, CA), pp. 1–4, July 2008.

[27] D. P. Fairchild and R. M. Narayanan, "Classification and modeling of human activities using empirical mode decomposition with S-band and millimeter-wave micro-Doppler radars," in *Proceedings of SPIE Conference on Radar Sensor Technology XVI*, (Baltimore, MD), April 2012. doi:10.1117/12.922448.

[28] P. H. Chen, M. C. Shastry, C. P. Lai, and R. M. Narayanan, "A portable real-time digital noise radar system for through-the-wall imaging," *IEEE Transactions on Geoscience and Remote Sensing*, vol. 50, no. 10, pt. 2, pp. 4123–4134, October 2012.

[29] A. S. Bugaev, V. V. Chapursky, S. I. Ivashov, V. V. Razevig, A. P. Sheyko, and I. A. Vasilyev, "Through wall sensing of human breathing and heart beating by monochromatic radar," in *Proceedings of the Tenth International Conference on Ground Penetrating Radar (GPR 2004),* vol. 1, (Delft, The Netherlands), pp. 291–294, June 2004.

[30] M. Y. W. Chia, S. W. Leong, C. K. Sim, and K. M. Chan, "Through-wall UWB radar operating within FCC's mask for sensing heart beat and breathing rate," in *Proceedings of European Radar Conference (EURAD 2005)*, (Paris, France), pp. 267–270, October 2005.

[31] S. Ivashov, V. Razevig, A. Sheyko, and L. Vasilyev, "Detection of human breathing and heartbeat by remote radar," in *Progress in Electromagnetic Research Symposium Proceedings*, (Pisa, Italy), pp. 663–666, March 2004.

[32] S. Jiping and L. Mingi, "Life detection and location methods using UWB impulse radar in a coal mine," *Mining Science and Technology (China)*, vol. 21, no. 5, pp. 687–691, September 2011.

[33] B. Levitas and J. Matuzas, "UWB radar for human being detection behind the wall," in *Proceedings of the 2006 International Radar Symposium*, (Krakow, Poland), pp. 1–3, May 2006.

[34] W. Li, X. Jing, Z. Li, and J. Wang, "A new algorithm for through wall human respiration monitoring using GPR," in *Proceedings of the 14th International Conference on Ground Penetrating Radar (GPR)*, (Shanghai, China), pp. 947–952, June 2012.

[35] S. Shirodkar, P. Barua, D. Anuradha, and R. Kuloor, "Heart-beat detection and ranging through a wall using ultra wide band radar," in *Proceedings of the 2011 International Conference on Communications and Signal Processing (ICCSP)*, (Calicut, India), pp. 579–583, February 2011.

[36] Z. Liu, L. Liu, and B. Barrowes, "The application of the Hilbert–Huang transform in through-wall life detection with UWB impulse radar," *PIERS Online*, vol. 6, no. 7, pp. 695–699, 2010.

[37] L. Liu, Z. Liu, and B. E. Barrowes, "Through-wall bio-radiolocation with UWB impulse radar: Observation, simulation and signal extraction," *IEEE Journal of Selected Topics in Applied Earth Observations and Remote Sensing*, vol. 4, no. 4, pp. 791–798, December 2011.

[38] V. M. Lubecke, O. Boric-Lubecke, A. Host-Madsen, and A. Fathy, "Through-the-wall radar life detection and monitoring," in *Proceedings of the 2007 IEEE/MTT-S International Microwave Symposium*, (Honolulu, HI), pp. 769–772, June 2007.

[39] E. Zaikov, "UWB radar for detection and localization of trapped people," in *Proceedings of the 11th International Radar Symposium (IRS)*, (Vilnius, Lithuania), pp. 1–4, June 2010.

[40] Q. Wang, Y. Li, J. Wu, and T. Zhang, "Life signal extraction in through-the-wall surveillance," in *Proceedings of the Annual International Conference of the IEEE Engineering in Medicine and Biology Society (EMBC 2009)*, (Minneapolis, MN), pp. 1343–1346, September 2009.

[41] Z. Zeng, J. Sun, J. Li, F. Liu, Q. Lu, and X. Chen, "The analysis of TWI data for human being's periodic motions," in 2011 *Proceedings of the 2011 IEEE International Geoscience and Remote Sensing Symposium* (IGARSS), (Vancouver, BC), pp. 862–865, July 2011.

[42] R. M. Narayanan, M. C. Shastry, P. H. Chen, and M. Levi, "Through-the-wall detection of stationary human targets using Doppler radar," *Progress In Electromagnetics Research B*, vol. 20, pp. 147–166, 2010.

[43] D. P. Fairchild and R. M. Narayanan, "Human activity classification using Hilbert-Huang Transform analysis of radar Doppler data," in *Proceedings of SPIE Conference on Radar Sensor Technology XV*, (Orlando, FL), April 2011. doi:10.1117/12.887348.

[44] P. H. Chen, R. M. Narayanan, C. P. Lai, and A. Davydov, "Through wall ranging and imaging using UWB random noise waveform: System design considerations and preliminary results," in *Digest of 2009 IEEE International Symposium on Antennas and Propagation* (Charleston, SC), June 2009. doi: 10.1109/APS.2009.5172369.

[45] G. Rilling and P. Goncalves, *EMD Toolbox for Matlab. Ecole Normale Superieure de Lyon, Lyon*, France, 2008. Available online at http://www.perso.enslyon.fr/patrick.flandrin/emd.html.

[46] C. C. Chang and C. J. Lin, *LIBSVM: a Library for Support Vector Machines*, 2001. Software available online at http://www.csie.ntu.edu.tw/ ~cjlin/libsvm.

[47] J. D. Bryan, J. Kwon, N. Lee, and Y. Kim, "Application of ultra-wide band radar for classification of human activities," *IET Radar, Sonar Navigation*, vol. 6, no. 3, pp. 172–179, March 2012.

Chapter 6
Identifying Human Movements Using Micro-Doppler Features

P. van Dorp

6.1 Introduction

Radar can be an extremely useful sensor technique to observe persons at critical infrastructures [1]. Radars became a common component of environment monitoring systems. They have properties not present in other sensors, such as insensitivity to darkness and light, penetration through common materials, and invisibility of their transmitted signals. A disadvantage of radar is that it transmits electromagnetic energy and that the measurements cannot be directly visualized like in the case of an electro-optical sensor. One way to visualize radar measurements is by showing them as a 3-D scene in a virtual environment. When observing humans, this requires the estimation of human motion features to animate a person. Detailed processing of the radar measurements gives range and radial velocity as a function of time of the person. These are used to visualize the scene in virtual reality. Real-time estimation of the human posture is an essential aspect in security application for proper countermeasures.

6.1.1 Radar Recognition

Radar recognition was primarily designed for battlefield surveillance radars and security radars applied for target classification. Hewis [2] and Nebabin [3] used the radial velocity distribution to classify the targets in humans, wheeled vehicles, tracked vehicles, helicopters and unknowns. The correct classification given is more than 80%. Another approach uses the physical dimensions of the target. Rohling [4] describes a radar system with range trilateration technique that estimates the object geometry. The classification between cars, cyclists, pedestrians, trees, traffic signs and a group of persons is more than 95% correct. The described recognition methods classify the human class but do not estimate the human posture. Estimation of the human posture is the next step to be carried out thereafter.

Recognition of human motion with radar is described in [5] and [6]. The approach in these references is based on radial velocity and radial acceleration estimation which give the individual human components. The acceleration estimation is time consuming and has errors due to the interfering reflections of individual human components.

If the object or a part of the object has rotation or translation in addition to the body motion, it might induce a frequency modulation and amplitude modulation on the return signal. These modulations give sidebands on both sides of the reflection of the human torso. This so-called micro-Doppler or micro-motion effect is observed in the 2-D time–velocity domain (see [7–9]). Additional range gives the 3-D range–velocity–time domain referred to the 3-D cubic micro-motion feature space (see [9–11]). These references give examples of micro-motion, time–frequency analyses algorithms and suggestions for future research. No recognition or parameter estimation techniques are described. Chen [7] and Mitra [12] give theoretical model studies of micro-motion. Another is the feature-based approach that estimates the parameters of a human walking model.

6.1.2 Human Motion Estimation

Analysis of radar measurements of human beings is concentrated on detection, tracking and classification of an object type. These are the main tasks of surveillance radars. The target signatures caused by human motion are observed and high-resolution time–range–frequency analysis is applied onto the measurements. Theoretical time–range–frequency responses are compared with real measurements and the results are in correspondence. The high-resolution analysis improves the interpretability of the measurements. As discussed previously the classification of a single human, a group of humans, wheeled vehicles, tracked vehicles and helicopters is possible. Classification and estimation of human body parts have not been examined so far.

In order to animate the human posture in a virtual world, time varying radar measurements, a human model and a radar model are required. Two approaches are possible. The first approach uses a parameterized model that incorporates the information of measured radar data with a corresponding state and a visual measurement. This approach has the disadvantage that both radar measurements and visual measurements are required with different scenarios. The second approach uses a description of the human with shapes, dimensions, radar cross section, and kinematics of the individual body parts and a radar model to simulate the radar response. Approaches which connect the radar model and human model can be broadly classified as model-based (top-down) and feature-based (bottom-up) (see [13]).

6.1.2.1 Model-Based

Model-based approaches estimate the parameters of a human walking model by minimizing a residual error between model predictions and radar measurements (see Figure 6.1). Three types of models are needed:

- a visualization model of the walking human (geometric model),
- a human walking model (kinematic model),
- the radar response of the walking human (radar model).

Different human model types can be distinguished with geometric, kinematic and radar parameters. The kinematic model is related to the geometric model due to the size of the human elements. Kinematic models are related to the human

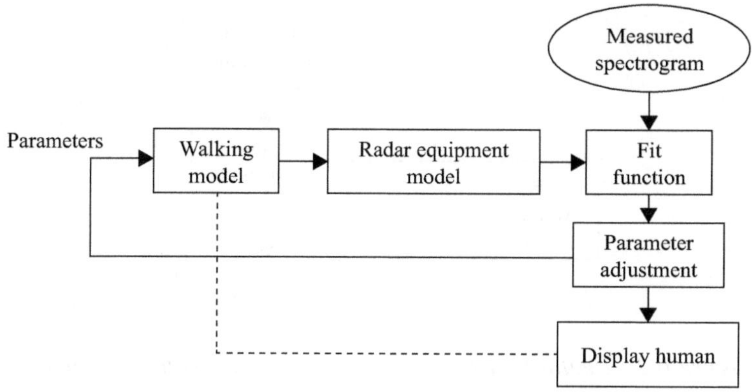

Figure 6.1 Model-based approach for human motion estimation

behaviour like walking and running or are successive human postures. Two methods are available, which connect the geometric and radar parameters of the human body:

- Surface human body model. This method assumes a precise human surface model used in CAD and other computer graphics software. The surface is described with regular surface elements like planes, cylinders and conic surfaces or with freeform surface models with segments and spline curves. All parts of the human model are adaptive in time. These models are highly realistic but limited in practical use due to their large number of model parameters. The radar cross section is estimated with physical optics (PO) and the surface is a perfect dielectric conductor (PEC).
- Physics-based human body model. This human body model reduces the number of parameters through the use of spheres, ellipsoids or point-based representations of the human body parts.

The modelling choice depends on many factors, for example the modelling accuracy, computation time, transmitter frequency and near field or far field range. The model is selected by comparing model results with measurements.

The advantages of the model-based approach are the direct relation between human geometric, kinematics and radar parameters, which solve the inverse function with radar measurement input and visualizing the human motion output. The disadvantages of the model-based approach are as follows: the iterative parameter minimization requires computation time for each human radar response calculation, the unknown minimization time, and each human motion requires a specific kinematic model. The model-based approach with database in which intermediate results are calculated offline and stored in a database solves the first two disadvantages.

6.1.2.2 Feature-Based

The feature-based (bottom-up) approach is from features to the human model parameters and is the inversion of the model-based approach (see Figure 6.2). Both the surface human body model and the physical-based human body model can be used in

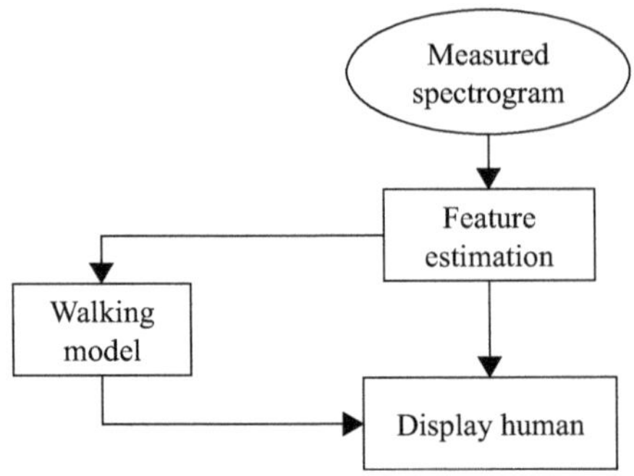

Figure 6.2 Feature-based approach for human motion estimation

the feature-based approach. The feature-based approaches extract local features like scatterer points and envelope from the measurements and track the corresponding features in each measurement of the sequence. The method requires a minimum number of features in order to solve the correspondence of the human model parameters and finds relations between the features and the human model parameters.

The advantage of the feature-based approach is fast computation. The disadvantage of the feature-based approach is the difficulty to find features with good correspondence with human motion parameters. This means that the inversion is difficult to obtain and that there is no connection with the human visualization model.

6.1.3 Chapter Overview

A short description of this chapter can be found below. The second section gives the human model used by the model-based approach and the feature-based approach. The human model contains shape and size of the human body part, Boulic kinematic model and the radar model. The third section describes the fit-function and parameter adjustment of the model-based approach. Linear and logarithmic fit functions are given and a fast parameter adjustment approach with knowledge of the kinematic model parameters is introduced. The approach enables estimation of human kinematic parameters but is time consuming. The feature-based approach is given in the fourth section. The feature estimation of this approach contains four steps. The first step segments the human response in the range–velocity domain. The second step estimates the velocity bounds and fits a model. The third step estimates the target centre and cycle frequency and fits a model. The last step filters all parameters. These parameters are inputs for the Boulic kinematic model and geometric model and enable the visualization of the virtual human. The approach enables real-time human motion estimation.

6.2 The Human Model

Computation of a realistic simulated spectrogram based on a simulated walking human is our goal. This requires detailed information about the kinematics, size and shape of the human body parts.

Human walking models have been developed in particular for realistic animation in virtual reality [14, 15]. To be suitable for animation, human models tend to have a large number of degrees of freedom. A small number of these are visible and distinguishable in radar measurements and could be estimated. Although there are many relations the most important relation is between the size of human body parts and the walking kinematics. Measurements show characteristics, which should be included in the model. The observed characteristics are

1. body, legs, and arms,
2. velocity specification of the walking model,
3. periodic behaviour of all human body parts,
4. smooth velocity relations of all human body parts,
5. directional information.

We investigated with the Walker animation system [16] model and Boulic model for our application. Both models have 12 human body parts. The Walker animation system model has fixed size and the Boulic model is scalable to the human length. The head is simulated with a sphere. The other human elements can be chosen between cylinders or rectangular cuboids. The Walker animation system model and Boulic model differ in the kinematics.

The Walker animation system model kinematics is specified with discrete sets of angles–time points of the upper arm, lower arm, upper leg and knee for human motion models like walking, running and jumping. Interpolation of the angle–time points gives a smooth curve. Other model kinematics could be specified. The walking velocity is varied by the cycle time. The observed limitations of this model are as follows:

1. The model gives discontinuities in the Doppler frequencies which is not a problem in animation but makes the model not usable for our application. The observed Doppler frequencies are smooth.
2. The human velocity is not related to the cycle time. The lack of this relation causes skidding or sliding feet.
3. The human motion is not related to the human size.
4. The model cycle time and model step length have to normalize with respect to the actual cycle time and step length. This normalization includes all human body parts and these functions are not known.

The Boulic kinematics is extracted from experimental data based on a wide range of normalized velocities including scalable human height. The model is related to experimental data and describes the kinematics of all 12 human body parts. The Boulic model enables generation of normal walking humans and generation of abnormal walking humans with unrealistic relations between the model parameters. Unrealistic relations give skidding or sliding feet. This Boulic model is suitable for detection of normal walking humans versus non-normal walking humans.

For these reasons, the Walker animation system is dropped, and based on this experience the Boulic walking model is selected. This section deals with the Boulic kinematics, Boulic size and shape of the human body parts.

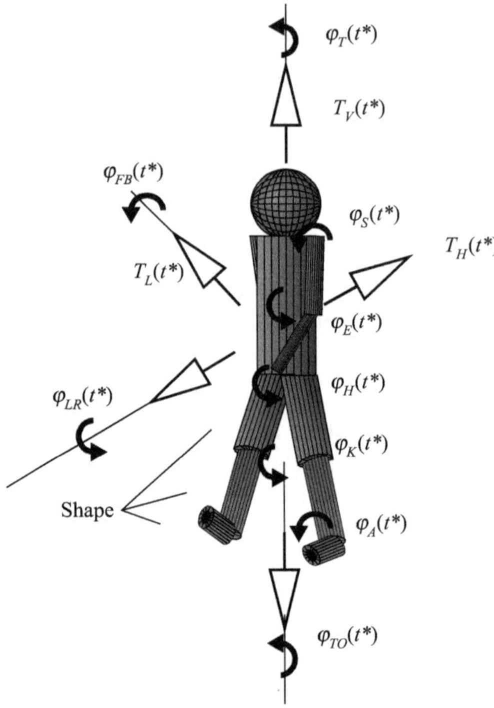

Figure 6.3 Human body model with 12 body parts. The Boulic model has three translations and nine rotation functions. The shape types are cylindrical and ellipsoid [17]

6.2.1 The Shape and Size of Human Body Parts

The human model is modelled as having 12 body parts (see Figure 6.3). The head is a sphere. The other human body parts are cylinders or ellipsoids. The radar cross section is estimated with physical optics (PO) and the surface is a perfect dielectric conductor (PEC). Cuboid shapes are unrealistic. The Walker animation system model has cuboid shapes which give sharp responses if the surface of a cuboid shape is perpendicular to the transmitted radar signal. These reflections do not agree with the smooth behaviour of measurements with different observation angles. Another aspect is that the cuboid shapes differ highly from real human shapes which have smooth shapes.

The sizes of the human body parts are based on average human body sizes. Table 6.1 presents the size of the human body parts.

The human body parts are related to each other by time-dependent translations and rotations (see Table 6.2). These translations and rotations depend on the human size. The positions of the human body parts are calculated using the Boulic model with quaternions.[1] The position of each human element is a multiplication of

[1] Quaternions have four dimensions, one real dimension and three imaginary dimensions. In this section quaternions represent 3-D translations and rotations.

Table 6.1 Human model shape parameters

Human element	Length symbol	Value, m	Radius symbol	Value, m	Shape
Head (H)	–	–	R_H	0.20	Sphere
Torso (T)	H_T	0.80	R_T	0.25	Cylindrical or ellipsoid
Upper leg (UL)	H_{UL}	0.50	R_{UL}	0.10	Cylindrical or ellipsoid
Lower leg (LL)	H_{LL}	0.50	R_{LL}	0.07	Cylindrical or ellipsoid
Feet (F)	H_F	0.20	R_F	0.07	Cylindrical or ellipsoid
Upper arm (UA)	H_{UA}	0.45	R_{UA}	0.05	Cylindrical or ellipsoid
Lower arm (LA)	H_{LA}	0.45	R_{LA}	0.04	Cylindrical or ellipsoid

Table 6.2 Trajectory models

Human element	Translations, m	Body rotations, rad	Left rotations, rad	Right rotations, rad
Lateral translation (L)	$T_L(t^*)$			
Horizontal translation (H)	$T_H(t^*)$			
Vertical translation (V)	$T_V(t^*)$			
Forward/backward angle (FB)		$\varphi_{FB}(t^*)$		
Left/right angle (LR)		$\varphi_{LR}(t^*)$		
Torsion angle (TO)		$\varphi_{TO}(t^*)$		
Hip angle (H)			$\varphi_H(t^*)$	$\varphi_H(t^* + 0.5)$
Knee angle (K)			$\varphi_K(t^*)$	$\varphi_K(t^* + 0.5)$
Ankle angle (A)			$\varphi_A(t^*)$	$\varphi_A(t^* + 0.5)$
Thorax angle (T)		$\varphi_T(t^*)$		
Shoulder angle (S)			$\varphi_S(t^*)$	$\varphi_S(t^* + 0.5)$
Elbow angle (E)			$\varphi_E(t^*)$	$\varphi_E(t^* + 0.5)$

translation quaternions and rotation quaternions. The remainder of this section gives information about the Boulic model.

The radar position is taken as the origin of the coordinate system and the person starts walking at a distance r_h [m] in front of the radar (see Figure 6.4). The reference point of the person is the centre of the human torso and the height corresponds with a point where the legs are connected to the human torso. The person walks with speed $s(t)$ [m/sec] in a direction $\alpha(t)$ [rad].

6.2.2 The Kinematics According to Boulic

The kinematics of the human body parts presented in this section is according to Boulic [18]. It normalized an observed person to a standard person with a leg length

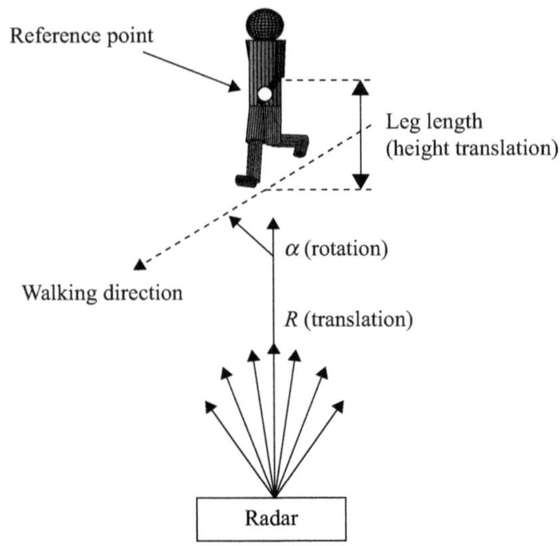

Figure 6.4 Person in front of the radar [19]

of 1 m. The relative normalized velocity v (m/sec) is expressed in the leg length and real body speed s [m/sec]. The leg length is the sum of the length of the upper leg h_{UL} [m], the length of the lower leg h_{LL} [m] and two times the foot radius R_F [m]. The constant b [m/m] depends on the size of the human and scales the real leg length of the observed person to a normalized leg length of a metre:

$$v = bs \tag{6.1}$$

and

$$b = \frac{1}{H_{LL} + H_{LL} + 2R_F} \tag{6.2}$$

The relative normalized velocity depends on the cycle length l^c [m] and the cycle frequency f^c [Hz] of the walking human:

$$v = l^c f^c \tag{6.3}$$

The cycle frequency is the number of cycles or steps in 1 sec. One complete cycle or step is the motion between two successive contacts of one heel with the floor. The cycle length is the distance between two contacts of the heel with the floor. Boulic gives an important empirical relationship between the relative velocity and the cycle length of a normal walking person. A normal walking person has natural arms, legs and body swing motion during human walking (see the diagonal line in Figure 6.5). Figure 6.5 presents the global borders of the normal human walking. If the relation between the arms, legs and body swing lie outside this region then the human is not walking normally. The normal walking person's empirical relationship between the relative velocity and the cycle length is

$$l^c = 1.346\sqrt{v} \tag{6.4}$$

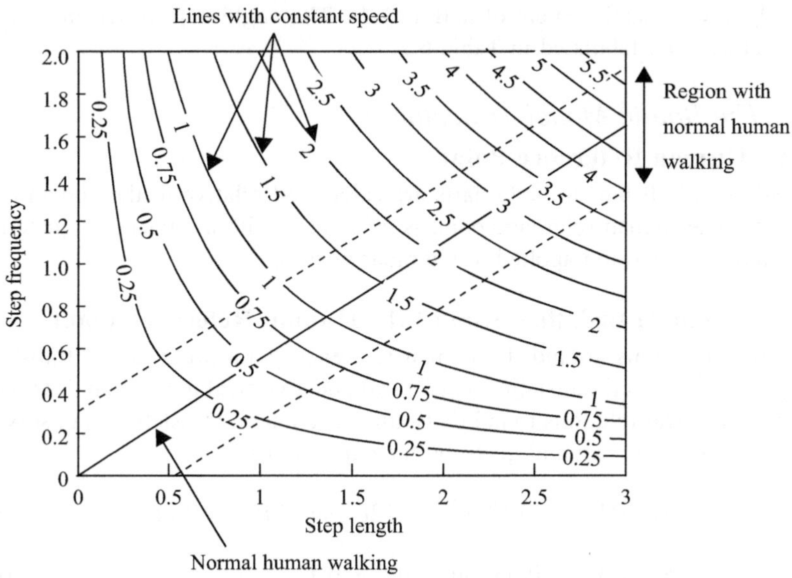

Figure 6.5 Relationship between relative velocity, cycle length and cycle frequency. The region with normal human walking lies on both sides of the line with normal human walking

Elimination of the relative velocity gives

$$f^c = \frac{l^c}{1.346^2} \tag{6.5}$$

This relationship is shown by the diagonal line in Figure 6.5. If the cycle frequency and cycle length don't fulfil this equation then the person is accelerating or decelerating. Figure 6.5 presents the relationships between the relative velocity, cycle frequency and cycle length. The figure presents the region with normal human walking behaviour. A human has non-normal walking behaviour if the parameters lie outside this region. Examples of non-normal motions are kicking, throwing and hobbling humans. Depending on the scenario this human gets special attention.

All angle trajectories and translation trajectories of a cyclic pattern can be expressed as a function of the relative time t^* [−] between zero and one within one human cycle: $t^* = |tf^c|_{\text{mod}1}$. The model assumes a constant relative velocity and so a constant cycle frequency. The position within one human cycle depends on the number of complete cycles and the start phase with cycle phase φ^c [−] between zero and one. The relative time with cycle phase is

$$t^* = |tf^c + \varphi^c|_{\text{mod}1} \tag{6.6}$$

The relative time lies between zero and one. Trajectories of the body parts are synchronized on the left human body parts, the relative time. The trajectories of the right body have a displacement of half a cycle. The angle trajectories and translation trajectories are tabulated in Table 6.2.

6.2.3 The Boulic Model Trajectories

6.2.3.1 Human Reference Point

The positions of all human body parts are calculated relative to the human reference point. The human reference point is calculated with a translation quaternion and a rotation quaternion applied to the radar position.

6.2.3.2 Torso Translation around the Human Reference Point

The torso translations around the human reference point are a lateral translation $T_L(t^*)$ [m], a horizontal translation $T_H(t^*)$ [m] and a vertical translation $T_V(t^*)$ [m]. These are empirical relations found by Boulic. The lateral translation is positive on the left side of the human and given by the expressions

$$For\ v \leq 0.5 T_L(t^*) = [-0.128v^2 + 0.128v]\sin(2\pi(t^* - 0.1)) \tag{6.7}$$

$$For\ v > 0.5 T_L(t^*) = -0.032\sin(2\pi(t^* - 0.1)) \tag{6.8}$$

The horizontal translation and vertical translation have equivalent expressions, which can be found in the Boulic's paper.

6.2.3.3 Torso Rotation around the Human Reference Point

The human body rotates around this reference point in three ways: forward/backward angle, left/right angle and a torsion angle.

6.2.3.4 Arms, Legs and Feet Positioning

The position and orientation of the arms, legs and feet are a combination of translations and rotations relative to the torso position and orientation. The procedure is equivalent to all human body parts and illustrated for the left lower arm.

The lower arm positioning is calculated after positioning of the upper arm. It is a continuation on this calculation. The position of the human's left lower arm is a translation to the left shoulder, a rotation of the left shoulder joint, a translation to the elbow, a rotation of the left elbow and a translation to the centre of the lower arm. The three translations depend on the human size and are time independent. The rotation of the left shoulder (in degrees) is

$$\varphi_S(t^*, v) = -3 - 9.88v[0.5 + \cos(2\pi t^*)] \tag{6.9}$$

where a rotation of zero lies parallel with the human torso. A positive value expresses a front swinging motion. The left shoulder has a $-2°$ rest on position. The position of the left elbow is a translation of H_{UA} in the direction of the elbow. The left elbow rotation is composed of three control points, which depend on the normalized velocity v. The control points defined the relative angles of the elbow

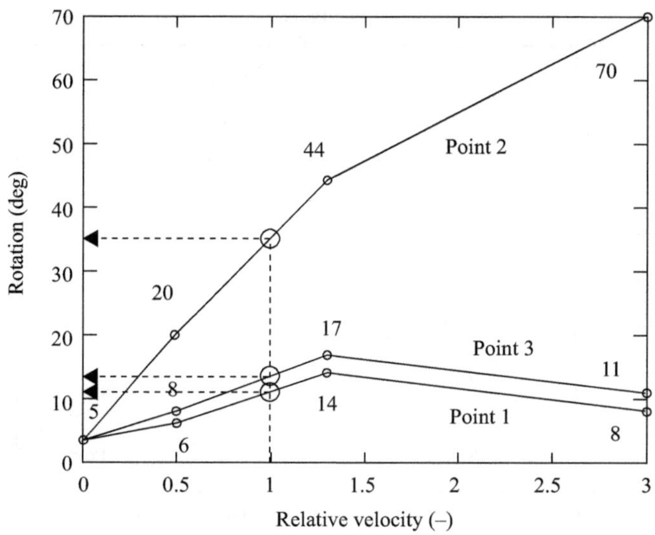

Figure 6.6 *Rotation control point function. The example control points are indicated with a circle*

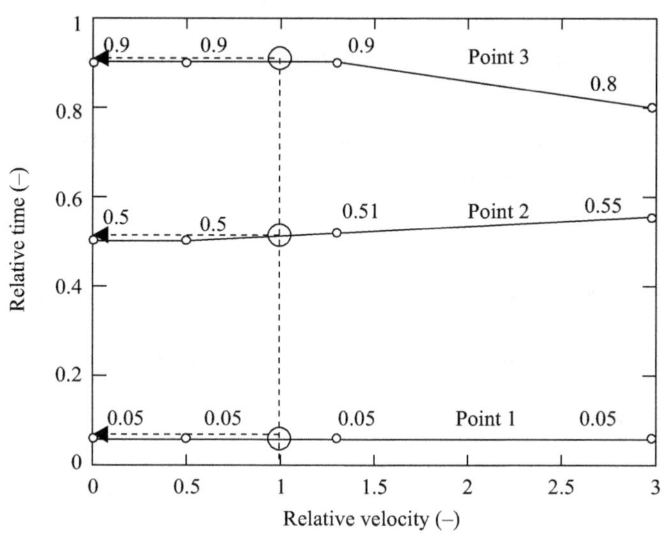

Figure 6.7 *Relative time control point function. The example control points are indicated with a circle*

rotation $\varphi_E(t^*, v)$ function as depicted in Figures 6.6–6.8. The trajectory presented in Figure 6.8 is constructed of three Hermit splines support functions that connect the three control points. The (v, t^*) function presented in Figure 6.7 gives the relative time coordinates of the three control points; the (v, φ_E) function presented in Figure 6.6 gives the rotation of the three control points. The two functions

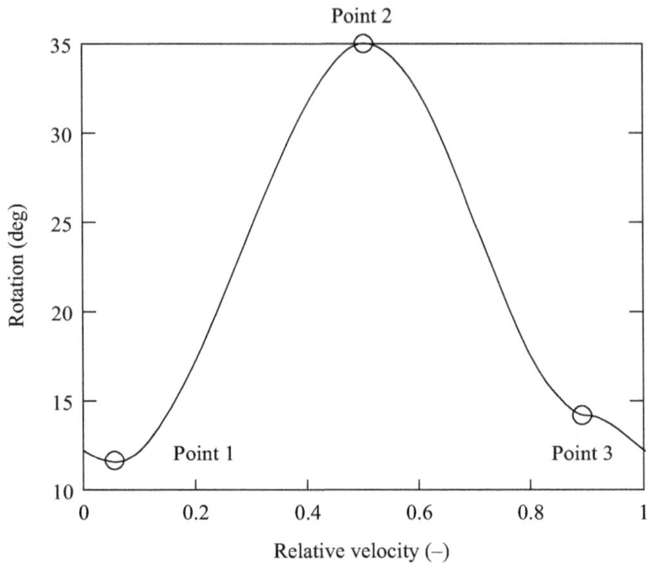

Figure 6.8 Example of the relative angle of the elbow function relative velocity of one. The example control points are indicated with a circle

are available in diagrams. The control points are a function of v in roughly three ranges:

$$\begin{aligned}
&\mathbf{A}: 0 \leq v < 0.5, \text{ start from the rest position to reach a slowly walking person,}\\
&\mathbf{B}: 0.5 \leq v < 1.3, \text{ almost constant velocity walking person,}\\
&\mathbf{C}: 1.3 \leq v < 3, \text{ fast walking person.}
\end{aligned} \tag{6.10}$$

The model is valid for relative velocities between 0.6 and 2.3. Below this range, the model extrapolates the human walking to the rest position. Above this range, the model extrapolates the human walking to a maximum relative velocity of 3.

6.2.4 The Radar Equipment Model

The radar equipment model contains the radar equation and the radar data processing model. The radar equation relates the range of a human body part, and it shapes to the radar characteristics. The human behaviour in time specifies the position of the human body parts. The radar data processing model calculates the simulated spectrogram and measured spectrogram.

The radar transmits an electromagnetic wave to the person. The incoming electromagnetic wave is reflected by the human body parts and received by the radar. The received signal is calculated with the radar equation [20]:

$$P_{\text{target}} = \frac{P_{\text{avg}} G^2 \lambda^2 \sigma}{(4\pi)^3 R^4 L} \tag{6.11}$$

where P_{target} [W] is the received signal energy, P_{avg} [W] the average transmitter power, G [−] the antenna gain, λ [m] the wavelength, σ [m^2] the radar cross section (RCS), R [m] the target range and L [−] the losses. The radar equation uses the RCS that describes the reflection by the human body parts. This reflection depends on the shape of the human body parts. The scattering amplitude and phase are indicated with the square root of the RCS $\sqrt{\sigma}$ [m]. The scattering amplitude and phase of a sphere with diameter R_s [m] are [21]

$$\sqrt{\sigma} = R_s\sqrt{\pi}\left[\left(1+\frac{\lambda}{j4\pi R_s}\right)e^{-j\frac{4\pi R_s}{\lambda}} - \frac{\lambda}{j4\pi R_s}\right] \tag{6.12}$$

The scattering amplitude and phase of a cylinder with diameter R_c [m], height H_c [m] and θ_c [rad], the angle from a broadside incidence, are [22]

$$\sqrt{\sigma} = -jH_c\sqrt{\cos(\theta_c)\frac{2\pi R_c}{\lambda}}\operatorname{sinc}\left(\frac{2\pi H_c}{\lambda}\sin(\theta_c)\right)e^{j\frac{2\pi}{\lambda}r}e^{j\frac{2\pi}{\lambda}R_c\cos(\theta_c)}e^{-\frac{\pi}{4}} \tag{6.13}$$

The scattering amplitude and phase of an ellipsoid are [21]

$$\sqrt{\sigma} = \frac{R_e^2\frac{1}{2}H_e\sqrt{\pi}}{(R_e\sin(\theta_e))^2 + \left(\frac{1}{2}H_e\cos(\theta_e)\right)^2}e^{j\frac{2\pi}{\lambda}2\Delta r} \tag{6.14}$$

where R_e [m] is the radius of the ellipsoid, the two equivalent sides, and H_e [m] the height of the ellipsoid, θ_e [rad] the angle with the height axis, and Δr [m] the distance between the centre of the ellipsoid and the point on the surface of the ellipsoid where the incoming wave makes first contact. The phase shift is an approximation according to [21]. Figure 6.9 presents the aspect angle dependencies

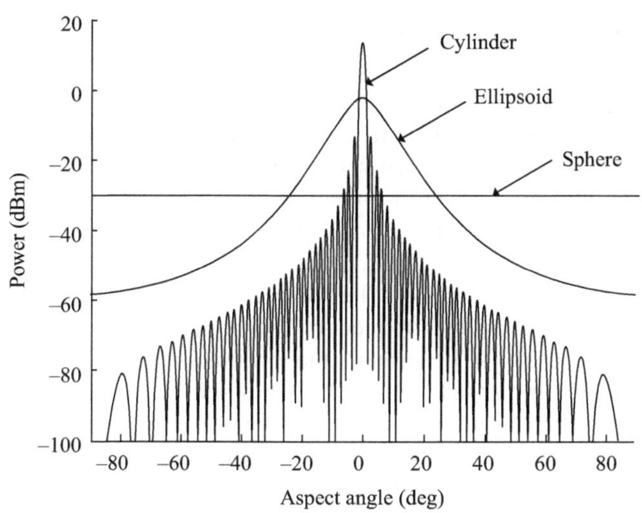

Figure 6.9 Sphere, cylinder and ellipsoid RCS aspect angle dependency

of these three shapes. The RCS of a sphere is independent of the aspect angle while the cylinder gives a strong reflection in case of a zero aspect angle. The ellipsoid RCS lies between that of the sphere and cylinder.

The given square root RCS values are the average values in time. In practice the echo signal of a human element varies in time due to meteorological conditions, equipment instability or variations in the human body parts RCS. The meteorological conditions and equipment instability are assumed insignificant compared with variation in the RCS of the human body parts. The Swerling model describes five statistically fluctuation target scattering models of the RCS in time (see [20]). Case 0 is a non-fluctuating target. Case 1 has constant amplitude for all detections on one scan but there is no correlation between echoes from scan to scan. This distribution applies to complex shapes with many similar reflection surfaces, for example an aircraft. Case 2 echoes vary from chirp to chirp as well as scan to scan. This distribution applies to rapid fluctuating targets. Cases 3 and 4 are the same as Cases 1 and 2, but they apply to targets, which have one dominant reflecting surface or scatter. Swerling Case 0 is applied that assumes a constant RCS throughout the measurement.

6.2.5 The Human Model Spectrogram Trajectories

Figure 6.10 gives a typical spectrogram of a walking human with the radial velocities of the Boulic model. The typical spectrogram belongs to an inbound walking human. The spectrogram depends on the aspect angle. Inbound and outbound humans give large Doppler frequency deviations while radial moving humans give small Doppler frequency deviations. These dependencies are all mitigated by the Boulic model. The human body centre point is the location used to calculate the radar response. We analyse the individual body components, the sum of all human body parts and the deviation with the Boulic trajectories.

6.2.5.1 Individual Body Parts

Figure 6.10(a) presents the spectrogram and Figure 6.10(b) shows all human components according to Boulic, and the remaining subplots show the projections of the Boulic components on the spectrogram. The projections agree reasonably with the spectrogram. The measured spectrogram in Figure 6.10(a) shows strong reflections caused by the motion of human components and zero velocity ground clutter reflections. The inbound walking human average velocity is 2 m/sec. The cycle frequency is 1 Hz, two forward foot movements. The torso, Figure 6.10(d), is the maximum or strong response. The torso has average velocity with superimposed sinusoidal with double frequency. The upper arms (Figure 6.10(e)) and upper legs (Figure 6.10(g)) lie around the maximum response. The backward movement of the lower arms (Figure 6.10(f)) has zero velocity while the forward movement is comparable with the forward upper leg movement (Figure 6.10(g)). The lower leg maximum forward movement (Figure 6.10(h)) is approximately two times the average velocity; the contact phase with the ground has approximately zero velocity. The feet (Figure 6.10(i)) give highest forward movement approximately three times the average velocity; the minimum velocity is zero.

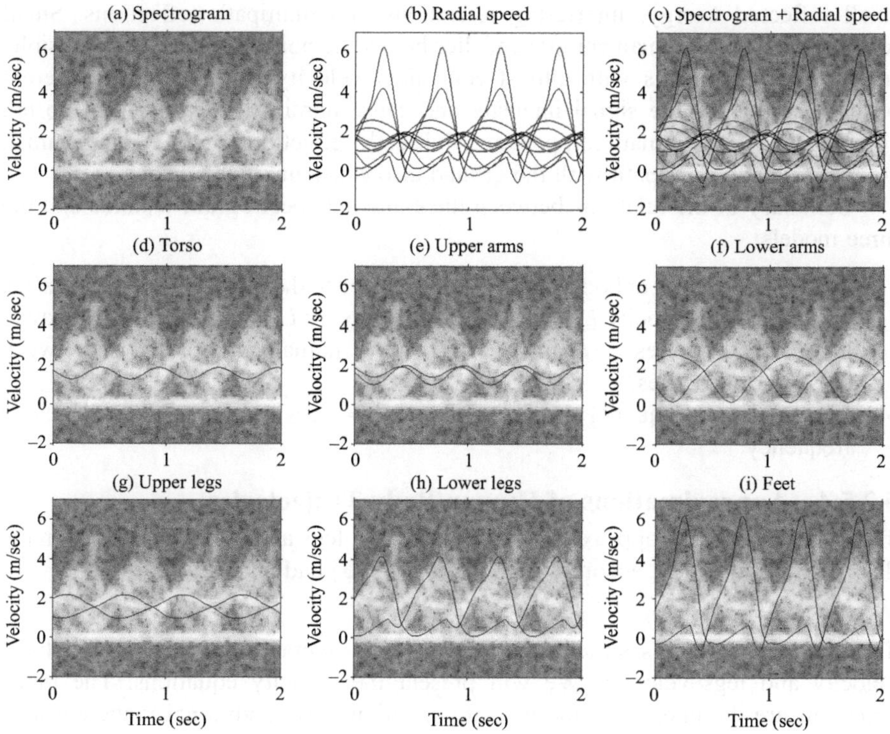

Figure 6.10 Typical spectrogram of a walking person with radial velocities of the Boulic model

6.2.5.2 Addition of All Human Components

Figure 6.10(c) presents all simulated human components and the spectrogram. The plot shows that all spectrogram velocities lie between the minimum and maximum simulated velocity. The lower legs and feet almost completely contribute to the velocity envelopes. The contribution of the feet is less than the contribution of the lower leg due to its small reflected energy. The torso, upper arms, lower arms and upper legs show sinusoidal amplitudes around the average velocity. The lower leg response looks like a sinusoidal component with additional component during the forward movement. This additional component is a second sinusoidal with higher frequency which gives the extra movement above on top of the other sinusoidal. All human components show a periodic behaviour with cycle frequency periodicity in a velocity-slice. The velocity-slices with the torso component show double cycle frequency.

6.2.5.3 Deviation from Boulic Trajectories

Deviations from the spectrogram and Boulic trajectories arise from imperfect parameter estimates and fluctuations of these in the recording window. The Boulic model is also only an approximation to reality. Other physical imperfections are the small reflected energy, interfering reflections and multipath reflections. Small reflected energy of human components lies below the noise level and is not visible. Interfering components with almost equivalent velocity and range but different positions influence the signal intensity and cause unreliable peaks which do not correspond with a human component. Multipath reflections of individual human components and multipath with the ground also cause unreliable peaks.

Summary of the relations between the extracted spectrogram features and the three models:

1. The centre velocity or body velocity $B(t)$ is related to the torso velocity model $s^T(t)$.
2. The minimum velocity $L(t)$ and maximum velocity $U(t)$ are related to the lower leg model velocities $s_U^L(t)$ and $s_L^L(t)$. The other human components lie between these two velocities.
3. The repetition frequency extracted of velocity-slices is a measure for the cycle frequency.

6.2.5.4 Approximations of Human Body Trajectories

This section gives the approximations of the torso, legs and arms trajectories in the direction of movement according to Boulic [18, 23] and Yam [24].

Torso Approximation

The walking pattern is structured in three parts: average human velocity, torso velocity and legs velocity. We will present the velocity equations. The range equations are the integral of the velocity equations with additional range constant provided by the human track information (see [20]). The human velocity is approximated with an average velocity S and acceleration $\dot{S}$ of the human body: $s(t) = S + \dot{S}t$. The torso moves forward and backward around the average human velocity. The forward and backward velocity of the torso indicated with superscript T is approximated with a linear average velocity of the human with a superimposed sinusoidal function with twice the cycle frequency:

$$s^T(t, \mathbf{z}^T) \approx S + \dot{S}t + S^T \cos(2\pi 2 f^c t + 2\pi\varphi^T) \tag{6.15}$$

where S^T is the sinusoidal amplitude, f^c the cycle frequency and $\mathbf{z}^T = (S^T, S, \dot{S}, f^c, \varphi^T)$ the parameter vector. Figure 6.10(d) presents the sinusoidal Boulic model torso movement which shows a good relation with this approximation. We use the centre velocity of each velocity-slice as an estimate of this torso velocity.

Arms and Legs Approximations

The forward and backward translations of the human arms and legs are superimposed on the torso motion and are approximated by sinusoidal functions according to Boulic and Yam. The arms result in an approximation with superimposed sinusoidal function with equivalent cycle frequency but other amplitude and cycle phase. The legs result in an approximation with a superimposed sinusoidal function with equivalent cycle frequency and two parts. The first part of the leg cycle corresponds with the swing movement and has a large amplitude. The other part which corresponds with the rest phase of the leg has small amplitude.

The strong amplitude and small amplitude are out of phase for each leg. The amplitude of the arms is smaller than the amplitude of the legs and the envelope belongs to the lower legs. We have mentioned that the leg motion is superimposed on the torso motion. The torso motion is small compared to the leg motion and could be ignored in the leg motion. Combining the two out of phase lower leg motions gives envelopes indicated with superscript L:

$$s_U^L(t, \mathbf{z}^L) \approx S + \dot{S}t + S^L\left|\cos(2\pi f^c t + 2\pi\varphi^L)\right| \tag{6.16}$$

$$s_L^L(t, \mathbf{z}^L) \approx S + \dot{S}t - S^L\left|\cos(2\pi f^c t + 2\pi\varphi^L)\right| \tag{6.17}$$

where $\mathbf{z}^L = (S,\ \dot{S},\ f^c,\ \varphi^L)$ is the lower leg parameter vector. Figure 6.10 shows Boulic's arms and legs trajectories. These trajectories differ from the sinusoidal approximations which have periodic behaviour with triangle shape behaviour. The measured spectrograms show that the bounds with strong reflections is almost sinusoidal and this justified the model.

The personification of the Boulic model with only three parameters originates from individual frequency, phase and amplitude parameters for the torso and leg. The personification torso frequency is f^T and leg frequency f^L.

6.3 Model-Based Human Motion Estimation[2]

6.3.1 Introduction

Radar measures the direction, distance and Doppler shift of the walking person as a function of time. The Doppler shift gives the radial velocity of the person. Detailed radar processing may extract more characteristics of the walking human. When human body parts do not move with constant radial velocity the Doppler signatures are time varying. Therefore time-frequency analysis techniques can be used to obtain more characteristics.

Different object parts induce different Doppler signatures, which enable the estimation of the behaviour of the elements of the object. The torso of a walking human induces a Doppler signature with constant velocity and RCS while the other human components induce sideband Doppler signatures on both sides of the torso. Doppler signatures of human beings are observed in the time–velocity domain (see [7–9], [25], and [26]). Additional range gives the time–velocity–range space (see [9–11] and [27]). These references give examples of Doppler signatures induced by time varying objects like helicopters and cars, time–velocity analyses algorithms and suggestions for future research. Chen [7] gives a theoretical model study of Doppler signatures induced by targets with vibration, rotation, tumbling and coning motions. Chen verifies the theoretical model study with high-resolution time-frequency transforms applied on real radar measurements. He compares models and real radar measurements. He confirms the effectiveness of the theoretical models.

[2] This section is based on P. van Dorp and F.C.A. Groen. 'Human walking estimation with radar'. *IEE Proc. Radar Sonar and Navigation, 150*(5), 356–365, October 2003.

Mitra [12] presents a theoretical study of Doppler signatures and presents radar parameter requirements in order to detect these Doppler signatures. The presented researches describe no detection, recognition or parameter estimation techniques applied on human motion or other extended target with multiple components. This section describes a model-based approach applied on human motion estimation.

6.3.2 Overview of the Method

The model-based approach minimizes the difference between simulated data and the measured data by adjusting the model parameters. The model-based approach is presented in Figure 6.1.

A simulated walking person shows the connected human body parts with a specification of their relative position as a function of time. One of the models describing these relative positions as a function of time for a small set of parameters for human walking is the Boulic model [18]. The Boulic parameters are the step frequency, step length, start phase and the type of movement. The kinematics of the human body parts is derived from the Boulic model. The kinematics model in combination with the geometric model that gives the shape and size of the human body parts describes the moving human. Two types of human gait are used in this section: that of the walker and of a person walking in a leisurely way (stroller or saunterer). The walker moves his arms during motion, whereas the stroller's arms hang by his sides. The radar equation is used to calculate the reflection of each human body part and adds the reflections at each time sample. The radar data processing model calculates the simulated spectrogram from it. The fit function calculates the difference between the measured spectrogram and the simulated spectrogram. The fit function to be used depends on the noise properties and reflected radar signal properties. The difference is a function of the parameters of the walking model. By minimizing the difference as a function of the parameters of the walking model, the best fit is found. The parameters that belong to the best fit describe the human walking behaviour. The scene of the walking person in a virtual environment is generated given the estimated walking parameters and the geometric model. The simulated walking person generates an image of the person at the measured position. Successive images of the animated person are shown in Figure 6.11.

6.3.3 The Fit Function

This section deals with the trade-off in the choice of the fit function (see [28–30]). The fit function calculates the error between the measured spectrogram and the simulated spectrogram. The choice of the fit function depends onto the error distribution. The spectrogram has no outliers and the errors are evenly distributed about the fit function. The least-squares error minimization gives reliable results with these assumptions and with Gaussian error distribution. If Gaussian error distribution is applicable it is forced by translation of the spectrogram data. The parameters of the simulated spectrogram are the kinematics parameters and the shape parameters. The kinematics parameters have to be estimated. The shape parameters are fixed. The kinematics parameters are the cycle frequency f^c, cycle

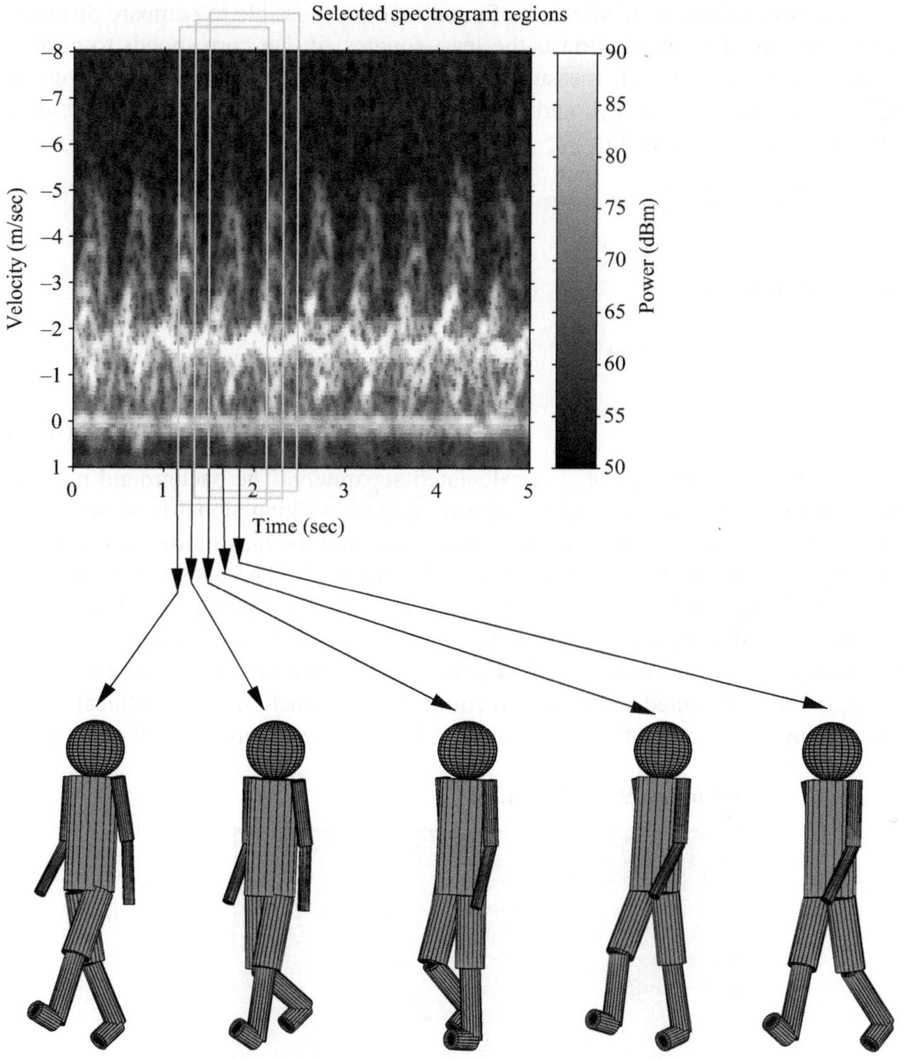

Figure 6.11 Human walking animation. The human motion is estimated for succeeding spectrogram slices. Successive estimated postures give the animated human

length l^c and cycle phase φ^c. The shape parameters are ellipsoid shape or cylindrical shape. The human body parts look like these two shapes.

6.3.3.1 Linear Least-Squares Fit

The measured spectrogram $\mathbf{X_m}(v_k, t_l)$ is normalized on power:

$$\mathbf{X}_{\mathbf{m}}^{\mathbf{Lin}}(v_k, t_l) = \frac{\mathbf{X_m}(v_k, t_l)}{\sqrt{\sum_{\text{all } v_k, t_l} \mathbf{X_m}(v_k, t_l)^2}} \tag{6.18}$$

The normalization stabilizes the fit and makes it possible to compare different measurements. The fit function is the least-squares fit that corresponds to normal distributed errors between measurement and model. The simulated spectrogram $\mathbf{X}_\mathbf{s}^{\mathbf{Lin}}$ has a scaling factor a_{Lin}, and a bias due to the background noise n_{Lin} for each velocity cell and time cell is

$$\mathbf{X}_\mathbf{s}^{\mathbf{Lin}}(v_k, t_l, f_{Lin}^c, l_{Lin}^c, \varphi_{Lin}^c, a_{Lin}, n_{Lin}) = a_{Lin}\mathbf{X}_\mathbf{s}(v_k, t_l, f_{Lin}^c, l_{Lin}^c, \varphi_{Lin}^c) + n_{Lin}(v_k, t_l) \tag{6.19}$$

and the fit function is

$$\chi^2(f_{Lin}^c, l_{Lin}^c, \varphi_{Lin}^c, a_{Lin}, n_{Lin}) = \sum_{\text{all } v_k, t_l} \Big|\mathbf{X}_\mathbf{m}^{\mathbf{Lin}}(v_k, t_l) - \mathbf{X}_\mathbf{s}^{\mathbf{Lin}}(v_k, t_l, f_{Lin}^c, l_{Lin}^c, \varphi_{Lin}^c, a_{Lin}, n_{Lin})\Big|^2 \tag{6.20}$$

The background noise n_{Lin} is calculated separately. The background noise is estimated from the velocity cells where no human walking signal is present. It is assumed that these velocity cells lie in a band around the maximum and minimum Doppler velocity in the spectrogram (see Figure 6.12). The background noise is fixed in the remainder of the fit. We only have to estimate the nonlinear parameters; the scaling factor a_{Lin} results from the fit function [31]. The gradient of the fit function with respect to a_{Lin} will be zero at the minimum. This gradient is solved for a_{Lin} and substituted in the fit function. The fit function is a nonlinear least-squares function; some of the parameters to be estimated cannot be solved linearly.

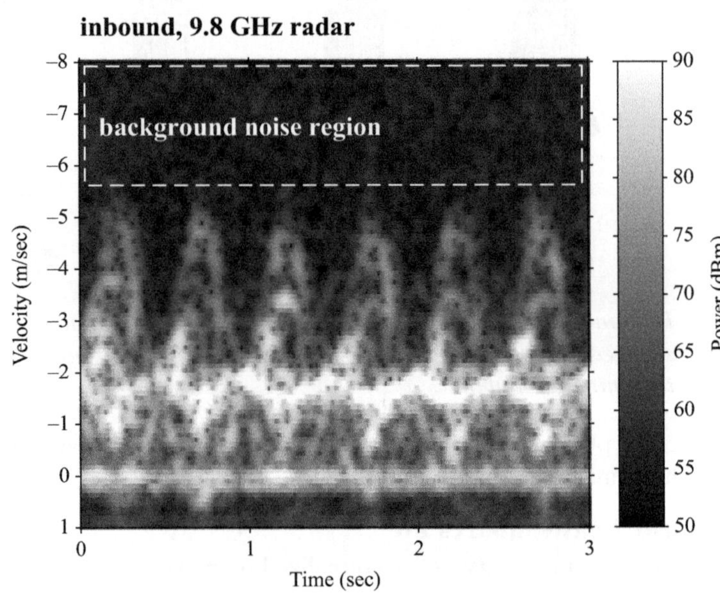

Figure 6.12 Spectrograms of an inbound walking person at 9.8 GHz. The region without human walking signal is indicated

6.3.3.2 Log Least-Squares Fit

In the linear least-squares fit small differences between the spectrograms can belong to large differences in the logarithms of the spectrograms in weak parts of the spectrogram [32]. The sum of squared differences is not an adequate measure. The sum of squared differences is an adequate measure if the error distribution function is a normal distribution that is not the case for the spectrogram. The probability density function of the RCS looks like an exponential distribution, which indicates Swerling Case 1 and Case 2 targets, and the multiplicative noise present in the radar. The log function forces the errors between the measurements and the model to a normal distribution. van den Heiden [33] presents the theory and justification of this step in radar data fit functions.

The measured spectrogram $\mathbf{X}_\mathbf{m}^\mathbf{n}$ is normalized as described by the least-squares fit:

$$\log(\mathbf{X}_\mathbf{m}^{\mathbf{Log}}(v_k, t_l)) = \frac{\log(\mathbf{X}_\mathbf{m}(v_k, t_l))}{\sqrt{\sum_{\text{all } v_k, t_l} \log(\mathbf{X}_\mathbf{m}(v_k, t_l)^2)}} \tag{6.21}$$

The transformation normalizes the measurement so that measurements with different dynamic ranges can be compared. The spectral distortion (SD) is a relative error measure that is defined by Broersen [34]:

$$\text{SD} = \frac{0.5}{2\pi} \int_{-\pi}^{\pi} \left| \ln(h(\omega)) - \ln(\hat{h}(\omega)) \right|^2 d\omega \tag{6.22}$$

where $h()$ is the true spectral density, and $\hat{h}()$ denotes the estimated spectral density. The SD is an accuracy measure in the frequency domain. Broersen [34] presents that the Parceval's relation gives the equivalence accuracy relation in time domain. This means that minimization of the SD has its equivalence in time domain. The applied minimization function is a scaled version of the SD with the log function of the measured spectrogram X_m^{Log} and simulated spectrogram X_s^{Log}:

$$\chi^2 = \sum_{\text{all } v_k, t_l} \left| \log(\mathbf{X}_\mathbf{m}^{\mathbf{Log}}(v_k, t_l)) - \log(\mathbf{X}_\mathbf{s}^{\mathbf{Log}}(v_k, t_l)) \right|^2 \tag{6.23}$$

The log fit function with unknown parameters is

$$\begin{aligned} \chi^2(f_{Log}^c, l_{Log}^c, \varphi_{Log}^c, a_{Log}, n_{Log}) = \sum_{\text{all } v_k, t_l} \big| &\log(\mathbf{X}_\mathbf{m}^{\mathbf{Log}}(v_k, t_l)) \\ &- \log(\mathbf{X}_\mathbf{s}^{\mathbf{Log}}(v_k, t_l, f_{Log}^c, l_{Log}^c, \varphi_{Log}^c, a_{Log}, n_{Log})) \big|^2 \end{aligned} \tag{6.24}$$

The background noise n_{Log} and the linear normalization a_{Log} constant are calculated separately.

Background Noise

The spectrogram regions without human signal are used to estimate the background noise with the standard deviation σ_n. The spectrogram regions without human signal are indicated with Ω_n. Spectrogram regions without human signal are the regions with high speed. The background noise conforms a lognormal distribution (see [20]). The background noise is estimated with the fit function:

$$\chi^2(n_{Log}) = \sum_{v_k, t_l \in \Omega_n} \left|\log(\mathbf{X}_{\mathbf{m}}^{\mathbf{Log}}(v_k, t_l)) - \log(n_{Log}(v_k, t_l))\right|^2 \tag{6.25}$$

This fit function is equivalent with the previous fit function without the human part. The background noise is fixed in the remainder of the fit.

Normalization Constant

The spectrogram region with a human signal which is much stronger than the background noise is selected for the normalization constant calculation. The background distribution is lognormal and normal after the log function. Values above a threshold are outliers in our case measurements that belong to the human response (see [31] for outlier selection). This spectrogram region, indicated with Ω_h, fulfils the following equation:

$$(v_k, t_l) \in \Omega_h \quad \text{for } a_{Log}\mathbf{X}_{\mathbf{s}}(v_k, t_l, f_{Log}^c, l_{Log}^c, \varphi_{Log}^c) > (n_{Log} + p\sigma_n) \tag{6.26}$$

when p equals 2 or 3 and the normalization constant a_{Log} is calculated as a linear fit. In the region Ω_h the background noise is ignored and the human signal remains. The log least-squares equation reduces to a function with linear log(a_{Log}):

$$\begin{aligned}
&\chi^2(f_{Log}^c, l_{Log}^c, \varphi_{Log}^c, a_{Log}) \\
&\quad = \sum_{v_k, t_l \in \Omega_h} \left|\log(\mathbf{X}_{\mathbf{m}}^{\mathbf{Log}}(v_k, t_l)) - \log(a_{Log}\mathbf{X}_{\mathbf{s}}(v_k, t_l, f_{Log}^c, l_{Log}^c, \varphi_{Log}^c))\right|^2 \\
&\quad = \sum_{v_k, t_l \in \Omega_h} \left|\log(\mathbf{X}_{\mathbf{m}}^{\mathbf{Log}}(v_k, t_l)) - \log(a_{Log}) - \log(\mathbf{X}_{\mathbf{s}}(v_k, t_l, f_{Log}^c, l_{Log}^c, \varphi_{Log}^c))\right|^2
\end{aligned} \tag{6.27}$$

Substitution of the estimated parameters in the log least-squares equation gives the fit.

6.3.3.3 Parameter Estimation

Minimization of the least-squares fit to find optimal values of the parameters using an exhaustive search over a grid is a computation intensive process. However, the grid gives insight into the behaviour of the fit, it has, for example, one minimum or multiple minima, and what can be good start values by numeric minimization. Numeric minimization of the least-squares fit with a minimum search algorithm requires start values (see Figure 6.13). We apply two line searches using *a priori* knowledge of the problem to find these starting values. The *a priori* information is that the human walking parameters lie in the neighbourhood of a normal walking

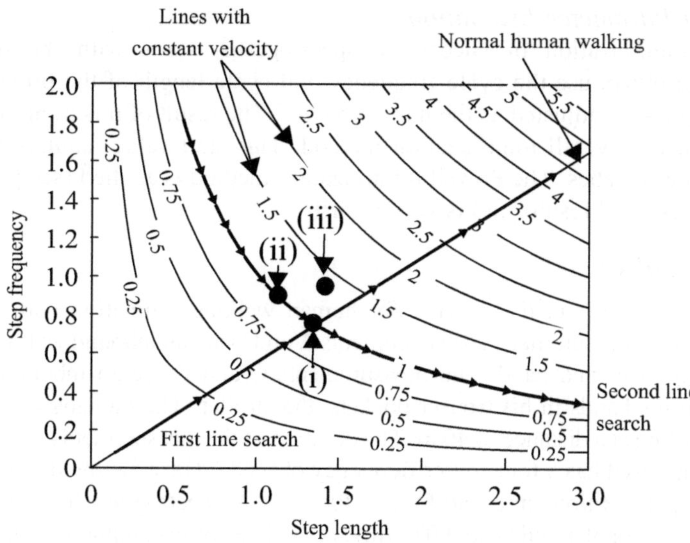

Figure 6.13 Parameter adjustment in the cycle length–cycle frequency plane. The parabolic lines are lines with constant human velocity. Point (i) is the minimum value of the first line fit, the normal walking line search. Point (ii) is the minimum value of the second line fit, the equivalent body velocity line search. Point (iii) is the absolute minimum after total minimization

person and the human body elements have equivalent velocity. This *a priori* knowledge is used in the parameter estimation.

Normal Walking Line Search

The minimization method used starts with a line search of the velocity of a normal walking person. We assume the walking human behaviour lies close to the normal walking human given by (6.4) and (6.5). The fit function is calculated for a discrete number of velocity values with corresponding length and frequency values onto the normal walking human curve. The minimum of the curve is selected. This result corresponds to the minimum cycle frequency f_n^c, cycle length l_n^c and cycle phase φ_n^c of a normal walking person. This result is indicated with the normal walking minimum in Figure 6.13.

Equivalent Body Velocity Line Search

The second line search is along the parabolic equation through this minimum:

$$f^c = \frac{f_n^c l_n^c}{l^c} \tag{6.28}$$

This line corresponds to equivalent human body Doppler velocities in the spectrogram. The result of this line search, the equivalent body velocity minimum in the figure, is the input of the overall Powell minimization method (see [31]). The result of the Powell minimization gives the absolute minimum.

Sequential Parameter Estimation

The first minimization of successive spectrograms starts with the described method; the others use the cycle frequency and cycle length of the previous one. The cycle phase is adjusted as the human moves. The result of this minimization is the input of the overall minimization method. This start value requires less time than two line searches. The Powell minimization method is applied (see [31]). This method gives good results and is very robust.

6.3.4 Results

In order to demonstrate the method for human walking estimation with radar we carried out four measurements. Measurements A and B are an inbound walking person with moving arms and hands and measurements C and D are an inbound walking person with his hands in his trouser pockets, the stroller. The humans walk straight inbound starting at a distance of 40 m. The duration of the measurements is 15 sec. The FMCW radar used has a transmitter frequency of 9.68 GHz, a bandwidth of 80 MHz and a sweep repetition time (SRT) of 340 μsec. A spectrogram has 256 × 3 072 elements, 256 Doppler cells and 3 072 time cells. The corresponding integration time is 1.13 sec. The range cell size is 1.87 m with 0.18 m/sec Doppler-cell size. The remainder of this section describes the results of these four measurements.

6.3.4.1 Measurements

Figure 6.14 presents example spectrograms of the four measurements with 1 sec duration. The spectrograms show strong reflections caused by the motion of the

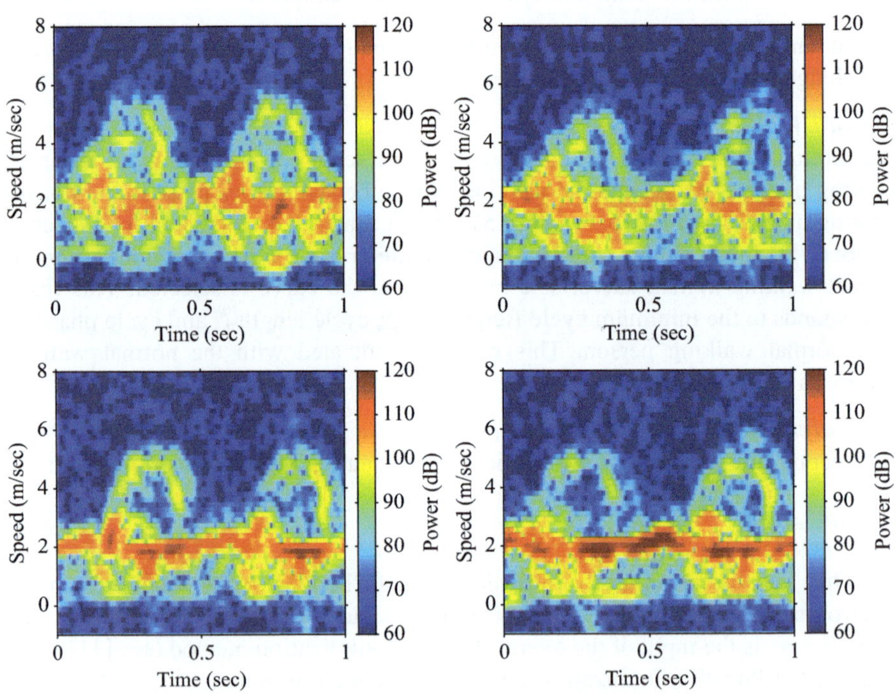

Figure 6.14 Measured spectrograms of human beings

human body parts. The average Doppler velocity is 2 m/sec; the cycle frequency is approximately 1 Hz. The average Doppler velocity belongs to the body of the human. The spectra show velocities belonging to other human body parts. These body parts move relative to the body with velocities greater and smaller than the average Doppler velocity. The spectrogram has a relatively high number of time elements if the range–velocity update is one chirp. It is wise to reduce the number of elements in order to speed up the calculations. A high Doppler velocity resolution is necessary to distinguish the individual human body parts. Reduction of the number of Doppler spectra by means of time averaging or sub-sampling could be applied. Both methods give equivalent results and as the sub-sampling requires less computation time it is applied. The reduction factor used is 32. Figure 6.15 presents four simulated spectrograms that correspond to the four models used with best fit. The figure presents walker- and stroller-simulated spectrograms for cylinder and ellipsoid shapes.

6.3.4.2 Fit Function

The structure of the linear and log fit functions for a typical example in the cycle length and cycle frequency plane is presented in Figure 6.16. The figure shows the minimum fit function for a given cycle length and cycle frequency. The typical example of the linear fit function has 35 minima. These local minima lie on the line with constant relative velocity. The absolute minimum is not always the correct minimum. The typical example of the log fit function has nine minima.

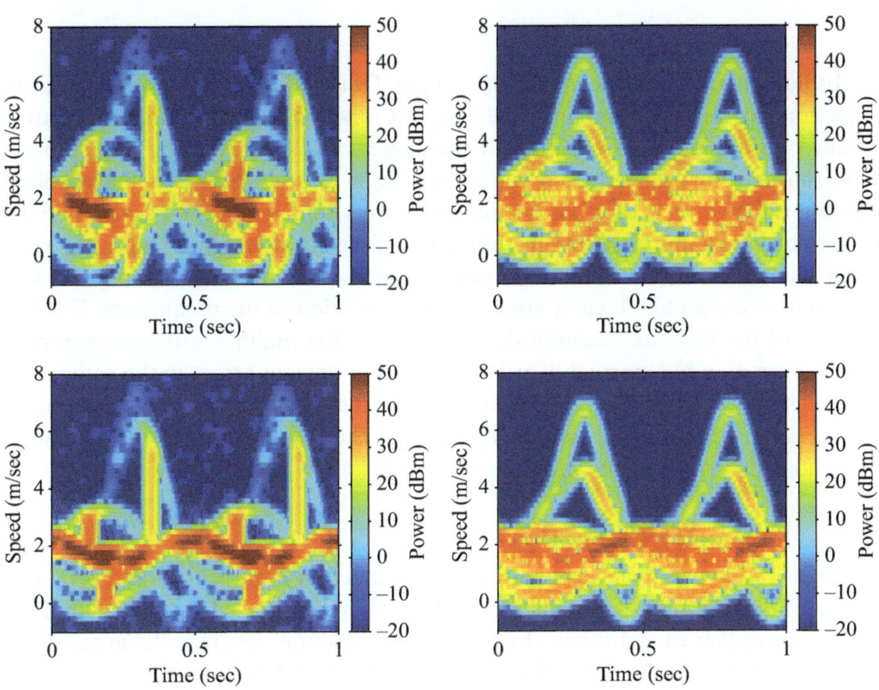

Figure 6.15 Simulated spectrograms of human beings

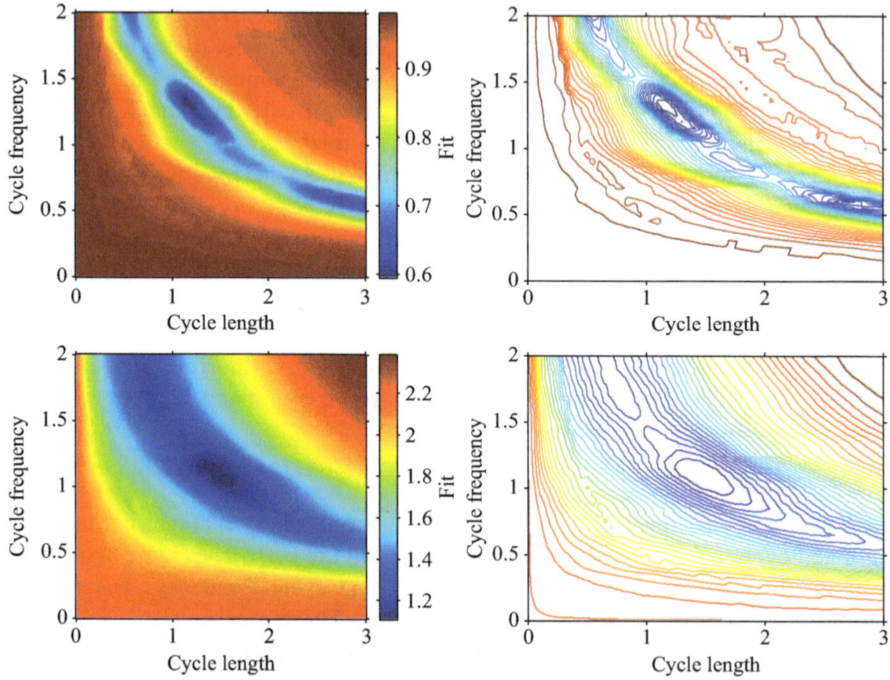

Figure 6.16 Fit functions

The absolute minimum is the correct minimum. Other minima lie far apart. The log fit function has fewer minima than the linear fit function. The absolute minimum is examined with a grid search onto all parameters and by restarting the numeric minimization with the estimated minimum. Both results show that the absolute minimum is the correct solution. The measurements contain normal human motions and the absolute minimum lies in this region. The local minima belong to non-normal human motions which lie outside the normal human motion region. The least-squares fit counted deviant errors in the estimation of the parameters. The log transform of the RCS exponential distribution and the multiplicative noise forced the log fit function to a normal distribution without deviant errors in the estimation of parameters. The absence of deviant errors gives fewer minima in the log fit function compared with the linear fit function. The results confirm the statistical assumptions described in the previous paragraphs.

6.3.4.3 Parameter Estimation

The results of the parameter estimation are presented in Figures 6.17 and 6.18 as a function of the four measurements. Four models have been used as presented in the legend. The walker model moves his arms during motion, whereas the arms of the stroller model hang by his sides. The mean and standard deviation, the error bars in the figures, are based on 400 overlapping spectrograms in 15 sec. The scene of the

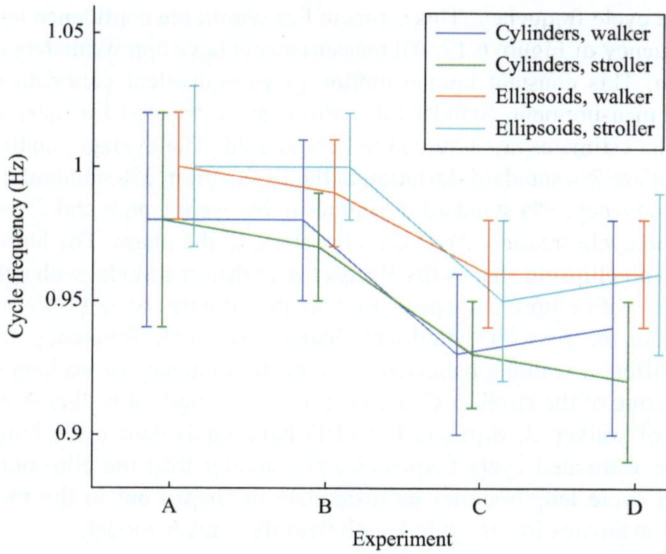

Figure 6.17 Estimated cycle frequency with error bars

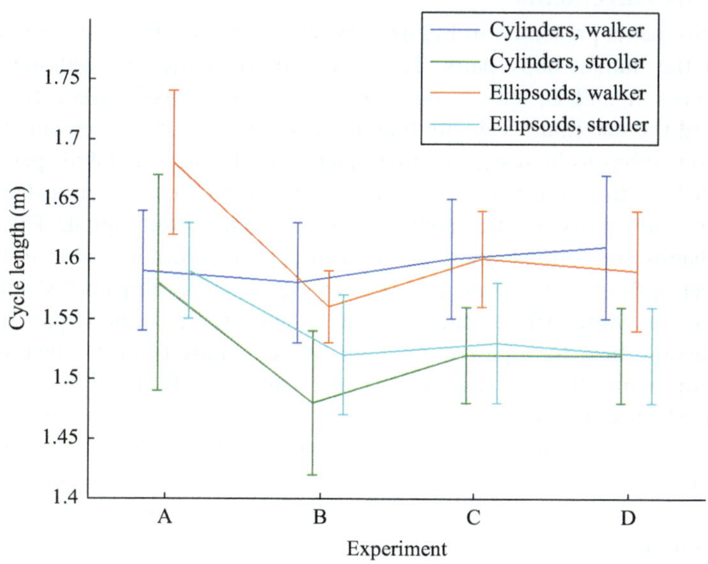

Figure 6.18 Estimated cycle length with error bars

walking human in a virtual environment is generated with the estimated parameters and the position of the human. A part of the generated scene is shown in Figure 6.11. The estimated cycle phase differs for each spectrogram. These phases lie on a wrapped line with an equivalent phase shift. The estimated phase is unwrapped and a straight line is fitted onto the data. The estimated slope is another estimate of the cycle frequency. This estimate lies within the confidence interval of the cycle frequency of Figure 6.17. All measurements have approximately constant human motion. This constant human motion gives equivalent parameters during one complete measurement. Standard deviations are estimated for these constant parameters. The estimates are assumed to be unbiased. The average relative errors of all estimates are 7% standard deviation in the fit function, 3% standard deviation in the cycle frequency, 3% standard deviation in the cycle length and 2% standard deviation in the cycle frequency based on the slope of the phase. The human body part's model with ellipsoid shapes fits always better than the model with cylindrical shapes. The fits with ellipsoid shapes based on all estimates have 11% error in the fit function than the fit with cylindrical shapes. The cycle frequency and cycle length show different walking behaviour. The cycle frequency of walkers A and B is greater than one of the strollers C and D. The cycle length of walker A is greater than the one of walker B. Strollers C and D have equivalent cycle length. The cylinder shape–estimated cycle frequencies are smaller than the ellipsoid shapes. The estimated cycle length shows no difference in shapes but in the model; the stroller model estimates lower cycle length than the walker model.

6.3.5 Discussion

6.3.5.1 Measurements

The measurements presented in Figure 6.14 show strong reflections caused by the motion of the human body parts. The spectrogram of the two walking persons, measurements A and B, shows more velocity components between the average velocity and the maximum velocity than the two strollers, measurements C and D. This difference has to be tested on other data as well. The simulated spectrograms in Figure 6.15 with cylindrical shapes show signals with broad spectral shape where the human body parts are perpendicular to the transmitted signal. These broad spectral shapes are vertical lines. These vertical lines are not visible in the measured spectrograms. The simulated spectrograms with ellipsoid shapes show smooth spectra. More diffuse scattering can be expected from body parts that look like an ellipsoid than for a perfect cylinder. We conclude from this that simulated spectrograms with ellipsoid shapes are better suited for fitting than the one with cylindrical shapes. The walker model shows extra frequency components around the frequency components caused by the human body. The upper arm frequency components cause the greatest velocity and are clearly visible. In the case of the stroller, the reflections of the arms add onto the body reflection and cause a stronger body reflection.

6.3.5.2 Parameter Estimation

As far as we know the minimum of the log fit function is always the absolute minimum. The half cycle frequency minimum and the double cycle frequency minimum belong to the locale minima. These two global minima lie far from the absolute minimum and give no problems in the numeric minimization. The log transform forced the measurement in normal distributed errors, and deviants errors are not counted in the estimation of parameters. The log fit function performs better than the linear fit function that agrees with the theory of Broersen [34, 35].

The numeric minimization of succeeding spectrograms follows the absolute minimum and does not jump to another minimum with double or half-cycle frequency. These results show that the estimated cycle frequencies and cycle lengths are around an average value, with some irregularities. The stochastic amplitude fluctuation of the measured signal and the transition of a range cell in another range cell cause these irregularities. Smoothing the estimated parameters in time removes these irregularities. The cycle frequency estimate and the slope of the phase estimate are equivalent. Their solutions lie within one standard deviation confidence interval. The human body part's model with ellipsoids always fits better than the cylindrical model. The cylindrical model causes a strong reflection where the body part is perpendicular to the transmitted radar signal, whereas the ellipsoid causes a smooth reflection. The clothes worn by the human cause a blurred RCS and look similar to the ellipsoid RCS. The strong reflection of the cylindrical model gives an extra error in the fit function. This error causes a smaller cycle frequency estimate in case of a cylindrical model. The cycle frequency and cycle length show different walking behaviour between different humans that can be used for human classification. The animated walking persons generated with the estimated parameters show a realistic look-alike of the real walking person. This is verified by visual inspection of video taken with the radar measurements. Although this was a quantitative comparison, the Boulic model is also applied onto human motion fits in captured video and the origin of the Boulic model corresponds to markers in captured video. The small variations in the estimated parameters do not affect the animation; smoothing is not necessary.

6.3.6 Conclusions

We have described a model-based approach of estimating human walking parameters from radar measurements and visualizing the results as a human walking scene in virtual reality. The approach minimizes the difference between simulated data and measured data. The human model used provides detailed simulated spectrograms. The model uses the Boulic kinematics model with three parameters: the cycle frequency, cycle length and the cycle phase. These are sufficient to simulate the walking human. The method is applied onto real radar measurements of inbound walking humans from a distance of 40 m. The log fit function gives an absolute minimum that corresponds to the correct solution that agrees with the theory of Broersen [35]. The linear fit function has more minima and the absolute minimum is not always the correct solution. The errors are 7% standard deviation in the fit function, 3% standard deviation in the cycle frequency, 3% standard deviation in the cycle length and 2% standard deviation in the cycle frequency based on the slope of the phase. The human body part's model with ellipsoid shapes fits always better than the model with cylindrical shapes; they have 11% error in the fit function. The ellipsoids produce more diffuse scattering. The cycle frequency and cycle length show different walking behaviour between different humans that can be used for classification. The described model-based approach opens possibilities in security, medicine, animation, military and other civil applications that need human motion. The approach is in principle also applicable to other moving objects like animals. Analysis of different behaviour with other human models and kinematics is one approach. Future research involves extending this approach to estimate the motion of each human body part.

6.4 Feature-Based Human Motion Estimation[3]

6.4.1 Introduction

Model-based approaches using the Boulic walking model for estimating human motion parameters have been described in the previous section. The animation showed motion personification originating from changes in parameters during the motion. The application of the Boulic model to other motion than walking is limited. Disadvantages are as follows: it requires different human kinematic models and the required computation time, and cannot be done in real-time. Although the model-based approach has a certain level of motion personification, not all personification of human motion is presented in the models, for example a different left/right step length. The feature-based approach we introduce in this section estimates the global Boulic parameters in a more efficient way making real-time posture estimation possible. The model-based approach that estimates the global Boulic parameters is the basis of our research. In this section we extract the global Boulic parameters (the basic level) and independent torso and leg features (personification) with a feature-based approach which can be applied in real-time applications. The Boulic model is described in section 6.2. Walking, running and jogging persons with different aspect angles are used. The scene of the moving person in a virtual environment is visualized with a geometric human model and estimated global Boulic parameters and personification, torso and leg parameters.

This section is organized as follows: Section 6.4.2 gives an overview of the method. Section 6.4.3 gives the feature extraction methods with Kalman filtering. Section 6.4.4 presents the experiments and section 6.4.5 discusses these. Finally, section 6.4.6 ends this section with conclusions.

6.4.2 Overview of the Method

An overview of the method with an example of a human spectrogram is presented in Figure 6.19. The spectrogram shows dominant torso reflections characterized by the average torso velocity, torso cycle frequency and torso amplitude. It shows also the average leg velocity, leg cycle frequency and a leg amplitude. The leg amplitude is related to the average leg velocity. Differences between the left and right leg are also observed in the two sinusoidal shaped lines. The results give the human posture that could be visualized in virtual reality.

[3] This section is based on P. van Dorp and F.C.A. Groen. 'Feature-based human motion parameter estimation with radar'. *Radar, Sonar and Navigation*, IET, 2, 135–145, 2008.

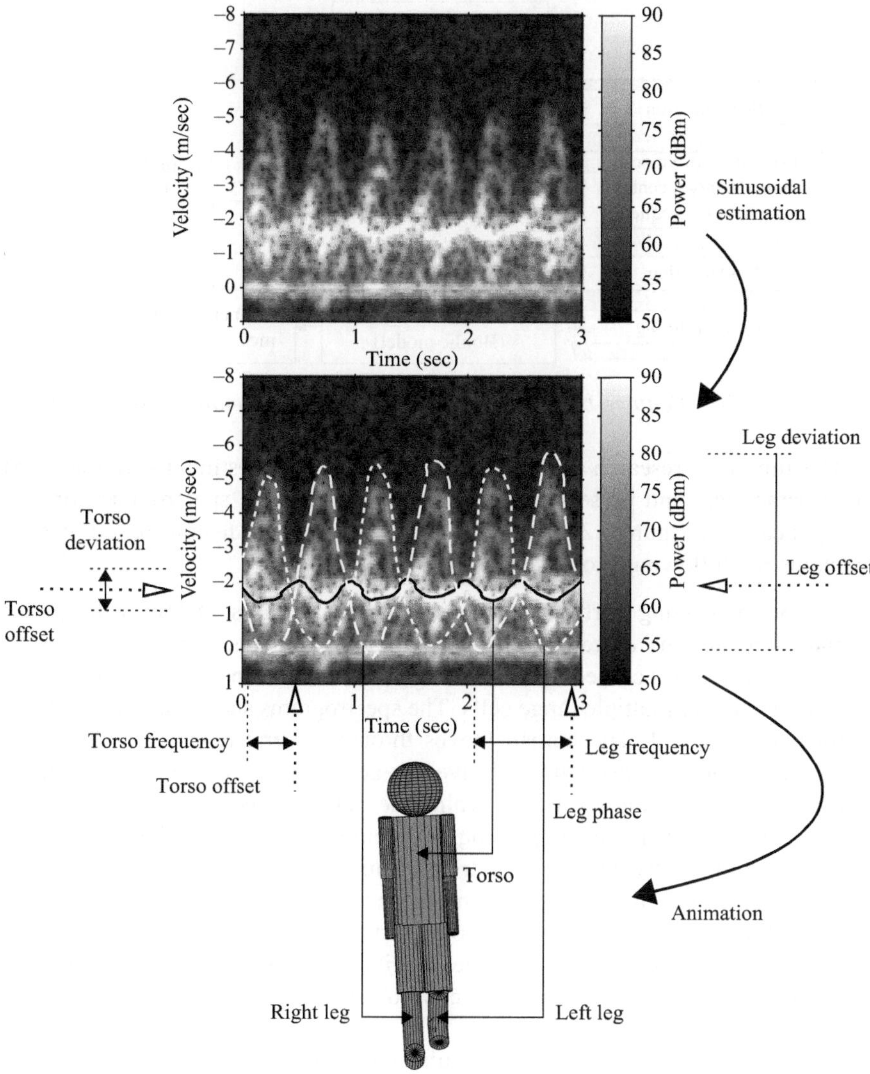

Figure 6.19 Method overview with schematic presentation of the steps. The human being is walking inbound

The Boulic model [11, 23] describes the global human walking kinematics with three global parameters. The motion of all human limbs can be formulated as global equations, based on these parameters. The Boulic model is a global model where the personification of the motion is removed by averaging over a large number of experimental measurements. The Boulic model is extended by Yam [24] who presents a hip and a knee rotation equation for human motions like strolling, walking and running.

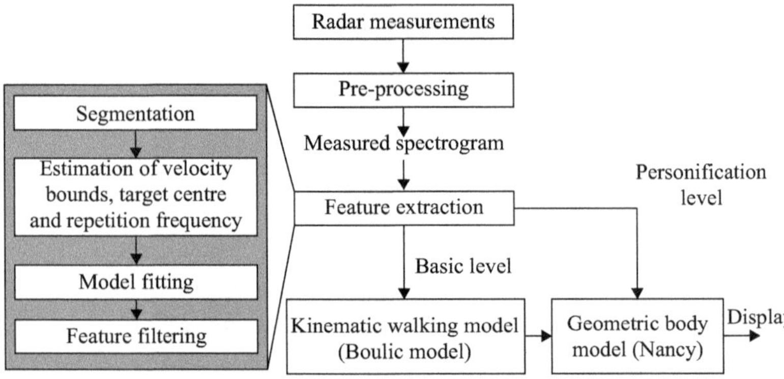

Figure 6.20 Feature-based approach for fast human motion estimation

The aim of the research described in this section is fast estimation of the global Boulic parameters and personification parameters from the spectrogram for real-time applications. Figure 6.20 presents our four-component feature-based method. It contains the following processes:

1. The pre-processing performs the radar processing tasks. It tracks the person and calculates the corresponding spectrogram. The spectrogram contains the velocity variations caused by the moving person. The response of a target can be distributed over multiple range cells. The spectrograms show views of the target for one range-cell. The human moves through several range-cells during the measurement. The target tracker gives an accurate target position. This target position has continuous range while the range cells have discrete range. Averaging adjacent range spectrograms remove the range migration effects. The estimation is based on the spectrogram and the range information is not used.
2. The feature extraction calculates the essential features from the spectrograms and estimates the global and personalized Boulic parameters from these. A least-squares fit of the centre frequency gives the sinusoidal parameters of the torso. A least-squares fit of the velocity bounds gives the sinusoidal parameters of the hip and knee flexing.
3. The walking kinematics model calculates the full posture of the person from the parameter values required for visualization.
4. The geometric body model component displays the posture of the human in virtual reality (see [36]).

6.4.3 Feature Extraction

The diagram in Figure 6.20 gives the four components of the feature extraction. The segmentation component determines the human response in the spectrogram. The next component extracts the maximum velocity $U(t)$, minimum velocity $L(t)$ and centre velocity $C(t)$ to estimate the leg and torso parameters of each time-slice. The methods are based on detections, percentiles and correlations. The cycle frequency or repetition frequency in the spectrogram is extracted. The model fitting component calculates the model features of the legs and torso. The feature filtering smoothes the features and estimates the global Boulic parameters.

6.4.3.1 Segmentation

The segmentation starts with the calculation of an adaptive threshold and decides on the basis of this threshold whether a spectrogram component belongs to the human response or not. The adaptive threshold V_T is calculated with the triangle histogram algorithm [37] or with the methods described in the detection algorithm in the model-based approach. The interval $-v_{\text{static}} \leq v \leq v_{\text{static}}$ is the region with stationary background clutter (see Figure 6.21). Velocity components smaller than v_{static} are removed because they contain reflections of static objects or ground clutter and the human response area is not around zero velocity.

The components with human response are clustered (see [38] and [39]), connected spectrogram components are selected; other targets in the scene are removed. The velocity bounds that contain the human response are between $v_{\min}$ and $v_{\max}$.

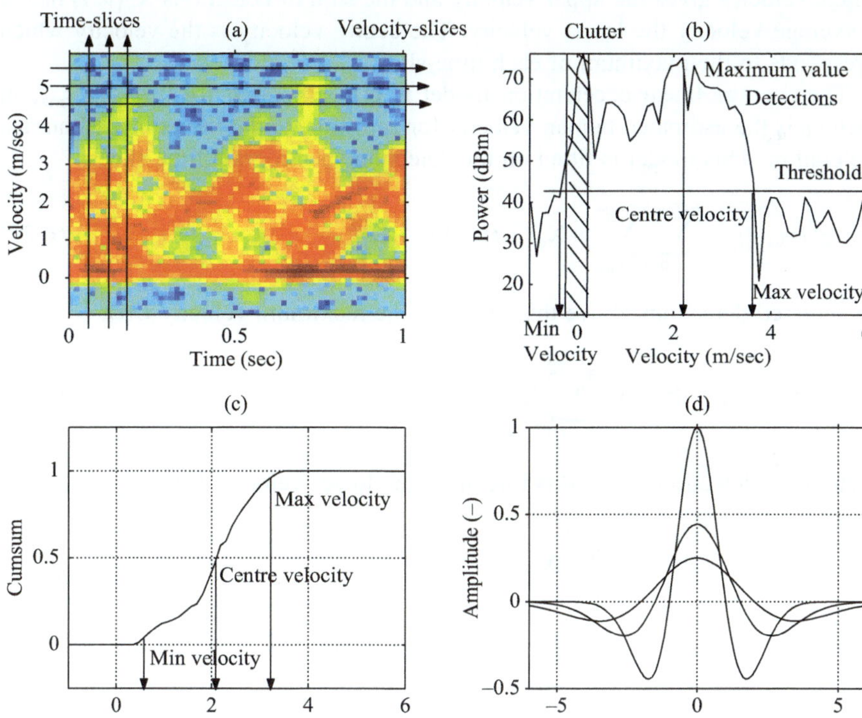

Figure 6.21 Three methods that extract the velocity bounds. (a) Spectrogram with time-slices and velocity-slices. (b) Velocity strength and detection threshold for time-slice 0.5 sec indicated in (a). (c) Percentile time-slice for time-slice 0.5 sec indicated in (a). (d) Three example Mexican hat wavelet functions with varying width

The spectrogram is indicated with $\mathbf{X}(v, t)$ with time window $0 \leq t \leq t_{max}$ where v is the velocity cell and t the time cell. The detection spectrogram $\mathbf{X}_d(v, t) = (\mathbf{X}(v, t) > V_T) \wedge (|v| > v_{static}) \wedge (v_{min} \leq v \leq v_{max})$ has ones for the detected human components.

6.4.3.2 Velocity Bounds, Object Centre and Cycle Frequency Estimation

This section describes three methods which estimate the velocity bounds and the object centre: the detection, percentile and correlation methods. These are common methods in image processing. The last paragraph presents the cycle frequency extraction method. Figure 6.21 presents the graphical picture of these methods.

Detection Method

This method input is the spectrogram (see Figure 6.21) and output the detection spectrogram where the amplitude lies above a threshold. Figure 6.21(b) presents one slice of the spectrogram. The method first extracts the average velocity. For each time-slice (see Figure 6.21(a)), the sum of detections $\mathbf{X}_d(v, t)$ above the average velocity gives the upper velocity and the sum of detections $\mathbf{X}_d(v, t)$ below the average velocity the lower velocity. The centre velocity is the velocity which corresponds to the maximum of each time-slice.

The average linear acceleration model is $\overline{v}(t) = \overline{v}_0 + \overline{a}t$ for $0 \leq t \leq t_{max}$ in which $\overline{v}_0$ is the estimated human velocity for zero time and $\bar{a}$ the estimated human acceleration. This model is fitted on the detection spectrogram time-slices:

$$\chi^2(\overline{v}_0, \overline{a}) = \sum_{t=0}^{t=t_{max}} \sum_{v=v_{min}}^{v=v_{max}} \left| v\, \mathbf{X}_d(v, t) - \overline{v}_0 - \overline{a}t \right|^2 \tag{6.29}$$

The summed detections above this line give the maximum velocity $U(t)$:

$$U(t) = \overline{v}_0 + \overline{a}t + \Delta v \sum_{v=\overline{v}(t)}^{v=v_{max}} \mathbf{X}_d(v, t) \tag{6.30}$$

the summed detection below this line gives the lower velocity $L(t)$:

$$L(t) = \overline{v}_0 + \overline{a}t - \Delta v \sum_{v=v_{min}}^{v=\overline{v}(t)} \mathbf{X}_d(v, t) \tag{6.31}$$

The centre velocity is defined as the maximum velocity response of each time-slice:

$$B(t) = \max_{v \in \mathbf{X}_d(v,t)} (\mathbf{X}(v, t)) \tag{6.32}$$

Percentile Method

This method calculates the cumulative amplitude distribution for each time-slice (see Figure 6.21(c)). The amplitude distribution gives the target versus background bounds. The background noise lies at the extremities and the target lies in the middle.

The cumulative amplitude distribution function $\mathbf{P}(v, t)$ of each time-slice is

$$\mathbf{P}(v,t) = \frac{\sum_{v=v_{\min}}^{v} \mathbf{X}_d(v,t)}{\sum_{v=v_{\min}}^{v=v_{\max}} \mathbf{X}_d(v,t)} \tag{6.33}$$

The percentile $\mathbf{P}(U(t), t) = 95\%$ gives the upper velocity, the percentile $\mathbf{P}(L(t), t) = 5\%$ the lower velocity and $\mathbf{P}(B(t), t) = 50\%$ the centre velocity.

Correlation Method

The correlation method estimates the width and centre of the human velocity distribution of a time-slice by correlation with a Mexican hat wavelet [40] with varying width σ and the best correlation is selected. The Mexican hat wavelet is given by (see Figure 6.21(d))

$$h(v,\sigma) = \frac{1}{\sqrt{2\pi\sigma^3}}\left[1 - \frac{v^2}{\sigma^2}\right] e^{-\frac{v^2}{2\sigma^2}} \tag{6.34}$$

The velocity response $\mathbf{X}(v_i, t_i)$ has a Poisson distribution; we use a log transformation to obtain Normal distributed background noise. Each log time-slice is correlated to $h(v, \sigma)$ for varying widths σ:

$$\mathbf{X}_c(v,t,\sigma) = \mathrm{corr}(h(v,\sigma), \log(\mathbf{X}(v,t))) \tag{6.35}$$

Each time-slice t_i has a maximum correlation with velocity v_{corr} and width σ_{corr}. The centre velocity is defined as the maximum correlation $B(t) = v_{\mathrm{corr}}$. The envelopes are twice the sigma $L(t) = v_{\mathrm{corr}} - 2\sigma_{\mathrm{corr}}$ and $U(t) = v_{\mathrm{corr}} + 2\sigma_{\mathrm{corr}}$.

Cycle Frequency

The Fourier transform to extract a repetition fails due to the short observation time and deformed sinusoidal shape. Other successful time-series estimate is the maximum entropy spectral analysis of Burg with autoregressive (AR) models [35]. We fit a sinusoidal with linear trend to the velocity-slice.

Before fitting the log transformation is applied to obtain Normal distributed background noise. The sinusoidal amplitude fit of velocity-slice v is

$$\chi^2(\mathbf{z}^A, v) = \sum_{t=0}^{t=t_{\max}} \left|\log(\mathbf{X}(v,t)) - A - \dot{A}t - \tilde{A}\cos(4\pi f^c t + 2\pi\varphi^A)\right|^2 \tag{6.36}$$

with $\mathbf{z}^A = (A, \dot{A}, \tilde{A}, f^c, \varphi^A)$ the parameters to be estimated. The repetition frequency is twice the cycle frequency due to the combined effect of two out of sync leg motions in a velocity-slice. The sinusoidal parameters are extracted with subtracted linear trend. We remove the velocity dependency of $\tilde{A}(f^c, v)$ by integrating over these:

$$\tilde{A}(f^c) = \sum_{v=v_{\min}}^{v=v_{\max}} \tilde{A}(f^c, v) \tag{6.37}$$

The cycle frequency corresponds with the frequency with maximum amplitude $f^A = \max_{f^c}(\tilde{A}(f^c))$.

6.4.3.3 Model Fit

This section describes the leg fit and torso fit.

Leg Fit

The leg fit has two stages. The first stage fits a sinusoidal curve to the difference of the upper and lower velocities:

$$\chi^2(\mathbf{z}^{AL}) = \sum_{t=0}^{t=t_{\max}} \left| [U(t) - L(t)] - \left[s_U^L(t, \mathbf{z}^L) - s_L^L(t, \mathbf{z}^L) \right] \right|^2 \tag{6.38}$$

with $\mathbf{z}^{AL} = (2\tilde{S}^L, f^L, \varphi^L)$ the estimated parameters. The minimum fit indicates the imbalance that results from the model error caused for example by an accelerating human. This fit is less sensitive to noise than the individual fits and independent of the upper and lower velocity.

The second stage extracts the linear trend and two off-phase cosine functions with fixed leg frequency f^L of the previous fit:

$$\chi^2(\mathbf{z}^L) = \sum_{t=0}^{t=t_{\max}} \left| L(t) - s_L^L(t, \mathbf{z}^L) \right|^2 + \sum_{t=0}^{t=t_{\max}} \left| U(t) - s_U^L(t, \mathbf{z}^L) \right|^2 \tag{6.39}$$

Torso Fit

The sinusoidal torso fit extracts the linear trend and sinusoidal of the centre velocity:

$$\chi^2(\mathbf{z}^T) = \sum_{t=0}^{t=t_{\max}} |B(t) - s^T(t, \mathbf{z}^T)|^2 \tag{6.40}$$

6.4.3.4 Filtering

Feature filtering with Kalman filters is applied to smooth the estimated parameters for imperfect estimates and estimates smoothed Boulic parameters in time. The filter input vectors used are

- torso $z^T = (S^T, \dot{S}^T, \tilde{S}^T, \varphi^T, f^T)$,
- leg $z^L = (S^L, \dot{S}^L, \tilde{S}^L, \varphi^L, f^L)$,
- global Boulic parameters with leg frequency and without cycle frequency = (v, l^c, f^L),
- global Boulic parameters with leg frequency and cycle frequency = (f^A, v, l^c, f^L).

6.4.4 Results

Two experiments are carried out applying the algorithm onto a test database and validation of the algorithm with the model-based approach presented in the first part of this section.

6.4.4.1 Database

We carried out 50 radar experiments of walking, jogging and running persons in normal terrain conditions. The azimuth aspect angles vary between 0° and 180°, the elevation between 0° and 30° (0° is horizontal view). The 9.68 GHz frequency-modulated continuous wave (FMCW) radar used has a variable bandwidth and 2 kHz sweep repetition frequency. Detailed description of FMCW radar processing and target tracking is covered in [38, 39, 41]. The spectrogram integration time is 1.024 sec. The time between successive spectrograms is 40 msec.

6.4.4.2 Feature Extraction

The methods are applied onto a typical walking human equivalent. The posture of the human in virtual reality is generated with the estimated parameters. This posture is compared with video observations. The comparison shows a good similarity between the radar-estimated motion and the video captured motion.

Detection Method

Figure 6.22 presents the results of the detection method. The velocity bounds in the spectrogram correspond with the bounds of the human response. A high signal-to-noise ratio gives a small number of gaps. The number of gaps increases with a low signal-to-noise ratio and the estimated velocity bounds lie dense at the torso velocity causing a systematic error with smaller amplitude. The torso response fluctuates and there are cases where other responses have higher intensity. There are occurrences where other human components interfere with the torso and reduce the torso amplitude, and another element with higher intensity is selected. Other occurrences show two interfering components that give a high intensity greater than the torso response.

Percentile Method

Figure 6.23 presents the percentile method with 5, 50 and 95 percentiles. The percentile method in upper and lower velocity shows equivalent behaviour with the detection method. The torso estimate gives a good correlation with the expected

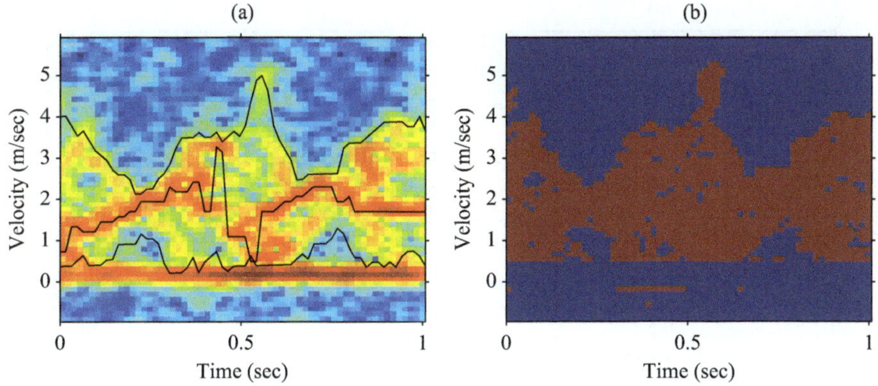

Figure 6.22 Detection method: left spectrogram with velocity bounds and centre velocity; right spectrogram with detections

torso in the spectrogram. The estimated sinusoidals show the cyclicity of the torso and legs. The torso frequency is twice the leg frequency. The estimated leg sinusoidal is less affected by the forward movement of the foot; near 0.5 sec it gives the average upper and lower velocities which correspond with the leg.

Correlation Method

The correlation results in Figure 6.24 presents the estimated centre and envelopes and the correlation time slice $\mathbf{X_c}(v, t=0, \sigma)$. The correlation results show a smoothed function with time varying maximum (v_{corr}, σ_{corr}). The velocity bounds are always symmetric around the centre velocity v_{corr}. The centre velocity deviates strongly with the expected torso. The velocity bounds and centre velocity are correlated with the Mexican hat function which differs from the real relation between these.

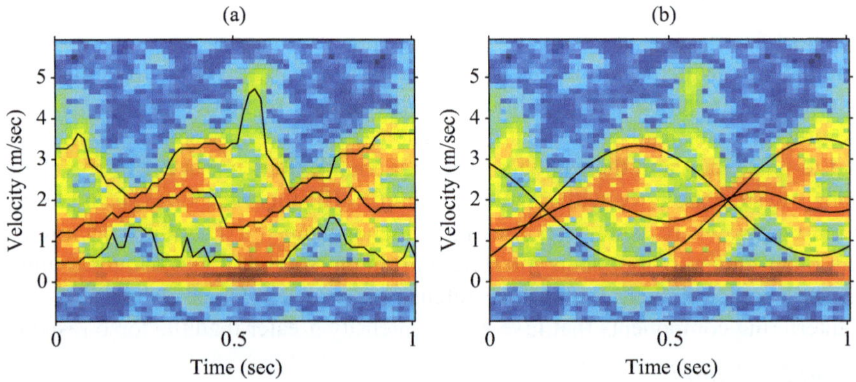

Figure 6.23 Percentile method: left spectrogram with velocity bounds and centre velocity; right spectrogram with estimated sinusoidals

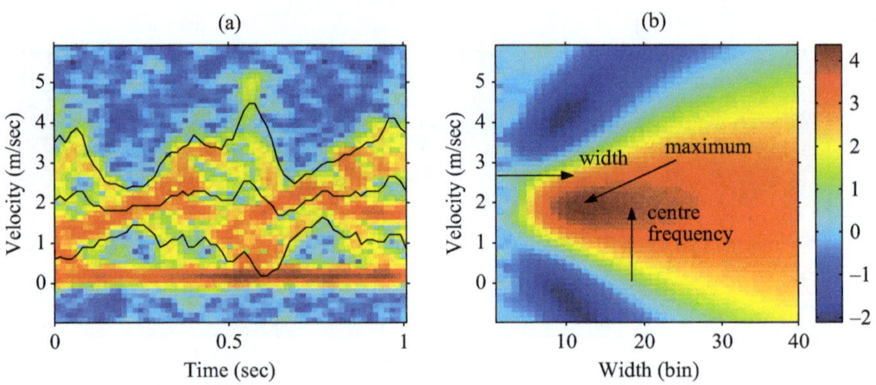

Figure 6.24 Correlation method: left spectrogram with velocity bounds and centre frequency; right example velocity–width space of one time-slice. The correlation results are the amplitude. The maximum correlation location is indicated with a circle

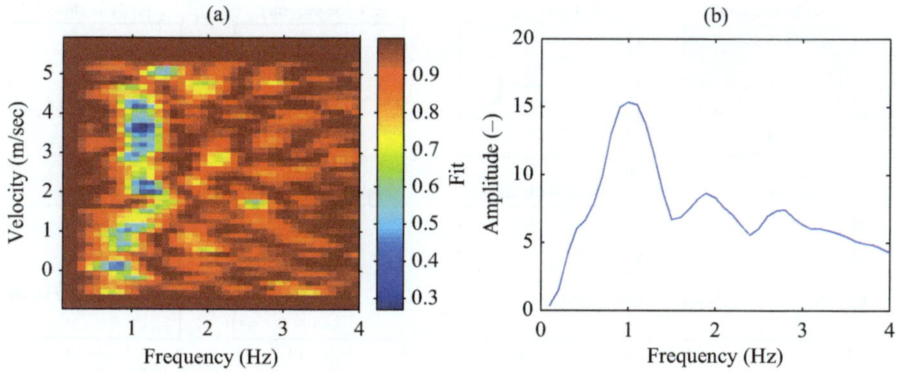

Figure 6.25 *Cycle frequency estimation: left, the amplitude fit function; right, the average cycle frequency strength*

Cycle Frequency Estimation

Figure 6.25 presents the results of the cycle frequency extraction. The fit function shows multiple minima on the line with 1 Hz cycle frequency. The torso component (2 m/sec velocity slice) shows a cycle frequency of 1.3 and 2.6 Hz in accordance with the double cycle frequency. The summed amplitudes show a strong peak with 1 Hz which corresponds with the average cycle frequency.

6.4.4.3 Estimated Parameters

The percentile result of a typical experiment of an inbound walking human is presented in Figure 6.26. Figure 6.26(a) presents the spectrogram. The human had four stages: the acceleration stage, the fast walking stage, the deceleration stage and the slow walking stage. The results of the parameter estimation are described in this section. Table 6.3 presents the corresponding average parameters in these stages.

Boulic Parameters

Figure 6.26(b–d) presents the three estimated global Boulic parameters $z^B = (f^c, l^c, \varphi^c)$. The cycle phase normalized between zero and one shows the periodic behaviour of the movement. The cycle frequency and cycle phase show some irregularities at the start and the end where the human has almost zero speed. The generated synthetic human with these parameters shows a very realistic moving human in which all individual human components show smooth behaviour. The start and end looks are also very realistic due to the small cycle length which corresponds with a slow walking human.

Leg Parameters

Figure 6.26(e) and 6.26(i) give the leg parameters $z^L = (S^L, \dot{S}^L, \tilde{S}^L, \varphi^L, f^L)$. The estimated sinusoidal is presented in Figure 6.26(j). The acceleration fluctuates during walking. These fluctuations give the personification of the moving human. The estimated sinusoidal of 1 ft in m/sec has a smoothed behaviour during the walking part, the beginning shows two small steps and the end shows four small steps. During walking the minimum velocity of the lower leg is 0.5 m/sec indicating no ground touch; the heel touched the ground.

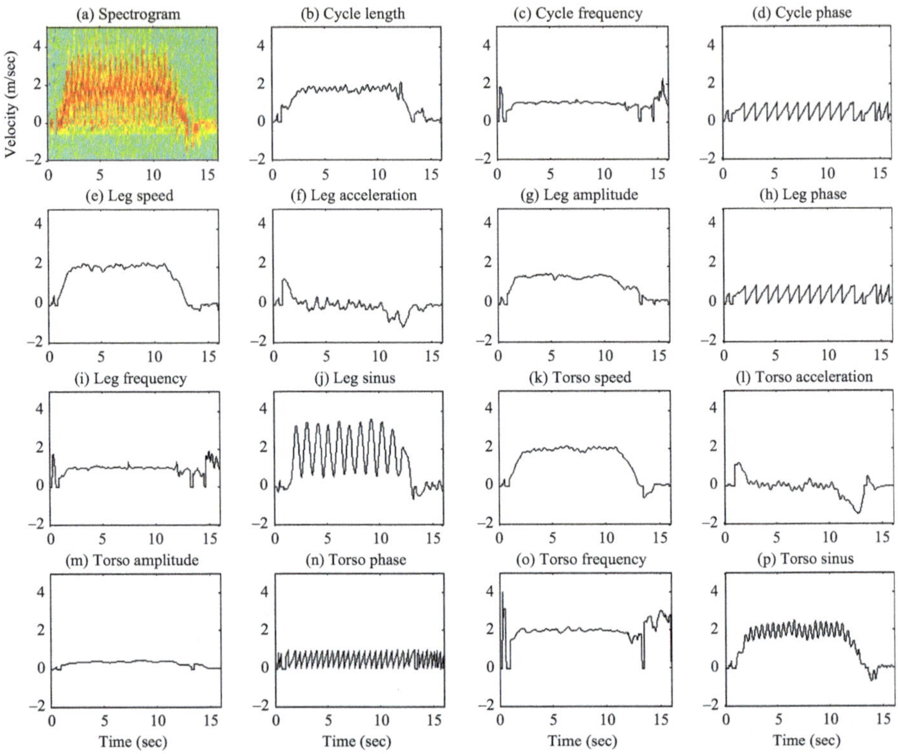

Figure 6.26 Estimated Kalman parameters of an inbound running human percentile method

Torso Parameters

Figure 6.26(k–o) give the torso parameters. The torso parameter vector is $z^T = (S^T, \dot{S}^T, \tilde{S}^T, \varphi^T, f^T)$. The estimated torso sinusoidal is presented in Figure 6.26(p). The velocity shows a ripple due to the sinusoidal approximation of the torso movement. The estimated amplitude has a parabolic shape with a maximum of $\tilde{S}^T = 0.5$ m/sec. During the walking part it increases and decreases during the movement. In the slow walking stage the human has some small steps with fast phase shift. The torso frequency f^T is twice the leg frequency and increases at the end due to the fast small steps. The estimated sinusoidal shows the torso behaviour around the average motion of the human.

The global Boulic parameters, and estimated leg and torso parameters show smooth behaviour with personification. The estimated personification differs from the leg and torso parameters calculated with the global Boulic parameters and results in a very realistic animated walking human.

Table 6.3 Average estimated percentile parameters inbound running human

Time interval	Parameter (range) units	Acceleration	Fast walking 2 m/sec	Deceleration	Slow walking 0.25 m/sec
	sec	0…4	4…11	11…13	13…15
Personification parameters					
Torso	S^T [0…3] m/sec	0	2	0	0.25
	$\dot{S}^T$ [0…3] m/sec^2	1.5	0	–1.5	0
	$\tilde{S}^T$ [0…3] m/sec	0	0.5	0	0.1
	φ^T [0…1] –	–	–	–	–
	f^T [0…2] Hz	3	2	3	3
Leg	S^L [0…3] m/sec	0	2	0	0.25
	$\dot{S}^L$ [0…3] m/sec^2	1.5	0	–1.5	0
	$\tilde{S}^L$ [0…3] m/sec	0	1.5	0	0.25
	φ^L [0…1] –	–	–	–	–
	f^L [0…2] Hz	0	1	0	1.5
Boulic parameters					
Global Boulic	l^c [0…3] m	0	2	0	0
parameters	f^c [0…2] Hz	0	1	0	2
with repetition frequency	φ^c [0…1] –	–	–	–	–

The time behaviour of the parameters is presented in Figure 6.26. Note that repetition frequency is interchangeable with cycle frequency.

6.4.4.4 Test on Database Measurements

The 50 experiments give 25 000 spectrograms belonging to a tracked person and the feature extraction is applied to it. Table 6.4 gives the Kalman filter error standard deviations of the extracted features. Figure 6.27 presents the average computation time of the Matlab implementation.

The results show that the three feature extraction methods give equivalent standard deviations. The leg parameters have higher standard deviation than the torso parameters. The cycle frequency standard deviation is smaller than the leg frequency standard deviation. Fusion of the cycle frequency and the leg frequency gives a 0.7 improvement in the standard deviation that agrees with uncorrelated parameters. The detection and percentile methods are six times faster than the correlation method. The computation time of the pre-processing is equivalent with the detection and percentile computation time and the cycle frequency extraction requires 1.2 times this time.

6.4.4.5 Comparison with the Model-Based Approach

Table 6.5 gives the standard deviations of the difference of the model-based approach and the feature-based approach. The cycle frequency and cycle length estimates have negligible zero offsets while the cycle phase has a small offset. The cycle frequency lies within 5% of the cycle frequency range, the cycle length lies

Table 6.4 Standard deviations estimated parameters based on the Kalman filter errors

		Parameter (range) units	Detection	Percentile	Correlation
Personification parameters					
Torso	σ^{S^T}	[0...3] m/sec	0.14	0.12	0.18
	$\sigma^{\dot{S}^T}$	[0...3] m/sec^2	0.26	0.19	0.25
	$\sigma^{\tilde{S}^T}$	[0...3] m/sec	0.05	0.04	0.05
	σ^{φ^T}	[0...1] –	0.17	0.17	0.17
	σ^{f^T}	[0...2] Hz	0.58	0.58	0.61
Leg	σ^{S^L}	[0...3] m/sec	0.12	0.09	0.12
	$\sigma^{\dot{S}^L}$	[0...3] m/sec^2	0.18	0.16	0.16
	$\sigma^{\tilde{S}^L}$	[0...3] m/sec	0.13	0.10	0.06
	σ^{φ^L}	[0...1] –	0.06	0.06	0.06
	σ^{f^L}	[0...2] Hz	0.22	0.24	0.23
General parameters					
Velocity	σ^V	[0...6] m/sec	0.10	0.10	0.10
Repetition frequency	σ^{f^A}	[0...2] Hz	0.19	0.19	0.20
Boulic parameters					
Global Boulic parameters without repetition frequency	σ^{l^c}	[0...3] m	0.35	0.32	0.26
	σ^{f^c}	[0...2] Hz	0.22	0.24	0.23
Global Boulic parameters with repetition frequency	σ^{l^c}	[0...3] m	0.20	0.20	0.21
	σ^{f^c}	[0...2] Hz	0.15	0.16	0.16

Note that repetition frequency is interchangeable with cycle frequency.

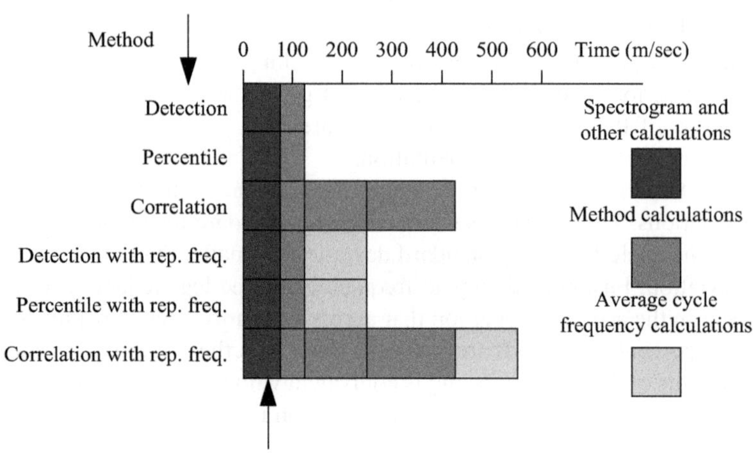

Figure 6.27 Average computation time

within 10% of the cycle length range and the cycle phase lies within 5% of the cycle phase range. The cycle phase has 15% offset. Detailed investigations show that the differences depend on the speed. High velocities result in small differences and low velocities result in greater differences. The reason is the flat envelope in case of low velocities.

Table 6.5 Comparison of the Boulic model-based approach and feature-based approach

Global Boulic parameters		Parameter (range) units	Detection	Percentile	Correlation
Frequency	σ^{f^c}	[0...2] Hz	0.07	0.07	0.06
Length	σ^{l^c}	[0...3 m	0.26	0.23	0.22
Phase	$\Delta\varphi^c$	[0...1] –	+0.14	+0.12	+0.10
	σ^{φ^c}	[0...1] –	0.05	0.04	0.03

The standard deviations σ^{f^c}, σ^{l^c}, σ^{φ^c} and phase offset $\Delta\varphi^c$ of the parameter error are presented. The cycle frequency offset is $\Delta f^c < 0.01$ Hz and cycle length offset is $\Delta l^c < 0.02$ m for all approaches.

6.4.5 Discussion

6.4.5.1 Methods

The detection method velocity bounds are affected by the number of missed detections and lie close to the centre velocity in case of a low signal-to-noise ratio. The percentile method is based on the cumulative sum of the signal; signals outside the velocity bounds affect the amplitude distribution. The correlation method gives an average symmetric velocity bounds based on the signal distribution.

The region in the spectrogram between the upper and lower bounds belongs to all scatterers of the human and these cause interfering signals. This interference causes strengthening and weakening of the received signals and the torso estimates deviate from the real location of the torso. Sometimes the maximum intensity differs from the torso but lies always close to the real value. The percentile torso estimate is very stable and gives a better estimate of the torso. The correlation method gives the worst torso estimate, an average between the upper and lower bound.

The cycle frequency extracted is very stable irrespective of the shape of the velocity bounds. The estimated standard deviation is smaller than 5%.

6.4.5.2 Parameter Estimation

The standard deviations presented in Table 6.4 show no essential difference between the extracted parameters. The cycle frequency standard deviation $\sigma^{f^A} = 0.19$ Hz is smaller than leg frequency $\sigma^{f^L} = 0.22$ Hz, fusion of these with a Kalman filter gives 0.7 standard deviation improvement $\sigma^{f^c} = 0.16$ Hz. This agrees with two uncorrelated parameters. The Kalman feature filtering gives smoothed features in time without irregularities.

The cycle frequency is not necessary. Both methods give a realistic look alike of the observed human motion. The generated human with global Boulic parameters shows personification and the results show that the addition of torso and leg parameters shows more personification.

Detailed investigation shows some differences:

1. The correlation method produces the best leg features. The correlation method centre frequency lies in the middle and has less relation with the torso frequency. Second best leg features give the percentile method.
2. The percentile method produces best torso features; the features are less affected by noise and interfering caused by other limbs.
3. The leg cycle frequency of all methods and the cycle frequency have equivalent standard deviation but the cycle frequency produces better features in case of flat velocity bounds. The flatter the velocity bounds the worse the quality of the leg frequency estimate.

The detection method and percentile method with and without cycle frequency require one sixth computation time of the correlation method. We advise for the percentile method with leg frequency and cycle frequency fusion if enough computation time is available, the cycle frequency could be ignored in a real-time application. At the current state-of-art a real-time solution is possible.

6.4.5.3 Difference Between the Model-Based and Feature-Based Approach

The model-based parameters and feature-based parameters correspond with each other with an exception of the offset of the cycle phase. The phases of the body components of the Boulic model in the model-based approach are all related to each other while the feature-based approach estimates different phase of the body and the legs. The model-based approach estimates an average cycle phase while the cycle phase of the feature-based approach belongs to the phase of the leg.

6.4.6 Conclusions

In this section we present a feature-based approach to estimate human motion parameters from radar spectrograms. The walking model of Boulic is used with personification information of the torso and leg. We have proposed a sinusoidal model for the torso and leg. The torso sinusoidal is related to the centre velocity and the leg sinusoidal is related to the lower and upper velocity bounds. Three methods are described which extract these velocities. Sinusoidal fits of velocity-slices give the cycle frequency that originates from periodic spectrogram components. Kalman filters smooth the features. The animated human generated with the features provide a realistic look-alike of the real motion of human irrespective of the walking model used to generate it.

The methods are applied to real radar measurements with different scenarios. The three methods give approximately equivalent results. The percentile method has best match with the velocity bounds and centre velocity. The computation time of the correlation method is about six times the other methods. We advise the percentile method with leg frequency and cycle frequency fusion. If enough computation time is available, the cycle frequency is not needed in real-time applications.

The comparison of the feature-based approach with the model-based approach shows the standard deviation cycle frequency f^c and cycle phase φ^c lie within 5% difference and the cycle length l^c within 10% of the parameter range. The cycle phase has a small offset due to the average cycle of the model-based approach and the leg phase in the feature-based approach. The accuracy of both approaches is equivalent but the feature-based approach is faster.

The feature-based approach brings the animation in the real-time domain and opens possibilities for person recognition besides animation.

References

[1] P. Withington, H. Fluhler, and S. Nag. Enhancing homeland security with advanced UWB sensors. *IEEE Microwave Magazine*, vol. 4, part 3: pp. 51–58, September 2003.

[2] M. Hewish. The last line of defense, reducing the vulnerability of fixed installations. *Janes's International Defense Review*, vol. 10:pp. 28–35, 1999.

[3] V.G. Nebabin. *Methods and Techniques of Radar Recognition*. Artech House, ISBN 0-89006-719-8, Norwood, MA, 1995.

[4] H. Rohling, F. Folster, F. Kruse, and M. Ahrholdt. Target classification based on a 24 GHz radar network. *Radar 2004, International Conference on Radar Systems*, 2004.

[5] A. Woltkiewicz, M. Nalecz, W. Kulpa, and K. Klembowski. Use of polynomial phase modeling to FMCW radar. Part C: Estimation of target acceleration in FMCW radars. *NATO Research and Technology Agency, Sensors and Electronics Technology Symposium on Passive and LPI Radio Frequency Sensors*, (paper no. 40C), 23–25 April 2001.

[6] M. Nalecz, R. Rytel-Andrianik, and A. Wojtkiewicz. Micro-Doppler analysis of signals received by FMCW radar. *International Radar Symposium*, pp. 651–656, 2003.

[7] V.C. Chen, R. Li, S.S. Ho, and H. Wechsler. Micro-Doppler effect in radar phenomenon, model, and simulation study. *IEEE Transactions on Aerospace and Electronic Systems*, vol. 42(no. 1):pp. 2–21, 2006.

[8] J.L. Geisheimer, E.F. Greneker, and W.S. Marshall. A high-resolution Doppler model of human gait. *Radar Sensor Technology and Data Visualization, Proceedings of SPIE*, vol. 4744:pp. 8–18, 2002. Komt de patent search voort van Kasper Haak uit, 2006.

[9] T. Thayaparan, S. Abrol, and E. Riseborough. Micro-Doppler radar signatures for intelligent target recognition. Technical Report, DRDC, Ottawa, TM 2004-170, Defence Research and Development Canada – Ottawa, September 2004.

[10] L. Marple. Sharping techniques for sensor feature enhancement. 26 May 2005.

[11] H. Ling. Micro-Doppler exploitation: Preliminary data collection and analysis. *NICOP Meeting London*, 7 June 2004.

[12] A.K. Mitra, M. Kobold, T. Lewis, and R. Williams. Theoretical radar-Doppler models for pivoting mechanical and biological objects-of-interest. *Proceeding SPIE, Algorithms for Synthetic Aperture Radar Imagery XIII*, vol. 6237:pp. 25–32, May 2006.

[13] C. Sminchisescu, D. Metaxas, and S. Dickinson. Incremental model-based estimation using geometric constraints. *IEEE Transactions on Pattern Analysis and Machine Intelligence*, vol. 27(no. 5), May 2005.

[14] F. Thomas and O. Johnston. *The Illusion of Life – Disney Animation*. ISBN 078686070. NY, Hyperion.

[15] T. White. *The Animator's Workbook*. ISBN 0823002292. Watson-Guptill, NY, 1988.

[16] P. Winston and K. Agarwal. Walking animation system. Class project at Harvey Mudd College, Claremont.

[17] P. van Dorp and F.C.A. Groen. Human walking estimation with radar. *IEE Proceedings in Radar Sonar and Navigation,* 150(5), 356–365, October 2003.

[18] R. Boulic, N. Magnenat-Thalmann, and D. Thalmann. A global human walking model with real time kinematic personification. *Visual Computing*, vol. 6(no. 6):pp. 344–356, December 1990.

[19] P. van Dorp and F.C.A. Groen. Feature-based human motion parameter estimation with radar. *Radar, Sonar and Navigation,* IET, 2, 135–145, 2008.

[20] M.I. Skolnik. *Radar Handbook,* 2nd edn. McGraw-Hill, NY, 1990.

[21] G.T. Ruck, D.E. Barrick, W.D. Stuart, and C.K. Krichbaum. *Radar Cross Section Handbook*, vol. 1. Plenum Press, New York, NY, 1970.

[22] E.F. Knott, J.F. Shaeffer, and M.T. Tuley. *Radar Cross Section*. 2nd edn. Artech House, Norwood, MA, 1993.

[23] R. Boulic, B. Ulicny, and D. Thalmann. Versatile walk engine. *Journal of Game Development*, vol. 1(no. 1):pp. 29–52, 2004.

[24] C.Y. Yam, M.S. Nixon, and J.N. Carter. Automated person recognition by walking and running via model-based approaches. *Pattern Recognition*, vol. 37(no. 5):pp. 1057–1072, May 2004.

[25] L. Cohen. *Time-Frequency Analysis*. Prentice Hall PTR, Upper Saddle River, NJ, 1995.

[26] V.C. Chen and H. Ling. *Time-Frequency Transforms for Radar Imaging and Signal Analysis*, ISBN 1-58053-288-8. Artech House, Norwood, MA, 2003.

[27] J. Liew and S.L. Marple. Three-dimensional lattice linear prediction parameter estimation method and fast algorithm. *IEEE Signal Processing Letters*, vol. 7(no. 6):pp. 164–168, 2000.

[28] S. Barbarossa and O. Lemoine. Analysis of nonlinear FM signals by pattern recognition of their time-frequency representation. *IEEE Signal Processing Letters,* vol. 3(no. 4):pp. 112–115, 1996.

[29] B. Boashash. *Time-Frequency Signal Analysis, Methods and Applications*. Longman Cheshire, Wiley, 1992.

[30] M.B. Priestley. *Spectral Analysis and Time Series*, vol. 1: Prediction and Control of Univariate Series; vol. 2: Prediction and Control of Multivariate Series, ISBN 0-12-564922-3, 6th edn. Academic Press, 1989.

[31] W.H. Press, S.A. Teukolsky, W.T. Vetterling, and B.P. Flannery. *Numerical Recipes in C*, 2nd edn. Cambridge University Press, 1992.

[32] P.M.T. Broersen. The performance of spectral quality measures. *IEEE Transactions on Instrumentation and Measurement*, vol. 50:pp. 813–818, 2001.

[33] R. van der Heiden. *Aircraft Recognition with Radar Range Profiles*. PhD thesis, University of Amsterdam, 27 October 1998.

[34] P.M.T. Broersen. The quality of lagged products and autoregressive Yule–Walker models as autocorrelation estimates. *IEEE Transactions on Instrumentation and Measurement*, vol. 58(no. 11):pp. 3867–3873, November 2009.

[35] P.M.T. Broersen. *Automatic Autocorrelation and Spectral Analysis*. Springer-Verlag New York, NY, 2006.

[36] C. Ballreich. VRML nancy. Humanoid Animation Working Group, http://www.h-anim.org, 1997.

[37] I.T. Young, J.J. Gerbrands, and L.J. van Vliet. *Fundamentals of Image Processing*. Delft University of Technology, The Netherlands, version 2.2, 1998.

[38] A. Farina and F.A. Studer. *Radar Data Processing, Volume I – Introduction and Tracking*. ISBN 0-471-90731-6. Research Studies Press, NY, 1985.

[39] A. Farina and F.A. Studer. *Radar Data Processing, Volume II – Advanced Topics and Applications*. Research Studies Press, NY, 1986.

[40] D.K. Mellinger and C.W. Clark. Recognizing transient low-frequency whale sounds by spectrogram correlation. *Journal of Acoustical Society of America*, vol. 107(no. 6):pp. 3518–3529, June 2000.

[41] B.R. Mahafza. *Radar Systems Analysis and Design Using MATLAB*, 2nd edn. Chapman & Hall/CRC, Boca Rotan, FL, USA, 2005.

Chapter 7

Micro-Doppler Signatures of Helicopter Rotor Blades

Karl Erik Olsen, Terje Johnsen, Øystein Lie-Svendsen, and Børge Torvik

7.1 Introduction

The distinguishing features of helicopters are their main and tail rotors. In addition, helicopters typically move at much lower speeds than many fixed wing aircraft. These two features can be used in radar measurements, both to identify an object as a helicopter and not a fixed wing aircraft, and to get information about the helicopter type. The speed of the tip of a rotor blade is typically much higher than the speed of the helicopter itself and will be close to the speed of sound. In a range-Doppler plot of the radar return from a helicopter, the main echo comes from the fuselage at low Doppler frequencies. The rotors typically appear as weaker returns at much higher Doppler frequencies and with a large spread in frequencies, which may contain both positive and negative frequencies depending on whether both advancing and receding rotor blades are seen. High time resolution measurements can be used to acquire further information about the type of helicopter, as we shall show in this chapter.

A weakness of this type of target classification is that the helicopter must be at fairly close range for the rotor signature to become visible. In particular, the radar return from the tail rotor is usually very weak.

In this chapter we present monostatic and bistatic measurements of helicopter radar signatures using the DiMuRa radar system, shown in sections 7.3.3.2 and 7.3.4.

In section 7.2 we provide background information useful in order to interpret the DiMuRa results: section 7.2.1 discusses the various helicopter classification methods. In section 7.2.2 we discuss the radar cross section (RCS) signature of rotor blades, based on electromagnetic calculations. In section 7.2.3 we show radar signature measurements of a small helicopter at different frequencies.

Section 7.3.1 gives a DiMuRa system description, while section 7.3.2 provides an overview of signal processing fundamentals with emphasis on the modifications necessary in order to identify and characterize rotor motion.

7.2 Background

7.2.1 Background Theory

The main rotor blades of a helicopter create lift as they rotate. The maximum rotational speed is limited to keep the airspeed around the advancing blade tip below the speed of sound, and thus avoid increased drag and vibration by the tips breaking the sound barrier. In practice this means that the sum of the helicopter forward air velocity and the blade tip velocity should never exceed approximately 330 m/s. This leads to a maximum main blade angular velocity given by

$$\omega_{\max} = \frac{v_s - v_h}{L}\,[\mathrm{rad/s}] \tag{7.1}$$

Here v_s is the speed of sound, v_h is the maximum speed of the helicopter, and L is the main rotor blade length.

Altitude and speed of the helicopter are controlled by changing the lift properties of the main blades. This is normally done by twisting the blades, and thereby changing the blades' angle of attack. During normal flight the rotational speed of the rotor is kept constant, which subsequently in the chapter will be shown to be useful for target recognition. As the engine of the helicopter turns the rotor against the air, the drag creates a torque that turns the body in the opposite direction. This is normally prevented by the use of a vertical tail rotor, which compensates the torque effect by pushing the tail. The thrust created by the tail rotor is controlled by changing the pitch angle of the rear rotor blades, and by that changing the heading of the helicopter. The signature of the tail rotor may also be used in a classification context.

7.2.1.1 Helicopter Classification Methods

For recognition of fixed wing aircraft, high range resolution techniques utilizing spatial attributes of the body are often used. These techniques build on the assumption that the target is a fixed body without dominant rotational parts. Although high-resolution techniques may contribute to recognition of helicopters, techniques based on analysis of time-varying scattering effects are more commonly used. These effects typically derive from rotational parts of the helicopters, such as main and rear rotor blades. From some aspect angles the rotational parts within the turbine of turbine driven helicopters may result in echoes with distinct spectral lines, which can be used for Jet Engine Modulation (JEM) analysis [1]. Nevertheless micro-Doppler effects from the rotors are the most widely exploited feature for recognition of helicopter by use of radar, and the rest of the chapter is limited to deal with these.

Classification Based on Main Rotor Flashes

Main rotor blades are long and thin, and for a basic understanding of the scattering mechanics, it can be modeled as a narrow flat plate with the dimensions of the blade [1]. However, for more detailed analysis of the scattering effects from the blades and the related RCS, an electromagnetic prediction code or compact range measurements should be used. In the context of target recognition, the RCS of the main and rear rotor blades are of interest in order to calculate the appropriate radar

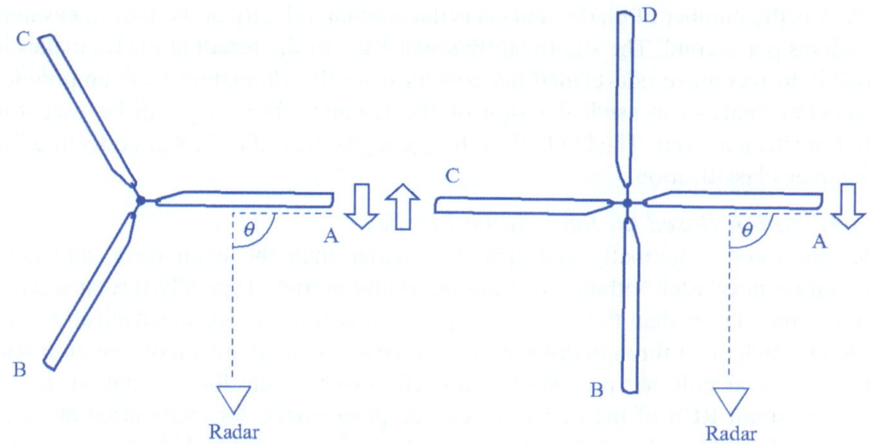

Figure 7.1 Odd and even numbered main rotors

system requirements for maximum effective recognition distance. A more detailed discussion on RCS characteristics of main rotor blades is found in section 7.2.2. Although the RCS values of the blades are not used as a feature for classification itself, the variation of RCS with aspect angle is very useful. The RCS of such long and thin objects has well-defined peaks around orthogonal incidence, which rapidly falls off with increasing angle. As the rotor rotates with a constant rotational speed, this results in what is commonly known as “blade flashes” with a constant rate. This blade flash frequency will obviously depend on the product of the rotational speed of the rotor and the number of blades. If flashes would originate only from the front side of the blades, the blades approaching the radar, the blade flash frequency would be the product of the number of blades and the rotational speed of the rotor alone. In practice flashes are also seen from the back side of the blades, even though the echoes from this side tend to be significantly lower. This level of backscattered energy from receding blades is dependent on the curvature and dimensions of the blade, and of the carrier frequency in use. This is discussed in more detail in section 7.2.2. From this, the blade flash frequency should be twice the earlier-mentioned product. However, on even bladed helicopters the flash from the approaching blades coincides with the receding blades in time. This is due to the symmetry and geometrical shape of the blades (see Figure 7.1).

In the odd numbered blade case, the approaching and receding blades are not positioned to provide a flash toward the radar at the same time. This means that every second flash will originate from the approaching and receding blades. This leads to the differentiation between odd- and even-bladed helicopters with regards to blade flash frequency:

$$f_{bf} = \begin{cases} N\dfrac{\omega_0}{2\pi} & \text{even-bladed helicopter} \\ 2N\dfrac{\omega_0}{2\pi} & \text{odd-bladed helicopter} \end{cases} \tag{7.2}$$

Here N is the number of blades and ω_0 is the angular velocity of the rotor measured in radians per second. The significantly lower RCS of the receding blades makes it possible to recognize odd-bladed helicopters from the alternating flash amplitude. If spectral analysis is used, the sign of the Doppler frequency can be used for differentiation as well. The blade flash frequency is proved to be a good feature for helicopter classification.

Classification Based on Rear Rotor Flashes

The rear rotor is normally significantly smaller than the main rotor, and may therefore rotate faster without breaking the sound barrier. Typically it rotates three to five times faster than the main rotor [1]. The two rotors are mechanically connected to each other through drive shafts and gearboxes, giving a fixed rotation rate ratio. This gear ratio has proved to be a useful discriminant for recognition. However, the small RCS of the rear rotor can be problematic for recognition at large distances. For spectral analysis, the aspect-dependent Doppler shifts from the rear blades may induce problems with regards to classification as the helicopter is seen from near broad side. RCS and polarization effects of rear rotor blades are discussed in section 7.2.3.

Classification Based on Spectral Analysis

In conventional pulse-Doppler signal processing the impulsive flashes tend to be averaged out into lower amplitudes in each of the Doppler bins [1], due to the high-frequency content of the rotor flashes and the short duration of the flashes compared to normal dwell times. This is the reason why helicopter recognition is commonly done in the time domain. However, valuable information can be extracted from the main rotor spectrum. Helicopter engine rotor modulation (HERM) detection technique utilizes the Doppler frequency spread of the main rotor. This spread is due to Doppler-shifted returns from approaching and receding blade tips. The width of the spread is given by two times this Doppler frequency and the spectrum is offset in the frequency domain by the Doppler frequency from the helicopter body. Candidate helicopters are the ones with rotational speed and rotor length that generate the measured spectrum. The conventional Fourier transform does not provide any time-dependent spectral description of a signal with time-varying spectral composition. As will be discussed in section 7.2.3, a spectrogram based on short-time Fourier transforms (STFTs) can give useful overview of the features available for helicopter recognition. This form of joint time–frequency analysis provides both spectral and temporal information about the signal. Within STFT-based spectrograms there is a trade-off between time and frequency resolution, both dependent on the time window size. Large window sizes give high-frequency resolution and poor time resolution, and vice versa [2]. This will be illustrated in section 7.3.3.2, where the same DiMuRa data are processed with different time window sizes (Figures 7.14–7.16).

7.2.1.2 Helicopter Classification Waveforms

In the design process of a helicopter recognition system, the waveform to be used requires some attention. In a conventional recognition system based on flash detection, the range resolution is of less importance and unmodulated pulses may be used. The most important requirement in this context would be that no blade flash will be lost between radar pulses. This requires that the pulse repetition frequency (PRF) of the system is high enough. Actually, in order to ensure that one complete pulse hits the blade within the flash duration period, the following requirement applies:

$$\tau_{PRI} + \tau_{\text{pulse}} \leq \tau_{bf} \tag{7.3}$$

This states that the sum of pulse repetition interval τ_{PRI} and pulse duration τ_{pulse} should be less than the main rotor blade flash duration. The flash duration time τ_{bf} can be estimated as the expected angular extent of the flash divided by the rotational speed ω:

$$\tau_{bf} \approx \frac{\lambda}{L\omega} \tag{7.4}$$

Here L is the length of the rotor blade and λ is the wavelength. For a normal sized helicopter, the flash rate may be a fraction of a millisecond, leading to a required PRF of several kilohertz. If a spectral analysis approach to recognition is used, an even higher PRF may be required to ensure that the Doppler frequency shift from the blade tips is sampled without ambiguities:

$$PRF_{\text{min}} = 2f_{\text{doppler}} = \frac{4v_{\text{radial}}}{\lambda} \tag{7.5}$$

7.2.2 Rotor Blade RCS

Since rotors are the main feature used in helicopter classification, it is useful to know the RCS of a rotor blade. We have carried out calculations of the monostatic RCS for the modeled rotor blade shown in Figure 7.2. The size and shape of the blade have been chosen to closely match the small UAV helicopter used in the measurements presented in section 7.2.3. The blade is assumed to be perfectly conducting metal (while the UAV blades are made of carbon fiber). The calculations are carried out at the frequencies used in the measurement – 1, 4, and 24 GHz – with the electromagnetic solver Efield [3], using the "multi-level fast multipole" method (MLFMM).

For the results we use a spherical coordinate system based on the Cartesian system presented in Figure 7.2 (in which the length of the blade is parallel to the z-axis). Placing the origin of the coordinate system in the rotor blade, we let the unit vector $\hat{\mathbf{r}}$ denote the direction from the blade to the radar. This vector makes an angle θ with the z-axis, and the projection of the vector into the xy-plane makes an angle ϕ with the x-axis. Hence the vector has the usual Cartesian components

$$\hat{r}_x = \cos\phi \sin\theta$$
$$\hat{r}_y = \cos\phi \sin\theta$$
$$\hat{r}_z = \cos\theta$$

Figure 7.2 The modeled rotor blade. The blade has a length (along the z-axis) of 79 cm, a width (along the x-axis) of 6 cm, and a thickness of 0.72 cm (see section 7.2.3)

For example, illumination of the rear (receding) edge corresponds to $\theta = 90°$, $\phi = 0°$.

Figure 7.3 shows a "global" view of the monostatic RCS at 24 GHz with a 6° resolution in θ and 12° in ϕ. The maximum RCS at $\theta = 90°$ and $\phi = 90°$ and 270° corresponds to broadside illumination. Edge-on illumination of the forward edge corresponds to $\theta = 90°$ and $\phi = 180°$. The narrow horizontal band at $\theta = 90°$ shows the rotor flash: a sharp and narrow increase in the backscatter from the rotor blade when illuminated perpendicularly to the rotor long axis (the z-axis).

The flash is much sharper than indicated in Figure 7.3; at 24 GHz it extends over only a fraction of a degree in the θ direction, as predicted by (7.4). We have therefore repeated the calculation with a much higher angular resolution for near frontside illumination.

Figure 7.4 shows horizontal and vertical "cuts" of the RCS. We note that the RCS drops by approximately a factor 10 with a 0.3° change in θ, both for horizontal and vertical polarization.

By contrast, in the ϕ direction, corresponding to a vertical cut, the RCS varies much more slowly, and in this direction the horizontal and vertical polarization behave quite differently. The slow ϕ variation implies that helicopter measurements are not sensitive to the rotor tilt as seen by the radar (a rotation of the blade around the z-axis), and hence will not be sensitive to the elevation angle of the helicopter as seen by the radar.

Figure 7.4 also shows that, with the modeled rotor blade in Figure 7.2 as input, the backscatter from vertical polarization is much smaller than from horizontal polarization, at least from (near) frontside illumination. As a function of rotor tilt angle (ϕ) the backscatter from vertical polarization is also much more variable than from horizontal polarization.

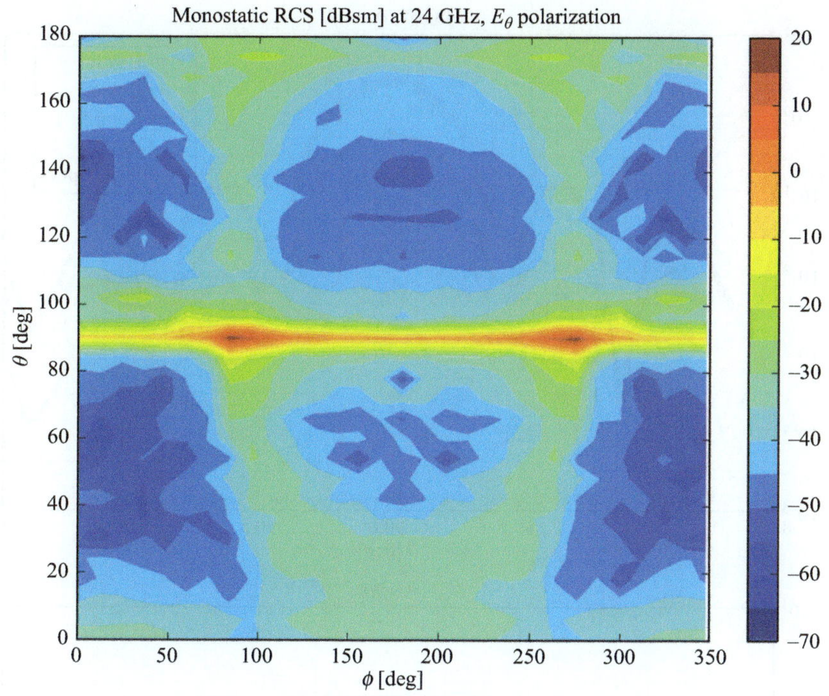

Figure 7.3 The monostatic RCS of the rotor blade with E_θ polarization. The coordinate system of Figure 7.2 is used, and θ and ϕ are the polar and azimuthal angles of the corresponding spherical coordinate system

Finally, Figure 7.4 compares the calculation using the actual rotor blade with the same calculation carried out assuming a rectangular box having exactly the same length, width, and thickness as the rotor blade (as given in the caption of Figure 7.2). Estimates of rotor RCS are frequently carried out approximating the rotor blade with a rectangular plate in the high frequency limit (the physical optics approximation) [1], for which a simple analytical expression for the RCS exists [4]. In our coordinate system with $\phi = 180°$ it reads

$$\sigma_{fp} = 4\pi \frac{L^2 H^2}{\lambda^2} \sin^2\theta \left(\frac{\sin((2\pi L/\lambda)\cos\theta)}{(2\pi L/\lambda)\cos\theta} \right)^2 \tag{7.6}$$

where λ is the wavelength, and L and H the length and height of the flat plate. We note from the top panel that the rectangular box and the actual rotor blade have the same overall shape of the RCS as a function of θ. However, the box approximation overestimates the RCS by approximately a factor 5. This is not surprising, given the curved shape of the rotor edge; at high frequency most of the radiation hitting the curved edge will be specularly reflected away from the radar so that the "effective" thickness of the rotor blade will be smaller than the physical thickness. The box approximation leads to very little difference between horizontal and vertical

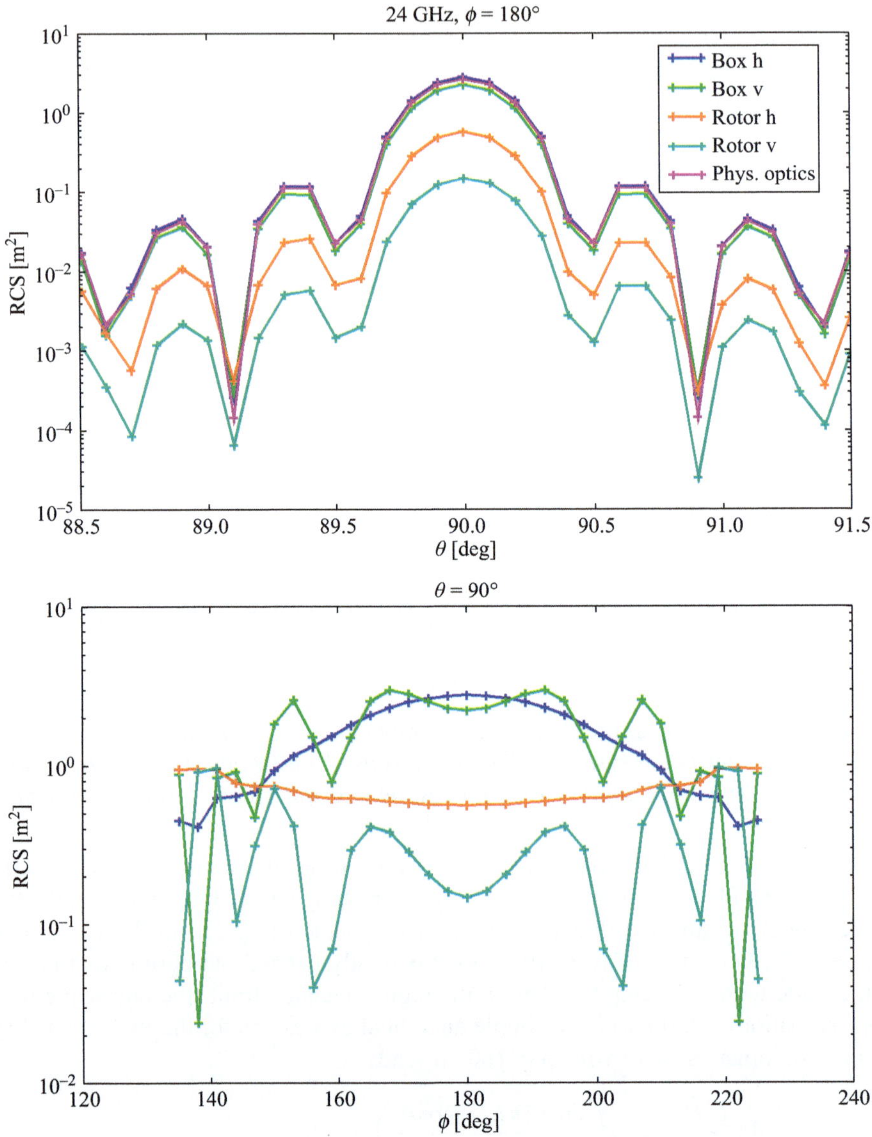

Figure 7.4 Monostatic RCS for front side illumination. Letters 'h' and 'v' indicate horizontal (E_θ) and vertical (E_ϕ) polarization, respectively. 'Box' is the result using a rectangular box with the same dimensions as the rotor blade of Figure 7.2. The purple curve shows the physical optics result from (7.6) for a rectangular plate with edge lengths equal to the length and thickness of the box

polarization, at least for edge-on incidence, while the actual rotor blade gave much less backscatter for vertical polarization. At lower frequencies the rectangular box approximation will be much better, as we shall see. Figure 7.4 also shows that the flat plate physical optics approximation is an excellent approximation to the full-scale calculation for the box at 24 GHz, although it strongly deviates from the actual rotor blade RCS.

Figure 7.5 compares the frontside (edge-on) RCS (also shown in Figure 7.4) with the RCS for backside (rear edge) illumination. From the top panel we note that the flash should be much weaker for backside illumination, with the RCS for horizontal polarization being about a factor 5 smaller than the front side RCS, and about a factor 10 reduction for vertical polarization. The bottom panel indicates that the magnitude of the flash for horizontal polarization should depend little on the angle of attack of the rotor blade or the elevation angle of the helicopter as seen by the radar. However, for vertical polarization the RCS does depend on ϕ. From these calculations we expect that the forward edge of the rotor should be more visible than the receding edge at high frequencies, and whether both flashes will be recorded depends on the radar sensitivity.

The rotor RCS at lower frequencies is shown in Figure 7.6 for front-side illumination and horizontal polarization. Note how the maximum RCS, at $\theta = 90°$, shows very little frequency dependence for the full electromagnetic calculation of the rotor blade signature, being in the range 0.25–0.5 m^2 for all three frequencies. At 1 GHz the flash is not very pronounced though – in this case the wavelength is comparable to the rotor size and hence the RCS remains large also at non-normal incidence, implying that the flash will have a long duration (and should thus no longer be labeled as a flash). The figure also shows that at 1 and 4 GHz, the rectangular box gives almost the same result as the actual rotor blade, showing that at low frequency the particular shape of the blade no longer matters, only its overall size.

Finally, we find it interesting to note that at none of these frequencies does the flat plate physical optics approximation give the correct RCS, deviating by approximately a factor 5 from the full calculation at 4 and 24 GHz, and underestimating it by a factor 100 at 1 GHz. At low frequency the flat plate (rectangular box) is a reasonable approximation, but physical optics is not applicable at low frequency (for this rotor size). At high frequency, physical optics is applicable, but the flat plate approximation is not (the shape of the blade has a large impact on the RCS).

7.2.3 Measurements on Scale Model

The results from two simple experiments with the UAV helicopter Kestrel 2000 can be used to illustrate the features available for helicopter recognition across different frequency bands. The first measurement was done with a K-HC1 24.0 GHz continuous wave radar from RFbeam Microwave GmbH. This unit has an aperture of approximately 5 cm and a transmitter power of 15 dBmW. Because of the low power, the target range was limited to 10 m in this case. The second experiment was performed using a N5245A network analyzer from Agilent Technologies, and

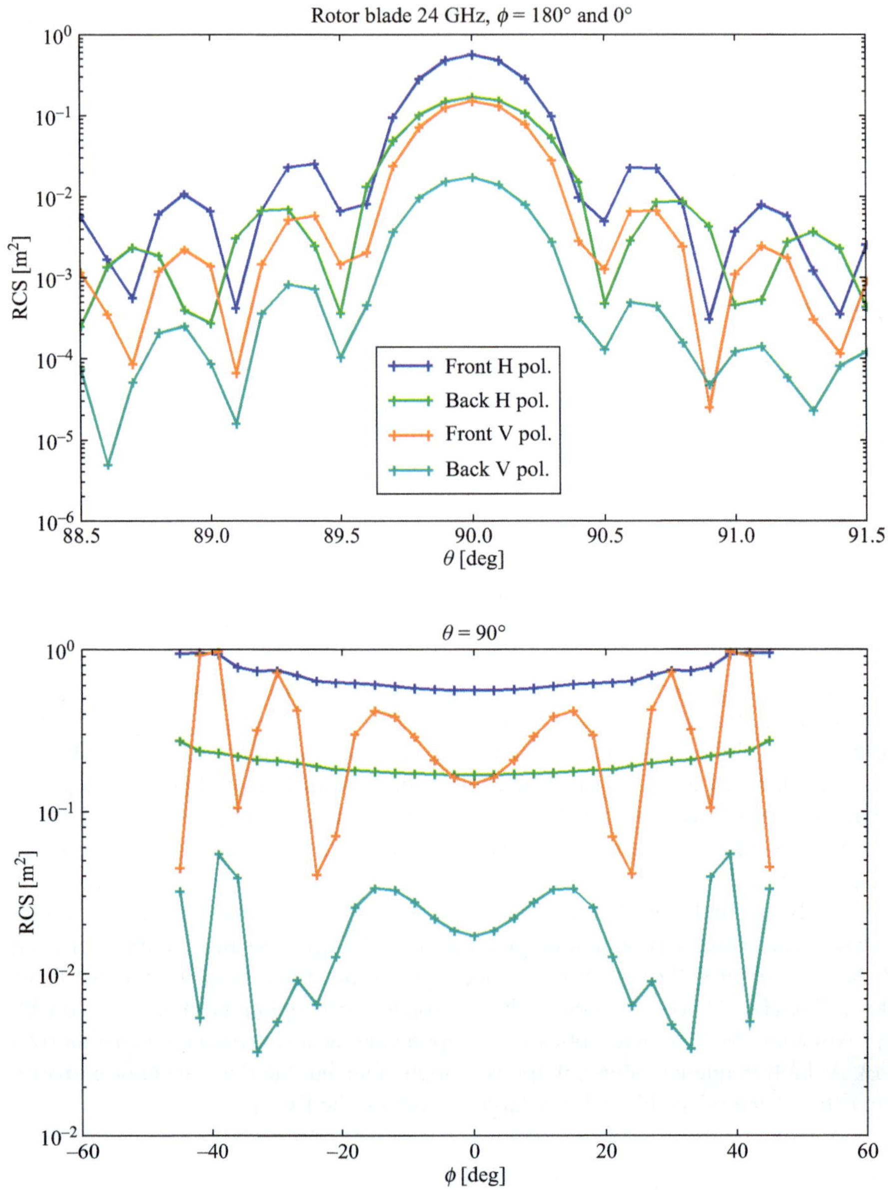

Figure 7.5 Comparison of the monostatic RCS for front- and back-side illumination

a wideband dual polarized reflector antenna from Q-PAR Angus Ltd. The diameter of this antenna is 70 cm. Maximum transmitter power in this setup was limited to 15 dBmW as well, but the target range could be increased to 30 m due to larger antenna gain. During the experiment each measurement was calibrated with

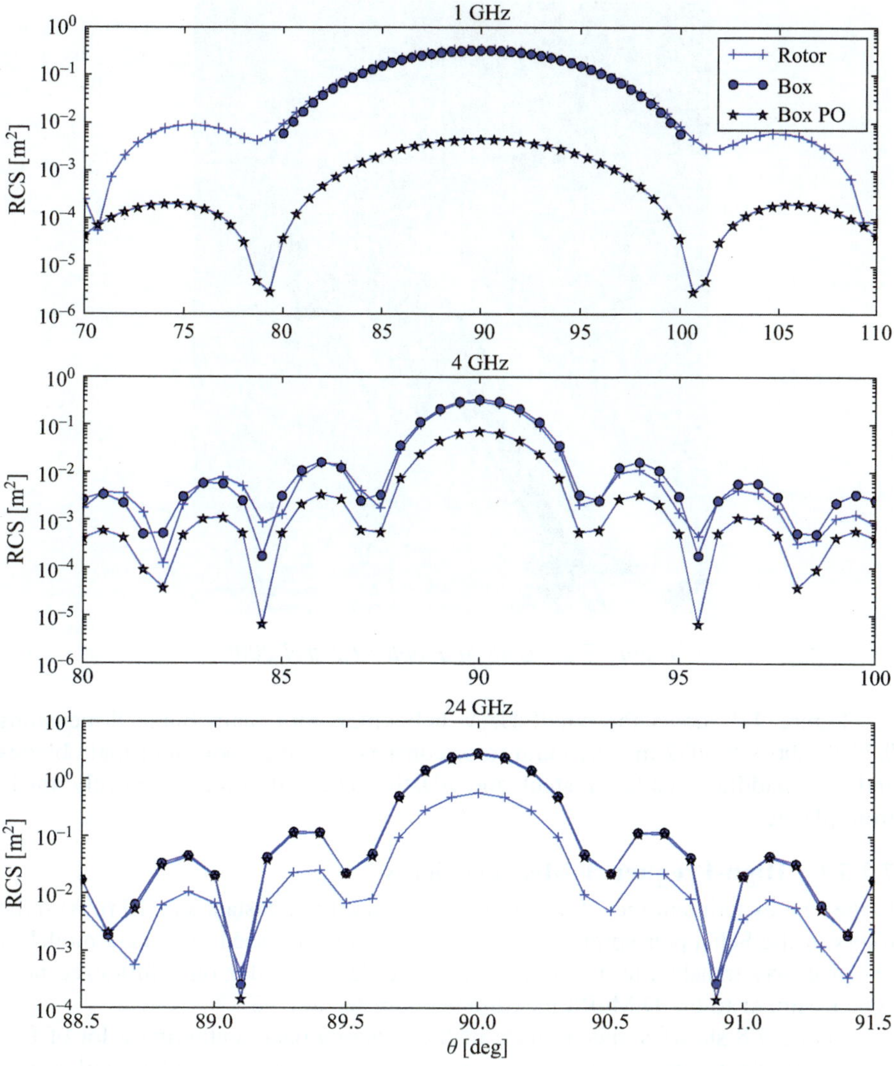

Figure 7.6 The RCS at 1, 4, and 24 GHz as a function of θ for horizontal (E_θ) polarization and for $\phi = 180°$. The same rotor and box dimensions as in Figure 7.4 were used. The bottom panel values are identical to those shown in Figure 7.4. "Box PO" denotes the physical optics result for the box, using (7.6)

measurements of a sphere with RCS of −10 dBsm in the optical region. In both experiments, the target was not considered to be in the far field. The measurements were conducted outdoors, and it must be emphasized that both experiments were carried out for demonstration purposes only.

Figure 7.7 UAS helicopter Kestrel 2000

Figure 7.7 shows the small UAV helicopter with main blade dimensions $0.79 \times 0.06 \times 0.0072$ m. The main rotor consists of two carbon fiber main blades and two paddlers made of steel and plastic. The tail rotor is entirely made from plastic.

7.2.3.1 High-Frequency Measurements

In the first experiment the 24.0 GHz radar was used at a distance of 10 m with the rotors of the helicopter spinning. The K-band frequency used at a scale model of this size may translate to a frequency in the order of 5 GHz on a full-sized helicopter, similar to the DiMuRa measurements in section 7.3.

Figure 7.8 shows a spectrogram of the gathered data containing a lot of features useful for target recognition. The spectrogram is based on STFTs with a time window somewhat longer than the main rotor flash duration, and maximum overlap. First of all it can be seen that the rotor has an even number of blades, as the flashes coincide in time from receding and approaching blades. See section 7.2.1 for more details. It should be noted that phase and amplitude inequalities in the quadrature demodulator of the receiver causes an unwanted mirror image in the spectrogram. Given the *a priori* knowledge of number of blades, the rotational speed and diameter of the main rotor can be found. In this experiment the rotation rate was estimated to be 950 rpm. Based on this estimate, and the maximum Doppler frequency shift of the blade tips found in the image, the distance from the blade tip to the rotation center can be estimated to 0.91 m. The correct length is 90 cm. The length of the shorter paddler can also be estimated to 31.5 cm.

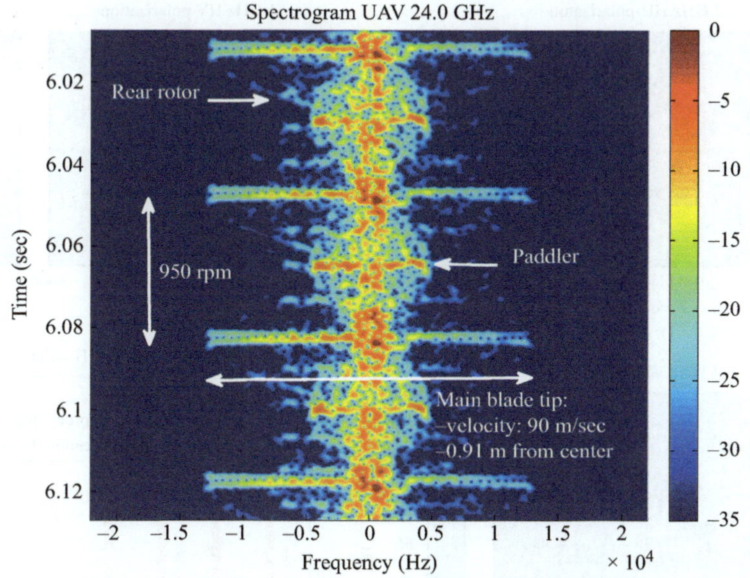

Figure 7.8 Spectrogram of UAV helicopter measured at 24.0 GHz

The actual length of this is 32 cm. The main blade tip speed in this example is measured to 90 m/s, which indicates that the engine is not running at full speed. Since the rear rotor is clearly visible in this experiment, the rear rotor flash rate could be measured to 4 275 rpm, and the gear ratio between main and rear rotor was then calculated to 4.2. The short distance between the target and radar implies that the electromagnetic wave front is curved across the target. Hence, a flash occurs at different positions along the blade as function of time. In fact the reflections from the parts closest to the hub are seen a measurable time before the tip, forming a distinct up-chirp in the spectrogram.

7.2.3.2 Low-Frequency Measurements

A different experiment was conducted for low-frequency measurements on the same UAV helicopter, using the mentioned network analyzer and larger antennas. The distance was increased to 30 m, resulting in a flatter wavefront across the target. Measurements were done from 1 to 6 GHz, with 1 GHz intervals. These frequencies used on such a small helicopter result in a low-frequency scattering regime on the blades (the wavelength is comparable to the length of the rotor blade). In the following results the returns from the hub of the rotor were filtered out by a high-pass filter, in order to show the returns from the main rotor blades only.

Figure 7.9 shows the spectrograms of different polarizations at 1 GHz. At this frequency the best visible feature is the main rotor blades measured at HH-polarization. At cross-polarizations the main rotor disappears completely. This indicates that the cross-polarization level is at least 25 dB lower than the flashes with HH-polarization. In the VV-polarization image, the main rotor blades cannot be identified. The rear

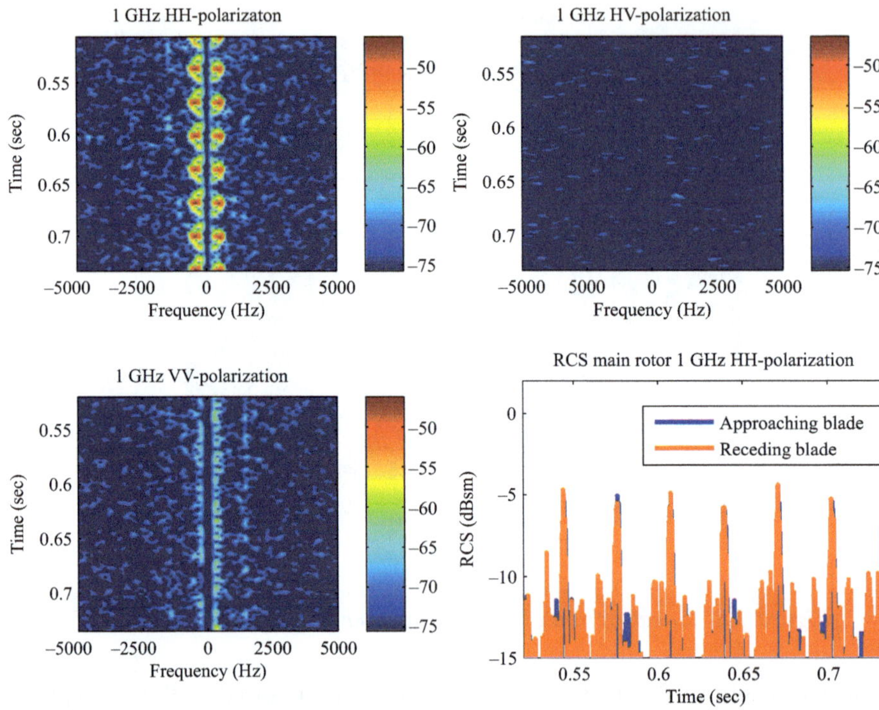

Figure 7.9 Measurements of UAV helicopter at 1 GHz with different polarizations

rotor is on the other hand visible at a level 15 dB below the main rotor flashes with HH-polarization. The lower right panel shows the calibrated RCS measurements for both approaching and receding blades. Note that the RCS at this frequency is equal from the front and the back, at a relatively constant level of −5 dBsm. The same effect is seen in the predictions discussed in section 7.2.2, and suggests that the shape of the rotor blade has no significance at this frequency. The absolute RCS values in this case match well with the predictions, but we reemphasize that the accuracy in such an experiment is low.

Figure 7.10 shows the spectrograms of different polarizations at 4 GHz. In the HH-polarization image, the main rotor is clearly visible with a large signal-to-noise ratio. The same applies to the paddler. For the cross-polarizations, there are some returns, from what is believed to be the paddler. The main rotor is not visible in this image, suggesting a suppression of more than 25 dB compared to the HH-polarization. In the VV-polarization image, both the approaching and receding main rotor blades are clearly visible. However, the strongest reflections are seen from the rear rotor. In the calibrated RCS plot in the lower right corner, it is clearly seen that the receding blade has lower RCS than the approaching ones. This is in contrast to the measurements at 1 GHz. The relative differences between receding and approaching blades match the predictions well, but the absolute values tend to be about 3 dB lower than predicted.

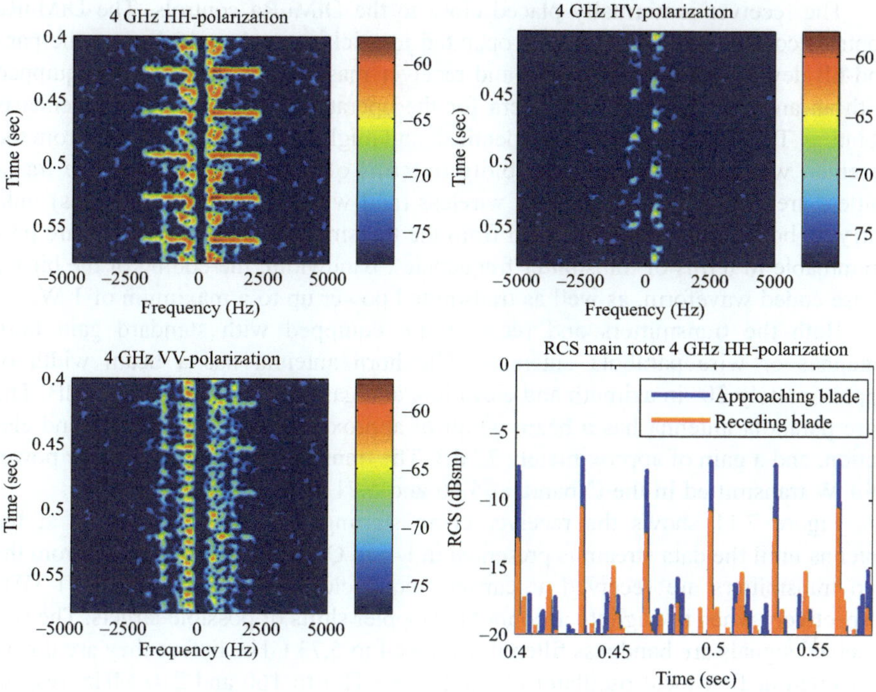

Figure 7.10 Measurements of UAV helicopter at 4 GHz with different polarizations

Although the calculations and measurements presented in this section were carried out for a scale model of a helicopter, the results are still applicable to a full-scale helicopter, provided the frequency is modified so that the ratio of wavelength to rotor blade length is approximately the same as in these calculations and provided that the rotor blade may (to a good approximation) be treated as a perfect electric conductor. The 24 GHz calculations and measurements correspond to $\lambda/L \approx 0.016$ (where again L is the rotor blade length). The results presented in section 7.3.3 correspond to $\lambda/L \approx 7 \times 10^{-3}$. Hence the actual helicopter measurements were carried out even closer to the high frequency (physical optics) regime than the results presented here.

7.3 Monostatic and Bistatic Helicopter Micro-Doppler

7.3.1 DiMuRa System Description

The Norwegian Defence Research Establishment has developed a binary phase coded experimental continuous wave multistatic radar (DiMuRa) consisting of two transmitters and one receiver. The radar has been used to detect a variety of air targets, and selected results have been published [5–12]. For the interested reader, in particular [12, pp. 730–744] might be a good starting point.

The receiver is normally placed close to the DiMuRa controls. The DiMuRa controls consist of three manually operated joystick controls, which steer the pan-and-tilt devices on the transmitter and receiver masts. Each mast is also equipped with an antenna and a video camera for the operator to see where the antenna is pointing. The two transmitters are identical, and might be placed up to 2 km from the receiver, which gives quite a flexibility in terms of bistatic operations. The transmitters are remotely controlled by wireless (and wired for shorter distances) link, carrying both control signals to and from the transmitters. The transmitters are programmable in terms of transmitter frequencies, bandwidth, the coding of the binary phase coded waveform, as well as transmitted power up to a maximum of 1 W.

Both the transmitters and receiver are equipped with standard gain horn antennas or wire parabola antennas. The horn antenna has a beam width of approximately 20° in azimuth and elevation, and gain of approximately 20 dB. The wire parabola antenna has a beam width of approximately 5° in azimuth and elevation, and a gain of approximately 27 dB. The transmitters have an average power of 1 W transmitted in the C-band, at 5.66 and 5.71 GHz, respectively.

Figure 7.11 shows the receiver chain starting with signal reception at the antenna until the data stream is presented in I- and Q-channels. The signals from the two transmitters are received at carrier frequencies around 5.66 and 5.71 GHz respectively when taking into account the Doppler shifts of possible targets. The two received signals are bandpass filtered from 5.64 to 5.73 GHz, before they are down-converted at first local oscillator (1. LO) (5.5 GHz) to 160 and 210 MHz, respectively. The receiver chain is then bandpass filtered (150–200 MHz) duplicated into two channels. The first channel is bandpass filtered 154–165 MHz leaving only the 160 MHz signal, while the other channel is bandpass filtered 204–215 MHz, leaving only the 210 MHz signal. The first channel is down-converted with 150 MHz at a second LO, leaving a signal at 10 MHz, while the other channel is down-converted with 200 MHz, also here leaving a signal at 10 MHz. Both channels are lowpass filtered before they are sampled at 37.5 MHz sampling rate. After sampling the signals are digitally down-converted and tuned leaving a two channel sampled signal (I and Q) with a bandwidth of 5 MHz sampled at a sampling rate of approximately 5 MHz. This means that the receiver may be interpreted as one receiver with two receiving chains inside. The two channels are digitally processed, displayed and if necessary stored in an ordinary PC. The signal processing will produce targets presented in range, Doppler and relative radar echo level (section 7.3.2).

7.3.2 Signal Processing Fundamentals

The results from this research emerges after the correlation process, and three of the most common ways to process the reference and surveillance channels follow. The two most straightforward algorithms are analogous to the calculation of the ambiguity function [13, pp. 297, 298],

$$|\Psi(\tau,\nu)| = \left|\sum_{n=0}^{N-1} e(n+\tau)d^*(n)e^{-j2\pi\nu n/N}\right| \tag{7.7}$$

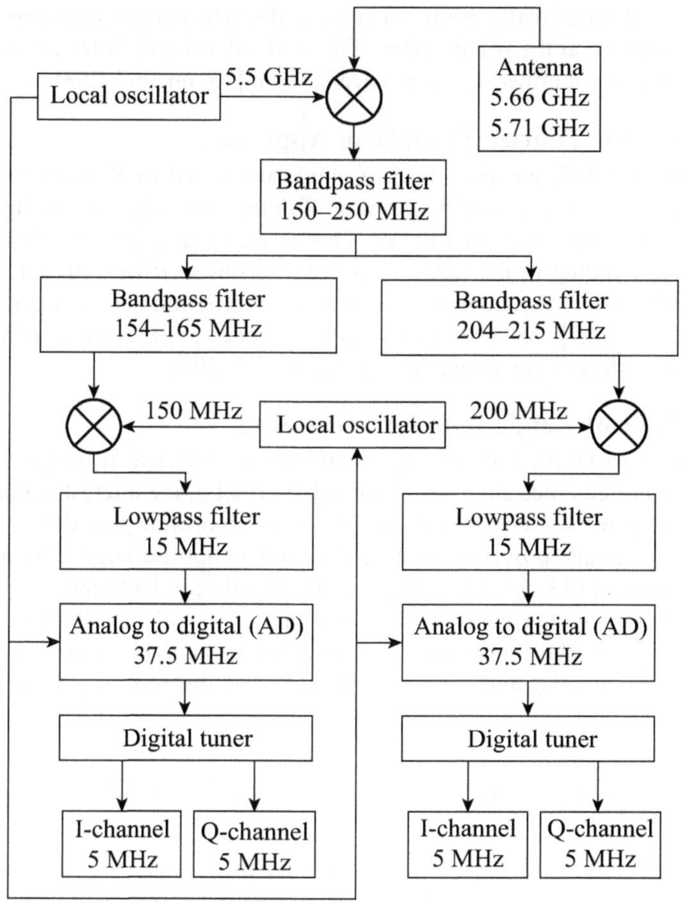

Figure 7.11 Schematic illustration of DiMuRa receiver hardware

where j is the complex constant $j = \sqrt{-1}$, Ψ denotes the discrete amplitude-range-Doppler surface, $e(n)$ is the sampled filtered echo signal, $d(n)$ is the sampled reference signal, τ is the time delay corresponding to the bistatic time difference of arrival (TDOA), and finally v is the Doppler shift of interest [13, pp. 297, 298]. There are two ways of interpreting this, as will be discussed in sections 7.3.2.1 and 7.3.2.2.

7.3.2.1 Cross-correlation Approach

Equation (7.7) [13, pp. 297, 298] may be interpreted as the correlation between $e(n)$ and $d(n)e^{j2\pi v(n/N)}$, where it is noted that the latter is a frequency-shifted version of $d(n)$, and this can be very efficiently obtained by cyclic shifting of the discrete Fourier transform (DFT)/fast Fourier Transform (FFT) transformed version of $d(n)$. The correlation for all ranges but one Doppler velocity is then calculated by multiplying each element of the DFT/FFT transformed $e(n)$ with each element of the shifted DFT/FFT transformed $d(n)$, and finally apply an inverse discrete Fourier transform (IDFT)/ inverse FFT (IFFT) on the result vector. This will calculate (7.7) for all ranges, but one Doppler value v. Further details are available in [13, pp. 297, 298].

7.3.2.2 Discrete Fourier Transform Approach

On the other hand [13, pp. 297, 298], if one is interested in all possible Doppler shifts, for one range, the most efficient way of doing this is by considering (7.7) as the Fourier transform of $e(n)d^*(n-\tau)$. This is easily achieved by element-wise rotating, corresponding to the delay τ, in order to obtain $d^*(n-\tau)$, and then calculate the DFT/FFT of the element-wise product of $e(n)$ and the rotated $d^*(n)$. This will calculate all the possible Doppler shifts for one range, corresponding to the delay τ. Further details are available in [13, pp. 297, 298].

7.3.2.3 Efficient Range Doppler Processing

Both the above-mentioned processing algorithms are efficient if only a few range cells or Doppler velocities shall be calculated. Normally, the interest is in a span of ranges and Doppler velocities, and the following algorithm provides an efficient way of achieving this, provided the Doppler shift is not too large. The method is briefly presented in [13, pp. 301–303], but full details can be found in [11, 12].

In the following the echo signal ($e(n)$ in (7.7)) will be referred to as the surveillance signal. The transmitted and reflected surveillance signal is received, down-converted, filtered and IQ sampled such that the resulting signal may be written[1]

$$U_{sur}(t_i)=\sum_{m=0}^{M-1}\xi_{sur}^{m}(t_i-v_m)u_{sur}(t_i-v_m)e^{j2\pi((f_c+f_D^m-f_{sur})t_i-f_c v_m-(t_R-t_L)f_D^m+\phi_c)} \quad (7.8)$$

where M is the number of targets (or the number of different scatterers within the same target), $u_{sur}(t_i-v_m)$ is the transmitted information carrying signal (in case of a passive system) or radar "code" (in case of an active radar), $\xi_{sur}^{m}(t_i-v_m)$ is a complex function accounting for time-varying amplitude and phase shift, both functions for the mth target.

From Figure 7.12 $v_m=t_T+t_R-t_L=(R_T+R_R-L)/c$, while f_D^m is the assumed constant Doppler shift of the mth target, and f_c is the carrier/center frequency of the broadcasted signal/channel where ϕ_c is the initial phase of the broadcasted carrier/center frequency. f_{sur} accounts for the down-conversion in the receiver, and the signal is sampled with sampling frequency f_s, yielding the sampling times $t_i=i/f_s$ for $i=0,1,2,\ldots,S-1$, where $S>0$ is the integer number of samples. For a target composed of multiple scatterers, $M>1$, the scatterers might be at the same range gate, with different Doppler shifts. This will be the case for helicopters, since the different parts of the rotors will scatter at different

[1] There exist several ways of realizing the reception, down-conversion (conditioning and adaptive), filtering and IQ conversion (if applied) of signals to baseband. The alternatives require different mathematical formulations in order to be precise, however the majority of them end up in a form like (7.8).

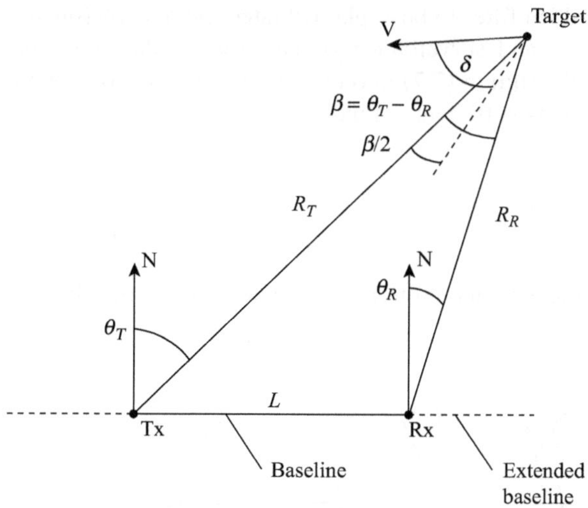

Figure 7.12 The bistatic geometry, north referenced, using the notation of Willis [14]

Doppler shifts f_D^m. By shifting u_{sur} by $t_T + t_R - t_L$, and not by $t_T + t_R$, $u_{sur}(t)$ is the transmitted signal as it would have appeared at the receiver Rx if it had traveled along the baseline shown in Figure 7.12. With this choice we do not have to introduce a similar shift in the reference signal below, essentially putting the transmitter at zero range.

In an active radar the reference signal $d(n)$, henceforth denoted U_{ref}, with which the surveillance signal will be matched in the matched filter, is the transmitted radar code (possibly filtered and shifted to the frequency band of U_{sur}). In the case of DiMuRa, U_{ref} is the actual code transmitted by the system. In a passive radar system, the reference signal is obtained either by a separate antenna aimed directly at the transmitter or by extracting the signal from the surveillance signal (implying that the direct signal is received in the sidelobe of the receiving antenna). For a passive system the reference signal is written, after reception, down-conversion (conditioning), filtering, and IQ sampling:

$$U_{ref}(t_i) = \xi_{ref}(t_i)u_r(t_i)e^{j2\pi((f_c - f_{ref})t_i + \phi_c)} \tag{7.9}$$

where $i = 0, 1, 2, \ldots, S-1$ for $S > 0$, an integer number of samples. $\xi_{ref}(t_i)$ plays the equivalent role for the direct path propagation as $\xi_{sur}(t_i)$ plays for the target reflected path, u_r is the transmitted reference signal (normally u_r and u_{sur} will be identical), and f_{ref} is the frequency down-conversion in the receiver, while the receiver is assumed to be at zero range (as mentioned above).

Olsen [15] showed how the cross-correlation (matched filter) (7.7) between the reference (7.9) and the surveillance signal (7.8) may be rearranged and written:

$$\Psi(i, f) = \sum_{s=0}^{A-1}\sum_{n=0}^{N-1} C_n U_{ref}^*(t_{n+sN})U_{sur}(t_{n+sN+i})e^{-j2\pi f t_{n+sN}} \tag{7.10}$$

where C_n is an added filter to be explained later and $S = AN$ (the total length of the recorded reference and surveillance signals used in the cross-correlation). As it stands, splitting the sum in (7.7) into two sums offers no computational advantage. However, if f or N is sufficiently small so that

$$ft_N = N\frac{f}{f_s} \ll 1 \tag{7.11}$$

(setting $t_0 = 0$), the exponential term will remain approximately constant during the sum over n:

$$\begin{aligned}\Psi(i, f) &\approx \sum_{s=0}^{A-1} e^{-j2\pi f t_{sN}} \sum_{n=0}^{N-1} C_n U^*_{ref}(t_{n+sN}) U_{sur}(t_{n+sN+i}) \\ &\equiv \sum_{s=0}^{A-1} e^{-j2\pi f t_{sN}} \hat{z}(i, s) = \sum_{s=0}^{A-1} \hat{z}(i, s) e^{-j2\pi s N f/f_s}\end{aligned} \tag{7.12}$$

Using the correlation theorem, the correlation $\hat{z}(i, s)$ may be efficiently computed by means of the FFT method, followed by an IFFT. This computational advantage is the main reason for splitting the sum into sums over n and s. Note also that the last expression in (7.12) is basically the discrete Fourier transform (DFT) of $\hat{z}(i, s)$, and hence this sum may also be efficiently evaluated using the FFT method. The value of i that maximizes $|\hat{z}(i, s)|$ will then give the target (bistatic) range, while the value of f that maximizes $|\Psi|$ gives the target radial velocity.

For a given sampling frequency f_s the choice of N dictates the maximum Doppler frequency that can be evaluated with the DFT, $\max(|f|) = f_s/(2N)$. Note that for these frequencies the requirement (7.11), needed to make the approximation in (7.12), is satisfied, although some degradation must be expected at the highest frequencies. We shall come back to this later (see (7.18) and Figure 7.13).

For large N this maximum frequency is much lower than the maximum frequency available in the signal, which from the sampling theorem is $\max(|f|) = f_s/2$. In order to capture the rotor motion of a helicopter, it is desirable to evaluate Ψ for frequencies higher than $f_s/(2N)$. In this case (7.12) cannot be used, and Ψ has to be evaluated from (7.10) using a much slower direct summation. We come back to this in section 7.3.2.4.

One could of course also increase the maximum frequency attainable in the DFT by lowering N (assuming f_s is given). However, the computational efficiency is reduced as N is lowered since the sum in $\hat{z}(i, s)$ is then short, reducing the gain of the FFT method compared with a direct summation, and for small N one is basically back to the methods outlined in sections 7.3.2.1 and 7.3.2.2.

Expressed in terms of the modeled surveillance and reference signals, (7.8) and (7.9), the cross-correlation reads

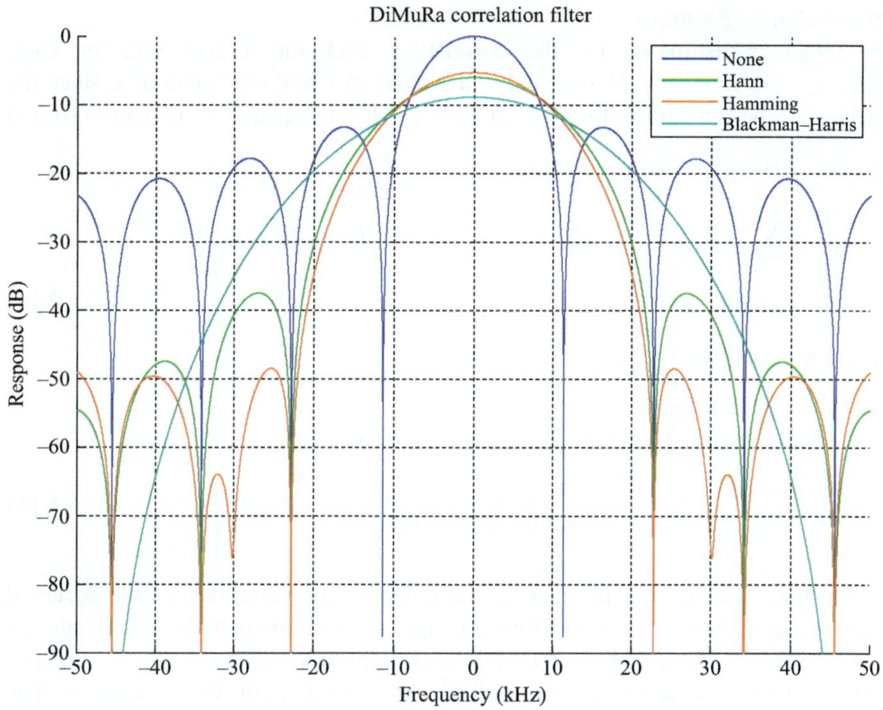

Figure 7.13 The filter effect, C_n, on the sum in (7.20), where the frequency on the horizontal axis refers to the Doppler shift, i.e., f_D^k. The DiMuRa set of parameters has been used, $f_s = 5$ MHz, $f_T = 5.7$ GHz, and requiring $v_{max} = \pm 300$ m/s, yields $f_s/N = 11\,400$ Hz, and $N = 439$

$$\hat{z}(i,s) = \sum_{m=0}^{M-1} e^{j2\pi(f_c - f_{sur} + f_D^m)t_i}\, e^{j2\pi(f_{ref} - f_{sur} + f_D^m)t_{sN}} \times \sum_{n=0}^{N-1} C_n \alpha_{ref}^*(t_{n+sN})\alpha_{sur}^m(t_{n+i+sN} - v_m) \times u_{ref}^*(t_{n+sN})u_{sur}(t_{n+i+sN} - v_m)e^{j2\pi(f_{ref} - f_{sur} + f_D^m)t_n} \tag{7.13}$$

for $i = 0, 1, 2, \ldots, N-1$ and $s = 0, 1, 2, \ldots, A-1$. $\alpha_{ref}(t)$ and $\alpha_{sur}^m(t)$ are the propagation factors ξ_{ref} and ξ_{sur} multiplied by constant phase factors (which do not matter since we eventually evaluate $|\Psi|$), and account for the propagation attenuation, multipath, initial phases and amplitudes. Full details are available in [15].

By inserting (7.13) into (7.12), a proper choice for f will minimize the oscillatory behavior caused by the f_D^m terms. However, for a target with many scatterers with different f_D^m, but at the same range gate, there is no single f that will cancel the oscillations from all the scatterers. Hence one should expect that after the correlation process the return from a target consisting of parts with different velocities will be lower than from a (non-rotating) rigid body of the same size and shape.

Target Range Estimate

The range is estimated by the correlation, and for a one reflector case, $\alpha^m_{sur}(t_{i+sN} - v_m) = 0$ for all $m \neq k$ for all values of i, and one value of s, since the situation will be comparable for all $s = 0, 1, 2, \ldots, A-1$. Equation (7.13) can be written

$$\begin{aligned}\hat{z}(i,s) &= e^{j2\pi(f_{ref}-f_{sur}+f^k_D)t_{sN}} e^{j2\pi(f_c-f_{sur}+f^k_D)t_i} \\ &\times \sum_{n=0}^{N-1} C_n \alpha^*_{ref}(t_{n+sN}) \alpha^k_{sur}(t_{n+i+sN} - v_k) u^*_{ref}(t_{n+sN}) u_{sur}(t_{n+i+sN} - v_k) \\ &\times e^{j2\pi(f_{ref}-f_{sur}+f^k_D)t_n} \end{aligned} \quad (7.14)$$

leading to the absolute value

$$\begin{aligned}|\hat{z}(i,s)| = \Bigg| &\sum_{n=0}^{N-1} C_n \alpha^*_{ref}(t_{n+sN}) \alpha^k_{sur}(t_{n+i+sN} - v_k) \\ &\times u^*_{ref}(t_{n+sN}) u_{sur}(t_{n+i+sN} - v_k)\, e^{j2\pi(f_{ref}-f_{sur}+f^k_D)\frac{n}{f_s}} \Bigg| \end{aligned} \quad (7.15)$$

Further, assume that there is no time-dependent variations in the received signal as a consequence of anomalous propagation, i.e., one path direct propagation from transmitter to receiver, and no phase or amplitude variations due to inhomogeneous medias, in which case α_{ref} and α^k_{sur} are assumed constants with time. The range to the target is found by maximizing the sum over the product of the two functions $u^*_{ref}(t_{n+sN})$ and $u_{sur}(t_{n+i+sN}-v_k)$ in (7.15) with respect to the index i.

A general result [16, p. 143] of the autocorrelation function is that it attains its maximum value at the origin. The product of the two functions $u^*_{ref}(t_{n+sN})$ and $u_{sur}(t_{n+i+sN}-v_k)$ can be thought of as time-shifted versions of each other, and thus the cross-correlation between them will be a time-shifted version of the autocorrelation, and will attain its maximum value at $t = v_k$ instead of at the origin. We assume $f^k_D = 0$ and $f_{ref} = f_{sur}$, and solve

$$\left| \sum_{n=0}^{N-1} u^*_{ref}(t_{n+sN}) u_{sur}(t_{n+i+sN} - v_k) \right|_{\max} = \omega_{\max}(s) \quad (7.16)$$

for i, which yields $i = [v_k f_s]$, where $[x]$ means the integer value of x. The error we make by the assumption $f^k_D = 0$ – or, equivalently, by using (7.12) instead of (7.10) – might be illustrated by the ambiguity function for the information signal u_{ref}, and depending on the waveform, it might be concluded that (7.16) also will hold for $f^k_D \neq 0$ and f^k_D of realistic target Doppler velocities. For passive radar, this is the case as seen in [17–24]. For the DiMuRa measurements shown in sections 7.3.3.2 and 7.3.4 it holds for the helicopter fuselage but not for the rotors. The correlation is now

$$|\hat{z}([v_k f_s], s)| \approx |\alpha^*_{ref}||\alpha^k_{sur}| \left| \sum_{n=0}^{N-1} C_n u^*_{ref}(t_{n+sN}) u_{sur}(t_{n+sN})\, e^{j2\pi(f_{ref}-f_{sur}+f^k_D)\frac{n}{f_s}} \right| \quad (7.17)$$

The informative exercise of choosing $C_n = 1$ and $u^*_{ref}(t_{n+sN})u^k_{sur}(t_{n+sN}) = 1$ for $n = 0, 1, 2, \ldots, N-1$ and $s = 0, 1, 2, \ldots, A-1$ results in the sum of (7.17) reducing to a geometrical sum which can be found in mathematical tables [25, 26] to be

$$\sum_{n=0}^{N-1}\left(e^{j2\pi(f_{ref}-f_{sur}+f_D^k)/f_s}\right)^n = \begin{cases} N & f = 0, \pm1, \pm2, \ldots \\ \dfrac{\sin N\pi f}{\sin \pi f}e^{j(N-1)\pi f} & f \neq 0, \pm1, \pm2, \ldots \end{cases} \quad (7.18)$$

for $f = (f_{ref} - f_{sur} + f_D^k)/f_s$. The above sum's zeros are given by

$$(f_{ref} - f_{sur} + f_D^k) = \frac{f_s}{N}p \quad p = 0, \pm1, \pm2, \ldots \quad \text{and} \quad p \neq mN, m = 0, \pm1, \pm2, \ldots \quad (7.19)$$

This exercise shows that the correlation between the reference and surveillance signal will be degraded by the target's Doppler shift, and even nulls may appear, which might to some extent be filtered away as shown in Figure 7.13.

The figure shows the effect of applying various filters, C_n for $n = 0,1,2,\ldots, N-1$, in the sum

$$\sum_{n=0}^{N-1} C_n\left(e^{j2\pi(f_{ref}-f_{sur}+f_D^k)/f_s}\right)^n \quad (7.20)$$

as a function of target Doppler shift f_D^k, for no filter, Hamming, Hann and Blackman–Harris [27]. This is in principle the situation of $\alpha^*_{ref}(t_{n+sN})\alpha^k_{sur}(t_{n+i+sN} - v_k) = 1$ and $u^*_{ref}(t_{n+sN})u_{sur}(t_{n+sN}) = 1$, $i = [v_k f_s]$, and $f_{ref} = f_{sur}$ of (7.17). As a DiMuRa example, the sampling frequency was around $f_s = 5$ MHz, the carrier frequency of the transmitted signal was $f_c = 5.7$ GHz, and it was required that maximum Doppler shift should be no higher than a velocity of $v_{max} = \pm300$ m/s with Doppler span $\Delta v = 600$ m/s, resulting in (the span PRF) $f_s/N = 11\,400$ Hz, and $N = 439$.

By carefully choosing the values of f_{ref} and f_{sur}, the response of the correlation may be put at a frequency of choice. That is, it might be handy to choose $f_{ref} - f_{sur} = 5\,700$ Hz which would place the correlation response window at $\pm5\,700$ Hz in Figure 7.13 for the DiMuRa example, in contrast to $f_{ref} = f_{sur}$, which means the window is ranging from 0 to 11 400 Hz, where the first null appears without filtering.

The fraction N/f_s is the coherent correlation interval for *one correlation* vector. The interval is determined by the parameter N, which determines the length of one row in the matrix to be N samples, which also might be considered to be the PRF = N/f_s of the processing, which also will determine the unambiguous range to be $R_{unamb} = c/\text{PRF} = cN/f_s$. In the correlation $2N$ samples are processed, which is a result of the fact that one correlated range gate is the result of a two channel correlation, each of N samples. The reference channel is defined in such a way that N filtered samples from the reference channel is followed by N trailing zeros, correlated against $2N$ samples from the surveillance channel, resulting in an N sample unambiguous range. The trailing zeros ensure that the N sample range is not interfered by the cyclic FFT/IFFT methods used to perform the correlation.

Target Doppler Estimate

In order to extract the target Doppler shift, $\alpha_{sur}^m(t_i - \nu_m) = 0$ for all $m \neq k$, (7.13) is recast into

$$\hat{z}(i,s) = e^{j2\pi(f_c - f_{sur} + f_D^k)t_i}\left[\sum_{n=0}^{N-1} C_n \alpha_{ref}^*(t_{n+sN})\alpha_{sur}^k(t_{n+i+sN} - \nu_k) \times u_{ref}^*(t_{n+sN})u_{sur}(t_{n+i+sN} - \nu_k)\, e^{j2\pi(f_{ref} - f_{sur} + f_D^k)t_n}\right] e^{j2\pi(f_{ref} - f_{sur} + f_D^k)\frac{N}{f_s}s} \quad (7.21)$$

for $i = 0, 1, 2, \ldots, 2, N - 1$ and $s = 0, 1, 2, \ldots, A - 1$, which in turn may be recast into

$$\hat{z}(i,s) = K(i,s,k)e^{j2\pi(f_{ref} - f_{sur} + f_D^k)\frac{N}{f_s}s} \quad (7.22)$$

where

$$K(i,s,k) = e^{j2\pi(f_c - f_{sur} + f_D^k)t_i}\left[\sum_{n=0}^{N-1} C_n \alpha_{ref}^*(t_{n+sN})\alpha_{sur}^k(t_{n+i+sN} - \nu_k) \times u_{ref}^*(t_{n+sN})u_{sur}(t_{n+i+sN} - \nu_k)e^{j2\pi(f_{ref} - f_{sur} + f_D^k)t_n}\right] \quad (7.23)$$

for $i = 0, 1, 2, \ldots, N - 1$ and $s = 0, 1, 2, \ldots, A - 1$. $K(i, s, k)$ depends on the correlation properties of $\alpha_{ref}(t_i)u_{ref}(t_i)$ and $\alpha_{sur}(t_i)u_{sur}(t_i)$ for $i = 0, 1, 2 \ldots, N - 1$, as well as the start time of the processing t_0. i.e., when it is t_0. For each range correlation, i.e., fixed value of s and varying values of i, the variation of $K(i, s, k)$ can be interpreted as the target's response in the correlation process described in the previous section, modified by the correlation properties of $u_{ref}(t_i)$, as well as the propagation contribution of $\alpha_{ref}(t_i)$ and $\alpha_{sur}(t_i)$. No time-varying electromagnetic conditions, and the fact that $i = [\nu_k f_s]$ is considered a constant parameter (one range bin is considered), show that the only time-varying component in (7.23) is $u_{ref}^*(t)u_{sur}(t)$. In the case of a periodic $u_{ref}(t)$, the PRF should be chosen PRF $= N/f_s$, in order to be equal to the periodicity of $u_{ref}(t)$. In that case $K(i, s, k)$ is constant over $s = 0, 1, 2, \ldots, A - 1$ for fixed values of i and k.

This was the case for DiMuRa [5, 6, 8, 10, 11], as discussed in section 7.3.2.4. For the general case, it will be assumed that the variations of $u_{ref}(t)$ over the chosen period is of such a character that $K(i,s,k)$ is relatively stable over $s = 0, 1, 2, \ldots, A - 1$. Equation (7.22) clearly shows the need to compensate $\exp(j2\pi(f_{ref} - f_{sur} + f_D^k)sN/f_s)$ in (7.22) with a term

$$\exp(-j2\pi(f_{ref} - f_{sur})sN/f_s)$$

which is readily implemented by a vector multiplication for all values of $s = 0, 1, 2, \ldots, A-1$.

The target Doppler frequency is now found through a FFT of (7.12) for each of the ranges $i = 0, 1, 2, \ldots, N-1$. It is important to keep in mind that now the index i corresponds to one range bin, and $s = 0, 1, 2, \ldots, A-1$ corresponds to all the Doppler bins, i.e., looking at all Doppler velocities at one range bin. This argument can obviously be made for all the i range bins, and thus all combinations of range and Doppler velocities are found, i.e., for all the N ranges of interest.

In order to lower the sidelobes of the FFT in the Doppler direction, a traditional window function/filter can be applied, as seen in Figure 7.13 for the range correlation as a function of target Doppler shift. Suitable filters are well described and the behavior is documented in the well-known work of Harris [27]. Mathematically, the filter will be denoted as D_s, and is written (7.22)

$$\hat{z}(i,s) = K(i,s,k) D_s \, e^{j2\pi (f_{ref} - f_{sur} + f_D^k) \frac{N}{f_s} s} e^{-j2\pi f_p \frac{N}{f_s} s} \tag{7.24}$$

for $s = 0, 1, 2, \ldots, A-1$ for all values of i, where the term $\exp(-j2\pi f_p sN/f_s)$ compensates for the term $\exp(j2\pi(f_{ref} - f_{sur})sN/f_s)$ of choice; normally this would be $f_p = f_{ref} - f_{sur}$.

7.3.2.4 DiMuRa Helicopter Rotor Processing

We now discuss the specific signal processing requirements for detecting rotor flashes with DiMuRa. We use the requirements for the monostatic results presented in section 7.3.3 as an example.

In this case the received surveillance signal from either channel was sampled at a sampling frequency $f_s = 4.69$ MHz. The transmitters emitted a continuously repeated signal $u_r(j)$, $j = 1 \ldots N_c$ (again down-converted and sampled at the frequency f_s), with a code length $N_c = 456$. The transmitted signal is a binary code, $u_r(j) = \pm 1$, with optimal autocorrelation properties.

Because of the periodicity of the reference signal, we should choose $N = N_c$ in the cross-correlation Ψ, as this simplifies the sum:

$$\Psi(i,f) = \sum_{s=0}^{A-1} \sum_{n=0}^{N-1} C_n U_{ref}^*(t_n) U_{sur}(t_{n+sN+i}) e^{-j2\pi f t_{n+sN}} \tag{7.25}$$

With the chosen parameters, the maximum frequency is $f_s/(2N) = 5.1$ kHz, provided we restrict ourselves to compute (7.10) using the FFT technique outlined above. This translates into a maximum Doppler speed $v_D = 135$ m/s. From section 7.3.3.1, the helicopter has a never exceed speed of 60 m/s, so this choice for $N = N_c$ is sufficient for the fast signal processing, using the approximation (7.12), to capture the motion of the helicopter fuselage. It is, however, not sufficient to capture the rotor motion, for which we want to compute Ψ for Doppler speeds up to about 300 m/s. In principle, we may compute (7.10) by a slow, direct summation for the high frequencies f required. However, $\Psi(i,f)$ is to be evaluated not only at all frequencies of interest, but also at all ranges, and a direct summation then becomes prohibitively time consuming. To avoid this, we take advantage of the fact that the

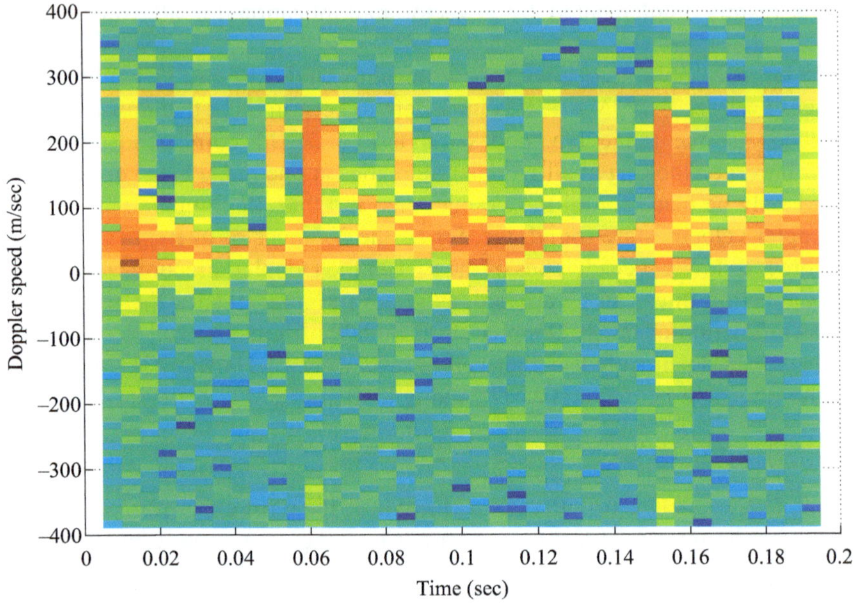

Figure 7.14 The cross-correlation $20\log_{10}|\Psi|$ processed with $A = 50$ for the helicopter at a bistatic range of 1.25 km

helicopter fuselage in general gives a much larger radar echo than the rotor blades, except at the times of rotor flashes. We therefore first do the fast range-Doppler processing, using (7.12) and FFTs, to find the range and radial velocity of the helicopter fuselage. Denoting the range found by range index i_t, we then recompute $\Psi(i_t, f)$ using the full expression (7.10) for the large range of frequencies needed to capture rotor motion. Since the summation is now only carried out at one range, $i = i_t$, we can afford to carry out the direct summation. (The chosen f_s leads to a range cell size of about 30, so the helicopter should only appear in one or two range cells.)

Although the transmitters operated continuously, of every second only 0.2 s of continuously sampled data were saved to disk, and this is obviously the maximum length of received data that can be coherently processed. Hence, we are not able to monitor rotor flashes for more than about one full main rotor rotation. The 0.2 s sample length corresponds to $A \leq 2\,055$. To be able to detect rotor flashes, A has to be chosen much shorter than this upper limit. For instance, the results presented in figure 7.14 were processed with $A = 50$, which allows for about 40 processing intervals within the 0.2 s.

It can be instructive to see how the periodicity of the emitted signal affects the correlation Ψ. Neglecting constant phase factors (which are irrelevant since we eventually compute the magnitude $|\Psi|$ of the cross-correlation), the surveillance signal from a point target located at range index i_t and with a Doppler shift f_t is

$$U_{sur}(t_{n+sN}) = \alpha_{sur} u_{ref}(t_{n-i_t})\, e^{j2\pi(n+sN)f_t/f_s} \tag{7.26}$$

where $u_{ref}(t_n)$ is the reference signal at the receiver location at time $t_n = n/f_s$ and we assume $f_{sur} = f_c$. As previously, α_{sur} accounts for attenuation and phase shift of the reference signal as it propagates to the target and to the receiver. We assume α_{sur} is constant within the 0.2 s sample length. The correlation (7.10) then becomes

$$\begin{aligned}\Psi(i_t, f) &\equiv \sum_{s=0}^{A-1}\sum_{n=0}^{N-1} C_n U_{ref}^*(t_n) U_{sur}(t_{n+sN+i_t}) e^{-j2\pi(n+sN)f/f_s} \\ &= \alpha_{sur}\left(\sum_{s=0}^{A-1} e^{-j2\pi sN(f-f_t)/f_s}\right)\sum_{n=0}^{N-1} C_n |u_{ref}(n/f_s)|^2 e^{-j2\pi n(f-f_t)/f_s} \\ &= \alpha_{sur}\left(\sum_{s=0}^{A-1} e^{-j2\pi sN(f-f_t)/f_s}\right)\sum_{n=0}^{N-1} C_n e^{-j2\pi n(f-f_t)/f_s} \end{aligned} \tag{7.27}$$

where we have seen that the DiMuRa radar emits a binary code with constant amplitude, $|u_{ref}(t_i)| = 1$. We recognize the first sum as the same geometric series encountered in (7.18):

$$\sum_{s=0}^{A-1} e^{-j2\pi sN(f-f_t)/f_s} = \begin{cases} A & \text{if } N(f-f_t)/f_s \text{ is an integer} \\ \dfrac{1-e^{-j2\pi AN(f-f_t)/f_s}}{1-e^{-j2\pi N(f-f_t)/f_s}} & \text{otherwise} \end{cases} \tag{7.28}$$

Hence we must expect that $|\Psi|$ will have maxima whenever $f - f_t$ is an integer times the PRF f_s/N. In addition to the true maximum at $f = f_t$, the first spurious maxima appear at $f = f_t \pm f_s/N$.

However, if we do not apply a windowing function, setting $C_n = 1$, the second sum in (7.27) also becomes a geometric series:

$$\begin{aligned}\sum_{n=0}^{N-1} e^{-j2\pi(f-f_t)/f_s n} &= \frac{1-e^{-j2\pi N(f-f_t)/f_s}}{1-e^{-j2\pi(f-f_t)/f_s}} \\ &= \begin{cases} 0 & \text{if } N(f-f_t)/f_s \text{ is a nonzero integer and} \\ & \quad (f-f_t)/f_s \text{ is not an integer} \\ N & \textit{if } f = f_t \textit{ or } (f-f_t)/f_s \text{ is an integer} \end{cases}\end{aligned} \tag{7.29}$$

In other words, if no windowing is used, a target should lead to a maximum in the cross-correlation $|\Psi|$ only at the true target velocity, while a non-constant windowing function C_n in general causes a single target to produce several spikes in $|\Psi|$, offset in velocity by

$$v_D = \frac{f_s}{2Nf_c} c \approx 270 \text{ m/s} \tag{7.30}$$

Hence filtration of the reference signal may cause the (weak) rotor flashes to become masked by "echoes" of, e.g., the signal from the fuselage. So the best strategy in this case is to employ no windowing function at all, $C_n = 1$. This is a consequence of the reference signal being constant in amplitude, only varying in phase. For an amplitude-modulated reference signal windowing may still be advantageous.

In hindsight we note that using a shorter code (smaller N) would have been preferrable as this would have increased the PRF f_s/N beyond the Doppler frequencies expected from rotor motion. We would thus avoid contamination from these spurious contributions.

The received signal will contain contributions from stationary clutter and the direct signal from the transmitter. Because of the periodicity of the emitted signal, this contamination will appear not only at zero Doppler, but also at multiples of the PRF f_s/N. Since f_s/N corresponds to a velocity $v_D \approx 270$ m/s, while the tip of the main rotor has a speed relative to the helicopter $v_m \approx 250$ m/s (section 7.3.3.1), the contributions from clutter and the direct signal will appear in the cross-correlation at typical helicopter rotor speeds. These signals are suppressed in the correlation function Ψ because they appear at a different range, but the signals themselves could be much stronger than, e.g., the weak echoes from the rotors, so the suppression from the processing will not be sufficient to mask them. The spurious signals will be quite apparent in the results shown in section 7.3.3.2, however, and we therefore make no attempt to remove them from the received signal.

7.3.3 Quasi-monostatic Signatures

7.3.3.1 Helicopter Parameters

We shall show results from DiMuRa measurements of the radar signature from a helicopter with twin blade main and tail rotors. The helicopter has a "never exceed speed" of about 60 m/s. The main rotor has a diameter of 14.6 m, nominally rotating at 324 rpm. Hence, the tip of the main rotor has a speed (in the helicopter rest frame) of

$$v_m = 248 \text{ m/s} \tag{7.31}$$

The tail rotor has a diameter of 2.6 m, rotating at 1 660 rpm, so that the tip of the tail rotor has a speed

$$v_t = 228 \text{ m/s} \tag{7.32}$$

With the helicopter flying at near the maximum speed, the radar signal processing thus needs to account for rotor speeds up to 310 m/s.

The ratio of the tail to main rotor rotation speed is $1660/324 \approx 5.1$. As we show later, we usually detect a flash only from the rotor blade moving toward the radar (both for the main and tail rotors). Hence, we would expect to see about five tail rotor flashes for every flash of the main rotor.

7.3.3.2 Results

For the monostatic results shown here, horizontal polarization was used for DiMuRa, both for the emitted signal and at reception.

Figure 7.14 shows typical results when both main and tail rotor flashes are seen for one data set (covering 0.2 s). As discussed in section 7.3.2.4, we have first processed this data set using the efficient algorithm presented in section 7.3.2.3 to find the range cell of the helicopter. Then the data are reprocessed for a much wider range of velocities, but only at this given range, to find the rotor flashes. In general, main rotor flashes could be seen out to distances of about 5 km, while the tail rotor could only be seen at close range (1–2 km). The helicopter fuselage appears as a wide band at approximately 30–50 m/s, with the sign convention being that positive velocities correspond to the helicopter approaching the radar site. The main rotor is seen as two flashes appearing at approximately 0.06 and 0.16 s. Note that the rotor blade moving toward the radar yields a much stronger signal than the receding rotor blade in agreement with the calculations in section 7.2.2. The 0.1 s time delay between the two flashes agrees well with the 0.093 s expected from the 324 rpm rotation speed. In this example, the maximum rotor speed seen in $|\Psi|$ is about 250 m/s, while the highest speed of the rotor tip (including the helicopter motion) would be about 280–300 m/s, indicating that the tip of the rotor blade is not visible in this particular example.

The tail rotor is seen as the regularly spaced flashes appearing between the main rotor flashes in Figure 7.14. We note that there are about five tail rotor flashes for every main rotor flash in accordance with the rotation speed of the main and tail rotors. From section 7.3.3.1 we expect the tip of the tail rotor to move at almost the same speed as the main rotor tip. The tail rotor flashes extend up to about 270 m/s. With $v_t = 228$ m/s (section 7.3.3.1) and a helicopter speed of 50 m/s, the maximum tail rotor speed should be about 280 m/s, so the results indicate that in this example the whole rotor blade contributes to the flash.

At approximately the mid-time between main rotor flashes, a pronounced signal is seen with speeds in the range 0–100 m/s. Similar signals are seen in many data sets. The range of speeds indicate that this return signal may come from areas near the hub of the main rotor.

In addition to these "real" signals, Figure 7.14 also shows a sharp enhancement at approximately 270 m/s, which we attribute to clutter and/or the direct signal, as discussed in section 7.3.2.4.

Figures 7.15 and 7.16 show the same data set as shown in Figure 7.14, but now processed using $A = 20$ and $A = 100$, respectively. By using a shorter integration time we obtain a finer time resolution at the expense of a coarser Doppler resolution and a smaller gain in the Doppler processing. We note from Figure 7.15 that both the main and tail rotor flashes have become much narrower in time, indicating that these are indeed short duration events. At the same time the main radar echo from the helicopter is smeared out over a wide range in velocities, indicating that the processing gain has been reduced with a shorter integration time. This effect is particularly striking when Figures 7.15 and 7.16 are compared, where we note that a long integration time, $A = 100$, leads to a much better separation of the fuselage from the background noise.

On the other hand, a long integration time is not advantageous for identifying rotor flashes. The flashes are of such short duration that nothing is gained by a long integration time, which instead causes the flashes to become smeared out in time.

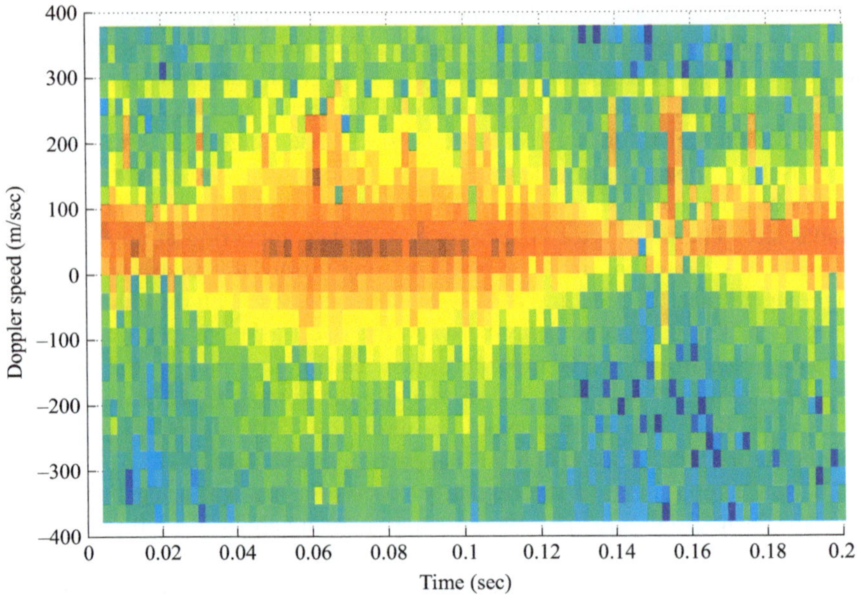

Figure 7.15 The same data as in Figure 7.14, but processed with A = 20

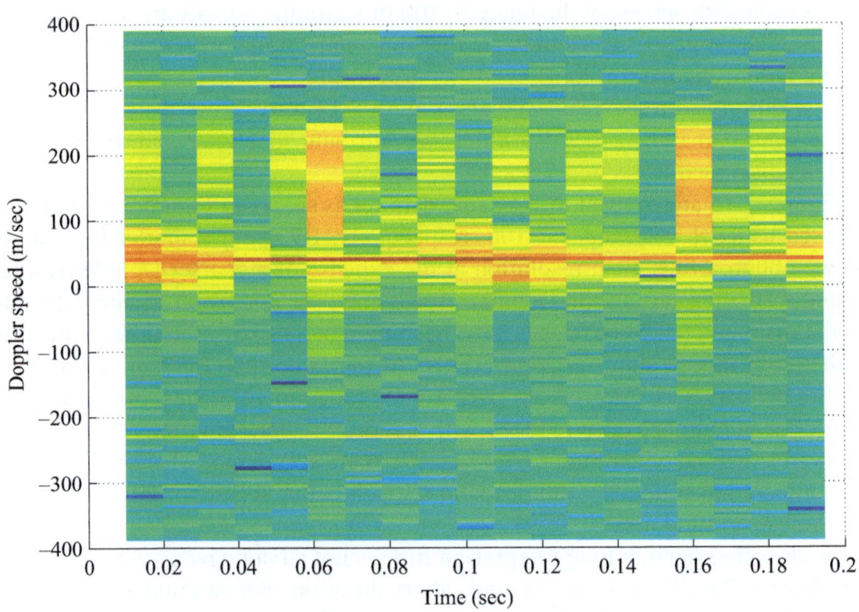

Figure 7.16 The same data as in Figure 7.14, but processed with A = 100

7.3.4 Bistatic Signatures

In bistatic or multistatic radar, the transmitter (Tx) and receiver (Rx) are separated as shown in Figure 7.12. Each Tx–Rx pair and helicopter position define a bistatic plane where the bistatic angle β of this pair is the angle between the vectors from the target to the receiver and transmitter. The bistatic bisector $\beta/2$ is defined along the half angle of the bistatic angle as shown in Figure 7.12.

In the simplest case the rotor rotates in the bistatic plane. This could be a good approximation for a hovering helicopter just above horizon in a relative flat landscape. For this geometry, a reflected Doppler-shifted signal originating from a flash of the main rotor blades appears when the rotor is perpendicular to a line along $\beta/2$. The velocity vector of the rotor is then parallel to $\beta/2$, which results in a maximum possible Doppler shift for this given bistatic configuration characterized by β. The bistatic Doppler shift of the signal can be written as

$$f_d = \left[\frac{dR_{Tx}}{dt} + \frac{dR_{Rx}}{dt}\right]/\lambda \tag{7.33}$$

where $\frac{dR_{Tx}}{dt}$ and $\frac{dR_{Rx}}{dt}$ are the velocity components toward Tx and Rx and λ is the wavelength of the transmitted radar signal. This equation can be written as

$$f_d = \frac{2V}{\lambda}\cos(\delta)\cos(\beta/2) \tag{7.34}$$

where V is the velocity projected onto the bistatic plane and δ is the angle separating this projected velocity vector and the bistatic bisector $\beta/2$ (see Figure 7.12).

If the rotor rotates out of the bistatic plane, a flash could be observed when the blade velocity component V has a lower value than when rotating in the bistatic plane. In bistatic radar, we can also form geometries where $\delta > 0$. According to (7.34) changes in V, δ, and β as a function of geometry result in different Doppler shifts. As we know the positions of radar Tx and Rx, and assuming the rotor rotation plane, different helicopter positions would result in a range of values for f_d.

To measure rotor blade flashes a series of measurements were recorded by use of bistatic radar at different geometries. This simulates one realization of multistatic radar with several Tx–Rx pairs. Vertical polarization was used for both transmitted and received signals.

A Eurocopter AS350 B2 helicopter, as shown in Figure 7.17, was used in the campaign. The main rotor of this helicopter has a diameter of $d = 10.69$ m and a rotation speed of $\omega_r \approx 400$ rpm, resulting in a maximum velocity of the rotor tip of $v_{max} = 224$ m/s. Generally, the rotor rarely rotates in the bistatic plane – there is an angle between the rotor rotation plane and the bistatic plane. This maximum velocity will then typically not be observed by radar since the projected rotor velocity toward the receiver and transmitter will be smaller due to bistatic configuration ($\beta > 0$) and rotation out of the bistatic plane.

The number of rotor blades for our test helicopter was $N_{main} = 3$ (main rotor) and $N_{tail} = 2$ (tail rotor). Flashes from rotor blades with positive and negative velocities will therefore alternate in time for the main rotor, whereas they will occur

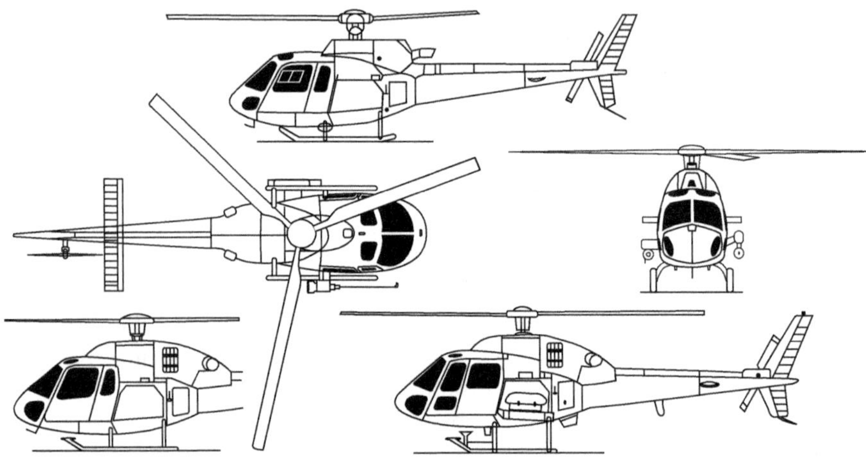

Figure 7.17 Helicopter of type Eurocopter AS350 B2 in different view angles. © 2003 IEEE. Reprinted, with permission, from IEEE 2003 Radar Conference Proceedings

simultaneously for the tail rotor. The period T between flashes for odd and even number bladed rotors is given by (see also section 7.2.1)

$$T_{\text{odd}} = \frac{60}{2N\omega_r}, \ T_{\text{even}} = \frac{60}{N\omega_r}$$

In Figures 7.18 and 7.19 typical results for a hovering helicopter are shown in quasi-monostatic and bistatic geometry, respectively. This processing can be used as a helicopter detector by observing the characteristic spreading in frequency from the rotor blades.

From Figure 7.18 it is evident that the monostatic configuration results in substantial clutter returns that bury the helicopter body. We observe that in the bistatic geometry of Figure 7.19, the clutter returns are significantly reduced and the body of the helicopter can be observed. With smaller common volume of the Tx and Rx beams, which is the case in a bistatic configuration, a reduced number of scatterers contribute to the clutter.

The velocity span is above ±200 m/s, which corresponds to calculated projected velocities of 217 and 207 m/s for the main rotor in the two geometries. In Figure 7.18, the upper blade flash is hardly visible. This is in contrast to the monostatic measurements presented in section 7.3.3.2, where the advancing blade gave a stronger return than the receding blade. It also disagrees with the calculations of section 7.2.2 which predict that the rear edge should not give a stronger return than the front edge (at low frequencies the two should be quite similar). In the bistatic geometry shown in Figure 7.19 both blade edges contribute significantly.

If instead, recorded data sets are presented as a function of time – by reducing the integration time A in (7.10) and thus computing Ψ for different time slices within the 300 ms data frame – additional quantitative information can be extracted.

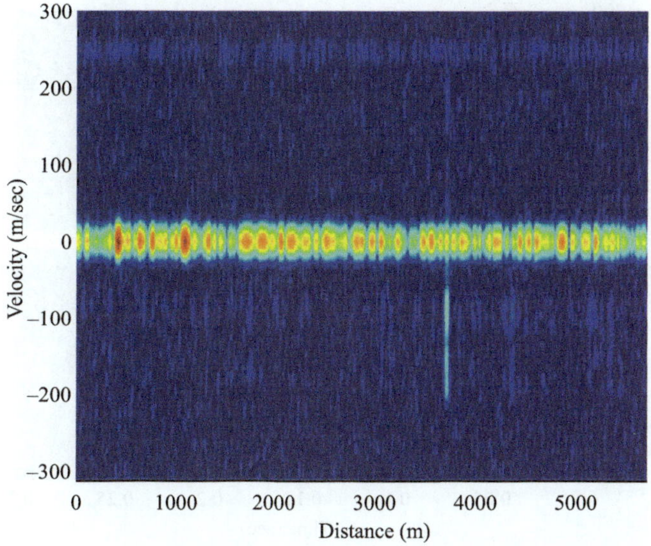

Figure 7.18 Quasi-monostatic measurement of a hovering helicopter integrated over 300 ms, showing the cross-correlation $20\log_{10}|\Psi|$ as a function of Doppler speed and range. © 2003 IEEE. Reprinted, with permission, from IEEE 2003 Radar Conference Proceedings

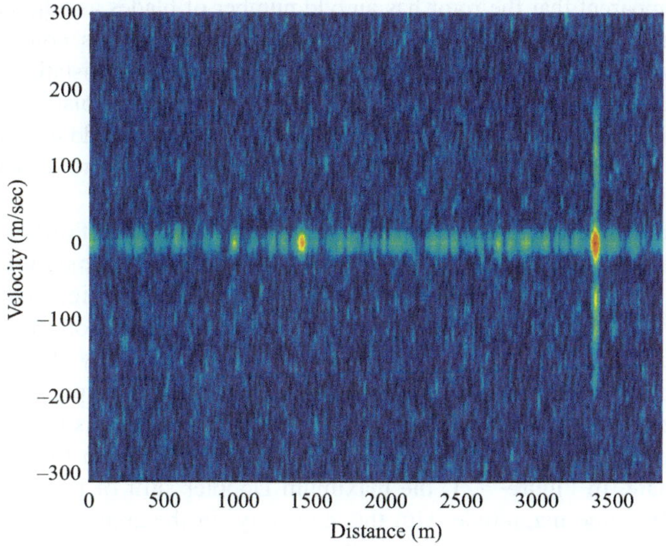

Figure 7.19 Bistatic measurement of a hovering helicopter with $\beta = 33°$ and integration time 300 ms. © 2003 IEEE. Reprinted, with permission, from IEEE 2003 Radar Conference Proceedings

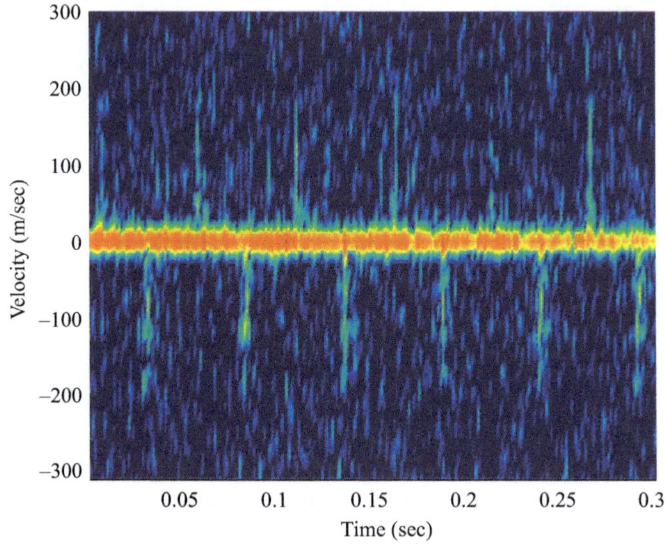

Figure 7.20 Hovering helicopter at $\beta = 33°$ plotted as a function of time for the range gate containing the helicopter. © 2003 IEEE. Reprinted, with permission, from IEEE 2003 Radar Conference Proceedings

In Figure 7.20 a time plot of the data used in Figure 7.19 shows each individual blade flash of the main rotor as the continuous wave radar continuously illuminates the target. It is apparent that the rotor has an odd number of blades as adjoined flashes have velocities of opposite sign. The period T between flashes could be easily extracted. In a measurement, the rotor velocity component in the bistatic plane could be estimated from measured helicopter track and knowledge of bistatic radar geometry. This information together with extraction of flash period, odd/even number of blades, and measured blade tip velocity could be used as input to a classification algorithm.

Systematic measurements have been carried out for a range of different bistatic angles and helicopter geometries. In Figures 7.21 and 7.22 two measurements are shown with approximately head and side view helicopter orientation and $\beta = 110°$. The flashes of the main rotor blades are marked with a circle in both figures. In the head orientation only the approaching blade flashes are clearly distinguishable, while both blade edges can be extracted in the side view. In between these flashes, the tail rotor contributes with flashes that appear symmetrically as there is an even number of rotor blades in the head view data. The flashes from the approaching blade dominate. In Figure 7.21, the maximum Doppler shift of the main and tail rotor is of the same magnitude with this geometry. In the approximate side view there seems to be some systematic contribution to the "pattern" that can be related to reflections of rotors and parts of the helicopter for the time period between flashes, i.e., when the rotors form two-bounce reflections. The cause of these contributions has not been investigated but they might add a property that can be

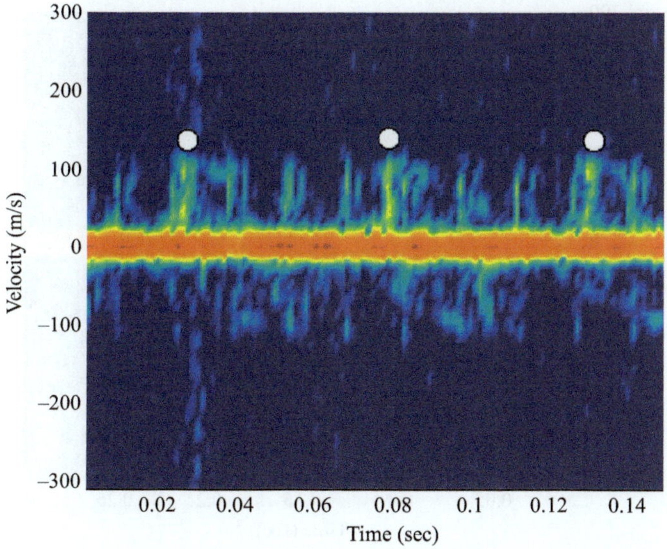

Figure 7.21 *Bistatic measurement at $\beta = 110°$ with an approximate head view of the helicopter. Main rotor flashes are marked with a circle. © 2003 IEEE. Reprinted, with permission, from IEEE 2003 Radar Conference Proceedings*

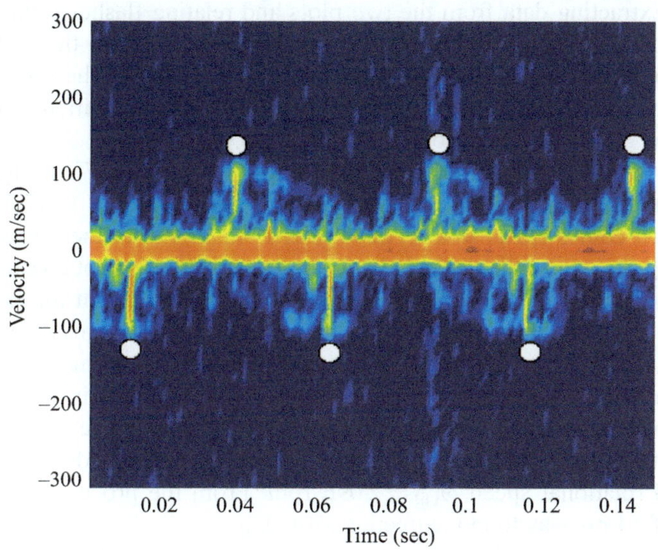

Figure 7.22 *Bistatic measurement at $\beta = 110°$ with an approximate side view of the helicopter. Main rotor flashes are marked with a circle. © 2003 IEEE. Reprinted, with permission, from IEEE 2003 Radar Conference Proceedings*

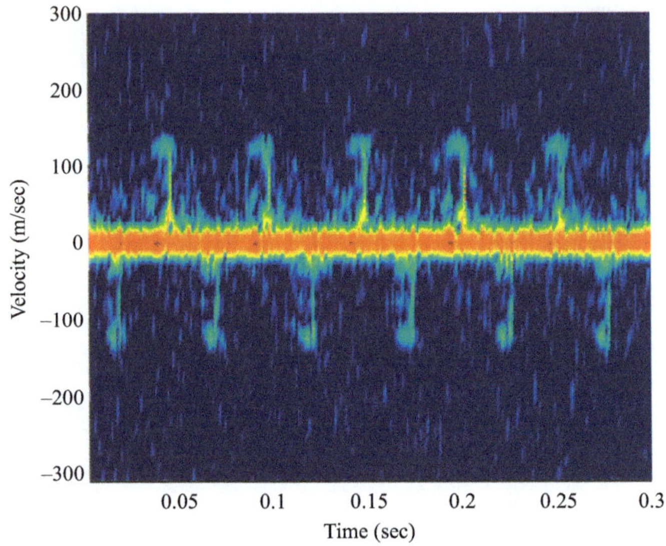

Figure 7.23 Bistatic measurement at $\beta = 86°$ with head view of helicopter. © 2003 IEEE. Reprinted, with permission, from IEEE 2003 Radar Conference Proceedings

exploited. Extracting data from the two plots and relating flashes to main and tail rotor should be possible. Multiple reflections could complicate the picture somewhat as shown for tail flashes in Figures 7.21 and 7.23 where there seems to be a double flash from the outer section of the main rotor blades shifted in time (rotation). Figure 7.21 also shows double flashes from the tail rotor.

As the bistatic angle is increased to $\beta = 86°$, 127°, and 137° in Figures 7.23–7.25, the Doppler shift of the main rotor decreases according to (7.34). This is expected as it comes closer to a forward scatter situation. For these measurements of increasing β in steps, the distance between Rx and Tx was kept constant at 2 km, resulting in a shorter range as β increased. It is also observed that for the latter two geometries, only the approaching rotor blade contributes, whereas the flash from the backside of the rotor blade is not detected. For these geometries the flashes from the tail rotor blades dominate. We observe differences in intensity of the receding and approaching blade for the two geometries, but extraction of flash characteristics for the tail rotor is possible. The tail rotor has a diameter of $L_{tail} = 1.86$ m and rotational speed $\omega_{tail} = 2\,096$ rpm. From the processed data a blade flash rate of 14 ms was found, agreeing with ω_{tail}.

The image in Figure 7.26 shows details of the tail rotor. It is evident that the rotor flash would be partly shadowed for given geometries. In Figure 7.24 the lower part of the spectrum is more visible, compared to Figure 7.25, which might be a result of shadowing. Even though the tail rotor dominates in the high β measurements, characteristics of the main rotor could still be extracted.

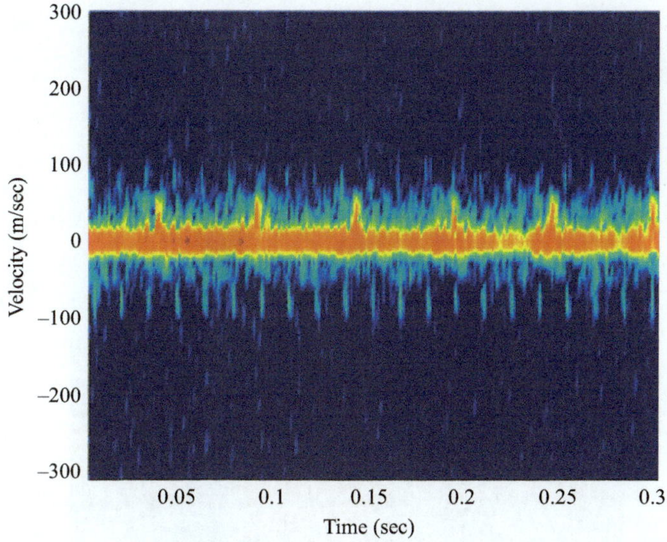

Figure 7.24 Bistatic measurement at $\beta = 127°$ with head view of helicopter. *© 2003 IEEE. Reprinted, with permission, from IEEE 2003 Radar Conference Proceedings*

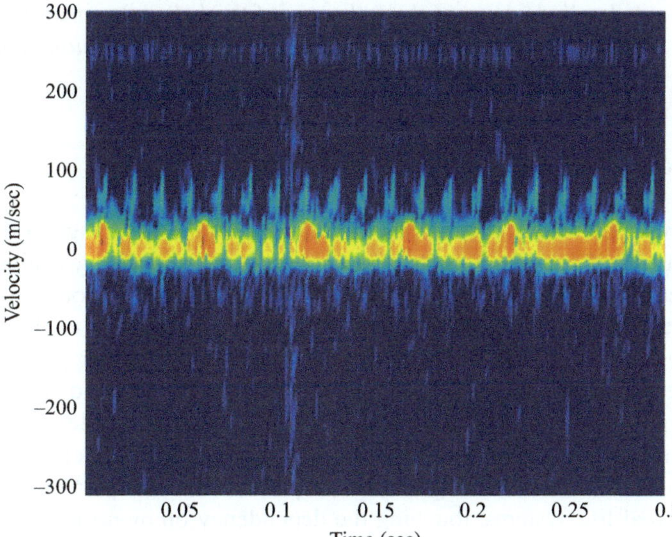

Figure 7.25 Bistatic measurement at $\beta = 137°$ with head view of helicopter. *© 2003 IEEE. Reprinted, with permission, from IEEE 2003 Radar Conference Proceedings*

Figure 7.26 A detailed image of the tail rotor of the helicopter drawn in Figure 7.17. © 2003 IEEE. Reprinted, with permission, from IEEE 2003 Radar Conference Proceedings

7.4 Conclusions

The work presented in this chapter has shown how to take advantage of the micro-Doppler features of the rotary parts of a helicopter. Traditionally, analysis of the amplitude variations of the target echo returns have been used in order to separate fixed wing and rotary wing aircraft. Modern radar systems are exploiting more signal processing, thus affording the extra costs of the Doppler processing. By applying the so-called short-time Fourier transform (STFT) to a data set, we can extract both amplitude variations (as a function of time) and Doppler contents from the signal, and thus more information will be available for target classification.

The work has been focusing on the properties of helicopter micro-Doppler results from real-life systems, touching the dependency on monostatic and bistatic systems, frequency, geometry, and waveform. RCS modeling results from generic helicopter blades have shown the effects of frequency choice as well as geometrical dependence of the flashes from both main and tail rotor. Common models ranging from a box-like rotor to a high-fidelity CAD model of a generic helicopter blade demonstrated the required model accuracy as a function of frequency.

Further, scaled helicopter measurements have been presented in order to demonstrate the effects of micro-Doppler from helicopters, as well as supporting the findings from the RCS modeling. A full mathematical algorithm for target detection in both range and Doppler was presented, before showing how to Doppler process the exactly same data set in order to classify as fixed or rotary wing target without changing the waveform or having to schedule a dedicated waveform.

Finally, examples of data sets from real targets have been presented. Both monostatic as well as bistatic measurements of helicopter micro-Doppler were presented and analyzed. The results showed and demonstrated the ability of such systems to discriminate between aircraft and helicopters. We showed how the STFT method can aid in the classification of the helicopter by providing information about the number of main and tail rotor blades, rotation speed of the rotor, speed of the rotor blade tip, and the ratio of the rotation speed of the main and tail rotors.

The results presented here, as well as the suggested algorithms demonstrated on real targets, are all well suited for modern radar systems capable of exploiting modern Doppler processing, facing the emerging challenge of non-cooperative target recognition.

References

[1] P. Tait, *Introduction to Radar Target Recognition*, ser. IEE Radar, Sonar, Navigation, and Avionics. The Institution of Engineering and Technology, 2005 [Online]. Available: http://books.google.no/books?id=hrp_QP-vz4sC

[2] V. Chen, *The Micro-Doppler Effect in Radar*, ser. Artech House Remote Sensing Library. Artech House, 2011 [Online]. Available: http://books.google.no/books?id=eJ7eMHpxt30C

[3] 'Efield electromagnetic solver suite', ESI Group [Online]. Available: http://www.efieldsolutions.com

[4] G. T. Ruck, D. E. Barrick, W. D. Stuart, and C. K. Krichbaum, *Radar Cross Section Handbook*. New York, NY: Plenum, 1970.

[5] K. E. Olsen, T. Johnsen, S. Johnsrud, R. Gundersen, H. B. I. Tansem, and P. Sornes, 'Results from an experimental continuous wave low probability of intercept bistatic radar – the first steps toward multistatic radar', in *Proceedings of the International Radar Conference, 2003*, September 2003.

[6] T. Johnsen, K. E. Olsen, and R. Gundersen, 'Hovering helicopter measured by bi-/multistatic CW radar', in *Radar Conference, 2003. Proceedings of the IEEE*, May 2003.

[7] T. Johnsen and K. E. Olsen, 'Bi- and multistatic radar', in *NATO Lectures Series SET-086 on Advanced Radar Signal and Data Processing*, 2005.

[8] T. Johnsen, K. E. Olsen, S. Johnsrud, and R. Skjerpeng, 'Simultaneous use of multiple pseudo random noise codes in multistatic CW radar', in *Radar Conference, 2004. Proceedings of the IEEE*, April 2004.

[9] T. Johnsen, R. Gundersen, K. E. Olsen, S. Johnsrud, I. Tansem, and P. Sørnes, Eds., *Bistatic Radar Measurements of Helicopter (in Norwegian)*.

FFI/Rapport-2002/01764, Norwegian Defence Research Establishment, 2002.

[10] K. E. Olsen, T. Johnsen, S. Johnsrud, I. Tansem, and P. Sornes, 'Multistatic and/or quasi monostatic radar measurements of propeller aircrafts', in *2007 IET International Conference on Radar Systems*, October 2007.

[11] K. E. Olsen, T. Johnsen, and S. Johnsrud, 'Details of the signal processing, simulations and results from the Norwegian multistatic radar DiMuRa', in *Waveform Diversity and Design Conference, Procedings of the IEEE*, February 2006.

[12] M. Wicks, E. Mokole, S. Blunt, R. Schneible, and V. Amuso, Eds., *Principles of Waveform Diversity and Design*, 1st edn. Raleigh, NC: Scitech, 2010.

[13] M. Cherniakov, Ed., *Bistatic Radar: Emerging Technology*. John Wiley & Sons, Chichester, England, 2008.

[14] N. J. Willis, *Bistatic Radar*. Silver Spring, MD: Technology Service Corporation, 1995.

[15] K. E. Olsen, 'Investigation of bandwidth utilisation methods to optimise performance in passive bistatic radar', doctoral thesis, University College London, 2011 [Online]. Available: http://discovery.ucl.ac.uk/1333270/

[16] E. W. Weisstein, Ed., *CRC Concise Encyclopedia of Mathematics*, 2nd edn. CRC Press LLC, Boca Raton, FL, 2003.

[17] H. D. Griffiths, 'From a different perspective: Principles, practice and potential of bistatic radar', in *Proceedings of the International Radar Conference*, September 2003.

[18] C. J. Baker and H. D. Griffiths, 'Bistatic and multistatic radar sensors for homeland security', in *NATO Advanced Study Institute, Advances in Sensing with Security Applications*, July 2005.

[19] H. D. Griffiths and C. J. Baker, 'Passive coherent location radar systems. Part 1: Performance prediction', *IEE Proceedings on Radar, Sonar and Navigation*, vol. 152, no. 3, pp. 153–159, June 2005.

[20] C. J. Baker, H. D. Griffiths, and I. Papoutsis, 'Passive coherent location radar systems. Part 2: Waveform properties', *IEE Proceedings on Radar, Sonar and Navigation*, vol. 152, no. 3, pp. 160–168, June 2005.

[21] H. D. Griffiths and C. J. Baker, 'The signal and interference environment in passive bistatic radar', in *Information, Decision and Control, IDC'07*, February 2007.

[22] R. Saini and M. Cherniakov, 'DTV signal ambiguity function analysis for radar application', *IEE Proceedings on Radar, Sonar and Navigation*, vol. 152, no. 3, pp. 133–142, June 2005.

[23] A. Lauri, F. Colone, R. Cardinali, C. Bongioanni, and P. Lombardo, 'Analysis and emulation of FM radio signals for passive radar', in *Aerospace Conference, 2007 IEEE*, March 2007.

[24] T. Tsao, M. Slamani, P. Varshney, D. Weiner, H. Schwarzlander, and S. Borek, 'Ambiguity function for a bistatic radar', *IEE Transactions on Aerospace and Electronic Systems*, vol. 22, no. 3, pp. 1041–1051, July 1997.

[25] K. Rottmann, Ed., *Mathematische Formelsammlung*, 4th edn. B.I. Wissenschaftsverlag, 1991.
[26] E. Weisstein, 'Exponential sum formulas. Mathworld – A Wolfram web resource, September 2011. [Online]. Available: http://mathworld.wolfram.com/ExponentialSumFormulas.html
[27] F. J. Harris, 'On the use of windows for harmonic analysis with the discrete Fourier transform', *IEEE Proceedings*, vol. 66, pp. 51–83, January 1978.

Chapter 8

Micro-Doppler Signatures of Small Boats

Victor C. Chen and David Tahmoush

8.1 Introduction

Radar detection and tracking of small boats in sea clutter is an important issue for both civilian and military applications. Rigid inflatable boats (RIBs) are commonly used in piracy, smuggling, and poaching activities; therefore, detecting small boats in the sea clutter environment becomes necessary. To define the strategies and improve the performance of detecting small boats in sea clutter, we need to understand characteristics of the sea or littoral clutter and the maneuverability and reflectivity of small boats in the sea.

In this chapter, we first use the small boats and sea clutter database collected by the Council for Scientific and Industrial Research (CSIR) of South Africa to demonstrate radar micro-Doppler signatures of small boats in sea clutter [1, 2]. Then, we compare these with micro-Doppler images of small boats in the littoral environment [6, 7].

8.2 Characterization and Modeling of Sea Clutter

Radars operating in an open environment will receive returns from the environment and the clutter. To develop suitable processing algorithms for radar detecting and tracking small boats in sea clutter, it is necessary to characterize sea clutter returns. Sea wave dynamics are determined by the sea state and the direction of sea waves. When wind and swell conditions change, sea state varies.

Statistic modeling of sea clutter has been investigated for decades [3–5]. Two types of sea surface waves – the capillarity wave and the gravity wave – have been modeled. The wind generates the capillarity wave, which has low amplitude and a short wavelength. The gravitational forces generate the gravity waves, which can also be triggered by the wind or other meteorological phenomena. The gravity wave is the main factor of sea wave energy and sea state. Sea waves are modeled by gravity waves driven by stable winds and short capillarity waves driven by turbulent gusts of wind near the surface.

High-resolution radar returns from sea clutter often appear very spiky. To understand detailed clutter characteristics and distinguish between spicks returned from clutter and that from small boats, spikes characteristics in Doppler domain can be very helpful. To effectively detect small boats in the sea clutter, these Doppler bins, where the small boat is located, must be extracted while Doppler spectral behavior of the sea clutter becomes a key characteristic.

8.2.1 Doppler Spectra Characteristics of Sea Clutter

Because individual scatterers of sea clutter are moving with respect to the radar, the phase of the returned signal from the clutter will vary with time. Its Doppler spectrum is asymmetric and time varying. It mean Doppler shift may not be zero Doppler and is dependent on the wind speed and direction. Figure 8.1 shows an example of the Doppler spectrum of sea clutter as function of time, derived from radar data collected by the CSIR. The CSIR radar is vertically polarized with a frequency of 9 GHz, a pulse repetition frequency of 5 kHz, and a range resolution of 15 m [1].

The spectrum width varies with time, with occasional extreme Doppler excursions, which could be the result of local wind gusting. The spectrum appears asymmetric in shape, with a nonzero mean Doppler shift. These features highlight the complexity of the relationship between the intensity modulation and the form of the spectrum, the former being dominated by the swell structure in the sea surface and the latter being additionally affected by the local gusting of the wind and the detailed scattering mechanism.

The sea/littoral clutter is generally nonstationary. However, during a coherent processing time interval, the clutter can be considered stationary and its Doppler characteristic constant. Figure 8.2(a) shows the power spectrum density function of the sea clutter at low sea state using a high-resolution radar with 0.33 m range resolution. The power spectrum density function in high-resolution sea clutter is narrow-banded as shown in Figure 8.2(a). Because of low sea states, the spiky sea waves are not strong. Figure 8.2(b) shows the power spectrum density function of

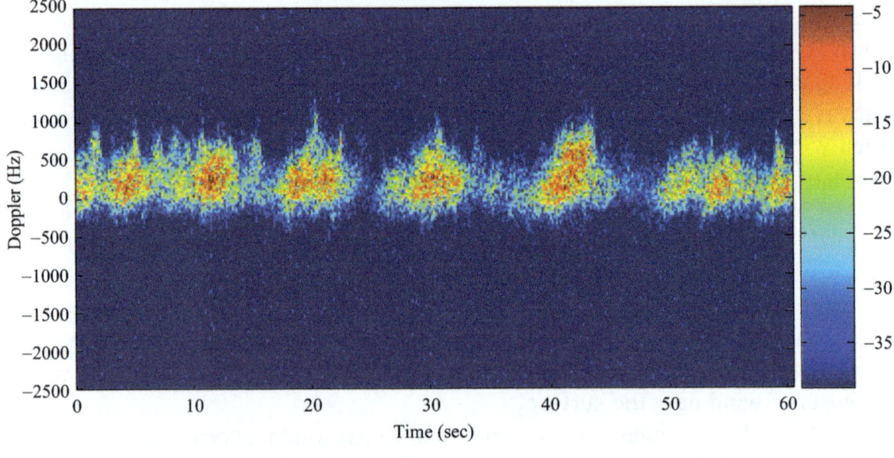

Figure 8.1 An example of the Doppler spectrum of sea clutter as function of time

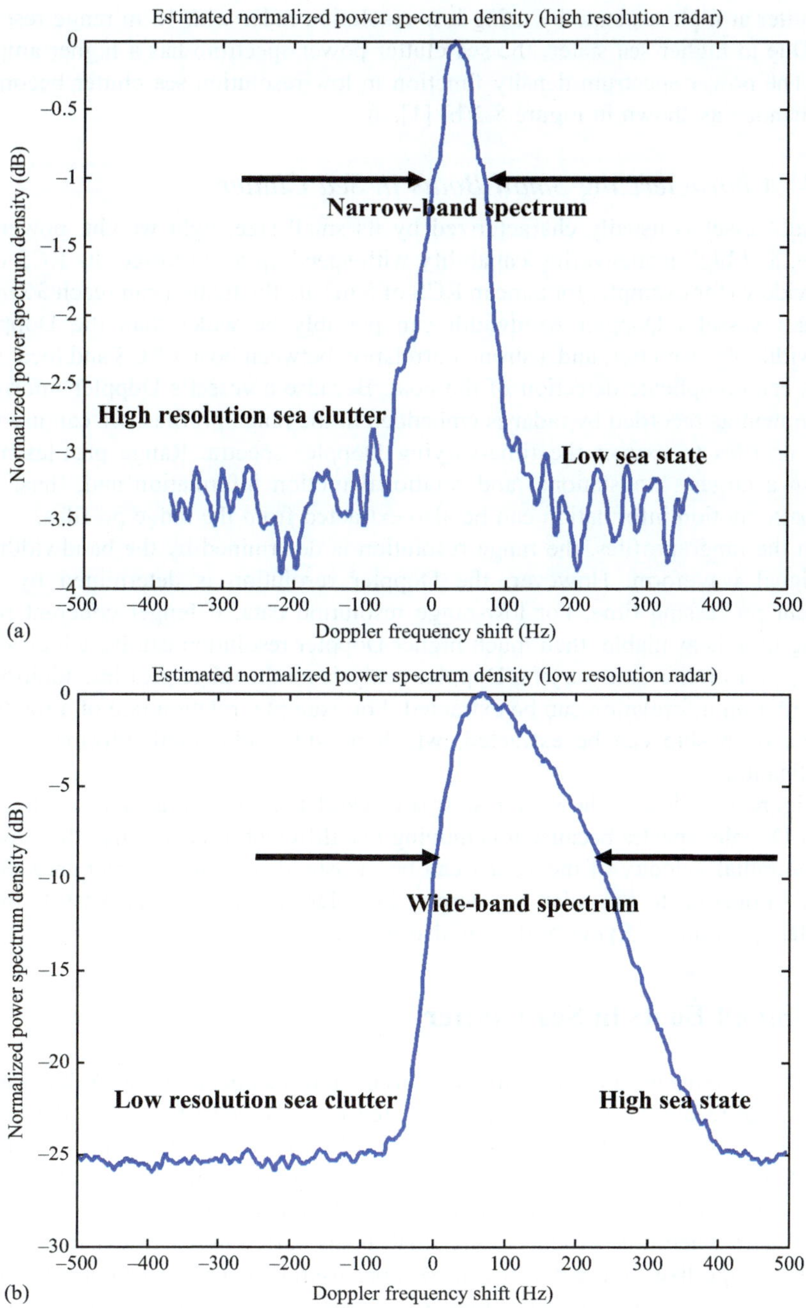

Figure 8.2 Sea clutter Doppler spectral behavior (a) power spectrum density with high-resolution radar (0.33 m resolution) at low sea state; (b) power spectrum density with low-resolution radar (15 m resolution) at high sea state

sea clutter at high sea state by using low-resolution radar with 15 m range resolution. Due to higher sea states, the sea clutter power spectrum has a higher amplitude. The power spectrum density function in low-resolution sea clutter becomes wide-banded as shown in Figure 8.2(b) [1].

8.2.2 Characterizing Small Boats in Sea Clutter

A small vessel is usually characterized by its small size, light weight, powerful engine, and high maneuvering capability with speed up to 20 m/sec. Its RCS can vary widely (for example, for a mean RCS of 5 m^2, its fluctuation can reach 35 m^2). A small vessel's Doppler bandwidth can possibly be wider than the Doppler bandwidth of sea clutter, and a strong correlation between boat's RCS and local sea waves can complicate detection of the boat. Because a vessel's Doppler information content as recorded by radar is embedded in the range profiles, we can use the range profiles to extract the time-varying Doppler spectra. Range profiles also contain a target's translational and rotational motion information and, thus, the kinematic motion information can be also extracted from the range profiles.

In the range profiles, the range resolution is determined by the bandwidth of the signal waveform. However, the Doppler resolution is determined by the coherent processing time. For low-range resolution data, if longer coherent processing time is available, then much higher Doppler resolution can be achieved.

By performing time-varying Doppler analysis on the range profiles, additional target motion information can be extracted. For example, rotation rate of a rotating antenna on a ship can be extracted, which provides additional information for classification.

Figure 8.3 clearly shows a fast small vessel that is separated from the sea clutter Doppler spectra because it is moving in a different direction than the waves, and the radial velocity of the vessel can be measured from its Doppler frequency shifts. Compared to the relatively strong Doppler spectrum of the vessel, weak Doppler spectrum of flying birds can also be seen [2].

8.3 Small Boats in Sea Clutter

The radar systems used in the CSIR measurement trials operate in the X-band with lower range resolution (15 m) and are located near the coast [1]. Since the range resolution used in the released database is low (15 m), the vessel occupies less than half of a range cell. Figure 8.4(a) is a typical example of range profiles and Figure 8.4(b) is the range-Doppler image of the boat and sea clutter.

In cases when it is difficult to see small boats buried in sea clutter in a range-Doppler image like Figure 8.4, it can be more useful to analyze Doppler features instead of range features. Because longer integration time is possible, high Doppler resolution (1 Hz or less) can be easily reached. Therefore, time-varying Doppler features of the small boat and sea clutter can be used for the purpose of detecting, discriminating, and tracking small boats in sea clutter.

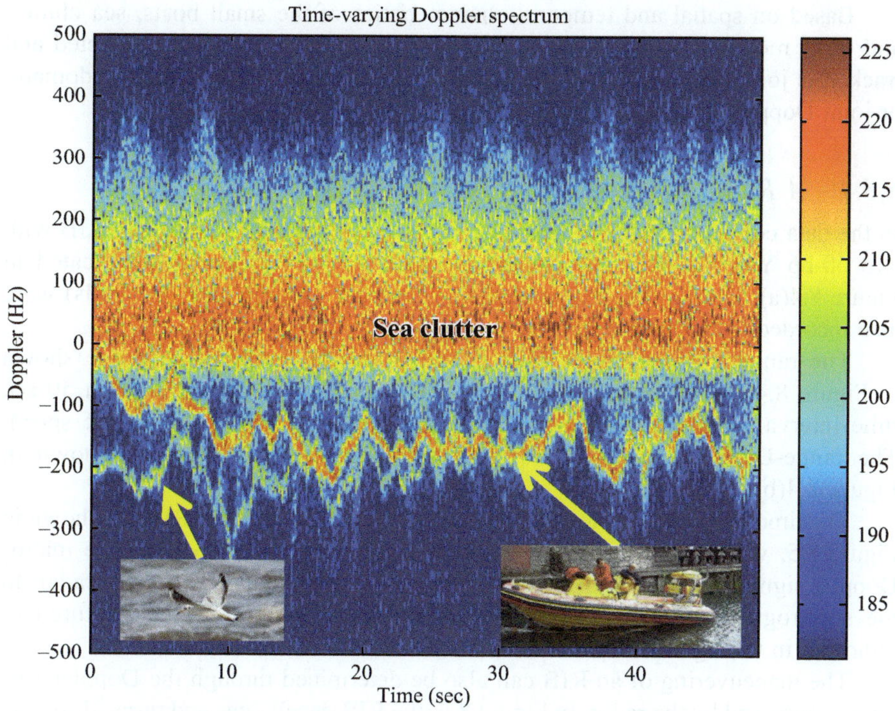

Figure 8.3 *Time-varying Doppler characteristics can be extracted from range profiles*

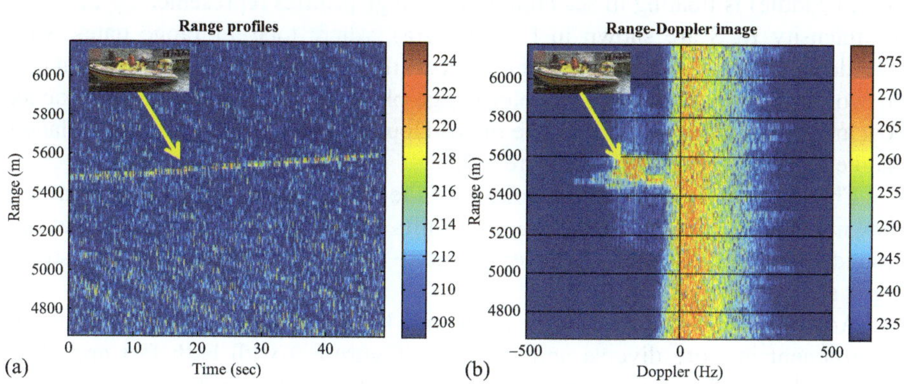

Figure 8.4 *(a) A typical example of range profiles of sea clutter with a moving RIB, and (b) the corresponding range-Doppler image of the sea clutter and the moving RIB*

Based on spatial and temporal characteristics of the small boats, sea clutter, and other moving objects (such as flying birds), small boats can be detected and tracked in joint range and slow-time (pulses) domain, joint range-Doppler domain, or joint Doppler and slow-time domain.

8.3.1 A Rigid Inflatable Boat in Sea Clutter

In the data collection, there is a Rotary Endeavour – Class 3 5.5 m long RIB with two 60-hp Yamaha outboard motors and a single VHF antenna as indicated in Figure 8.4(a). During the data collection, a number of large birds (seagulls) were also recorded in the dataset [1].

The range profiles represented by dBm^2 of the intensity RCS are shown in Figure 8.4(b), where the RCS of the RIB can be seen during about a 50 sec time interval. The RIB is moving away from the radar with 2–3 m/sec speed. The range-Doppler image plot of the sea clutter and the RIB are shown in Figure 8.4(b).

The time-varying Doppler spectrum of the sea clutter and the RIB are shown in Figure 8.5, where the Doppler shift of the sea clutter is positive and the micro-Doppler signature of the RIB is negative because it is moving away from radar. In the spectrogram, micro-Doppler signature of flying birds was also captured as indicated in the figure.

The maneuvering of an RIB can also be determined through the Doppler variations observed by the radar. In Figure 8.6, the RIB decelerates and turns, throwing out a wake [6].

8.3.2 A Paddled Kayak in Sea Clutter

In the data collection, a kayak boat (a small human-powered boat with double-bladed paddle) is floating in sea clutter. The range profiles represented by dBm^2 of the intensity RCS are shown in Figure 8.7(a), where total 11 range gates were recorded and during about 40 sec time interval the kayak boat did not move out one or two range gate. The range-Doppler image plot of the sea clutter and the kayak boat are shown in Figure 8.7(b). The micro-Doppler signature of the double-bladed paddle stroking is captured in Figure 8.8.

The time-varying Doppler spectrum of the sea clutter and the kayak boat are shown in Figure 8.7, where the Doppler shifts of the sea clutter and the boat are around zero Doppler. Micro-Doppler signature of the double-bladed paddle stroking is also captured in the figure.

Additional measurements were made on a kayak in littoral clutter. The littoral environment is more diverse and shown in Figure 8.9 with both fast and slow paddling by the kayaker in the spectrogram [6]. Another look at a kayaker is shown in Figure 8.10, where the paddling stops and the speed of the kayak slows at around 80 sec [7].

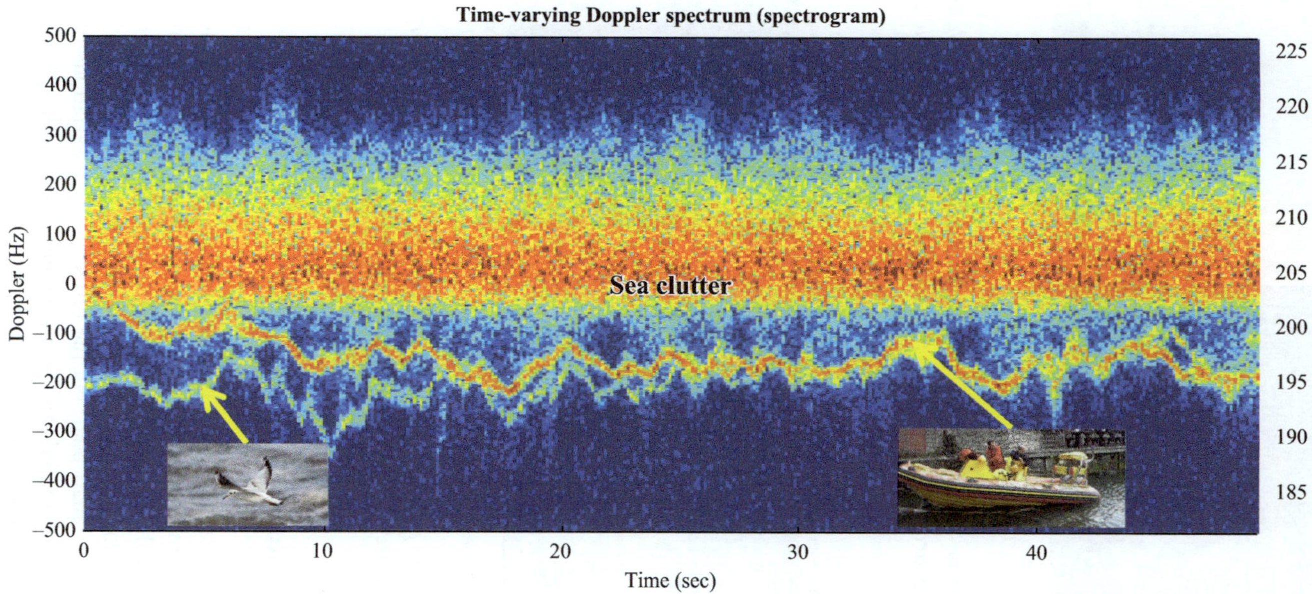

Figure 8.5 The time-varying Doppler spectrum of the RIB, sea clutter and flying birds can be seen

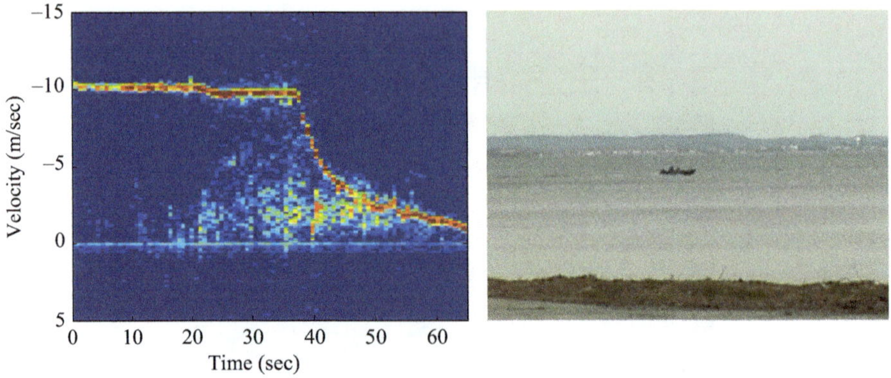

Figure 8.6 Zodiac motorboat Doppler (left) and image (right). Note the appearance of wakes in the Zodiac spectrogram as the zodiac turns

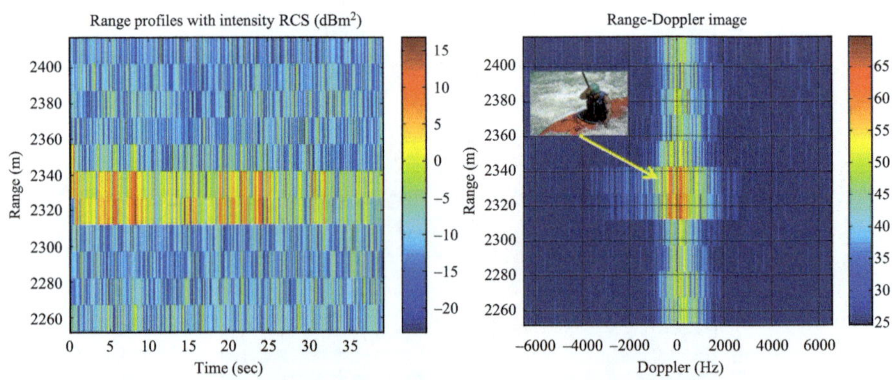

Figure 8.7 Range profiles and range-Doppler plot of the kayak in sea clutter collected CSIR in the Signal Hill trial

8.4 Small Vessels Conclusions

Direct detection of small vessels is complicated by their relatively low RCS and the sea clutter. As shown, small vessels often disappear from the range profiles and thus become difficult to track. The range-Doppler does show a detection with better signal-to-noise ratio than the range profiles, but the micro-Doppler approach detects not only the small boat, but the birds as well. Variations in the motion are discernable like when a rapid turn is carried out. Even paddling is observable with a kayaker, and variations in paddling show up as well. The use of micro-Doppler techniques allows the measurement of exoclutter signals like the paddle motion of the kayak even when the Doppler of the kayak is within the clutter.

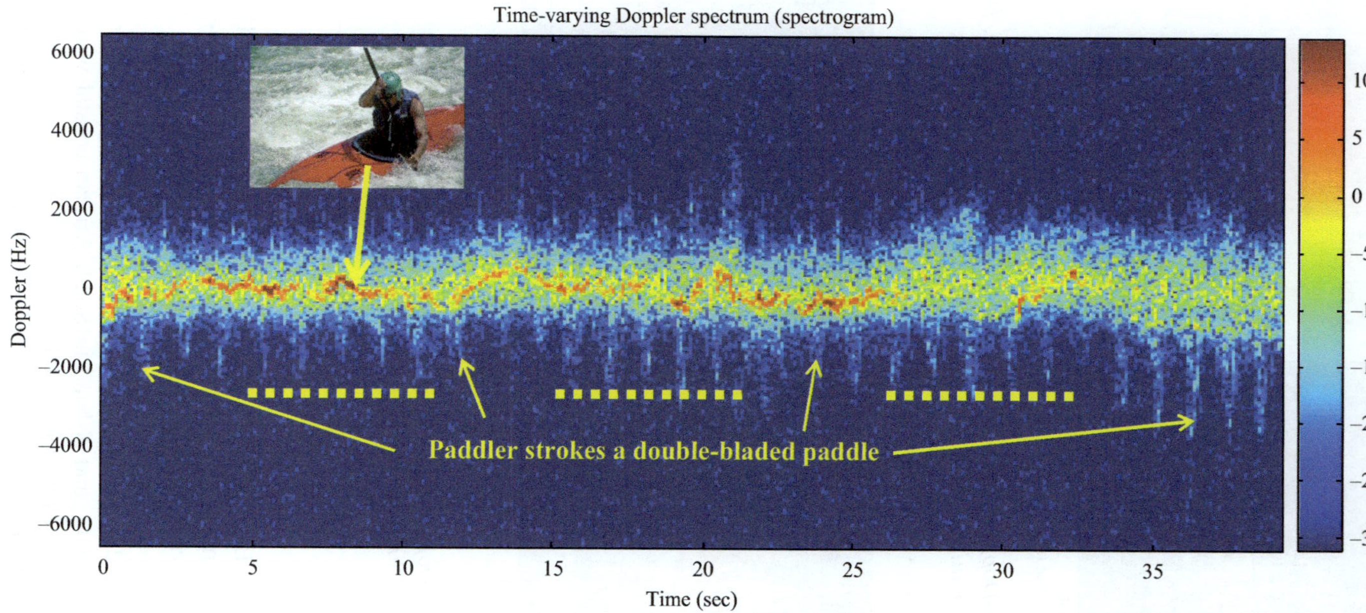

Figure 8.8 Time-varying Doppler spectrum of the CSIR Signal Hill trial, where the micro-Doppler signature of the double-bladed paddle stroking can be seen

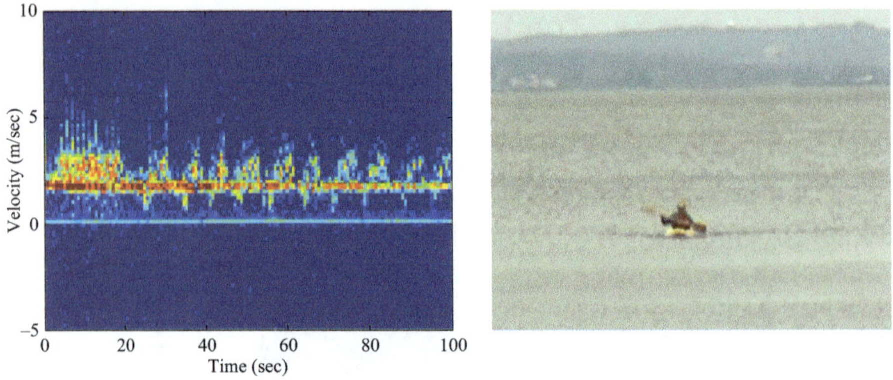

Figure 8.9 Micro-Doppler features of a kayak (left) and image (right). Note the kayaker initially paddles quickly, then does a slower stroke

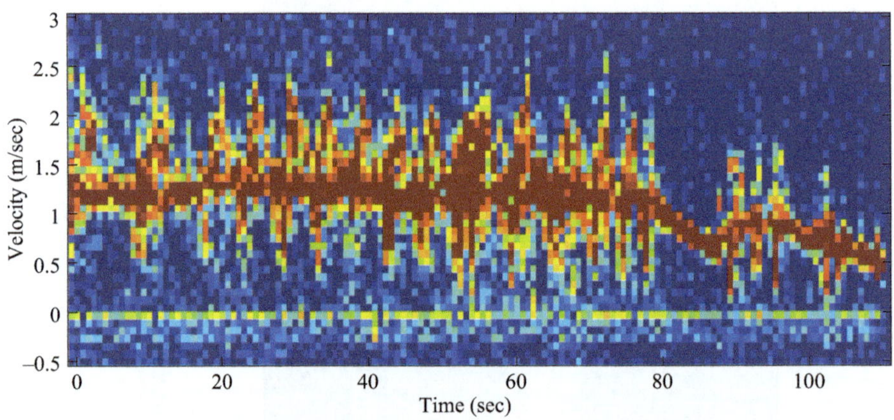

Figure 8.10 Micro-Doppler features of a kayak with the paddle swings stopping around 80 sec and the resulting slowing of the kayak

References

[1] P. Herselman, C. Baker, and H.J. de Wind, "An analysis of X-band calibrated sea clutter and small boat reflectivity at medium-to-low grazing angles," *International Journal of Navigation and Observation*, vol. 2008, Article ID 347518.

[2] P. Herselman and C. Baker, "Analysis of calibrated sea clutter and boat reflectivity data at C- and X-band in South African coastal waters," *IEEE International Radar Conference*, 2007.

[3] K. Ward, "Compound representation of high resolution sea clutter," *Electron. Lett.*, vol. 17, no. 16, pp. 561–563, August 6, 1981.

[4] K. Ward, C.J. Baker, and S. Watts, "Maritime surveillance radar Part I: radar scattering from the ocean surface," *IEE Proc.*, vol. 137, Pt. F, no. 2, pp. 51–62, April 1990.

[5] K. Ward, R.J.A. Tough, and S. Watts, *Sea Clutter, Scattering, the K Distribution and Radar Performance*, vol. 20, *IET Radar, Sonar, and Navigation Series*, IET, London, UK, 2006.

[6] J. Silvious and D. Tahmoush, "Radar tracking and classification of littoral targets," *Proceedings of SPIE*, vol. 8361, April 2012.

[7] D. Tahmoush and J. Silvious, "Pattern of life for radar port and river security," *IEEE Conference on Technologies for Homeland Security*, November 2012.

Chapter 9

Multistatic Micro-Doppler Signature Processing

Graeme E. Smith and Christopher J. Baker

9.1 Introduction

The exploitation of micro-Doppler radar signatures is a maturing area that aims to extract micro-motion information intrinsic to a given target with the ultimate aim of enhanced classification. Micro-Doppler has been used with considerable success in both the natural and synthetic sensing worlds. For example, bats use micro-Doppler in order to measure wing beat frequencies associated with flying insects of different types [1]. In radar, micro-Doppler measurement of human gait signatures enables different forms of movement to be identified [2] (although identification of different humans has yet to be achieved). There is also evidence in the natural world that bats use flight trajectories that deliberately gain information from a multiplicity of aspects [3]. Using data derived from observations at multiple aspects improves classification in radar systems [4–7]. It is an approach that can be exploited by distributed or multistatic radar systems, including those using MIMO and related processing techniques. In this chapter the generation and exploitation of multistatic micro-Doppler radar target signatures are examined with the ultimate aim of improving the classification of targets.

The superior classification performance of the multistatic approach is a result of the additional information that comes from viewing a target at multiple aspects. (It's a lot easier to recognize a person from the front of their head than it is from the back of their head and even easier using both perspectives). Thus the combination of a distributed, multisite network and the measurement of micro-Doppler signatures makes for a powerful means of acquiring information from target echoes.

There is an important distinction to be made between multiple aspects and a multistatic or multisite geometry. Multiple aspects provide extra information via the different viewing angles. This is termed "angular diversity." Thus multiple aspects can be obtained by moving a traditional monostatic radar system from one viewing position to another. By making measurements of the micro-Doppler at each different position, multi-aspect, micro-Doppler information can be gathered and exploited. In other words a series of time separated, multi-aspect, micro-Doppler target signatures are acquired. This results in extra information for target classification. However, this approach also imposes some limitations that can be

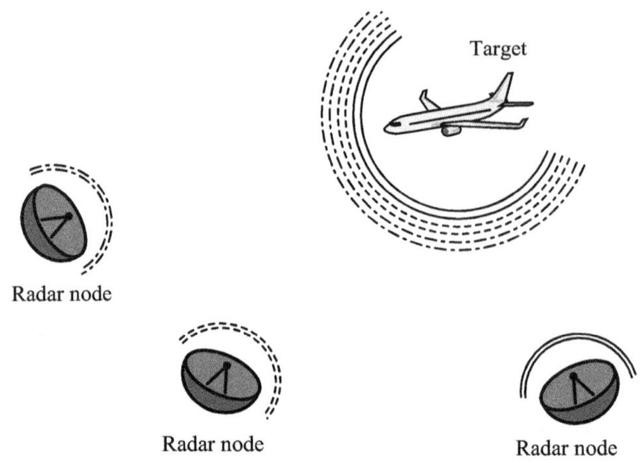

Figure 9.1 Multistatic geometry of a collection of monostatic radar systems illustrating different viewing angles obtained of a target (spatial diversity)

overcome using a multisite network of radar system that allows simultaneous measurements over both monostatic and bistatic geometries.

First, a multistatic geometry can be considered as consisting of a set of distributed monostatic radars, i.e., equivalent to placing a single radar at each site or node in the network. Figure 9.1 shows a multistatic network of monostatic radar systems each viewing an aircraft target from a different aspect. A fully coherent distributed radar network can be constructed that operates from a common timing signal such that there is phase coherency at each receiver with respect to each transmitter. This system can subsequently be further generalized by removing the requirement for the radars to be monostatic. That is to say, there can be co-location of transmitters and receivers or separate location of transmitters and receivers or a mixture of both. The term "node" will henceforth be used for the location of either a transmitter and/or a receiver.

Thus all the receivers can measure scattered radiation caused by illumination from all of the transmitters. This allows monostatic and bistatic radar theory to be applied to the network with the total response being the sum of the individual monostatic and bistatic responses. Figure 9.2 shows, in schematic form, a fully coherent multistatic radar system comprising two transmitters and two receivers in the form of two cooperative monostatic radar systems located at two spatially separated nodes.

Figure 9.2 shows that there are four separate "radar systems" that can be individually identified: two monostatic systems and two bistatic systems. Thus there are three different geometries (as the two bistatic cases will be reciprocal with one another) with which the target is viewed.

If the multistatic network is fully coherent, there will be $N_{tx} \times M_{rx}$ bistatic/monostatic radar systems comprising the network, where N_{tx} is the number of transmitter nodes and M_{rx} is the number of receiver nodes. In the example shown in Figure 9.2 there are two transmitters and two receivers making a total of four radar

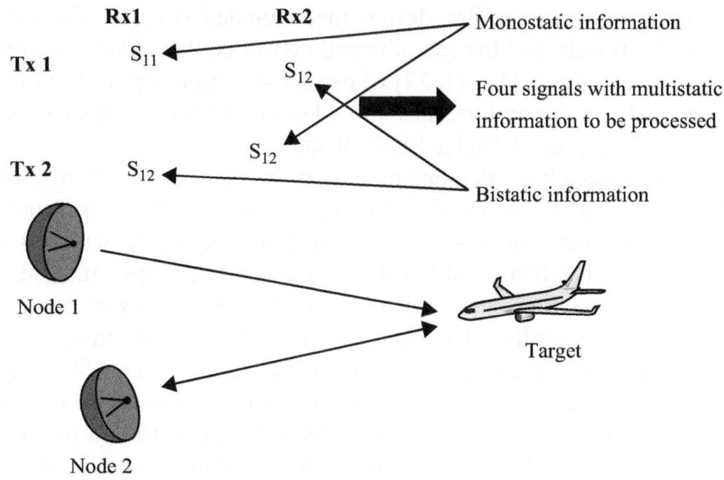

Figure 9.2 A fully coherent multistatic radar geometry

systems (two transmitters multiplied by two receivers). This allows a multiplicity of micro-Doppler signatures to be measured simultaneously. This is quite different to the case where a single monostatic radar system is moved, over time, from one viewing location to another.

As will be seen later in this chapter, the properties measured by a multistatic radar system depend both on the target being viewed and the angles subtended between the nodes (as measured at the target). This dependence will modify the target signature regardless of whether it is a range profile, two-dimensional image, or a micro-Doppler measurement. Monostatic radar can be considered a special case where the angle between nodes at the target is 0°. However, even for monostatic geometries the target signature is a very sensitive function of viewing angle.

Thus each node of a multistatic radar system may be a transmitter, receiver, or transceiver (both transmitter and receiver). The single node of the monostatic radar system is a transceiver. Spatial separation of the multistatic radar nodes provides simultaneous observation of the scene of interest from multiple angles. The increase in the number of viewing angles increases the amount of target information available and ultimately leads to greater accuracy when estimating target parameters. In turn, estimating parameters more accurately, leads to improvements in detection, tracking, and discrimination performance [4–7].

The micro-Doppler signature has been shown to be a powerful discriminator of radar targets. Initial studies of micro-Doppler signatures are generally attributed to Chen *et al.* who provide an analytic analysis of the phenomena as well as time-frequency analysis of real data [8–11]. Immediately apparent from this work is that different types of micro-motion have different forms when observed in time-frequency plots and that these different forms are visible to the eye. This led to further research in which automated recognition of targets based on their micro-Doppler signature was reported by several authors [2, 12–22]. Much of the emphasis of this research was on classifier design that included dynamic time warping, Gaussian mixture models, and the use of neural networks. Particularly, significant is the work of Stove and Sykes [12, 21, 22] whose classification approach, a mixture of linear discriminant analysis and heurists, provides the target recognition capabilities of the operationally deployed Thales MSTAR radar.

While these studies have demonstrated that the micro-Doppler signature is a potent contributor to target recognition, the recognition task has most usually been reserved for cases where there are just a small number of target classes. This restricts utility and hence there is still a strong need for improvement. As discussed above, one potential way to achieve this is to increase the number of target aspects. Indeed multi-aspect recognition has been shown to dramatically improve classification performance for high range resolution profile based systems [4–7]. Here we explore multistatic micro-Doppler signatures and how they enhance the information gathered from a target with the aim of improving classification performance.

In the remainder of this chapter we examine multi-aspect, micro-motion behavior of radar targets. While the resulting signature contains additional information and can help overcome problems such as self-occlusion (parts of the target obscuring other parts of the target along the radar receiver line of sight), it also contains some redundancy. However, by formulating a proper understanding of this redundancy we will see that multi-aspect, micro-Doppler has the potential to reduce the complexity of target recognition algorithms and thence improve performance.

In the next section micro-Doppler radar is briefly reviewed in order to set the scene for extension to the multistatic case. Section 9.3 formulates a fundamental description of the properties of bistatic radar that can be used to assemble those of a multistatic radar. Section 9.4 describes the formation of the key micro-Doppler parameters under a multistatic geometry. In section 9.5 the range of measurements that can be made with a multistatic radar system is examined. Lastly in, section 9.6, the relationship between signatures of real world multistatic measurements of human subjects and the distribution of transmitters and receivers is investigated.

9.2 Background

Radio frequency (RF) micro-Doppler signatures of targets composed of moving components have been exploited in various ways over many years. The vast majority of reported research is for monostatic radar illumination where a single transmitter and a single receiver are colocated. Here, the case where there are many transmitters and many receivers, all arbitrarily located, is considered. Thus the geometry of the multistatic network is itself a design tool that can be used to maximize the information contained in the multiple micro-Doppler signatures.

The history of multistatic micro-Doppler can be traced back to the original research into target classification. In early research the micro-Doppler signatures used for classification were those resulting from propeller engine modulation (PEM). As the name implies, PEM originates from the propellers on aircraft [23, 24]. Propellers are significant radar reflectors that move at relatively high speeds with a regularity determined by the design of the propeller blades and their rotation rate.

They generate transient components that form part over the overall target radar cross-section (RCS). These transients are known as blade flashes. Blade flashes have also been exploited in the more modern counterpart to PEM, jet engine modulation (JEM). The transients result in Doppler shifts much larger than those arising from the aircraft fuselage. Such distinctive signatures were detectable even with early radar systems. As the sensitivity of receivers and the power of transmitters increased, it became possible to discern physical target scattering features that had smaller and smaller magnitudes of backscattering. The contribution to this backscattered echo signal from the physical parts of a target that exhibit micro-motions (i.e., those that vibrate, oscillate, and rotate) along with small motions of the target itself (such as precession of warheads or the bouncing of a ground vehicle chassis on the suspension), resulted in PEM being generalized to form part of the broader set of motions embraced by the term micro-Doppler [9–11]. However, these micro-Doppler signatures were gathered using monostatic radar systems only. More recently, the monostatic case has been broadened to include bistatic micro-Doppler [25–28]. A logical extension to bistatic micro-Doppler geometries is the multistatic geometry. In fact multistatic systems can be decomposed into a co-operating set of bistatic components, which therefore makes a natural building block for analysis and assessment of multistatic micro-Doppler radar systems.

Chen *et al.* were the first to analyze bistatic micro-Doppler signatures and use them as part of target recognition. The inverse synthetic aperture radar (ISAR) technique was used to observe and exploit such signatures [8, 29]. The case was examined where the micro-motions of a target occurred along a line tangential to the radar line of site and therefore produced little radial motion that would contribute to a micro-Doppler signature. They showed that by using a bistatic geometry a micro-Doppler signature could always be observed. Subsequently the analysis of Chen *et al.* has been expanded by a number of authors to consider the case when the target is illuminated bistatically by applying synthetic aperture radar (SAR) processing [27, 30–33]. Here the authors have typically assumed a fixed transmitter and a moving receiver and go on to contrast the bistatic performance with that of monostatic through a bistatic factor (see section 9.3.3). The bistatic factor varies between 0 and 1, with 1 being the case when a pseudo monostatic geometry is obtained. In other words the monostatic case provides the upper bound on the magnitude of the micro-Doppler shift. However, other characteristics that result from using a bistatic geometry, such as counter stealth, reduced clutter, reduced dynamic range requirements make it inherently appealing. Further, the complexity of target scattering from irregularly shaped objects means that there are occasions where the bistatic micro-Doppler signature is more prominent than that for the monostatic case. The notion of ground clutter reduction, for example, was demonstrated by Zhang *et al.* who combine displaced phase center array (DPCA) techniques with bistatic micro-Doppler to extract micro-motion parameters for targets in stationary clutter [25, 34].

Of the reported research into bistatic imaging, only Ai *et al.* [32] consider how moving to a multistatic system might enhance target information still further. They consider adding a bistatic receiver node to a separately located monostatic transceiver node to obtain two observations of a target. Their motivation is to more accurately estimate the magnitude of a vibration at some angle to the plane of observation (which includes the bistatic bisector that is the bistatic expansion to the LOS). By exploiting micro-motion measurements of a target using both monostatic and bistatic geometries the angular offset can be compensated and a more accurate estimate made. Smith *et al.* have also considered the impact of a multistatic geometry but this time to deliberately enhance the quality of information contained in the backscattered micro-Doppler signature so that the performance of automated target recognition is improved [26, 28, 35]. This research shows that multistatic micro-Doppler signatures do indeed contain more target information, but that there is also redundancy. We discuss these results in more detail toward the end of this chapter. However, first the foundations of multistatic radar are reviewed.

9.3 Bistatic and Multistatic Radar Properties

In this section the theory of bistatic and multistatic radar is introduced placing emphasis on those aspects that directly affect the resulting form of measured micro-Doppler signatures. A multistatic radar system can be decomposed into a series of interacting bistatic components. In this way we can use the relatively well known properties of bistatic radar and then integrate these to build up to the multistatic case. For a comprehensive treatment of a bistatic radar, refer to the excellent text by Willis [36].

Figure 9.3 shows a schematic of an arbitrary bistatic radar geometry identifying the major geometrical relationships [37].

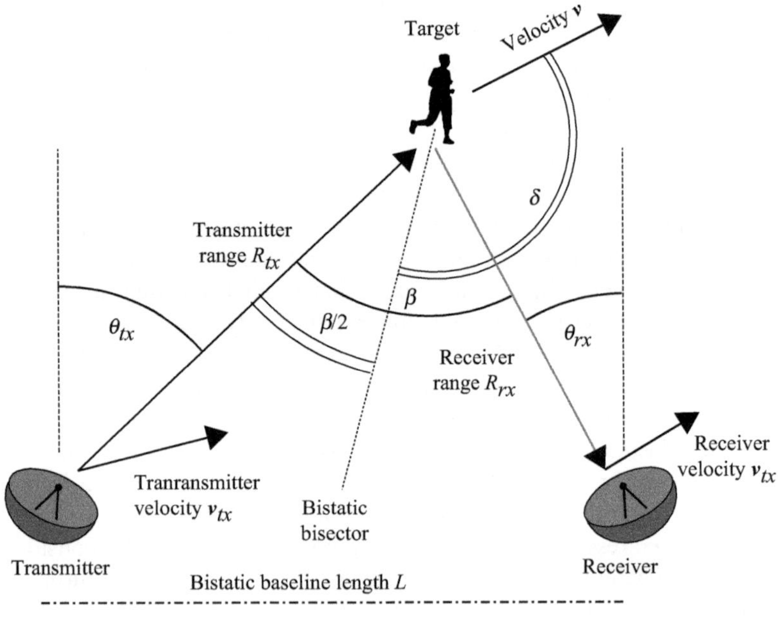

Figure 9.3 The geometry of a bisatic radar system

The transmitter and receiver are separated by a distance L, where L is referred to as the baseline. R_{tx} is the range from the transmitter to the target and R_{rx} is the range from the target to the receiver. The angle subtended by the target with respect to the locations of the transmitter and receiver is known as the bistatic angle β, and the line evenly dividing β is known as the bistatic bisector. θ_{tx} and θ_{rx} are the transmitter and receiver angles respectively and are usually referenced to a common orientation (such as North or the vertical as shown in the figure). Lastly we can define δ as the angle between the bistatic bisector and the direction of motion (or micro-motion), v, of the target. For completeness the figure shows the transmitter and receiver as having arbitrary velocities, v_{tx} and v_{rx} respectively, but to simplify the treatment here we will assume these to be zero.

As a result of the separation of transmitter and receiver, some of the basic equations related to radar performance have significantly different forms to their monostatic counterparts.

9.3.1 Range

The round trip delay in a bistatic radar, τ, is due to the sum of R_{tx} and R_{rx}, that is

$$\tau = (R_{tx} + R_{rx})/c \tag{9.1}$$

where c is the speed of propagation of the electromagnetic (EM) wave. Note that for the case when $\beta = 0°$, $R_{tx} = R_{rx} = R$ and (9.1) reverts to the monostatic case for round trip delay ($\tau = 2R/c$). Since the radar measures the round trip delay, τ, it is only possible to derive the sum $R_{tx} + R_{rx}$ not the ratio between them. Lines of equi-range form a series of concentric ellipses with the transmitter and receiver at their focal points. Assuming the location of the transmitter and receiver is known, the location of the target is constrained to lie on an ellipse as described by

$$(R_{tx} + R_{rx}) = \sqrt{(x - x_{rx})^2 + (y - y_{rx})} + \sqrt{(x - x_{tx})^2 + (y - y_{tx})} \tag{9.2}$$

where the receiver node is at (x_{rx}, y_{rx}) and the transmitter is at (x_{tx}, y_{tx}). An example showing the variation of iso-range contours is shown in Figure 9.4. If both the transmitter and receiver antennas are omni-directional there is no way of knowing where the target is located on any given range ellipse.

9.3.2 Location

An estimate of the target location can be achieved through two main methods: (i) Use of directional antennas and (ii) multistatic geometries. Directional antennas may be used at both the transmitter and receiver nodes. The intersection of transmit and receive beams will reveal the portion of the iso-range ellipse at which the target is located thus greatly reducing the positional ambiguity. Targets are usually located with reference to the receiver position. The target to receiver range can be computed using the following equation:

$$R_{rx} = \left\{(R_{tx} + R_{rx})^2 - L\right\}/\left\{2(R_{tx} + R_{rx} + L\sin\theta_{rx})\right\} \tag{9.3}$$

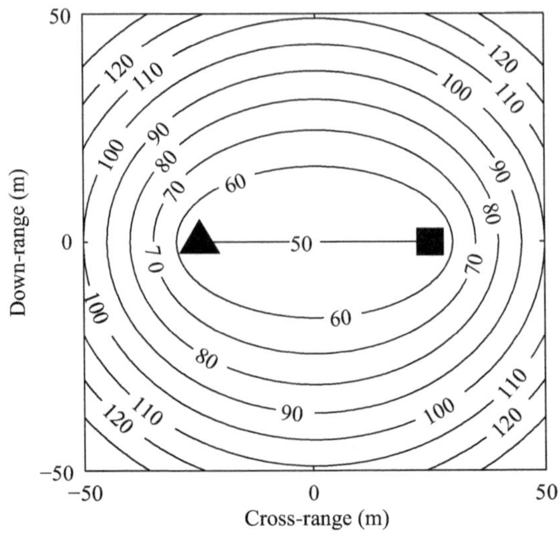

Figure 9.4 Range contours are ellipses in bistatic radar

which assumes the angle between the receive path and the reference direction, θ_{rx}, is known, see Figure 9.3. After calculating R_{rx} both the range and the angle from the receiver are known and the true target location can be determined to an accuracy governed by the bistatic radar design parameters.

Operation with two directional antennas requires their pointing direction to be both carefully controlled and synchronized such that they illuminate a common area as they each scan over an angular ambit of 360°. If both antennas are not pointing at the target at the same time then the transmitter to receiver radar path will not be complete and the target will be missed. Such coordination can be challenging, especially if mechanically scanned antennas are used. Two alternatives are possible. If an isotropic or broad beamwidth transmission is used, the receiver requires little or no synchronization. However, this restricts both sensitivity and location accuracy as the transmit power is not being concentrated on a target (as it would be if a high gain antenna were used). Alternatively high gain electronic scanning allows for the required nonlinear scan patterns to be more easily mapped out. This improves ultimate accuracy and sensitivity but comes at greater system complexity and cost.

The second method for target localization uses a multistatic geometry. For example, if the radar has three nodes, then, depending on the ratio of transmitters to receivers, at least two transmit–receive pairs can be formed and the target will be located on two iso-range contours. Solving for where the contours cross provides the target location. A disadvantage of this approach is that the solution is ambiguous with a crossing point in front of and behind the baseline; however, inclusion of further nodes or using directional antennas can overcome this ambiguity. It is, of course, possible to combine these two approaches.

9.3.3 Range Resolution

The ability to discriminate targets in range is an important component of multistatic micro-Doppler radar. At a coarse level it allows a target to be better separated from background clutter. At a finer level, over resolving a target can enable the micro-Doppler signature for individual parts of a target to be measured. By considering, again, the multistatic micro-Doppler system to be made up of individual bistatic pairs the effects of geometry on range resolution can be examined. The range resolution of a monostatic radar is given by

$$\Delta R = c/(2B) \tag{9.4}$$

where c is the speed of propagation (approx. 3×10^8 msec^{-1}) and B is the transmitted waveform bandwidth.

The width of a bistatic range bin may be defined as the separation ΔR_B between two concentric iso-range ellipsoids $(R_{tx} + R_{rx})$. Thus the resolution will be a function of the target, transmitter, and receiver geometry. This is illustrated schematically in Figure 9.5.

The resolution is highest on an extension of the bistatic baseline L and degrades to a worst case situation at an angle of 90° to the baseline. Willis [36] provides an approximation to the bistatic range resolution

$$\Delta R_{\mathrm{bi}} = c/(2B \cos\{\beta/2\}) \tag{9.5}$$

which represents an upper bound.

Note that when the bistatic angle is zero the geometry reverts to the monostatic case and the range resolution is $c/2B$. This, therefore, represents the best that can be achieved. In other words separation of the transmitter and receiver can only reduce achievable range resolution. The expression $\cos(\beta/2)$ is commonly referred to as the bistatic factor and relates the bistatic parameter to an equivalent monosatic measurement made along the bistatic bisector. For bistatic geometries (9.5) is the minimum separation between two iso-range contours on which targets can reside and still be individually detected. For bistatic angles of $\beta > 0°$, (9.5) will always

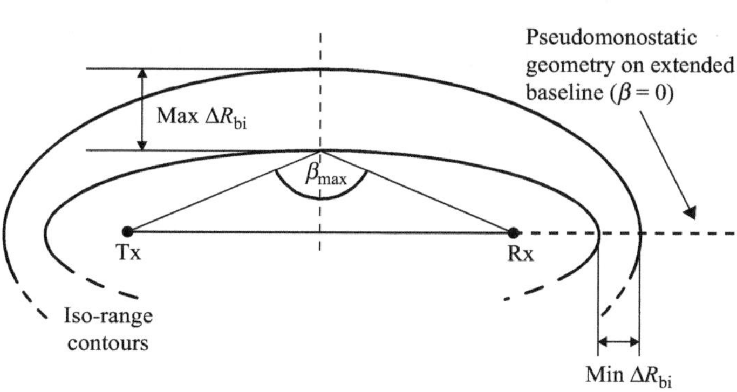

Figure 9.5 Definition of bistatic range resolution as separation between two iso-range contours

give a value of ΔR that is greater (coarser) than the monostatic range resolution (for an equivalent bandwidth). Furthermore, when $\beta = 180°$ the target lies on the bistatic baseline and the radar is considered to be operating in a forward scatter geometry. Forward scatter is a special case where $\Delta R = \infty$ and it is not possible to separate targets based upon range. Thus, overall, with a multistatic geometry there are a multiplicity of bisatic angles and hence a multiplicity of range resolutions.

9.3.4 Detection Performance

The bistatic radar range equation has a similar form to that of monostatic radar but differs in one important respect. In bistatic radars the signal-to-noise ratio (SNR) for a target is a function of the product of the square of the transmitter to target range (R_{tx}) and the square of the target to receiver range (R_{rx}). Thus SNR is given by [38]

$$SNR = \frac{P_{tx} G_{tx} G_{rx} \lambda^2 \sigma F_{tx}^2 F_{rx}^2}{(4\pi)^3 R_{tx}^2 R_{rx}^2 kTBNL_{tx} L_{rx}} \tag{9.6}$$

where P_{tx} is the transmitted power, G_{tx} and G_{rx} the transmit and receive antenna gains, λ the carrier wavelength, σ the target bistatic RCS, F_{tx} and F_{rx} the path propagation factors for the transmit and receive paths, k the Boltzmann's constant, T the operating temperature, B the system bandwidth, N the noise figure, and L_{tx} and L_{rx} the losses for the transmitter and receiver hardware. The form of (9.6) means that the iso-SNR contours are sets of ovals (ovals of Cassini [38, 39]), as shown in Figure 9.6. Note that this is in contrast to the elliptical iso-range lines.

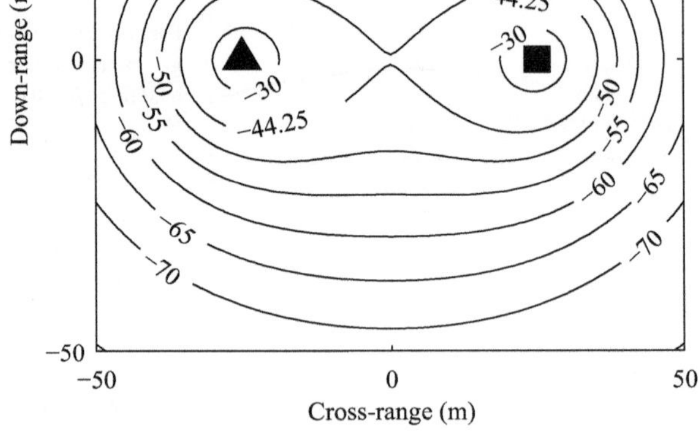

Figure 9.6 Ovals of Cassini describe contours of constant SNR in bistatic radar geometries

This means that the SNR and therefore the detection performance of the radar system are not necessarily constant for two targets at the same receiver range but at different locations (i.e., on different parts of the same iso-range contour). Further, at extended bistatic ranges, the contours look like a regular oval but at shorter bistatic ranges they have a figure-of-eight shape. If the detection range of the radar is specified as an SNR limit, then it follows from (9.2), (9.6), and Figures 9.4 and 9.6 that the largest bistatic range at which a target may be detected will vary as a function of the direction to the target. Again, this is quite different to the monostatic radar case where iso-SNR contours are coincident with iso-range contours and the detection range is independent of target direction. For the multistatic case detection range can be evaluated by summing the individual bistatic components.

9.3.5 Doppler

The Doppler shift, just as in monostatic radar, is given by the time rate of change of the total path between transmitter, target, and receiver. The total path length is $R_{tx} + R_{rx}$ and the Doppler frequency and following the treatment in Willis [36] can be expressed as

$$f_D = \frac{1}{\lambda}\left[\frac{d}{dt}(R_{tx} + R_{rx})\right] = \frac{1}{\lambda}\left[\frac{dR_{tx}}{dt} + \frac{dR_{rx}}{dt}\right] \tag{9.7}$$

Keeping the transmitter and receiver stationary and assuming the target has a finite velocity, dR_{tx}/dt is the projection of the target velocity onto the transmitter to target LOS, i.e.,

$$\frac{dR_{tx}}{dt} = V\cos(\delta - \beta/2) \tag{9.8}$$

In the same way dR_{rx}/dt is the projection of the target velocity onto the receiver LOS, i.e.,

$$\frac{dR_{rx}}{dt} = V\cos(\delta - \beta/2) \tag{9.9}$$

Inserting (9.8) and (9.9) into (9.7) we get an expression for bistatic Doppler

$$f_D = (2v/\lambda)\cos(\beta/2)\cos(\delta) \tag{9.10}$$

where v is the target speed and other parameters are as defined above. $\beta = 0°$, corresponding to the monostatic case and the Doppler is $\frac{2v}{\lambda}\cos\delta$. Critically, (9.10) shows that the bistatic Doppler depends on the bistatic angle, which in turn depends on the target location. The bistatic factor term, $\cos(\beta/2)$, is now in the numerator compared to its appearance in the denominator when considering range resolution.

If the angle between the bistatic bisector and the target heading, δ, is held constant, the Doppler shift will vary as a function of bistatic angle (target location) alone. An example set of iso-Doppler contours, in Hz, is shown in Figure 9.7 for a case where the radar wavelength is 12.5 cm, the target speed is 1 msec^{-1}, and $\delta = 45°$. Key points to observe from the figure are that (a) the Doppler shift along the

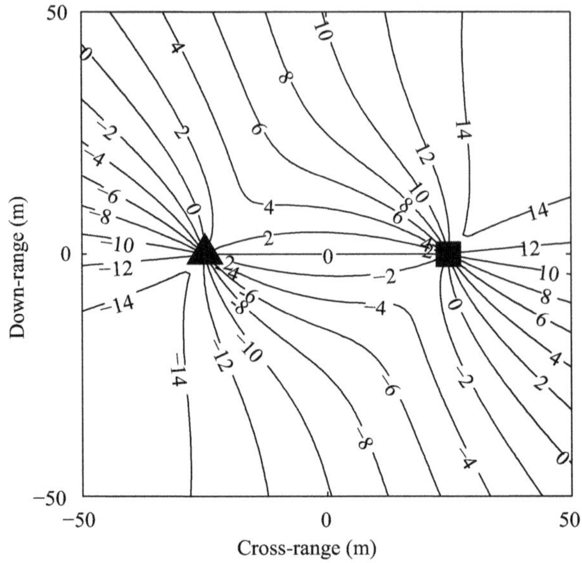

Figure 9.7 Iso-Doppler contours, in Hz, for an S-band radar. Target moves at 1 msec^{-1} and 45° to the bistatic bisector

bistatic baseline is 0 Hz and that (b) the Doppler shift is a maximum along the line of the bistatic baseline, but only when not between the nodes. The former observation tells us that in truly forward scatter geometries it is not possible to measure the Doppler shift. While the latter indicates that in "over the shoulder" geometries (where one node must look past, or over the shoulder of, the other) the performance is equivalent to a monostatic system.

The reduction in Doppler shift of a bistatic channel compared to a monostatic channel is of course, particularly significant for the study of micro-Doppler signatures. The signature can be thought of as comprising multiple components, each having their own time-frequency properties and each depending on the target micro-motions, see Chapter 2. As the bistatic angle increases the Doppler frequencies of each of these components will reduce and the overall signature bandwidth will fall. In turn, this will reduce the ability to separate the components and potentially limit the usefulness of the signature. However, as will be seen in section 9.4 there are gains when observing the micro-Doppler signature multistatically and, so long as β does not become too large, these provide valuable advantages.

Thus the basic foundations by which a bistatic radar is able to measure the micro-Doppler signatures of targets have been laid. By treating multistatic systems as a series of monostatic and bistatic channels, the multistatic signature will be similarly composed of its monostatic and bistatic constituents.

For a monostatic radar to measure Doppler it has to be coherent such that both phase and amplitude are captured. This is achieved by having a common and accurate timing signal that allows echo reception to be synchronized to transmission such that phase from pulse to pulse is preserved and measured. Bistatic and multistatic radar must also be phase coherent. However, as the transmitter and receiver are physically separated this becomes more challenging. It is possible to distribute a clock signal over cable to the nodes of multistatic radar or even to have multiple atomic clocks; however, these solutions become undesirable if the node separation is significant or if long-term operation is desired. Fortunately, the signals of global navigation satellite systems (GNSSes) may be exploited to obtain coherence [40–42]. The GNSS signals provide a 1 Hz reference signal to which a local oscillator is phase-locked. While there are quantization errors on 1 Hz signal they can be removed through averaging and coherence achieved. The GNSS signals can also be used to control the timing between the radar nodes, such that pulses are transmitted at known times [41, 42].

9.4 The Multistatic Micro-Doppler Signature

In this section the theory, and specifically the mathematics, of the micro-Doppler signature is expanded from the monostatic to the multistatic case. Detailed consideration of the monostatic micro-Doppler signature is given in Chapter 2, with which the reader is already assumed to be familiar. Here, a multistatic micro-Doppler signature is assumed to comprise multiple components arising from the multiple channels of the underlying radar system. First, it is considered how to expand the monostatic micro-Doppler theory described in Chapter 2 to the bistatic case. Second, the meaning of a signature that comprises multiple channels is examined, where each represents a different observation of the target. Finally, some basic worked examples to illustrate the significance of multistatic micro-Doppler are outlined.

9.4.1 Multistatic Micro-Doppler Mathematics

As already discussed, multistatic radar comprises multiple transmitter and receiver nodes, or even multiple transceiver nodes or a mixture thereof, which are spatially separated but still able to operate in concert. Figure 9.8 shows, in outline, the principal components and typical geometry of a multistatic system. The signal transmitted by the transceiver node (shown as a solid black line) is scattered from the target (solid grey lines) and received both by itself and by the receiver only node. This, therefore, defines one monostatic and one bistatic channel. The transmit only node also broadcasts a signal toward the target (dashed black line) that is again scattered by the target (dashed grey lines) and received at the transceiver and receiver nodes defining a further two bistatic channels. The scattering angle in each of these channels is different, resulting in changes to measured properties such as received power and Doppler shift. Additionally, the changes in illumination and scatter angle result in different self-occlusions (when some components of the target cause other target components to be shadowed and hence not observed) in each channel.

In this way it may be seen that the separation of the nodes in a multistatic radar, results in the different channels observing different target properties. This provides the basis by which more information may be obtained about the target. The channels are either monostatic or bistatic, with the latter being a generalization of the former.

Figure 9.9 shows the geometry of a target exhibiting micro-motion being observed by a single channel of a multistatic radar. The target is represented by the

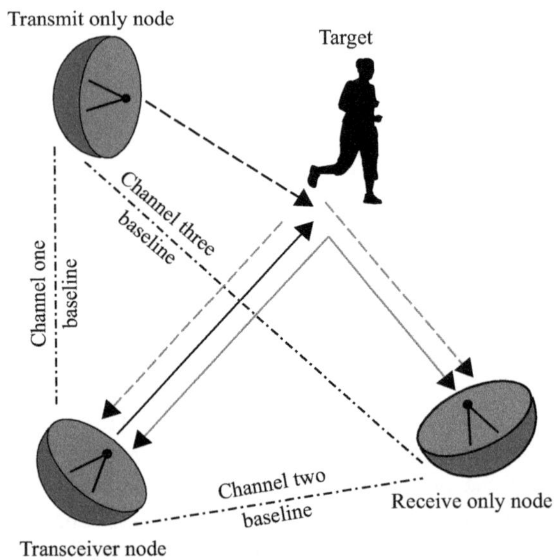

Figure 9.8 A generic geometry showing the principal features of a multistatic radar system

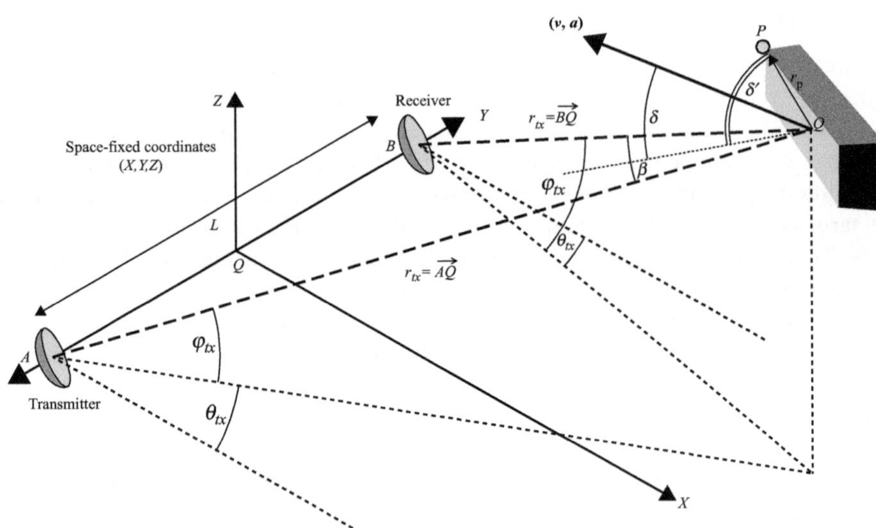

Figure 9.9 Geometry of a bistatic channel observing a target exhibiting micro-motions

grey block that has its center at Q. It has a gross motion described by velocity and acceleration vectors, $\boldsymbol{v}$ and $\boldsymbol{a}$. The target is considered to comprise a number of point scatterers. An example is shown in Figure 9.9 by the single scatterer at location P. The target is considered to be spinning about its center, Q, with a motion described by the Euler angular velocity $\boldsymbol{\omega} = [\,\omega_x \quad \omega_y \quad \omega_z\,]$ where ω_x, ω_y, and ω_z are the angular velocities about the X, Y, and Z axes respectively. If $\boldsymbol{r}_{P,0} = \overrightarrow{QP}$ is the initial vector from the target center to the scatterer then after a time t the scatterer will have moved to a new location described by

$$\begin{aligned} Q(t) &= Q_0 + t\boldsymbol{v} + \frac{1}{2}\boldsymbol{a}t^2 \\ P(t) &= Q(t) + \boldsymbol{\Omega}_t \boldsymbol{r}_{P,0} \end{aligned} \tag{9.11}$$

where Q_0 is the initial target center position and $\boldsymbol{\Omega}_t$ is a rotation matrix obtained from the angular velocity vector $\boldsymbol{\omega}$. The form of $\boldsymbol{\Omega}_t$ is described in Appendix A.

The transmitter and receiver are at points A and B respectively and the bistatic baseline has a length L defined as $L = \|\overrightarrow{AB}\|$, where $\|\cdot\|_2$ is the Euclidean norm of a vector. The bistatic plane is that which contains the three points A, B, and Q and the bistatic angle, β, is the angle $\angle AQB$, which is the angle between the LOSes from the transmitter and receiver. The transmit and receive ranges are, respectively, $r_{tx} = \|\overrightarrow{AQ}\|_2$ and $r_{tx} = \|\overrightarrow{BQ}\|_2$ while the transmitter and receiver look angles (azimuth and elevation) are, again respectively, (θ_{tx}, ϕ_t) and (θ_{rx}, ϕ_{rx}) and are measured from the direction of the x-axis.

The vectors from the radar nodes to the target can be defined as $\boldsymbol{r}_{tx} = \overrightarrow{AQ}$ for the transmitter and $\boldsymbol{r}_{rx} = \overrightarrow{BQ}$ for the receiver while the vector from the target center to the point scatterer is $\boldsymbol{r}_P = \overrightarrow{QP}$. The backscatter signal received from the scatterer at P can be described by

$$s_{rx}^P(t) = \sigma_P \exp\left(\frac{j2\pi}{\lambda}\|\boldsymbol{r}_{tx}(t) + \boldsymbol{r}_P(t)\| + \|\boldsymbol{r}_{rx}(t) + \boldsymbol{r}_P(t)\|\right) \tag{9.12}$$

where λ is the wavelength of the carrier frequency, σ_p is the reflectivity of the scatter at P. A time dependence has been introduced to the vectors in (9.12) to indicate that the vectors change over time in a manner consistent with (9.11). The received signal from the entire target is then

$$s_{rx}(t) = \iiint_{\text{all } P} \sigma_P \exp\left(j\frac{j2\pi}{\lambda}\|\boldsymbol{r}_{ts}(t) + \boldsymbol{r}_P(t)\| + \|\boldsymbol{r}_{rx}(t) + \boldsymbol{r}_P(t)\|\right) d\boldsymbol{r}_P \tag{9.13}$$

In (9.13) the signal $s_{rx}(t)$ is the progression of the complex amplitude reflected back from the target. There is no consideration given to the range extent of the target so we assume that either the target is being illuminated by a continuous wave (CW) radar or, if modulation is employed, the target remains within a single range bin for the duration of the observation.

The phase term of signal received from a single point scatterer on the target is

$$\Phi_P(t) = \frac{2\pi}{\lambda}\left(\|\boldsymbol{r}_{tx}(t) + \boldsymbol{r}_P(t)\|_2 + \|\boldsymbol{r}_{tx}(t) + \boldsymbol{r}_P(t)\|_2\right) \tag{9.14}$$

This expression can be broken into two parts, one for the linear translation motion and the other for the micro-motion

$$\mathbf{\Phi}_P(t) = \mathbf{\Phi}_{trans}(t) + \mathbf{\Phi}_{micro}(t) \tag{9.15}$$

where

$$\mathbf{\Phi}_{trans} = \frac{2\pi}{\lambda}\left(r_{tx}(t) + r_{rx}(t)\right) \tag{9.16}$$

and

$$\mathbf{\Phi}_{micro}(t) = \frac{2\pi}{\lambda} 2r_P(t)\cos\frac{\beta}{2}\cos(\delta') \tag{9.17}$$

For the derivation of (9.16) and (9.17) see Appendix A and note that in the equations the non-bold face variables represent the magnitudes of their vector counterparts, e.g., $a = \|\boldsymbol{a}\|_2$.

The Doppler shift is the time derivative of the phase function so

$$f_{D_{trans}}(t) = \frac{2\pi}{\lambda}(v + at)\cos(\beta/2)\cos(\delta) \tag{9.18}$$

and

$$f_{D_{micro}}(t) = \frac{4\pi}{\lambda}\cos\frac{\beta}{2}\cos(\delta')\|\dot{\mathbf{\Omega}}_t \mathbf{r}_{P,0}\| \tag{9.19}$$

where $\dot{\mathbf{\Omega}}_t$ is the time derivative of $\mathbf{\Omega}_{tx}$ and the full derivation is provided in Appendix A. Since the target is a collection of many scatterers the final expression for Doppler may be written as

$$f_D(t) = f_{D_{trans}}(t) + \iiint_{all\ P} f_{D_{micro}}(t)\mathrm{d}P \tag{9.20}$$

Note that in (9.20) the translational Doppler shift, which corresponds to the gross motion of the target, is outside the integral. This is consistent with the derivation in Appendix A, since this term relates to the center of the target that is the same for all of the scatterers.

The above summary, and the full derivation in Appendix A, is for a single channel but a multistatic radar comprises many channels. If there are N_{tx} transmit nodes and N_{rx} receive nodes then the total number of channels in radar, and therefore in the multistatic micro-Doppler signature, will be $N_{tx} \times N_{rx}$. (Note here we consider a transceiver as a collocated transmitter and receiver, so a system with two transceiver nodes would have $2 \times 2 = 4$ channels.) This represents the principal difference between multistatic and monostatic micro-Doppler signatures: multistatic signatures contain many channels while monostatic signatures contain just one channel.

9.4.2 Multistatic Micro-Doppler Simulation Example

Here, the simulated results of a micro-Doppler signature as observed by a three channel multistatic system are presented. The outputs of this simulation, time-frequency representations (TFRs) of the signature in each channel, are used to illustrate the key prominent features of multistatic micro-Doppler signatures (note that the simulation does not include noise or power variation with the R^{-4} law).

Figure 9.10(a) shows the target to be composed of two scatterers, one in linear motion representing bulk velocity, the other with a sinusoidal motion representing a micro-Doppler component. The stars indicate the location of the two scatterers at time $t = 0$ sec and the circles show their progression over time. The initial vector from the origin to the target center is $\begin{bmatrix}-2\times10^3 & 5\times10^3 & 500\end{bmatrix}^T$. The scatterer with linear motion only is located at the target center and the micro-motion scatterer is initially at a vector $[1 \quad 0 \quad 0]^T$ from the center. The micro-motion is described by the Euler angular velocity vector $[0 \quad 0 \quad 720]^T$, measured in degrees per second, and the linear motion is described by velocity vector $[0 \quad -20 \quad 0]^T$, measured in msec^{-1}. We therefore have a target traveling at 20 msec^{-1} along a path parallel to the y-axis. This is composed of two scatterers the second of which orbits the first at a distance of 1 m and makes two rotations per second. Although extremely simple, the target might be considered as a crude model of a helicopter: the linear motion scatterer representing the fuselage and the rotating scatterer the rotor blades.

The radar-target geometry is shown in Figure 9.10(b) and comprises three nodes. The first is a transceiver node located at $\begin{bmatrix}-2\times10^3 & 0 & 0\end{bmatrix}^T$, which operates with a 5 GHz carrier, and the remaining two nodes are receivers and are located at $\begin{bmatrix}2\times10^3 & 0 & 0\end{bmatrix}^T$ and $\begin{bmatrix}0 & 2\times10^3 & 5\times10^3\end{bmatrix}^T$, respectively. It follows that the radar has three channels and these are summarized in Table 9.1. The angles given in Table 9.1 are for the target center at its initial location. The

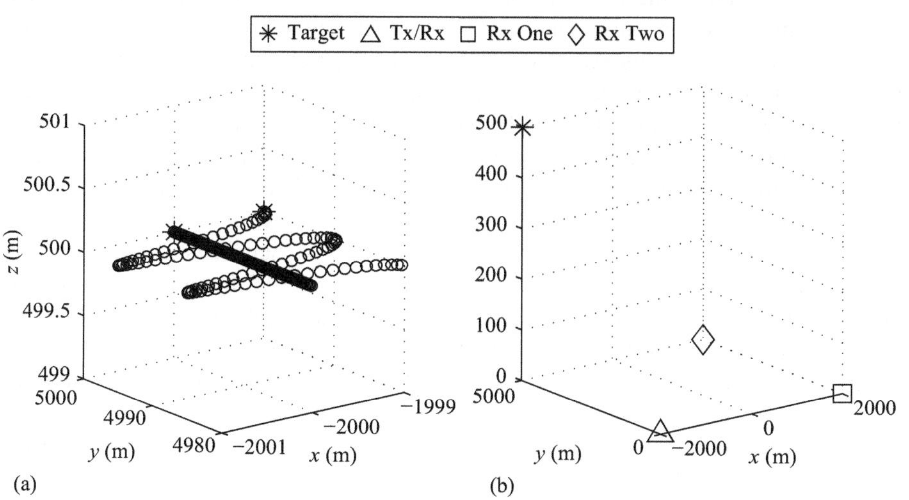

Figure 9.10 Details of the multistatic micro-Doppler simulation (a) progression of target components with time and (b) the radar-target geometry

Table 9.1 Parameters of the multistatic micro-Doppler simulation calculated using (9.10)

Channel	Baseline L (km)	Bistatic Angle β (°)	Velocity Angle δ (°)	Predicted Lin. Doppler (Hz)	Predicted Micro-Doppler (Hz)	
					min	max
Tx/Rx (Monostatic)	0	0	5.7	663.8	246.7	1081
Tx/Rx to Rx 1 (Bistatic 1)	4	38.5	20.1	591.5	219.8	936.1
Tx/Rx to Rx 2 (Bistatic 2)	6.4	89.3	45.6	332.1	62.7	601.

simulation is run for 1 sec, so the target moves forward by 20 m, which is not far enough to allow appreciable change to the angles given a bistatic range of the order of 10 km. Also presented in Table 9.1 are the predicted maximum and minimum Doppler values for the target signature.

The multistatic micro-Doppler signature output from the simulation is shown in Figure 9.11 as a set of TFRs. The TFR employed was a short time Fourier transform with a hamming window applied to reduce the sidelobes [43]. In each subplot of Figure 9.11 two signal components are visible one with a constant frequency, which corresponds to the scatterer with linear motion only, and the other with a sinusoidal frequency variation, which corresponds to the rotating scatterer. The linear component is commonly referred to as the main Doppler line (MDL), since it typically has higher power than the oscillating components and for a real target it represents the contribution of all of the components that do not exhibit a micro-motion and move with the gross motion only. The other component, which exhibits the sinusoidal frequency variation in this example, is the micro-Doppler component and represents the contribution for the target components that do exhibit micro-motion. For a real target there would be several micro-Doppler components and they would have more complex forms. The subplots, (a) through (c), of Figure 9.11 represent, respectively, the monostatic, first bistatic, and second bistatic channels of the radar.

The signatures presented in Figure 9.11 are consistent with the predicted results of Table 9.1. The linear Doppler shift in each channel is at the predicted frequency, and the minimum and maximum frequencies of the micro-Doppler are also as predicted to a reasonable accuracy. There is a slight variation between the predicted minimum and maximum values and those observable in Figure 9.11 due to rounding when calculating the angle δ between the micro-motion velocity and the bistatic bisector.

The initial frequency of the micro-Doppler component of the signatures varies between the channels. This variation is a result of the projection of the rotating scatterer's velocity onto the bistatic bisector for the different channels. The initial conditions are such that for the monostatic channel the peak micro-Doppler

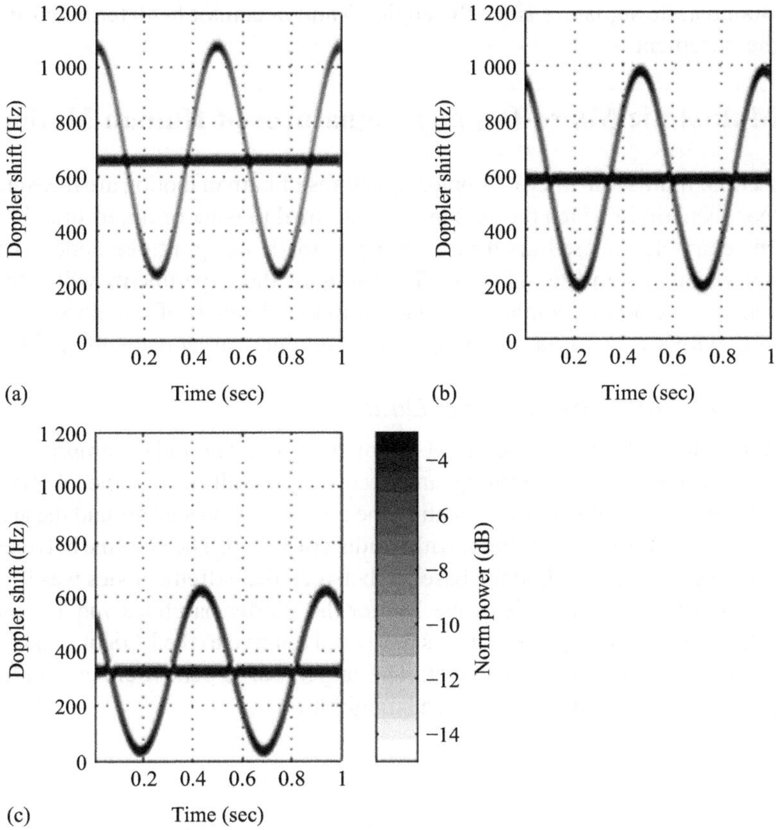

Figure 9.11 Time-frequency representations of the signature in the three channels of the simulation (a) the monostatic channel, (b) the first bistatic channel, and (c) the second bistatic channel

frequency is observed at time $t = 0$; however, this is not true for the other two channels and their initial frequencies are less than the maxima expected. It should be noted that there is no reason the micro-Doppler should be at a maxima for time $t = 0$; rather, the magnitude of the micro-Doppler shift when the target is first observed will be a function of the state of the micro-motions at that time. Despite the changes in velocity projection angle, the period of the micro-Doppler component is the same in all channels. Two clear periods are always visible indicating that the rotating target has moved through two cycles of its motion; since the observation interval is 1 sec we note that this is consistent with the 720 sec^{-1} rate set for the rotating scatterer.

In summary the multistatic geometry allows multiple micro-Doppler signature to be measured. The additional channels are simultaneous observations that can be used to solve for the underlying properties of the target motion. This is the approach taken by Ai *et al.* [32] to determine the rotational radius of a point scatterers based on a multistatic micro-Doppler signature and cannot be performed using a monostatic signature since the angles δ and δ' cannot be determined from a single measurement.

9.5 Multistatic Micro-Doppler Signatures of Human Motion

In this section, multistatic micro-Doppler signatures of human motion are measured in an outdoor scenario in which the person is constrained to walk or run in straight lines only. This example is based upon measurements made using a three node 2.4 GHz multistatic radar system [26, 44, 45]. The results of the experiments will inform a discussion on how best to combine the data from the channels of a multistatic radar, particularly with a view to supporting automatic or aided target recognition (ATR).

9.5.1 Gathering Experimental Data

Figure 9.12 shows the radar and target geometry used. The radar comprised three nodes, each capable of transmitting and receiving, resulting in nine observation channels: three monostatic and six bistatic. The nodes were spaced around the arc of a circle, centered on the target position, with a radius of 120 m. The baseline between the adjacent nodes was 50 m, while the baseline between the extreme nodes was 97.6 m. The corresponding bistatic angles were 24° for the shorter baselines and 48° for the longer. The direction of target travel was measured relative to the LOS of node 3, i.e., the line between node 3 and the target. The target was a person either running or walking, and the motion was always in a straight line.

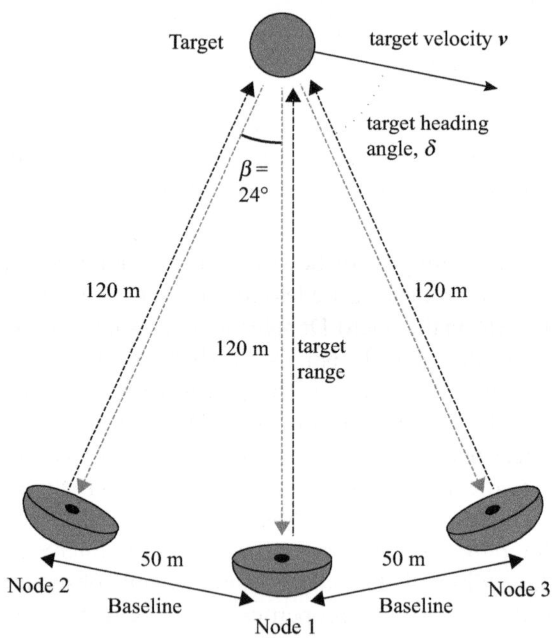

Figure 9.12 The radar and target geometry for the multistatic measurements

Detailed technical details of the multistatic radar system can be found in the references of Derham and of Doughty [44, 45]. The system's parameters, as they were set for the experiments, are detailed in Table 9.2. When considering the PRF it is important to appreciate that pulse interleaving was used: that is to say, each node of the radar had a PRF of 20 kHz, but only one node transmits at a time while all three receive. Strictly therefore, the three nodes do not make simultaneous observations of the target. The interleaving is uniform so the time delay between the measurements when node 1 transmits and when node 3 transmits is just 33 μsec that was considered insignificant compared to the durations of micro-motions of the target. As such, the data was processed as if all three nodes transmitted simultaneously. A limiting factor, however, was the system dynamic range that was only 35 dB, much smaller than the dynamic range of the micro-Doppler signature anticipated from [46, 47]. Despite this, the micro-Doppler signature of a human gait was observable but with some loss of detail.

The volunteer whose gait was being recorded moved at two velocities walking pace, ≈ 1.5 msec^{-1}, and running, ≈ 4.5 msec^{-1}. These velocities were estimated by measuring the volunteer under controlled conditions. When the volunteer was running, his direction of motion was at 0° to the LOS of node 3, an approaching geometry; when he was walking his motion was at 135° to node 3's LOS, a receding geometry. The expected frequency of the main Doppler line in each of the radar's channels can be readily calculated from (9.10) and are listed in Table 9.3. Each row represents a different channel identified by the transmitting and receiving node numbers. Associated with each channel are a bistatic angle, β, and an angle to the target direction of motion, δ. The latter parameter varies between the running and walking cases. It can be seen that the frequency of the main Doppler line is greatest when β is close to 0° and when δ is close to either 0° or 180°. This is because under these conditions the channel is monostatic and the target is either traveling head-on or tail-on to the bistatic bisector (see section 9.3).

The data processing employed extracted a sampled CW signal from the original range profiles. For each channel the sampled data was matched filtered with the transmitted chirp to provide a range profile. The range bins containing the volunteer's backscatter signal were identified in the range profiles by a simple process that relied on prior knowledge of his location and the fact that he was the only target. The volunteer migrated between range bins over the 60 000 pulses collected. To compensate for this in every range profile the range bins in which the volunteer's signal

Table 9.2 The operating parameters of the NetRAD multistatic radar

Parameter	Unit	Value	Parameter	Unit	Value
Number of transceiver nodes	–	3	Pulse width	μsec	0.6
Tx Power	mW	200	BW	MHz	40
Antenna gain	dBi	21	Waveform	–	Up-chirp
Pulse interleaving	–	True	Capture time	μsec	1.28
Pulse repetition frequency	kHz	20	No pulses	–	60 000

Table 9.3 Frequency of the main Doppler line for personnel target multistatic micro-Doppler signatures.

Tx Node	Rx Node	β (°)	Target running 0° to LOS of Node 3		Target walking 135° to LOS of Node 3	
			δ (°)	Main Doppler Line Freq. (Hz)	δ (°)	Main Doppler Line Freq. (Hz)
1	1	0	24	66.3	159	−22.6
1	2	24	36	57.5	171	−23.4
1	3	24	12	69.5	147	−19.9
2	1	24	36	57.5	171	−23.4
2	2	0	48	48.6	183	−24.2
2	3	48	24	60.6	159	−20.6
3	1	24	12	69.5	147	−19.9
3	2	48	24	60.6	159	−20.6
3	3	0	0	72.6	135	−17.1

was observed, across all pulses, were coherently summed to create a single large range bin. The complex samples from this expanded range bin for each pulse were then taken to be a sampled CW signal. Techniques, such as calculating the STFT were then applied to the sampled CW signal to observe the micro-Doppler signature. By repeating this process on each channel, nine time-synchronized signals were analyzed representing the multistatic micro-Doppler signature.

9.5.2 Experimental Results and Their Implications

Results from the experiments are shown in Figure 9.13. The two parts of the figure are each divided into nine sub-plots, one for each channel of the multistatic system. Each sub-plot shows the spectrograms for the echo signal in that channel and is titled "Tx N, Rx M" to indicate which node is transmitting and which is receiving. The monostatic channels are along the leading diagonal of the figures and the bistatic channels are to either side.

All nodes of the multistatic network can both transmit and receive. The number of channels increases as the product of the number of transmit and receive nodes. Under these conditions the challenge of data fusion becomes apparent: how best can the information in the different channels be exploited? It is also important to consider if these extra channels actually provide new target information [48].

Between the sub-plots of Figure 9.13, the frequency of the main Doppler line varies, in accordance with Table 9.3. It can also be seen that the bandwidth of the signature varies between channels. The maximum bandwidth was observed in the monostatic channel node 3 Tx, node 3 Rx, where the target is moving along the LOS. Again, this result is in agreement with the key aspect angle dependencies as predicted in section 9.3.

The oscillation rates of the signature components that represent the target's limbs are the same in the different channels. This is to be expected because the nine channels are simultaneous observations of the target. When, for example, a leg is

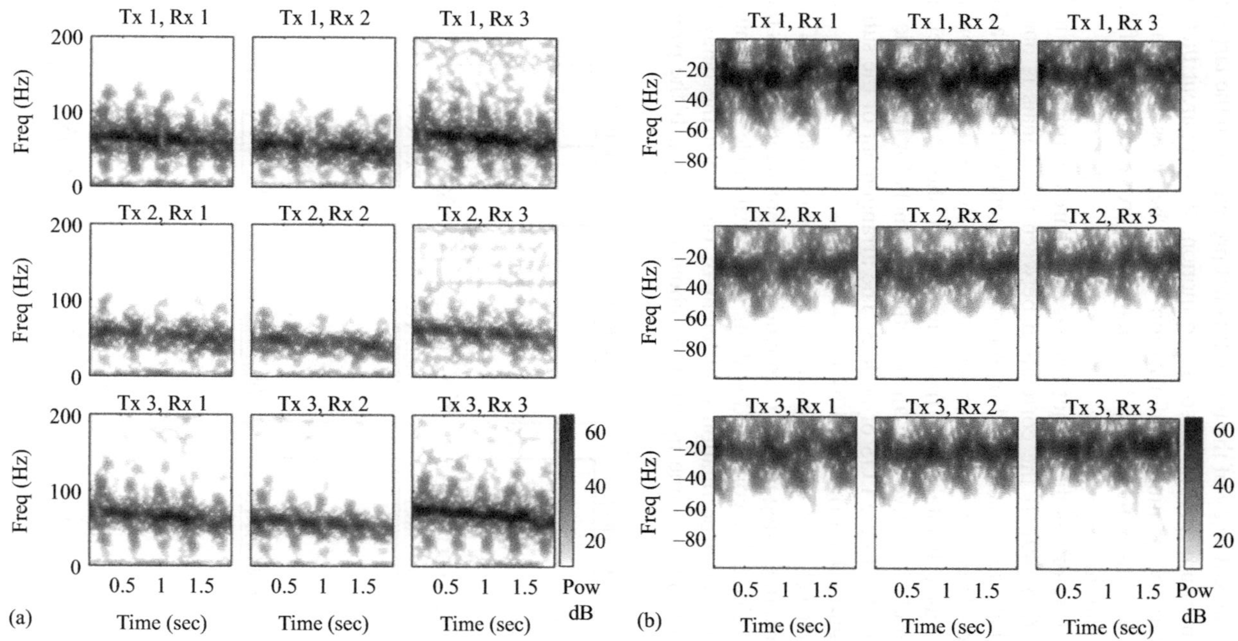

Figure 9.13 Multistatic micro-Doppler signature of a person (a) running and (b) walking

swinging forward in one of the channels, it is swinging forward in all the channels. It is also possible to observe a slight oscillation in the main Doppler line of the signature component relating to the torso when the target is walking. This is a result of the forward–backward sway of the torso. It is not present when the target runs as the increased momentum inhibits torso micro-motion. It is apparent that the different channels carry very similar information and it becomes important to ask which parts of a multistatic micro-Doppler signature contain useful, complementary information.

A simple measure of whether or not two signals share information is to look at their cross-correlation [49]. If the correlation between the signals is one at zero delay, we know they are the same and therefore the second contains no additional information; if they do not correlate then they contain different information, and the second channel is as valuable as the first. Figure 9.14 shows the correlation for node 1 transmit, node 3 receive with all the other channels for the simulation of a person walking at 135° to the LOS of node 3. The arrangement of the sub-plots is the same as in Figures 9.13(a) and 9.13(b).

Clearly when the channel is correlated with itself, the correlation coefficient at a delay of zero is one, indicating perfect correlation. Of more significance is the result when the selected channel is correlated with its reciprocal, node 3 transmit, node 1 receiver. Again we see a correlation coefficient of one when the delay is zero, suggesting they contain the same information about the target. If the path node 3

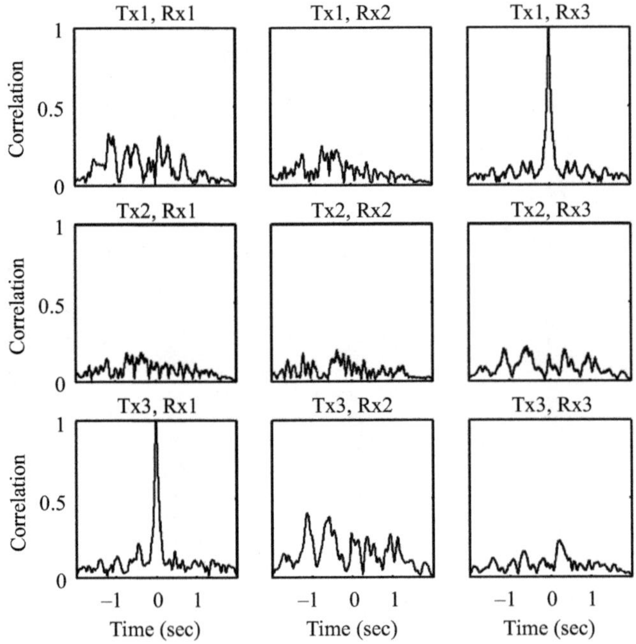

Figure 9.14 Cross-correlation of channel node 1 Tx, node 3 Rx with all other channels for the real person walking at 135° to node 3's LOS

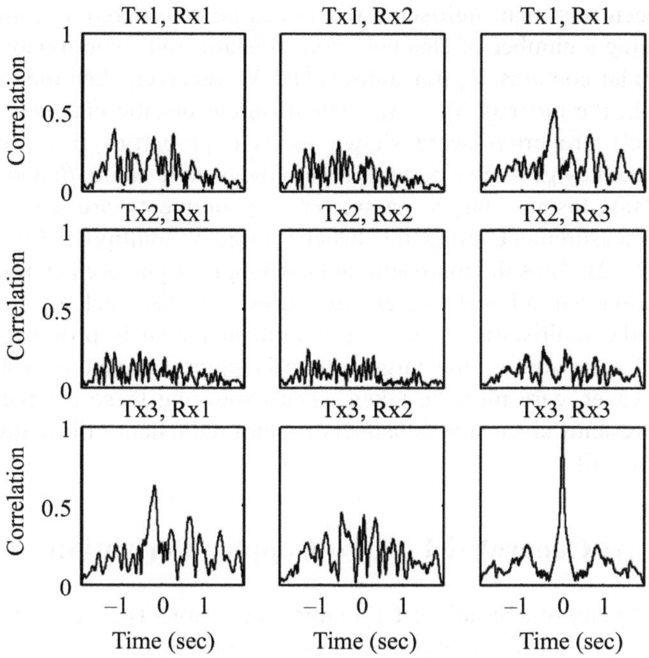

Figure 9.15 Cross-correlation of channel node 3 Tx, node 3 Rx with all other channels for the real person walking at 135° to node 3's LOS

transmit, node 1 receive is considered the reciprocal of node 1 transmit, node 3 receive, then it can be concluded that the reciprocal path contains no additional target information. In the other channels shown in Figure 9.14 the correlation is always low. Similar results were obtained for all bistatic channels: the reciprocal always correlated perfectly indicating that it contained no additional information.

The situation for the three monostatic channels is different. Figure 9.15 shows the results from cross-correlating channel node 3 transmit, node 3 receiver with all other channels. Only the autocorrelation case shows a high correlation indicating that new and useful target information is present in the other channels. Making this additional information available to a classifier would be expected to improve radar ATR performance, as has been reported for HRRP based systems, although this is not always the case [4–7].

9.6 Summary and Conclusions

In this chapter the theory of multistatic micro-Doppler signature measurement has been introduced and developed to explain how additional information is generated. This additional information aids target classification. It has been demonstrated in the research literature that observing a target from multiple aspects improves the probability of correctly classifying it.

It has been seen that multistatic radars can be considered as composite systems containing a number of channels, some bistatic and some monostatic. If the multistatic radar contains N_{tx} transmitters and N_{rx} receivers then the total number of channels in the radar is $N_{tx} \times N_{rx}$. The different bistatic channels and monostatic channels properties were shown to have properties that are geometry dependent. The key geometry parameters are the bistatic angle, β, and the bistatic bisector. Many bistatic target parameter measurements are equivalent to a monostatic measurement along the bistatic bisector multiplied by the bistatic factor of cos(β/2). Thus the multisatic micro-Doppler signature is represented as a series of individual micro-Doppler signatures, one for each of the $N_{tx} \times N_{rx}$ channels in the multistatic radar. The additional channels provide alternative, simultaneous estimates of the target micro-Doppler parameters from different aspects. However, care must be taken when exploiting these channels in target recognition systems since some channels contain redundant information that will provide no benefit.

Appendix A: Generalized Micro-Doppler Derivation

A general geometry of a bistatic radar illuminating a target is provided in Figure 9.9, here an expression for the Doppler shift is derived. For the derivation a constant target RCS and negligible noise are assumed. Thus the analysis focuses on the derivation of the phase expression. The time derivative of phase gives the expression for Doppler.

The notation used in the derivation is as follows:

- Vectors are noted using either lower case bold face or an arrow above two points, e.g., $\boldsymbol{a}$ is a vector and $\overrightarrow{AB}$ is the vector between points A and B.
- Matrices are noted in bold face upper case, e.g., $\boldsymbol{M}$.
- The magnitude of a vector is indicated using lower case regular face, e.g., $a = \|\boldsymbol{a}\|_2$ where $\|\cdot\|_2$ is the Euclidean norm.
- A unit vector is represented using a hat above the vector, e.g., $\hat{\boldsymbol{a}} = \boldsymbol{a}/a$ is a unit vector along $\boldsymbol{a}$.
- The scalar, or dot, product between vectors $\boldsymbol{a}$ and $\boldsymbol{b}$ is denoted $\langle \boldsymbol{a}, \boldsymbol{b} \rangle$.
- The vector that bisects the vectors $\boldsymbol{a}$ and $\boldsymbol{b}$ is $b\boldsymbol{a} + a\boldsymbol{b} = ab(\hat{\boldsymbol{a}} + \hat{\boldsymbol{b}})$.

In the derivation we also exploit the following identities and expansions:

- The scalar product result $\langle \boldsymbol{a}, \boldsymbol{a} \rangle = a^2$.
- The Taylor expansion $\sqrt{1 + 2x} = 1 + x + \mathcal{O}(x^2)$ which is valid for $|x| < \frac{1}{2}$.
- The trigonometric identity $\cos(\alpha) = 2\cos^2\left(\frac{\alpha}{2}\right) + 1$.

The scatterer at P is some distance from Q, the center of the target. The vector between the two points is given by

$$\boldsymbol{r}_P = \overrightarrow{QP} \tag{9.A-1}$$

The transmitter and receiver are at points A and B, respectively, and we define the vectors from them to the target center as

$$\begin{aligned} \boldsymbol{r}_{tx} &= \overrightarrow{AQ} \\ \boldsymbol{r}_{rx} &= \overrightarrow{BQ} \end{aligned} \tag{9.A-2}$$

Note that the angle between $\boldsymbol{r}_{tx}$ and $\boldsymbol{r}_{rx}$ is the bistatic angle β. The vector from the transmitter and receiver to the scatterer at P may be written as

$$\begin{aligned} \overrightarrow{AP} &= \boldsymbol{r}_{tx} + \boldsymbol{r}_p \\ \overrightarrow{BP} &= \boldsymbol{r}_{tx} + \boldsymbol{r}_p \end{aligned} \tag{9.A-3}$$

Assuming the carrier signal wavelength is λ the phase of the signal received from the scatterer at P is then given by [26, 38, 39]

$$\boldsymbol{\Phi}_P = \frac{2\pi}{\lambda}\left(\|\overrightarrow{AP}\|_2 + \|\overrightarrow{BP}\|_2\right) = \frac{2\pi}{\lambda}\left(\|\boldsymbol{r}_{tx} + \boldsymbol{r}_P\|_2 + \|\boldsymbol{r}_{rx} + \boldsymbol{r}_P\|_2\right) \tag{9.A-4}$$

The term $\|\boldsymbol{r}_{tx} + \boldsymbol{r}_P\|_2$ can be simplified as follows:

$$\begin{aligned} \|\boldsymbol{r}_{tx} + \boldsymbol{r}_P\|_2 &= \sqrt{\langle \boldsymbol{r}_{tx} + \boldsymbol{r}_P, \boldsymbol{r}_{tx} + \boldsymbol{r}_P\rangle} \\ &= \sqrt{r_{tx}^2 + r_P^2 + 2\langle \boldsymbol{r}_{tx}, \boldsymbol{r}_P\rangle} \\ &= r_{tx}\sqrt{1 + r_P^2/r_{tx}^2 + 2\langle \boldsymbol{r}_{tx}, \boldsymbol{r}_P\rangle/r_{tx}^2} \end{aligned} \tag{9.A-5}$$

assuming that $r_{tx} \gg r_P$ then $r_p^2/r_{tx}^2 \approx 0$:

$$\begin{aligned} \|\boldsymbol{r}_{tx} + \boldsymbol{r}_P\|_2 &= r_{tx}\sqrt{1 + 2\langle \boldsymbol{r}_{tx}, \boldsymbol{r}_P\rangle/r_{tx}^2} \\ &= r_{tx}(1 + \langle \boldsymbol{r}_{tx}, \boldsymbol{r}_P\rangle/r_{tx}^2) \end{aligned} \tag{9.A-6}$$

using the Taylors series expansion of $\sqrt{1+2x}$ to the first order. Finally,

$$\|\boldsymbol{r}_{tx} + \boldsymbol{r}_P\|_2 = r_{tx}(1 + \langle \boldsymbol{r}_{tx}, \boldsymbol{r}_P\rangle/r_{tx}^2) = r_{tx} + \langle \hat{\boldsymbol{r}}_{tx}, \boldsymbol{r}_P\rangle \tag{9.A-7}$$

A similar derivation can be made for the $\|\boldsymbol{r}_{rx} + \boldsymbol{r}_P\|_2$ term allowing (9.A-4) to be written as

$$\begin{aligned} \boldsymbol{\Phi}_P &= \frac{2\pi}{\lambda}(r_{tx} + r_{rx} + \langle \hat{\boldsymbol{r}}_{tx}, \boldsymbol{r}_P\rangle + \langle \hat{\boldsymbol{r}}_{rx}, \boldsymbol{r}_P\rangle) \\ &= \frac{2\pi}{\lambda}(r_{tx} + r_{rx} + \langle \hat{\boldsymbol{r}}_{tx} + \hat{\boldsymbol{r}}_{rx}, \boldsymbol{r}_P\rangle) \\ &= \frac{2\pi}{\lambda}(r_{tx} + r_{rx}) + \frac{2\pi}{\lambda}(\langle \hat{\boldsymbol{r}}_{tx} + \hat{\boldsymbol{r}}_{rx}, \boldsymbol{r}_P\rangle) \\ &= \boldsymbol{\Phi}_{trans} + \boldsymbol{\Phi}_{micro} \end{aligned} \tag{9.A-8}$$

where $\boldsymbol{\Phi}_{trans} = \frac{2\pi}{\lambda}(r_{tx} + r_{rx})$ is the phase relating to the round trip path to the target center, Q, and $\boldsymbol{\Phi}_{micro} = \frac{2\pi}{\lambda}(\langle \hat{\boldsymbol{r}}_{tx} + \hat{\boldsymbol{r}}_{rx}, \boldsymbol{r}_P\rangle)$ is the additional phase shift resulting from the extra distance between Q and P.

The Doppler shift can be obtained by taking the time derivative of (9.A-8). Considering the translation component of phase first,

$$f_{D_{trans}} = \frac{d}{dt}\Phi_{trans} = \frac{2\pi}{\lambda}\left(\frac{dr_{tx}}{dt} + \frac{dr_{tx}}{dt}\right) \tag{9.A-9}$$

and the same derivation and arguments made in section 9.3.5 to obtain

$$f_{D_{trans}} = \frac{2\pi}{\lambda}(v + at)\cos(\beta/2)\cos(\delta) \tag{9.A-10}$$

The velocity and acceleration here are the magnitude of the vectors, $\boldsymbol{v}$ and $\boldsymbol{a}$, shown in Figure 9.9.

For the micro-Doppler component of (9.A-8) first the scalar product is expanded to give

$$\Phi_{micro} = \frac{2\pi}{\lambda}\|\hat{\boldsymbol{r}}_{tx} + \hat{\boldsymbol{r}}_{rx}\|_2 r_p \cos(\delta') \tag{9.A-11}$$

where knowledge that $\hat{\boldsymbol{r}}_{tx} + \hat{\boldsymbol{r}}_{rx}$ is along the bistatic bisector has been exploited so that the angle between $\hat{\boldsymbol{r}}_{tx} + \hat{\boldsymbol{r}}_{rx}$ and $\boldsymbol{r}_p$ must be δ' as indicated in Figure 9.9. Since $\hat{\boldsymbol{r}}_{tx}$ and $\hat{\boldsymbol{r}}_{rx}$ are unit vectors, an expansion of $\|\hat{\boldsymbol{r}}_{tx} + \hat{\boldsymbol{r}}_{rx}\|_2$ can be carried out similar to that performed above, remembering that $\hat{\boldsymbol{r}}_{tx} = \hat{\boldsymbol{r}}_{rx} = 1$ and that the angle between the vectors is the bistatic angle, β, to give

$$\|\hat{\boldsymbol{r}}_{tx} + \hat{\boldsymbol{r}}_{rx}\|_2 = \sqrt{2 + 2\cos\beta} \tag{9.A-12}$$

Applying the trigonometric identity highlighted at the start of this appendix leads to

$$\|\hat{\boldsymbol{r}}_{tx} + \hat{\boldsymbol{r}}_{rx}\|_2 = \sqrt{2 + 4\cos^2\frac{\beta}{2} - 2} = 2\cos\frac{\beta}{2} \tag{9.A-13}$$

which can be substituted back into (9.A-11) to give

$$\Phi_{micro} = \frac{2\pi}{\lambda} 2r_P \cos\frac{\beta}{2}\cos(\delta') \tag{9.A-14}$$

Again, to obtain the micro-Doppler shift the time derivative of phase is taken giving

$$f_{D_{micro}} = \frac{4\pi}{\lambda}\cos\frac{\beta}{2}\cos(\delta')\frac{dr_P}{dt} \tag{9.A-15}$$

where the observation interval is assumed to be short enough that δ' can be considered constant over time.

The final part of the derivation is to consider $\frac{dr_P}{dt}$, which is the progression of $\boldsymbol{r}_p$ with time, i.e., the micro-motion. The model of the micro-motion depends on the target being considered. If the target is tumbling in addition to its linear velocity and acceleration the model for the vector $\boldsymbol{r}_p$ may be written as

$$\boldsymbol{r}_p = \boldsymbol{\Omega}_t \boldsymbol{r}_{P,0} \tag{9.A-16}$$

where t is time, $\boldsymbol{r}_{p,0}$ is the vector to P at $t=0$ sec, and $\boldsymbol{\Omega}_t$ is a time dependent rotation matrix obtained from a Euler angular velocity vector $\boldsymbol{\omega}$ that defines the rotational velocity about the X, Y, Z axes, e.g.,

$$\boldsymbol{\omega} = [\omega_X \quad \omega_Y \quad \omega_Z]^{\mathrm{T}}$$

$$\boldsymbol{\Omega}_t = \begin{bmatrix} \cos\omega_Y t\cos\omega_Z t & \cos\omega_X t\sin\omega_Z t + \sin\omega_X t\sin\omega_Y t\cos\omega_Z t & \sin\omega_X t\sin\omega_Z t - \cos\omega_X t\sin\omega_Y t\cos\omega_Z t \\ -\cos\omega_Y t\sin\omega_Z t & \cos\omega_X t\cos\omega_Z t - \sin\omega_X t\sin\omega_Y t\sin\omega_Z t & \sin\omega_X t\cos\omega_Z t + \cos\omega_X t\sin\omega_Y t\sin\omega_Z t \\ \sin\omega_Y t & -\sin\omega_X t\cos\omega_Y t & \cos\omega_X t\cos\omega_Y t \end{bmatrix} \tag{9.A-17}$$

This motion model is chosen for the point P since it can describe a large number of motions depending on the elements of $\boldsymbol{\omega}$. The time derivative of (9.A-16) is then

$$\frac{dr_P}{dt} = \frac{d\|\boldsymbol{r}_P\|_2}{dt} = \|\dot{\boldsymbol{\Omega}}_t \boldsymbol{r}_{P,0}\| \tag{9.A-18}$$

where $\dot{\boldsymbol{\Omega}}_t$ is the time derivative of $\boldsymbol{\Omega}_t$. Substituting (9.A-18) into (9.A-15) gives

$$f_{D_{micro}} = \frac{4\pi}{\lambda}\cos\frac{\beta}{2}\cos(\delta')\|\dot{\boldsymbol{\Omega}}_t \boldsymbol{r}_{P,0}\| \tag{9.A-19}$$

The final Doppler signature returned from scatterer at P is therefore

$$\begin{aligned} f_D &= f_{D_{trans}} + f_{D_{micro}} \\ &= \frac{2\pi}{\lambda}\cos\frac{\beta}{2}\left((v+at)\cos(\delta) + 2\cos(\delta')\|\dot{\boldsymbol{\Omega}}_t \boldsymbol{r}_{P,0}\|\right) \end{aligned} \tag{9.A-20}$$

which is a time varying signature, since the rotation matrix derivative $\dot{\boldsymbol{\Omega}}_t$ depends on time. Note that if the transmitter and receiver were collocated then $\beta = 0°$, $\cos\frac{\beta}{2} = 1$ and (9.A-20) becomes the expression of the monostatic micro-Doppler, see Chapter 2. Further, note that the closer P is to the target center, Q, the smaller $r_{p,0}$ will be and the value of $f_{D_{micro}}$ will tend toward 0 Hz. In the extreme case where $r_{p,0} = 0$ m, the scatterer is at the target center, so the tumbling motion would not affect it, and (9.A-20) be the Doppler shift of a target with no micro-motion.

References

[1] C. J. Baker, M. Vespe, and G. J. Jones, "Target classification by echo locating animals," in *Waveform Diversity and Design Conference, 2007. International*, pp. 348–352, 4–8 June 2007.

[2] Y. Kim and H. Ling, "Human activity classification based on micro-Doppler signatures using a support vector machine," *Transactions on Geoscience and Remote Sensing, IEEE*, vol. 47, no. 5, pp. 1328–1337, 2009.

[3] J. A. Simmons, P. A. Saillant, and S. P. Dear, "Through a bat's ear," *Spectrum, IEEE*, vol. 29, no. 3, pp. 46–48, 1992.

[4] M. Vespe, C. J. Baker, and H. D. Griffiths, "Radar target classification using multiple perspectives," *IET Radar, Sonar & Navigation*, vol. 1, no. 4, p. 300, 2007.

[5] M. Vespe, C. J. Baker, and H. D. Griffiths, "Automatic target recognition using multi-diversity radar," *Radar, Sonar Navigation, IET*, vol. 1, no. 6, pp. 470–478, 2007.

[6] F. Zhu, X.-D. Zhang, Y. Hu, and D. Xie, "Nonstationary hidden Markov models for multiaspect discriminative feature extraction from radar targets," *Transactions on Signal Processing, IEEE*, vol. 55, no. 5, pp. 2203–2214, May 2007.

[7] S. Ji, X. Liao and L. Carin, "Adaptive multi-aspect target classification and detection with hidden Markov models," in *Conference on Acoustics, Speech, and Signal Processing, 2004. Proceedings. (ICASSP '04). IEEE International*, 2004, vol. 2, pp. ii–125-8.

[8] V. C. Chen, *The Micro-Doppler Effect in Radar*. Artech House, 2011.

[9] V. C. Chen, F. Li, S.-S. Ho, and H. Wechsler, "Analysis of micro-Doppler signatures," *Radar, Sonar and Navigation, IEE Proceedings*, vol. 150, no. 4, pp. 271–276, 2003.

[10] V. C. C. Chen, H. Wechsler, F. Li, and S.-S. Ho, "Micro-Doppler effect in radar: phenomenon, model, and simulation study," *Transactions on Aerospace and Electronic Systems, IEEE*, vol. 42, no. 1, pp. 2–21, Jan. 2006.

[11] V. C. Chen, "Doppler signatures of radar backscattering from objects with micro-motions," *Signal Processing, IET*, vol. 2, no. 3, pp. 291–300, Sep. 2008.

[12] A. G. Stove and D. Classifier, "A Doppler-Based Target Classifier Using Linear Discriminants and Principal Components," in *The Institution of Engineering and Technology Seminar on High Resolution Imaging and Target Classification, 2006*, pp. 107–125, 2006.

[13] G. E. Smith, K. Woodbridge, and C. J. Baker, "Radar micro-doppler signature classification using dynamic time warping," *IEEE Transactions on Aerospace and Electronic Systems*, vol. 46, no. 3, pp. 1078–1096, 2010.

[14] A. Balleri, K. Chetty, and K. Woodbridge, "Classification of personnel targets by acoustic micro-doppler signatures," *Radar, Sonar Navigation, IET*, vol. 5, no. 9, pp. 943–951, 2011.

[15] I. Bilik and P. Khomchuk, "Minimum divergence approaches for robust classification of ground moving targets," *Transactions on Aerospace and Electronic Systems, IEEE*, vol. 48, no. 1, pp. 581–603, 2012.

[16] I. Bilik and J. Tabrikian, "Radar target classification using Doppler signatures of human locomotion models," *Transactions on Aerospace and Electronic Systems, IEEE*, vol. 43, no. 4, pp. 1510–1522, 2007.

[17] I. Bilik, J. Tabrikian, and A. Cohen, "GMM-based target classification for ground surveillance Doppler radar," *Transactions on Aerospace and Electronic Systems, IEEE*, vol. 42, no. 1, pp. 267–278, 2006.

[18] T. Damarla, M. Bradley, A. Mehmood, and J. M. Sabatier, "Classification of animals and people ultrasonic signatures," *Sensors Journal, IEEE*, vol. 13, no. 5, pp. 1464–1472, 2013.

[19] Y. Li, L. Du, and H. Liu, "Hierarchical classification of moving vehicles based on empirical mode decomposition of micro-Doppler signatures," *Transactions on Geoscience and Remote Sensing, IEEE*, vol. 51, no. 5, pp. 3001–3013, 2013.

[20] G. E. Smith, K. Woodbridge, and C. J. Baker, "Naive Bayesian radar micro-doppler recognition," in *Conference on Radar, 2008 International*, 2008, pp. 111–116.

[21] A. G. Stove and S. R. Sykes, "A Doppler-based automatic target classifier for a battlefield surveillance radar," in *RADAR 2002*, pp. 419–423, 2002.

[22] A. G. Stove and S. R. Sykes, "A Doppler-based target classifier using linear discriminants and principal components," in *Radar Conference, 2003. Proceedings of the International*, pp. 171–176, 2003.

[23] M. R. Bell and R. A. Grubbs, "JEM modeling and measurement for radar target identification," *IEEE Transactions on Aerospace and Electronic Systems*, vol. 29, no. 1, pp. 73–87, 1993.

[24] J. Martin, B. Mulgrew, "Analysis of the theoretical radar return signal form aircraft propeller blades," in *Radar Conference, 1990. Record of the IEEE 1990 International*, pp. 569–572, 7–10 May 1990.

[25] W. Zhang, C. Tong, Q. Zhang, Y. Zhang, and X. Zhang, "Extraction of vibrating features with dual-channel fixed-receiver bistatic SAR," *Geoscience and Remote Sensing Letters, IEEE*, vol. 9, no. 3, pp. 507–511, May 2012.

[26] G. E. Smith, K. Woodbridge, C. J. Baker, and H. Griffiths, "Multistatic micro-Doppler radar signatures of personnel targets," *IET Signal Processing*, vol. 4, no. 3, pp. 224–233, 2010.

[27] C. Clemente and J. J. Soraghan, "Vibrating target micro-Doppler signature in bistatic SAR with a fixed receiver," *Transactions on Geoscience and Remote Sensing, IEEE*, vol. 50, no. 8, pp. 3219–3227, 2012.

[28] G. E. Smith, K. Woodbridge, and C. J. Baker, "multistatic micro-Doppler signature of personnel," in *2008 IEEE Radar Conference, RADAR 2008*, 2008.

[29] V. C. Chen, A. des Rosiers, and R. Lipps, "Bi-static ISAR range-Doppler imaging and resolution analysis," in *Radar Conference, 2009 IEEE*, pp. 1–5, 2009.

[30] C. Clemente and J. J. Soraghan, "On the effect of vibrating target in bistatic SAR images," in *IMA Conference on Mathematics in Defence*, pp. 1–5, 2011.

[31] C. Clemente and J. J. Soraghan, "Characterization of vibrating targets in bistatic SAR," in *Sensor Signal Processing for Defence (SSPD 2011)*, pp. 1–5, 2011.

[32] X. Ai, X. Zou, J. Yang, J. Liu, and Y. Li, "Feature extraction of rotating target based on bistatic micro-Doppler analysis," in *Conference on Radar (Radar), 2011 IEEE CIE International*, vol. 1, pp. 609–612, 2011.

[33] R. Zhu, Q. Zhang, X. Zhu, and Y. Luo, "Micro-Doppler analysis of vibrating target in bistatic radar," in *Conference on Synthetic Aperture Radar, 2009. APSAR 2009. 2nd Asian-Pacific*, pp. 981–984, 2009.

[34] W. Zhang, C. Tong, L. Ren, X. Chen, and X. Zhang, "Bistatic micro-Doppler radar signatures of vibrating targets with along-track

interferometry," in *Microwave, Antenna, Propagation, and EMC Technologies for Wireless Communications (MAPE), 2011 IEEE 4th International Symposium*, pp. 505–508, 2011.
[35] G. E. Smith, K. Woodbridge, and C. J. Baker, "Multiperspective micro-Doppler signature classification," in *Conference on Radar Systems, 2007 IET International*, pp. 1–5, 2007.
[36] N. J. Willis, *Bistatic Radar*. SciTech Publishing, 2005.
[37] M. C. Jackson, "The geometry of bistatic radar systems," *Commun. Radar Signal Process. IEE Proc. F,* vol. 133, no. 7, pp. 604 –612, 1986.
[38] N. J. Willis and H. D. Griffiths, *Advances in Bistatic Radar.* SciTech, 2007.
[39] M. Cherniakov, Ed., *Bistatic Radars: Emerging Technology*. Wiley, 2008.
[40] M. Inggs, G. Inggs, S. Sandenbergh, W. Al-Ashwal, K. Woodbridge, and H. Griffiths, "Multistatic networked radar for sea clutter measurements," in *Geoscience and Remote Sensing Symposium (IGARSS), 2011 IEEE International*, pp. 4449–4452, 2011.
[41] J. S. Sandenbergh, M. R. Inggs, and W. A. Al-Ashwal, "Evaluation of coherent netted radar carrier stability while synchronised with GPS-disciplined oscillators," in *Radar Conference (RADAR), 2011 IEEE*, pp. 1100–1105, 23–27 May 2011.
[42] J. S. S. Sandenbergh and M. R. R. Inggs, "A common view GPSDO to synchronize netted radar," in *Conference on Radar Systems, 2007 IET International*, pp. 1–5, 2007.
[43] V. C. Chen and H. Ling, *Time-Frequency Transforms for Radar Imaging and Signal Analysis*. Artech House, 2002.
[44] S. R. Doughty, "PhD thesis: Development and performance evaluation of a multistatic radar system," University of London, 2008.
[45] T. E. Derham, S. R. Doughty, K. Woodbridge, and C. J. Baker, "Design and evaluation of a low-cost multistatic netted radar system," *Radar, Sonar Navigation, IET,* vol. 1, no. 5, pp. 362–368, Oct. 2007.
[46] S. L. Marple, Jr., "Sharpening and bandwidth extrapolation techniques for radar micro-Doppler feature extraction," in *Radar Conference, 2003. Proceedings of the International*, pp. 166–170, 2003.
[47] S. L. Marple, Jr., "Large dynamic range time-frequency signal analysis with application to helicopter Doppler radar data," in *Signal Processing and its Applications, Sixth International, Symposium on. 2001*, vol. 1, pp. 260–263, 2001.
[48] R. O. Duda, P. E. Hart, and D. G. Stork, *Pattern Classification*, 2nd ed. Wiley, 2006.
[49] T. M. Cover and J. A. Thomas, *Elements of Information Theory*, 2nd ed. Wiley, 2006.

Chapter 10

Signal Decomposition of Micro-Doppler Signatures

L. Stanković, M. Daković, T. Thayaparan, and V. Popović-Bugarin

10.1 Introduction

Micro-Doppler effects appear in the inverse synthetic aperture radar (ISAR) imaging when a target has one or more fast-moving parts [1–7]. Similar effects appear in the synthetic aperture radar (SAR) imaging as well [8]. These effects may decrease readability of radar images. Namely, the spectral content of the micro-Doppler signal changes over time in a wide range. Therefore, the micro-Doppler may cover a rigid body and make it difficult to detect. On the other hand, the micro-Doppler effects, at the same time, carry useful information about features (type, velocity, size, etc.) of the fast-moving parts of the rigid body [9]. It is easier to estimate these features if the micro-Doppler effect is separated from the rigid body part of the radar image. Thus, the extraction of the micro-Doppler effects is an important problem in the radar imaging. Since the spectral content of signal corresponding to the micro-Doppler effects is time-varying, time-frequency (TF) analysis techniques are an appropriate tool for the micro-Doppler features analysis and extraction.

Based on time-frequency representation (TFR) of the radar signal, several methods for the micro-Doppler effects analysis are presented here. The most common and simplest form of the signal that corresponds to the micro-Doppler effect is a sinusoidal frequency-modulated (FM) signal. This kind of the micro-Doppler signatures can be extracted based on the inverse Radon transform of its TFR [10]. This transform projects a two-dimensional (2D) sinusoidal signature into a single point whose position corresponds to the parameters of the sinusoidal signature. The technique based on the inverse Radon transform can be used for other non-sinusoidal periodic micro-Doppler forms. In general, the micro-Doppler effect, as a nonstationary signal with time-varying spectral content, can be separated from the rigid body by using the L-statistics on the TF signal representations [10, 11]. The L-statistics-based separation of the micro-Doppler effects from the rigid body is very efficient and simple for any micro-Doppler form. The Viterbi algorithm can improve the micro-Doppler tracking in high noise environments [12].

In section 10.2 a model of radar signal for the fast-moving reflectors is reviewed. The drawbacks of the Fourier transform (FT), as a common tool for the SAR/ISAR imaging, in the case of fast-moving reflectors imaging is explained. The most commonly used TFRs, which are a suitable tool for the nonstationary signals analysis, are shortly reviewed in this section as well. A technique for modulation parameters estimation of a sinusoidal FM related micro-Doppler signal, based on the inverse Radon transform, is presented in section 10.3. Its application to the cases, with a fraction of period of the micro-Doppler signal being available, is presented. Since, the micro-Doppler is not necessary of the form of a sinusoidal FM signal, a simple algorithm for period estimation of a general form of the periodic signals is presented in this section, as well. The micro-Doppler must be separated from a rigid body in order to be further analyzed, thus the L-statistics-based algorithm for the micro-Doppler separation is presented in section 10.4. In section 10.5, a Viterbi-based algorithm for the micro-Doppler tracking in high noise is given.

10.2 Micro-Doppler Signal Model

Consider continuous wave radar that transmits signal in a form of coherent series of chirps $\exp(j\pi Bt^2/T_r)$, for $0 \leq t \leq T_r$. In one revisit, the transmitted signal consists of M such chirps. The repetition time of a single chirp is denoted by T_r. The coherent integration time (CIT) is $T_c = MT_r$.

The received signal of a point target is delayed, changed in amplitude, and possibly shifted in frequency. After appropriate dechirping and filtering, it is of the form

$$r(t) = \sigma \exp(j\pi Bd(t)t/(cT_r)) \exp(j2d(t)\omega_0/c) \tag{10.1}$$

where σ is the reflection coefficient of the target, $d(t)$ is the target distance from radar, ω_0 is the radar operating frequency, while c is the speed of light.

In numerical realizations, radar signal is discretized. Each chirp contains N samples. The obtained radar signal is two-dimensional, where chirp's repetition time is related to so called "slow-time", while sampling rate within one chirp is related to "fast-time."

Assuming that the distance is constant during T_r then the FT along fast-time will, in an ideal case, be represented by a delta pulse at frequency $\pi Bd(t)/(cT_r)$, being proportional to the range location $d(t) = d$ of the reflecting point. Thus, in order to analyze the cross-range representation in the radar imaging, we will consider only the Doppler part of the received signal of a point target, in the continuous dwell time,

$$s(t) = \sigma \exp(j2d(t)\omega_0/c) \tag{10.2}$$

Here the whole CIT will be considered. Thus, we will not assume that $d(t)$ is constant within the CIT.

10.2.1 ISAR Setup

In the ISAR case, the aim is to obtain a high-resolution image of a target based on the change in the target viewing angle with respect to the fixed radar. The illustration of a target and the relevant quantities in the ISAR system is given in Figure 10.1. Suppose that a target consists of K point scatterers, whose locations are (x_i, y_i), $i = 1, 2, \ldots, K$ in the coordinate system with origin in the center of target rotation. The coordinate axis in the direction of the line radar-target (range coordinate) is denoted by x, while y is the coordinate in the direction normal to this line (cross-range coordinate). The target's total movement can be considered as a superposition of translation and rotation.

Assume that during the considered time interval, point scatterers can be described by a rotation with $\theta(t) = \omega_R t$. The new coordinates of a scatterer are

$$\begin{bmatrix} x_i'(t) \\ y_i'(t) \end{bmatrix} = \begin{bmatrix} \cos(\theta(t)) & \sin(\theta(t)) \\ -\sin(\theta(t)) & \cos(\theta(t)) \end{bmatrix} \begin{bmatrix} x_i \\ y_i \end{bmatrix} \tag{10.3}$$

The range position of the i-th reflector is

$$d_i(t) = \sqrt{(R + x_i'(t))^2 + y_i'^2(t)} \cong R + x_i'(t) = R + x_i \cos(\theta(t)) + y_i \sin(\theta(t)) \tag{10.4}$$

since the distance R is much larger than $x_i'(t)$ and $y_i'(t)$.

The received signal, for a system of point scatterers, can be modeled as a sum of individual responses. In addition, common ISAR imaging models assume that all point scatterers share the same angular motion, Figure 10.1. Then, the Doppler part of the received signal (10.2), within the m-th radar sweep, corresponding to K rigid body points, can be written as [13]

$$s_B(t) = \sum_{i=1}^{K} \sigma_{Bi} \exp(j2[R_B(t) + x_{Bi} \cos(\theta_B(t)) + y_{Bi} \sin(\theta_B(t))]\omega_0/c) \tag{10.5}$$

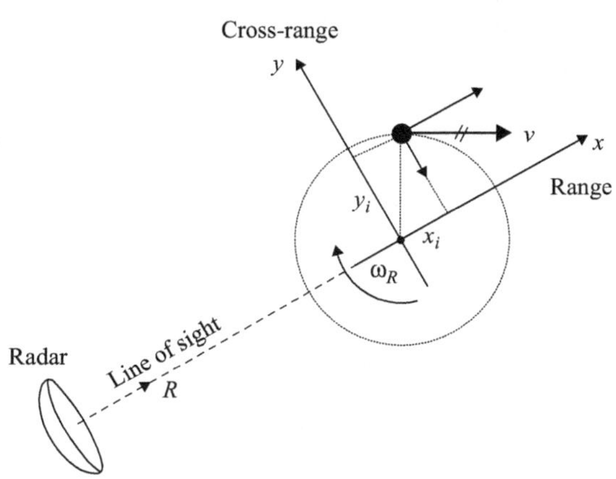

Figure 10.1 Geometry of the ISAR system

where the target's translation and angular motion are denoted by $R(t)$ and $\theta(t)$, with indices B corresponding to the rigid body points. The initial locations of points, in the coordinate system whose origin is in the center of the target rotation, are (x_{Bi}, y_{Bi}).

For the rigid body points $|\theta_B(t)| \ll 1$ holds during the CIT, following in $\cos(\theta_B(t)) \approx 1$ and $\sin\theta_B(t) \approx \theta_B(t) = \omega_B t$, where ω_B is the effective rigid body rotation rate. Motion compensation techniques [14, 15] are employed to remove the influence of the translation motion, thus we can also neglect the remaining factor $R_B(t) + x_{Bi}$, so that we have

$$s_B(t) = \sum_{i=1}^{K} \sigma_{Bi} \exp\left(j\frac{2\omega_0}{c} y_{Bi}\omega_B t\right) \tag{10.6}$$

The previous approximations cannot be used for fast-rotating (moving) micro-Doppler points, since their angular position $\theta_R(t)$ can significantly change during the CIT. Index R is used for fast-moving micro-Doppler points. Assume that there are P fast rotating points, which rotate around their central points (x_{R0i}, y_{R0i}) with radii A_{Ri}. The coordinates of i-th point are

$$x_{Ri}(t) = x_{R0i} + A_{Ri} \sin(\theta_{Ri}(t))$$
$$y_{Ri}(t) = y_{R0i} + A_{Ri} \cos(\theta_{Ri}(t))$$

According to (10.3), at a time instant t, for rotating points we have

$$\begin{bmatrix} x'_{Ri}(t) \\ y'_{Ri}(t) \end{bmatrix} = \begin{bmatrix} \cos(\theta_B(t)) & \sin(\theta_B(t)) \\ -\sin(\theta_B(t)) & \cos(\theta_B(t)) \end{bmatrix} \begin{bmatrix} x_{R0i} + A_{Ri} \sin(\theta_{Ri}(t)) \\ y_{R0i} + A_{Ri} \cos(\theta_{Ri}(t)) \end{bmatrix}$$

where $\theta_B(t)$ represents angular velocity of the rigid body. Assume now that the rotations speed of the i-th fast rotating point is ω_{Ri}, as well as that the initial angle is zero, then $\theta_{Ri}(t) = \omega_{Ri}t$. After compensation for R in (10.4), the distance from the radar to the i-th fast rotating reflector is

$$\begin{aligned} d_{Ri}(t) \cong x'_{Ri}(t) &= [x_{R0i} + A_{Ri} \sin(\theta_{Ri}(t))]\cos(\theta_B(t)) + [y_{R0i} + A_{Ri} \cos(\theta_{Ri}(t))]\sin(\theta_B(t)) \\ &= x_{R0i} \cos(\theta_B(t)) + y_{R0i} \sin(\theta_B(t)) + A_{Ri} \sin(\theta_{Ri}(t) + \theta_B(t)) \end{aligned}$$

In the case of $|\theta_B(t)| \ll 1$, $\cos(\theta_B(t)) \approx 1$ and $\sin(\theta_B(t)) \approx \theta_B(t) = \omega_B t$,

$$d_{Ri}(t) \cong [x_{R0i} + y_{R0i}\omega_B t] + A_{Ri} \sin((\omega_{Ri} + \omega_B)t)$$

After the distance compensation for x_{R0i} and by taking $\omega_{Ri} \gg \omega_B$, we have

$$d_{Ri}(t) \cong y_{R0i}\omega_B t + A_{Ri} \sin(\omega_{Ri}t) \tag{10.7}$$

The received signal, including both the rigid body points and the fast rotating micro-Doppler points, can be written as

$$s(t) = \sum_{i=1}^{K} \sigma_{Bi} \exp\left(j\frac{2\omega_0}{c} y_{Bi}\, \omega_B t\right) + \sum_{i=1}^{P} \sigma_{Ri} \exp\left(j\frac{2\omega_0}{c}\left[y_{R0i}\, \omega_B t + A_{Ri} \sin(\omega_{Ri} t)\right]\right) \tag{10.8}$$

Terms in the first sum in (10.8) represent complex sinusoids at frequencies proportional to the cross-range positions of the reflectors y_{Bi}, $i \in [1, K]$ and the rotating rate ω_B.

A similar form of the received signal is obtained in the case of vibrating points. If a reflecting point vibrates, around a central point (x_{V0i}, y_{V0i}), along a line parallel to the line-of-sight, with frequency ω_{Vi}, reaching the maximum amplitude A_{Vi} from the central point, we get the form

$$s(t) = \sum_{i=1}^{K} \sigma_{Bi} \exp\left(j\frac{2\omega_0}{c} y_{Bi}\, \omega_B t\right) + \sum_{i=1}^{P} \sigma_{Vi} \exp\left(j\frac{2\omega_0}{c}\left[y_{V0i}\, \omega_B t + A_{Vi} \sin(\omega_{Vi} t)\right]\right) \tag{10.9}$$

Thus, the vibrations can be analyzed in the same way as the rotations.

The micro-Doppler points may not make a whole period of motion within the CIT, or their motion may be aperiodic in nature. Analytical form of the micro-Doppler part of the signal for any other arbitrary motion can be easily obtained by using

$$x_{Ri}(t) = x_{R0i} + x_{arb}(t)$$
$$y_{Ri}(t) = y_{R0i} + y_{arb}(t)$$

in (10.3), and by following the presented procedure. The Doppler part of the radar signal that corresponds to an arbitrary moving micro-Doppler point is then a FM signal with the instantaneous frequency (IF)

$$\Omega_{Ria}(t) = \frac{2\omega_0}{c}\left[y_{R0i}\, \omega_B + \frac{dx_{arb}(t)}{dt}\right] \tag{10.10}$$

In practice, we can also have a situation when the reflection coefficients of, for example a rotating point, are highly dependent on the position (angle to the line of sight)

$$s(t) = \sum_{i=1}^{K} \sigma_{Bi} \exp\left(j\frac{2\omega_0}{c} y_{Bi}\, \omega_B t\right) + \sum_{i=1}^{P} \sigma_{Ri}(\theta_{Ri}) \exp\left(j\frac{2\omega_0}{c}\left[y_{R0i}\, \omega_B t + A_{Ri} \sin(\omega_{Ri} t)\right]\right)$$

If we have a small fast rotating plane plate, then its reflectivity for all other angles, except for $\theta_{Ri} = 0$, could be negligible. This case can be modeled as

$$\sigma_{Ri}(\theta_{Ri}) = \sigma_{Ri} \exp(-S|\sin \theta_{Ri}(t)|)$$

where S is a large constant such that we can approximate $\sigma_{Ri}\exp(-S|\sin\theta_{Ri}(t)|)$ with $\sigma_{Ri}\delta_c(S\sin\theta_{Ri}(t))$, where

$$\delta_c(t) = \begin{cases} 1 & \text{for } t = 0 \\ 0 & \text{elsewhere} \end{cases}$$

In this case, the micro-Doppler effect manifests as a series of flashes.

10.2.2 Spectral Analysis

Since we consider only the Doppler part in the received signal, the analysis of the radar signal reduces to one-dimensional signal (10.2) and its FT,

$$S(\Omega) = FT\{s(t)\} = \int_{-\infty}^{\infty} s(t)e^{-j\Omega t}dt$$

analysis. In the discrete form, the FT reads:

$$S(k) = \sum_{m=0}^{M-1} s(m)\exp(-j2\pi mk/M) \tag{10.11}$$

Its squared absolute value $|S(k)|^2$ is called the periodogram. If we calculate the FT of the signal corresponding to one point of the rigid body only, we get

$$S_{B1}(\Omega) = FT\left\{\sigma_{B1}\exp\left(j\frac{2\omega_0}{c}y_{B1}\,\omega_B t\right)\right\} = 2\pi\sigma_{B1}\delta\left(\Omega - \frac{2\omega_0}{c}y_{B1}\,\omega_B\right)$$

The FT is a delta pulse at the position proportional to the cross-range coordinate y_{B1}. The delta pulse position depends not only on the radar parameters, that are constant, but also on the rotation speed ω_B. The basic property in spectral analysis is that a higher concentration of the FT is achieved by using longer signal sequences. However, within longer time intervals, the target point velocity may change. These changes cause corresponding frequency changes that spread the FT and may decrease radar efficiency.

The Doppler part of the radar signal that corresponds to a rotating reflector is a sinusoidal FM signal, with the IF:

$$\Omega_{Ri}(t) = \frac{2\omega_0}{c}\left[y_{R0i}\,\omega_B + A_{Ri}\omega_{Ri}\cos(\omega_{Ri}t)\right] \tag{10.12}$$

It can be concluded that the location of a rotating scattering points in the cross-range coordinate is proportional to its true location multiplied by the rigid body rotation speed. In addition, its resulting rotating rate $\omega_{Ri} = \theta'_{Ri}(t)$ is significantly higher than the rotating rate of the rigid body ω_B. Therefore, the radar image of rotating reflectors is not concentrated at one frequency. It is spread over the interval proportional to its rotating rate and the radius of rotation. The FT of a sinusoidal FM signal is known in the literature. Its coefficients are in the form of Bessel functions. Obviously, the FT is not a good tool for imaging and analyzing the micro-Doppler part of this signal.

Time variations of the spectral content may be tracked by using a TFR. A general form of TFR of this kind of signals would produce

$$TFR_i(t,\Omega) \sim \sigma_i \delta(\Omega - \Omega_{Ri}(t)) \tag{10.13}$$

with some additional spreading factors. Thus, the TFR of this part of the radar signal, within the CIT would be concentrated around the IF $\Omega_{Ri}(t)$ (10.12) in the TF plane [16].

A simplest way to localize the signal behavior in shorter time intervals, within the CIT, and to produce a TFR is in applying a window function to the standard FT. It results in the short-time Fourier transform (STFT) [16, 17]

$$STFT(t,\Omega) = \int_{-\infty}^{\infty} s(t+\tau)w(\tau)e^{-j\Omega\tau}d\tau \tag{10.14}$$

where $w(\tau)$ is a window function used to truncate the considered signal. The main drawback of the STFT is a high dependence between time and frequency resolutions. The increase of resolution in one domain (for example in frequency domain, by increasing window width) causes the decrease of resolution in the other domain. The optimal window width can be calculated by using the concentration measures [18]. The squared absolute value of the STFT is called the spectrogram.

The Wigner distribution (WD) is the basic quadratic TFR. It achieves a higher concentration of TFR than the spectrogram [16, 17]

$$WD(t,\Omega) = \int_{-\infty}^{\infty} s\left(t+\frac{\tau}{2}\right)s^*\left(t-\frac{\tau}{2}\right)w\left(\frac{\tau}{2}\right)w^*\left(-\frac{\tau}{2}\right)e^{-j\Omega\tau}d\tau \tag{10.15}$$

Since the WD is quadratic signal representation, it suffers from the undesired cross-terms, appearing as a result of cross-product of different signal components when the analyzed signal has more than one component (scatterer).

The S-method (SM) is a TFR that can produce significant improvement in imaging of radar targets (as in the WD) without introducing cross-terms (as in the spectrogram). It is based on the standard STFT (or the FT) and its "correction" to achieve a high concentration as in the WD

$$SM(t,\Omega) = \int_{-\infty}^{\infty} P(\theta)STFT(t,\Omega+\theta)STFT^*(t,\Omega-\theta)d\theta$$

A discrete form of the SM is [16]

$$\begin{aligned} SM(m,k) &= \sum_{i=-L}^{L} STFT(m,k+i)STFT^*(m,k-i) \\ &= |STFT(m,k)|^2 + 2\mathrm{Re}\left\{\sum_{i=1}^{L} STFT(m,k+i)STFT^*(m,k-i)\right\} \end{aligned} \tag{10.16}$$

where L is the number of correcting terms. Usually, just few of them are sufficient to improve the STFT-based concentration, avoiding the cross-terms [19]. The SM can be applied in focusing the whole FT-based radar image, without the STFT application. Then its form is

$$\begin{aligned} SM(m,k) &= \sum_{i=-L}^{L} S(m,k+i)S^*(m,k-i) \\ &= |S(m,k)|^2 + 2\mathrm{Re}\left\{\sum_{i=1}^{L} S(m,k+i)S^*(m,k-i)\right\} \end{aligned} \tag{10.17}$$

where $|S(m,k)|^2$ is the standard 2D FT-based radar image and $S(m,k+i)S^*(m,k-i)$ are the correcting terms, while in this case m is range and k is cross-range.

10.2.3 SAR Setup

Consider a SAR setup with geometry and corresponding quantities as in Figure 10.2. First, assume that the radar movement is parallel to the imaged terrain. This position of the radar will be described by the elevation angle $\theta = 0$. Denote the radar speed as $\vec{v}_R$ and assume that it moves in a direction of the x_0 axis of the coordinate system, $\vec{v}_R = v_R \vec{i}_{x_0}$. Assume that the line-of-sight is in the direction of y_0 axis, with unity vector $\vec{i}_{y_0}$. The distance of the radar to the center of the imaged terrain is denoted by $\vec{R}_0 = R_0 \vec{i}_{y_0}$. At one instant, the radar movement corresponds to the rotation of the terrain, with respect to the radar, with $\vec{v}_R = \vec{\omega}_B \times \left(-\vec{R}_0\right)$.

Now assume that there is only one stationary reflecting point on the ground, located at (x_{0i}, y_{0i}, z_{0i}) from the center of image. Taking into account that the image plane with the reflecting point rotates with $\vec{\omega}_B$ around the central point, movement (trajectory) of a reflector can be described by a simple rotation matrix (10.3), as in the ISAR case, resulting in

$$\begin{aligned} x'_{0i}(t) &= x_{0i}\cos(\omega_B t) + y_{0i}\sin(\omega_B t) \\ y'_{0i}(t) &= -x_{0i}\sin(\omega_B t) + y_{0i}\cos(\omega_B t) \\ z'_{0i}(t) &= z_{0i} \end{aligned} \tag{10.18}$$

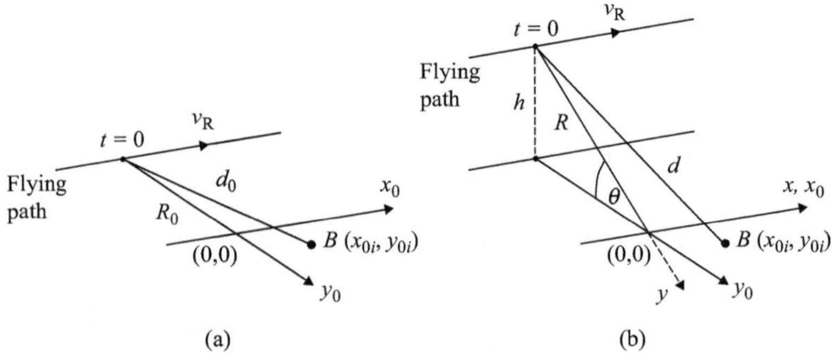

Figure 10.2 Geometry of SAR systems: (a) $\theta = 0$, (b) $\theta \neq 0$

The signal $s_i(t)$ reflected from this point, as a function of its distance from the radar $d_i(t)$ (Figure 10.2(a)), is of the form

$$s_i(t) = \sigma_i \exp\left(j\frac{\omega_0}{c}2d_i(t)\right) = \sigma_i \exp\left(j\frac{2\omega_0}{c}\sqrt{x_{0i}'^2(t) + (R_0 + y_{0i}'(t))^2 + z_{0i}'^2(t)}\right)$$

Since the distance R_0 is much greater than any point coordinate, we may write

$$s_i(t) \cong \sigma_i \exp\left(j\frac{2\omega_0}{c}(R_0 + y_{0i}'(t))\right)$$

$$s_i(t) = \sigma_i \exp\left(j\frac{2\omega_0}{c}(-x_{0i}\sin(\omega_B t) + y_{0i}\cos(\omega_B t))\right) \tag{10.19}$$

where the constant phase term $\exp(j2\omega_0 R_0/c)$ is omitted and we returned to the equality sign for simplicity of further notation.

In general, if the radar is at $\theta \neq 0$, we have a new coordinate system, related to the radar, in which the initial positions of the reflecting points are rotated for θ angle around the x_0 axis, Figure 10.2(b), as

$$\begin{aligned} x_{0i}(\theta) &= x_{0i} \\ y_{0i}(\theta) &= y_{0i}\cos(\theta) - z_{0i}\sin(\theta) \\ z_{0i}(\theta) &= y_{0i}\sin(\theta) + z_{0i}\cos(\theta) \end{aligned} \tag{10.20}$$

The rotation in this coordinate system then remains the same as for $\theta = 0$, with the rotation vector

$$\vec{v}_R = \vec{\omega}_B \times \left(-\vec{R}\right)$$

$$v_R = \frac{\omega_B}{\cos(\theta)}R_0 = \omega_{B\theta}R_0$$

The resulting rotation from (10.18), using the coordinates in (10.20), can be now written as

$$\begin{aligned} x_i'(t) &= x_{0i}\cos(\omega_{B\theta}t) + y_{0i}(\theta)\sin(\omega_{B\theta}t) \\ y_i'(t) &= -x_{0i}\sin(\omega_{B\theta}t) + y_{0i}(\theta)\cos(\omega_{B\theta}t) \\ z_i'(t) &= z_{0i}(\theta) \end{aligned}$$

The signal $s_i(t)$, with the same assumption as for (10.19), is then of the form

$$\begin{aligned} s_i(t) &= \sigma_i \exp(j2y_i'(t)\omega_0/c) \\ &= \sigma_i \exp(j2(-x_{0i}\sin(\omega_{B\theta}t) + [y_{0i}\cos(\theta) - z_{0i}\sin(\theta)]\cos(\omega_{B\theta}t))\omega_0/c) \end{aligned} \tag{10.21}$$

Assume now that, in addition to the pure rotation caused by radar movement, the point rotates around its central point (x_{0i}, y_{0i}, z_{0i}) in the plane x_0Oy_0, then, with $z_{0i} = 0$. This rotation is described

$$\begin{aligned} x_{0i}(t) &= x_{0i} + A_i\sin(\omega_{R_i}t) \\ y_{0i}(t) &= y_{0i} + A_i\cos(\omega_{R_i}t) \end{aligned}$$

Thus, we have

$$x'_t(t) = [x_{0i} + A_i \sin(\omega_{R_i}t)]\cos(\omega_{B\theta}t) + [y_{0i} + A_i \cos(\omega_{R_i}t)]\cos(\theta)\sin(\omega_{B\theta}t)$$
$$y'_t(t) = -[x_{0i} + A_i \sin(\omega_{R_i}t)]\sin(\omega_{B\theta}t) + [y_{0i} + A_i \cos(\omega_{R_i}t)]\cos(\theta)\cos(\omega_{B\theta}t)$$
$$z'_t(t) = [y_{0i} + A_i \cos(\omega_{R_i}t)]\sin(\theta)$$

The resulting received radar signal is

$$s_{Ri}(t) = \sigma_i \exp\left(j\frac{2w_0}{c}(y'_i(t))\right) = \sigma_i \exp\left(j\frac{2\omega_0}{c}(-[x_{0i} + A_i \sin(\omega_{R_i}t)]\sin(\omega_{B\theta}t))\right)$$
$$\times \exp\left(j\frac{2\omega_0}{c}([y_{0i} + A_i \cos(\omega_{R_i}t)]\cos(\theta)\cos(\omega_{B\theta}(t)))\right)$$

Here, the approximations $\omega_{B_\theta}t \ll 1$ and $A_i \sin(\omega_{R_i}t) \ll x_{0_i}$ holds, and

$$s_{Ri}(t) \cong \sigma_i \exp\left(j\frac{2\omega_0}{c}[-x_{0i}\omega_{B\theta}t + A_i \cos(\omega_{R_i}t)\cos(\theta)]\right) \tag{10.22}$$

where the constant term $\exp(jy_{0i}\cos(\theta)2\omega_0/c)$ is omitted. Corresponding IF, in the spectrum of signal, is

$$\Omega_{Ri}(t) = \frac{2\omega_0\omega_{B\theta}}{c}\left[-x_{0i} - A_i\frac{\omega_{R_i}}{\omega_{B\theta}}\sin(\omega_{R_i}t)\cos(\theta)\right] \tag{10.23}$$

Generally, in addition to the pure rotation, caused by radar movement, the point may have a fast arbitrary movement around (x_{0i}, y_{0i}, z_{0i}) described by $x_{0i}(t) = x_{0i} + x_{arb}(t)$, $y_{0i}(t) = y_{0i} + y_{arb}(t)$ and $z_{0i}(t) = z_{0i} + z_{arb}(t)$, where index $_{arb}$ is used to denote such a movement. With approximation $\omega_{B\theta}t \ll 1$, appropriate distance compensation and by neglecting constant phase terms, we get

$$s_{Ri}(t) \cong \sigma_i \exp\left(j\frac{2\omega_0}{c}\left[-x_{0i}\omega_{B\theta}t + [y_{arb}(t)\cos(\theta) - z_{arb}(t)\sin(\theta)]\right]\right) \tag{10.24}$$

In this way, we can model, for example, an experiment with a pendulum caused micro-Doppler in the SAR, when $\theta_i(t) = \theta_{0i}\ \cos(\sqrt{g/l}t)$, with θ_{0i} being the pendulum maximal angle and g is the gravity acceleration. For a pendulum with the center located at $(0,0,l)$ oscillating in yOz plane we can derive $y_{arb}(t) = l\ \sin\theta_i(t)$, $z_{arb}(t) = l - l\ \cos\theta_i(t)$ and $x_{arb}(t) = 0$.

In simulations, the original form of the received radar signal (10.21), without approximations, may produce more accurate results.

We can conclude that the ISAR and the SAR setup can be considered in a unified way. Thus, all considerations regarding FT and TFR driven for the ISAR, hold for the SAR signals, as well. The micro-Doppler effect is modeled as non-stationary FM signal. For analysis of that class of signals many TF signal analysis methods are developed.

10.3 Inverse Radon Transform Based Micro-Doppler Parameters Estimation

The micro-Doppler can be used to determine properties of a target [13, 20, 21], since the micro-Doppler parameters are related to the parameters of the corresponding micro-motions. Therefore, the radar micro-Doppler signatures are of a great potential for identifying properties of unknown targets. In the previous section, we have shown that the micro-Doppler may assume various complex forms. However, the most common form of the micro-Doppler is a form of sinusoidal FM signal. Thus, in this section we first focus on this simple, common, and specific form of the micro-Doppler. The analysis of this form of the micro-Doppler is based on an interesting property of the inverse Radon transform, to project sinusoidal 2D structures into a point at the position that corresponds to the parameters of this sinusoid. This approach may work even if we have only a fraction of period in the analyzed signal. A method for detection of period for a periodic TF structure form is presented as well. When period of a periodic TF from is known, the complexity of the inverse Radon transform calculation is reduced.

10.3.1 The Inverse-Radon Transform Review

Let us shortly review the Radon and the inverse Radon transform [22]. A projection of a 2D function $f(x, y)$ onto x axis is

$$R_f(x,0) = \int_{-\infty}^{\infty} f(x,y)dy$$

The FT of the projection onto x is equal to the slice of the 2D FT along corresponding u axis, since

$$\int_{-\infty}^{\infty} R_f(x,0)e^{-jxu}dx = \int_{-\infty}^{\infty}\left(\int_{-\infty}^{\infty} f(x,y)dy\right)e^{-jxu}dx$$
$$= \int_{-\infty}^{\infty}\int_{-\infty}^{\infty} f(x,y)e^{-jxu}e^{-jy0}dxdy = F(u,0)$$

The same holds for the projection onto y axis.

A rotated version of the 2D function may be described in new, rotated coordinate system, by a coordinate rotation transform. For an angle α, it reads

$$\begin{bmatrix} x \\ y \end{bmatrix} = \begin{bmatrix} \cos(\alpha) & -\sin(\alpha) \\ \sin(\alpha) & \cos(\alpha) \end{bmatrix} \begin{bmatrix} \xi \\ \zeta \end{bmatrix}$$

$$\begin{bmatrix} \xi \\ \zeta \end{bmatrix} = \begin{bmatrix} \cos(\alpha) & \sin(\alpha) \\ -\sin(\alpha) & \cos(\alpha) \end{bmatrix} \begin{bmatrix} x \\ y \end{bmatrix}$$

Projection of a function $f(x, y)$ onto ξ, with a varying rotation angle α, is called the Radon transform of $f(x, y)$:

$$R_f(\xi,\alpha) = \int_{-\infty}^{\infty} f(\xi,\zeta)d\zeta = \int_{-\infty}^{\infty}\int_{-\infty}^{\infty} f(\xi,\zeta)\delta(\lambda-\xi)d\lambda d\zeta$$

If the Radon transform $R_f(\xi, \alpha)$ is given, then 2D function $f(x, y)$ producing Radon transform is the inverse Radon transform.

Example 1 Find the Radon transform of a function $f(x, y) = \delta(x - x_0)\delta(y - y_0)$ in x, y domain. Projection of $f(x, y)$ onto x axis is

$$R_f(x,0) = \int_{-\infty}^{\infty} f(x,y)dy = \delta(x-x_0)$$

For an arbitrary direction $\xi_0 = x_0\cos(\alpha) + y_0\sin(\alpha)$, $\zeta_0 = -x_0\sin(\alpha) + y_0\cos(\alpha)$ the function

$$f(\xi,\zeta) = \delta(\xi-\xi_0)\delta(\zeta-\zeta_0)$$

results in the Radon transform

$$R_f(\xi,\alpha) = \int_{-\infty}^{\infty} f(\xi,\zeta)d\zeta = \delta(\xi-\xi_0) = \delta(\xi-(x_0\cos(\alpha)+y_0\sin(\alpha)))$$

Note that this is a sinusoidal pattern along the line

$$\xi = x_0\cos(\alpha) + y_0\sin(\alpha) = \sqrt{x_0^2+y_0^2}\cos(\alpha + \arctan(y_0/x_0))$$

in the 2D (ξ, α) domain, with the amplitude $\sqrt{x_0^2+y_0^2}$ and the phase $\psi = \arctan(y_0/x_0)$. Of course, the Radon transform is periodic in α with 2π. Projections for $0 \leq \alpha < \pi$ are sufficient to calculate all transform values. **Thus, a point in (x, y) domain transforms to a sinusoidal pattern in Radon transform domain. Therefore, the inverse Radon transform of a sinusoidal 2D pattern is a point.** This relation will be used to transform a sinusoidal pattern into a point via inverse Radon transform.

10.3.2 Parameters Estimation

Bearing in mind the previous theory we may conclude that the TFR of sinusoid FM signal can be concentrated in a single point by applying the inverse Radon transform. The Radon transform of a point with position (x_r, y_r) is

$$R_f(\xi,\alpha) = \delta(\xi - A_r\cos(\alpha+\psi_r)) \tag{10.25}$$

where $A_r = \sqrt{x_r^2+y_r^2}$ and $\psi_r = \arctan(y_r/x_r)$. Period of the Radon transform in α is 2π. The inverse Radon transform of a sinusoidal pattern in (α, ξ) plane is a point in

(x_r, y_r) domain. Therefore, since the rotating (vibrating) micro-Doppler points produce sinusoidal patterns in the TF domain (10.12) and (10.23), the inverse Radon transform performed on the TFR of the micro-Doppler part of the radar signal will have peaks at the positions corresponding to the FM parameters of the micro-Doppler points. It is easy to conclude that a transformation which concentrates whole energy of the desired signal into a single point, while keeping disturbances being spread, is a good detector.

For the inverse Radon transform calculation, we should know the period of sinusoidal pattern, which is not known in advance. To solve this problem we may assign the considered time interval (in TFR) to a set of possible periods in α. The whole TFR is then considered as periodic in time containing $\alpha/(2\pi)$ periods. For example, we assign the whole duration of the TF micro-Doppler representation to possible periods from $\pi/2$ up to 4π with step of 0.1π. Then we perform inverse Radon transform with all of these assumed periods. When the true periodicity of the micro-Doppler representation coincides with this parameter, then the inverse Radon transform reaches its maximal concentration. The final estimation of α is the one that gives the minimal measure (maximal concentration), calculated as [18]

$$H_{IR}(\alpha) = \frac{\sum_{k=0}^{M-1} \sum_{k=0}^{M-1} |IR_\alpha(x,y)|}{\sqrt{\sum_{k=0}^{M-1} \sum_{k=0}^{M-1} |IR_\alpha(x,y)|^2}} \tag{10.26}$$

Once α is estimated, modulation frequency f_{Ri} can be now estimated as $\hat{f}_{Ri} = \hat{\alpha}/(2\pi)$.

When the assumed period α matches the true one, a peak in the inverse Radon transform is detected at a position (x_r, y_r). This position depends on the parameters A_r and ψ_r of the fast rotating micro-Doppler point. Thus, after finding the peak position in the inverse Radon transform, we may estimate the micro-Doppler parameters as $A_r = \sqrt{x_r^2 + y_r^2}$ and $\psi_r = \arctan(y_r/x_r)$.

Example 2 Consider $N = 129$ samples of non-noisy FM signal:

$$s_R(t) = \sigma_R \exp\left(j\frac{A_R}{f_R}\sin(2\pi f_R t + \psi_R)\right) \tag{10.27}$$

sampled with $t = n\Delta t$, $\Delta t = 1/128$, $n = 0, 1, \ldots, N-1$. The signal parameters are $\sigma_R = 1$, $f_R = 1.3$, $A_R = 50$, and $\psi_R = 45°$. The spectrogram and the S-method of the considered signal are presented in Figure 10.3(a) and 10.3(b). Spectrogram is calculated with a 17 point Hann window, while the S-method is calculated by using (10.16) with a 55 point Hann window and $L = 8$. In both cases, the TFR is calculated at each available time instant. The parameter α is varied from 0.2 to 15 with step 0.2. For each α the inverse Radon transform along with the corresponding concentration measure is calculated. The concentration measure $\mu(\alpha)$ is presented in Figure 10.3(c) and 10.3(d). The minimum of measure (i.e., highest concentration) is depicted with a circle.

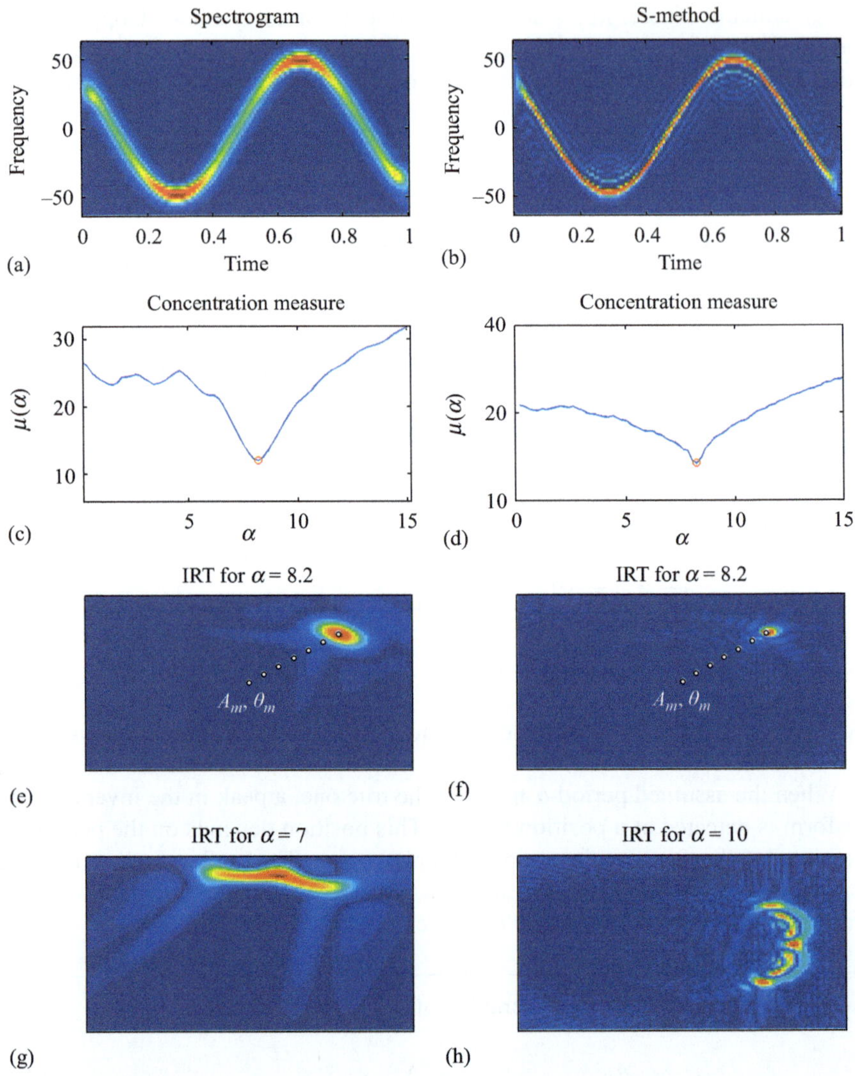

Figure 10.3 Modulation parameters estimation for a sinusoidal FM signal. TFR (a), (b); concentration measures (c), (d); the inverse Radon transform with highest (optimal) concentration (e), (f); and the inverse Radon transform when parameter α is not optimally chosen (g), (h)

Figure 10.3(e) and 10.3(f) present inverse Radon transform obtained for $\alpha = 8.2$. Maximum position in the inverse Radon transform is determined and modulation parameters are estimated in the spectrogram case as

$$\hat{f}_R = \frac{8.2}{2\pi} \approx 1.3, \quad \hat{A}_R = 49.4, \quad \hat{\Psi}_R = 45.0^\circ$$

and in the S-method case as

$$\hat{f}_m = \frac{8.2}{2\pi} \approx 1.3, \quad \hat{A}_m = 47.7, \quad \hat{\Psi}_m = 47.0^\circ$$

As we can see, in both cases modulation parameters are very close to the true values.

The inverse Radon transform for $\alpha = 7$ (in the spectrogram case) and for $\alpha = 10$ (in the S-method case) are given in Figure 10.3(g) and 10.3(h), respectively. From these subplots, we can see that the inverse Radon transform for the optimal α (subplots (e) and (f)) is better concentrated than the inverse Radon transform with other α.

Assume that we have P rotating reflectors (reflecting micro-Doppler points), so that the received radar signal is of the form (10.8), (10.22):

$$s_R(t) = \sum_{i=1}^{P} \sigma_{Ri} \exp\left(j\frac{A_{Ri}}{f_{Ri}} \sin(2\pi f_{Ri} t + \psi_{Ri})\right) + \varepsilon(t) \qquad (10.28)$$

where $A_{Ri} \cos(2\pi f_{Ri} t + \psi_{Ri})$ is the Doppler frequency caused by vibrating or rotating parts. Here, f_{Ri} denotes the frequency of the rotation (or vibration), and ψ_{Ri} is an arbitrary phase angle, while $\varepsilon(t)$ denotes disturbing components and noise.

If there are more than one rotating reflectors, with different or the same rotation speed, two scenarios are possible. One is that the concentration measure μ of the obtained inverse Radon transform produces simultaneously several or all P values of α, with visible and distinguishable concentration measure peaks. However, due to different amplitudes and different number of periods in the TF plane, usually only the strongest component is visible in the concentration measure. Thus, its parameter α and Doppler frequency shift f_{Ri} are estimated first. For the obtained value of α the maximum position in the inverse Radon transform should be found, leading to the estimation of A_{Ri} and ψ_{Ri} for the corresponding reflector (for example, $i = P$). Since we know f_{Ri} from the estimated period in α, the radar signal that corresponds to the estimated micro-Doppler parameters may be formed and the initial signal can be demodulated:

$$s'_R(t) = \sum_{i=1}^{P} \sigma_{Ri} \exp\left(j\frac{A_{Ri}}{f_{Ri}} \sin(2\pi f_{Ri} t + \psi_{Ri})\right) \exp\left(-j\frac{A_{R_P}}{f_{R_P}} \sin(2\pi f_{R_P} t + \psi_{R_P})\right)$$

In this way, P-th component of the analyzed signal is demodulated do the DC, and it can be easily filtered-out by setting zeros on a couple of samples around DC in the FT of signal $s'_R(t)$. Then, the filtered signal should be multiplied by $\exp(jA_{R_P}/f_{R_P} \sin(2\pi f_{R_P} t + \psi_{R_P}))$ in order to obtain the original form of the remaining $P-1$ micro-Doppler components. The procedure of calculating the best concentrated inverse Radon transform, estimating the micro-Doppler parameters and removing the reconstructed component should be repeated for each micro-Doppler component.

Example 3 Let us consider a multicomponent noisy signal (10.28) consisted of $P = 3$ sinusoidally FM components. Signal parameters are $\sigma_{R_1} = 1$, $f_{R_1} = 1.3$, $A_{R_1} = 50$, $\Psi_{R_1} = 45°$, $\sigma_{R_2} = 0.6$, $f_{R_2} = 1$, $A_{R_2} = 35.7$, $\Psi_{R_2} = -60°$, $\sigma_{R_3} = 0.7$, $f_{R_3} = 0.8$, $A_{R_3} = 28.6$, and $\Psi_{R_3} = 180°$, while the noise $\varepsilon(n)$ is a complex white Gaussian noise with variance 0.25.

In Figure 10.4(b) the concentration measure μ of the inverse Radon transform is shown. Here, only the strongest component is visible in the concentration measure.

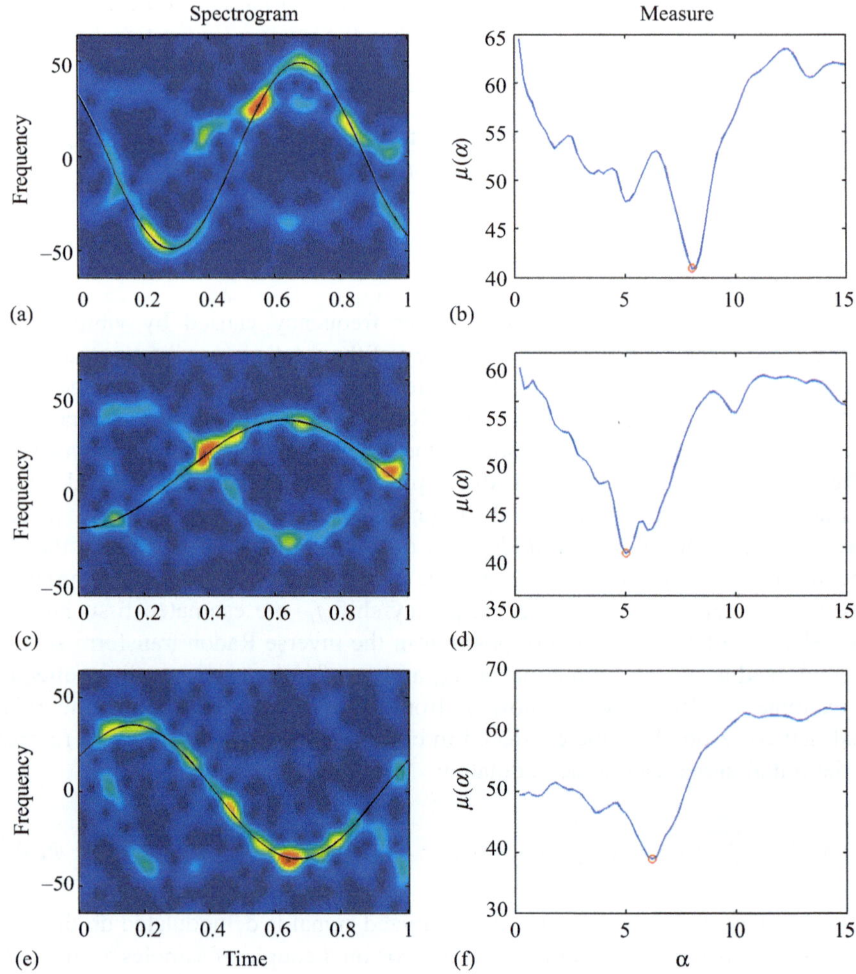

Figure 10.4 Multicomponent signal: Estimation of the first component (a), (b); second component (c), (d); and third component (e) and (f). Each component is removed from the signal after estimation, according to the described procedure, prior to the next component estimation

The corresponding parameter α is presented by red circle. The other parameters of the corresponding sinusoidal FM component are estimated from the IRD calculated with this value of α. The estimated parameters are $\hat{f}_{R_1} \approx 1.3$, $\hat{A}_{R_1} \approx 49.3$, $\hat{\Psi}_{R_1} = 49.0°$ (first component). Sinusoid that corresponds to the first estimated component which is formed by using the estimated parameters is depicted by the solid curve in Figure 10.4(a). Now, the strongest component is removed and the same procedure is repeated for the remaining components, until the energy of the remaining signal, as compared to the initial signal, is negligible. Estimated values from the other components are $\hat{f}_{R_2} \approx 1$, $\hat{A}_{R_2} = 35.33$, $\hat{\Psi}_{R_2} = -57.45°$, $\hat{f}_{R_3} \approx 0.8$, $\hat{A}_{R_3} = 28.5$, and $\hat{\Psi}_{R_3} = 180.0°$. From Figure 10.4(c) and 10.4(e), we can see that the estimated modulation parameters highly correspond to the components IFs. Moreover, from Figure 10.4(c) we can see that the proposed procedure can be successfully used even in the case when a shorter interval than a period of sinusoidal FM component is available.

Example 4 Multicomponent signal composed of a sinusoidal FM component, a linear FM component, and a constant frequency component

$$s(t) = \exp\left(j\frac{50}{1.3}\sin(2\pi 1.3t + \pi/4)\right) + 0.6\exp\left(j40\pi(t-0.9)^2\right) + 0.6\exp(j50\pi t)$$

is considered in this example. Results obtained with the presented procedure are shown in Figure 10.5(a) and 10.5(b). The estimated parameters are $\hat{f}_R = 8.2/(2\pi) \approx 1.3$, $\hat{A}_R = 49.3$, $\hat{\Psi}_R = 44.0°$. From this example, we see that the presented method is robust to the interferences.

The presented estimation procedure could be used even if the analyzed signal is periodic, but not sinusoidal FM signal. As an illustration a non-sinusoidal periodic modulated signal $x_1(t)$ with a varying amplitude

$$x_1(t) = A(t)\exp\left(j\int_0^t 200\sqrt[3]{\arcsin(\cos(1.8\pi u))}du\right)$$

is estimated.

Although this method is derived having in mind a sinusoidal modulation, the results presented in Figure 10.5(c) and 10.5(d) clearly show that the applicability of the method is not limited to the sinusoidal modulation patterns only. The estimated modulation parameters for signal $x_1(t)$ are $\hat{f}_R = 11.4/2\pi \approx 1.8$, $\hat{A}_R = 54.5$, $\hat{\Psi}_R = 1.3°$. The closest estimated sinusoid is presented in Figure 10.5(c) as well.

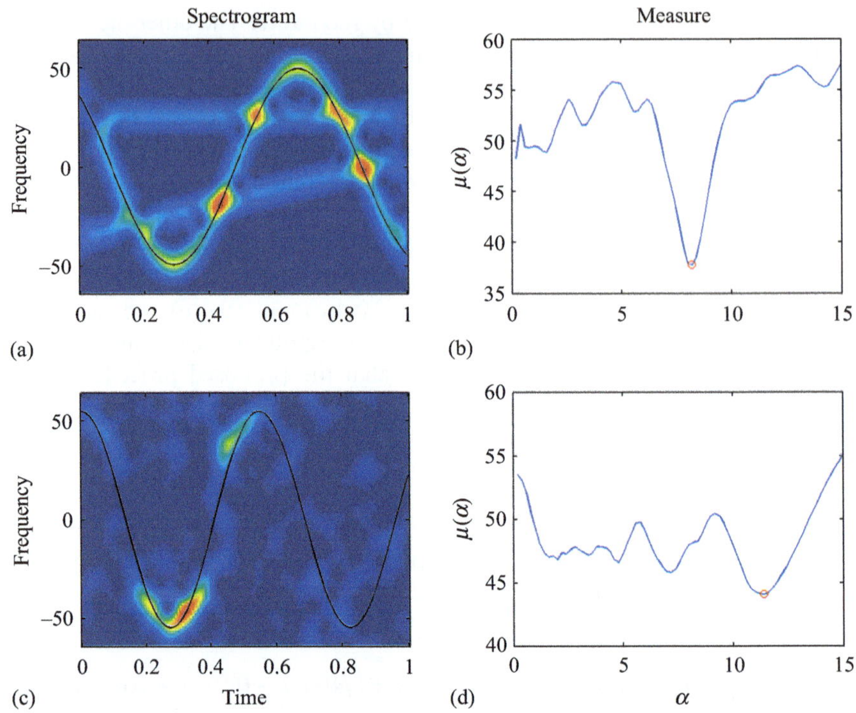

Figure 10.5 Modulation parameters estimation for multi-component signal consisted of constant frequency, linear FM and sinusoidal FM component: TFR (a); concentration measure (b). Modulation parameters estimation for signal with nonsinusoidal modulation and varying amplitude: TFR (c); concentration measure (d)

Example 5 In some cases all signal samples are not available. Consequently, we can calculate the spectrogram only at time instants/intervals when signal samples $x(n)$ are available. This procedure will be illustrated by considering the same signal as in the example with a multicomponent signal, assuming that the samples from the intervals 20–28, 50–80, and 95–110 are missing. Since the total number of samples is 129 we have 43% of missing samples. The estimation results obtained by using the available signal values are presented in Figure 10.6. Regions with unavailable samples are presented with white color in this figure. The parameters estimated using available samples are: $\hat{f}_{R_1} \approx 1.3$, $\hat{A}_{R_1} = 49.3$, $\hat{\Psi}_{R_1} = 45.0°$, $\hat{f}_{R_2} \approx 0.8$, $\hat{A}_{R_2} = 27.9$, $\hat{\Psi}_{R_2} = 178.7°$, $\hat{f}_{R_3} \approx 0.9$, $\hat{A}_{R_3} = 34.9$, $\hat{\Psi}_{R_3} = -54.6°$.

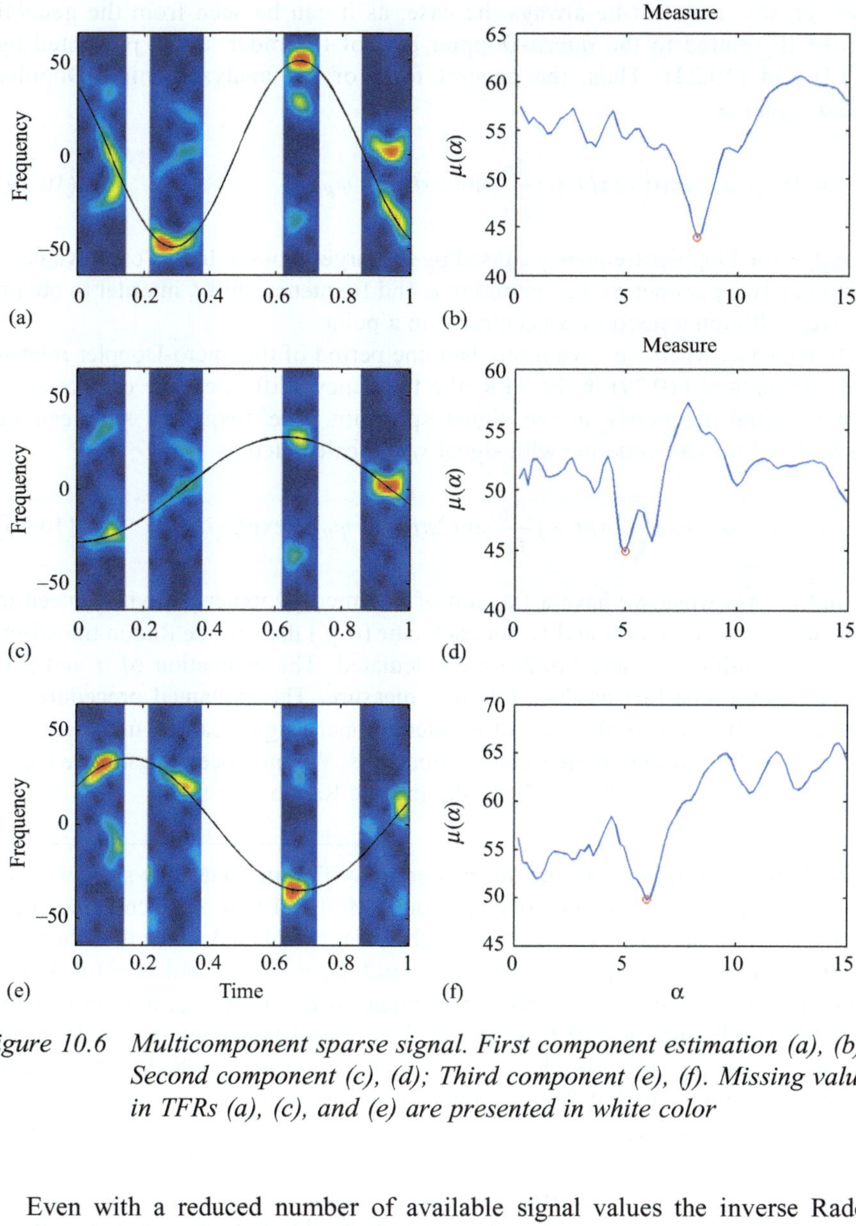

Figure 10.6 Multicomponent sparse signal. First component estimation (a), (b); Second component (c), (d); Third component (e), (f). Missing values in TFRs (a), (c), and (e) are presented in white color

Even with a reduced number of available signal values the inverse Radon transform based method produced accurate estimates.

In the previously explained procedure, we assumed that the TF pattern of the micro-Doppler signal is centered on middle (zero) frequency (cross-range).

However, this may not be always the case, as it can be seen from the general form of IF related to the micro-Doppler part of the radar signal presented by (10.12) and (10.23). Thus, the general form of the analyzed micro-Doppler related signal is

$$s_{Ri}(t) = \sigma_{Ri} \exp\left(j2\pi f_i t + j\frac{A_{Ri}}{f_{Ri}}\sin(2\pi f_{Ri}t + \psi_{Ri})\right) \tag{10.29}$$

where f_i is the Doppler frequency caused by the target motion. In this case, we need to estimate two parameters, i.e., period in α and frequency shift f_i in order to obtain the inverse Radon transform concentrated in a point.

In the case, when we have more than one period of the micro-Doppler related sinusoidal pattern (10.29) in the TFR, the frequency shift $\hat{f}_i$ can be estimated by finding central frequency in the signal spectrum. The frequency shift can be removed and we can continue with signal $s_{fR_i}(t)$ calculated as

$$s_{Ri}(t) = \sigma_{Ri} \exp\left(j2\pi f_i t + j\frac{A_{Ri}}{f_{Ri}}\sin(2\pi f_{Ri}t + \psi_{Ri})\right)\exp(-j2\pi\hat{f}_i t) \tag{10.30}$$

In the case when we have a fraction of the micro-Doppler period, we need to perform 2D search, over α and f_i. For each pair (α, f_i) the inverse Radon transform and corresponding measure (10.26) are calculated. The estimation of α and f_i is obtained as the one that produce minimal measure. The explained procedure for parameters estimation in the case of multicomponent signal can be used with the frequency shift $\hat{f}_i$ present in the signal components. We only need to estimate (α, f_i) for each component in order to obtain the inverse Radon transform.

Example 6 In order to illustrate this procedure of finding frequency shift, we will analyze the signal composed of one sinusoid FM signal that represents one component related to the micro-Doppler part of the radar signal, and which is of the form (10.29), with $\sigma_{R_1} = 10$, $A_{R_1}/f_{R_1} = 96$, $\psi_1 = -\pi/2$, $f_{R_1} = 1/512$, and $M = 1\,024$. The signal IF is not centered along zero cross-range in the TF plane, it is shifted for $f_1 = -60$. The frequency shift f_1 is estimated as central frequency in signal's spectrum and removed by (10.30). The inverse Radon transform is applied on the TFR of the obtained signal. Period T_{R_1} is estimated as the value that produces the best concentration of the inverse Radon transform, while A_{R_1} and ψ_1 are estimated based on the position of the point in the most concentrated inverse Radon transform. The black solid curve in Figure 10.7 presents the sinusoid reconstructed by using the estimated values. Frequency shift estimated by the presented procedure is $\hat{f}_1 = -60$, the estimated period is $T_{R_1} = 512.8026$ s, leading to the $\hat{f}_{R_1} = 0.0019501$ Hz, $\psi_1 = -0.5022\pi$, while normalized amplitude is 0.37327. The estimate sinusoid follows the TFR representation of the analyzed micro-Doppler related signal.

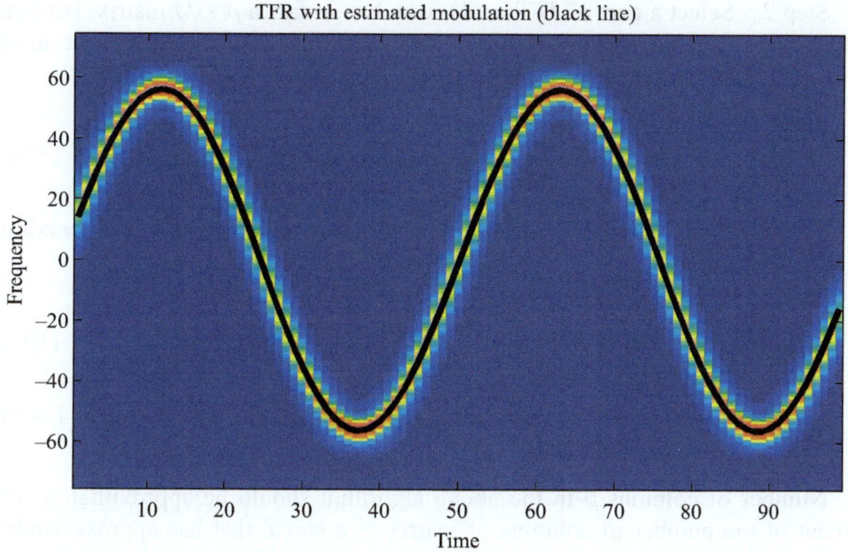

Figure 10.7 TFR of the m-D part with estimated modulation (black thick curve)

10.3.3 The Micro-Doppler Period Estimation

The micro-Doppler effect's period is considered as a parameter in the inverse Radon transform method. By estimating the period of the micro-Doppler effect, we would simplify the previous method. Here, an algorithm for the micro-Doppler period estimation is presented. The algorithm is based on the TF signal analysis concepts. The basic idea of the algorithm lies in the fact that, since the micro-Doppler is a periodic function, its parts are periodically repeated over time, and that this repetition might be detected directly from the TF plane.

One way to detect this repetition is to use a part of the TF plane, shift it across time index m, and compare with every part of the TFR with equivalent length, detecting the time points of its values repetition in the TFR. To this aim, the shifted part is multiplied with the corresponding part of the original TFR for the given m and the sum of the obtained product is used as a similarity measure between these two parts of the TFR. The sum is largest for time index m for which the values in the compared parts of the STFT are the same (or almost the same).

The explained procedure may be presented in a form of an algorithm:

Step 1. Calculate the TFR of signal $s(m)$ and denote is values by $TFR(m, k)$. This $TFR(m, k)$ is a $(M - M_w) \times M$ matrix, where M_w is time window width used for the TFR calculation.

Step 2. Select a part of *TFR*(*m*, *k*), which is in fact a $p \times M$ matrix, consisted of *p* first columns of the matrix *TFR*(*m*, *k*). For every instant *m*, calculate the similarity measure:

$$M_e(m) = \sum_{l=0}^{p-1} \sum_{k=0}^{M-1} TFR(l,k)TFR(m+l,k) \tag{10.31}$$

Step 3. Find maxima *Max* in $M_e(m)$ using the threshold THR_{Me} with condition given by

$$Max > THR_{Me} \sum_{l=0}^{p-1} \sum_{k=0}^{M-1} TFR(l,k) \tag{10.32}$$

Step 4. The distance between two successive maxima gives the micro-Doppler discrete period T_R.

Number of columns *p* in the above algorithm should be approximately 25/*L* percent of the number of columns of matrix of a signal that has approximately *L* periods. That value is obtained heuristically. It is easy to obtain the approximate number of periods by directly observing the considered TFR. Then, if the number of periods of IF in the considered TFR (for signal of length *M*) is for example about 2, the parameter *p* should be approximately 10% of the number of columns in matrix *TFR*(*m*, *k*).

Threshold THR_{Me} selection depends on signal-to-noise ratio (SNR) of the noisy signal. Larger values of threshold can be used for low SNRs, while it should be decreased with the increase of SNR. It is used in order to avoid false maxima in $M_e(m)$ (10.31).

Example 7 Multicomponent noisy FM signal $x(m) = x_1(m) + x_2(m) + x_3(m) + \varepsilon(m)$ is considered in this example. The signal is consisted of one linear FM signal whose IF is nonperiodic:s

$$x_1(m) = 0.6\ \exp(-j\pi 0.6m^2/512 + j0.5\pi m) \tag{10.33}$$

one complex sinusoid with frequency 0 (constant IF – $x_2(t)$) and one component with periodic IF:

$$x_3(m) = \exp\left(j\frac{0.6}{2f_R}\sin(2\pi f_R m + \pi/2) + j\frac{0.6}{36f_R}\sin(6\pi f_R m + \pi/2) + j0.2\pi m\right) \tag{10.34}$$

where $f_R = 1/120$. Noise is white, complex, Gaussian, and SNR = 10 dB. The STFT of the analyzed signal is presented in Figure 10.8(a), while the normalized similarity measure is shown on Figure 10.8(b). In addition to the sinusoidally FM signal, we have here disturbances caused by the other two nonperiodic components. As it is expected, nonperiodic components of signal *x*(*m*) distorted maxima of

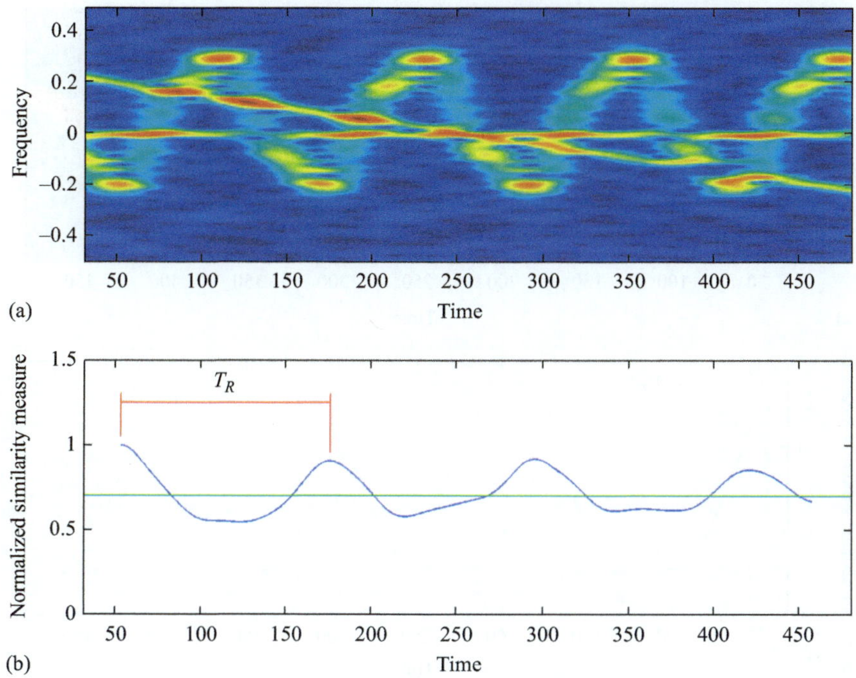

Figure 10.8 Micro-Doppler period estimation of multicomponent signal in white noise environment: (a) STFT; (b) Obtained normalized similarity measure

similarity measure, but, still with a suitable threshold selection, the algorithm is able to estimate the period. Estimated micro-Doppler period is $T_R = 120.0247$ s.

Example 8 Consider the discrete time-limited chirp signal:

$$x_1(m) = \begin{cases} \exp(-j\pi 0.5 f_R m^2/512 - j0.6\pi m) & \text{for } 0 \leq m \leq 120 \\ 0 & \text{elsewhere} \end{cases} \tag{10.35}$$

Discrete signal $x(m)$, defined as a periodic noisy array consisted of 5 signals $x_1(m)$:

$$x(m) = \sum_{l=0}^{4} x_1(m - lT_R) + \varepsilon(m) \tag{10.36}$$

with $T_R = 120$ is analyzed. Noise is white, complex, Gaussian, with SNR = 10 dB. Figure 10.9(a) shows the STFT of signal (10.36) in first 512 points, while Figure 10.9(b) shows normalized similarity measure. The micro-Doppler period is estimated as a distance between two successive maxima as $T_R = 120.3488$.

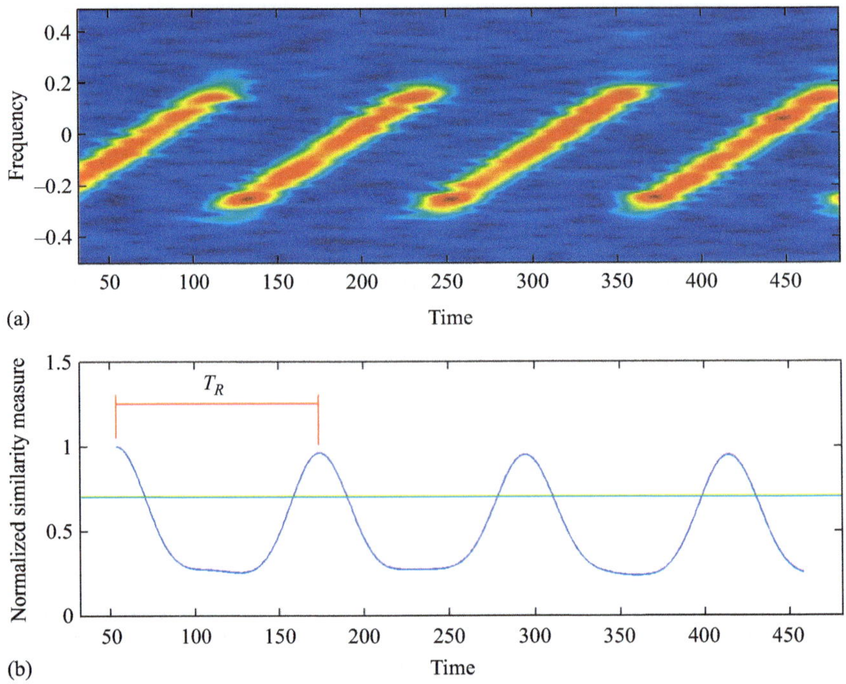

Figure 10.9 The m-D period estimation of signal (10.36): (a) STFT; (b) Obtained normalized similarity measure

10.3.4 Sparsity Domain of the Micro-Doppler Signal

The micro-Doppler signal in the form

$$s_R(t) = \sum_{i=1}^{P} \sigma_{Ri} \exp\left(j\frac{A_{Ri}}{f_{Ri}} \sin(2\pi f_{Ri}t + \psi_{Ri})\right)$$

could be considered as a sparse signal in the inverse Radon transform domain. In an ideal case, its inverse Radon transform is a sum of delta pulses at

$$IR_i(x,y) = \sum_{i=1}^{P} \sigma_{Ri}\delta(\mathrm{T} - T_{Ri}, x - x_{Ri}, y - y_{Ri})$$

It means that it can be represented by P coefficients only, with their positions. This number is much smaller than the total number of observation in TF domain. Thus, the micro-Doppler signal caused by pure oscillations is sparse in the inverse Radon transform domain. In order to reconstruct this kind of signal we can use a reduced (compressive sensed) set of observations. In the case of one micro-Doppler component only three observations are sufficient to reconstruct the whole micro-Doppler signal, as it will be shown next.

10.3.5 The Micro-Doppler Analysis from a Reduced Data Set

Real-world scenarios may involve signal analysis based on short duration or reduced data sets of available signal values [23–25]. Then, it is required to detect and estimate the micro-Doppler parameters based on the signals, containing only some parts or a fraction of a micro-Doppler signature [26]. We have already shown that the inverse Radon transform may be efficiently used to process this kind of signals. Here, more theory on this topic is provided along with a presentation of the simplest solution, requiring only three instants to recover sinusoidal micro-Doppler signature parameters.

In general, the micro-Doppler signature of a rotating or vibrating point may be described in the TF domain by

$$TFR_i(t,\Omega) \sim \sigma_i \delta(\Omega - A\cos(\omega_R\, t + \varphi))$$

Discrete-time form of this micro-Doppler signature, with a normalized time step, is

$$\Omega(m) = A\cos(\omega_R m + \varphi)$$

In the case, when the micro-Doppler period T is known and associated to 2π (by $\alpha = 2\pi m/T$) then only two values would be sufficient to detect and recover this micro-Doppler. For example, if we know the micro-Doppler signature position at the instant t_1 (in discrete domain at m_1) corresponding to α_1 then we can write

$$\Omega(m_1) = \xi_1 = x_0 \cos(\alpha_1) + y_0 \sin(\alpha_1)$$

Just one more equation

$$\Omega(m_2) = \xi_2 = x_0 \cos(\alpha_2) + y_0 \sin(\alpha_2)$$

is needed to find x_0 and y_0 producing the amplitude $\sqrt{x_0^2 + y_0^2}$ and the phase $\psi = \arctan(y_0/x_0)$ of this micro-Doppler signature. Note that from the instants corresponding to $\alpha_1 = 0$ and $\alpha_2 = \pi/2$ we would be able to read $\Omega(0) = x_0$ and $\Omega(m_2 = T/4) = y_0$, directly.

If the period is unknown, we need three independent instants to recover parameters of this kind of the micro-Doppler effect. We may write

$$\begin{aligned}
\Omega(m_1) &= \xi_1 = x_0 \cos(2\pi m_1/T) + y_0 \sin(2\pi m_1/T)\\
\Omega(m_2) &= \xi_2 = x_0 \cos(2\pi m_2/T) + y_0 \sin(2\pi m_2/T)\\
\Omega(m_3) &= \xi_3 = x_0 \cos(2\pi m_3/T) + y_0 \sin(2\pi m_3/T)
\end{aligned}$$

where T is the unknown period in time (with normalized step). In general, this system can be solved by varying parameter T and finding its value when the above system of three linear equations has a solution (x_0, y_0). A simple solution of the previous system is obtained by considering $m_1 = m + k$, $m_2 = m - k$, $m_3 = m$, when we may write

$$\begin{aligned}
\frac{\Omega(m+k) + \Omega(m-k)}{2\Omega(m)} &= \frac{A\cos(\omega_R m + \varphi + \omega_R k) + A\cos(\omega_R m + \varphi - \omega_R k)}{2A\cos(\omega_R m + \varphi)}\\
&= \frac{2A\cos(\omega_R m + \varphi)\cos(\omega_R k)}{2A\cos(\omega_R m + \varphi)} = \cos(\omega_R\, k)
\end{aligned}$$

Thus, the micro-Doppler oscillating frequency is

$$\omega_R = \frac{1}{k}\arccos\left\{\frac{\Omega(m+k)+\Omega(m-k)}{2\Omega(m)}\right\}$$

for $0 \leq \omega_R \leq \pi/k$. This relation limits the values of lag k to several samples. It is not so important since the considered time interval, in this case, is much smaller than a half of the period. After the frequency is estimated the other two parameters A and φ easily follow.

Example 9 The micro-Doppler analysis from the reduced data set is illustrated by experimental data. Experimental trials using an X-band radar were conducted in order to investigate and determine the micro-Doppler radar signatures of targets. The target used for this experimental trial was a spinning blade with corner reflectors attached. The reflectors were designed to reflect electromagnetic radiation with minimal loss. These controlled experiments can simulate the rotating types of objects,

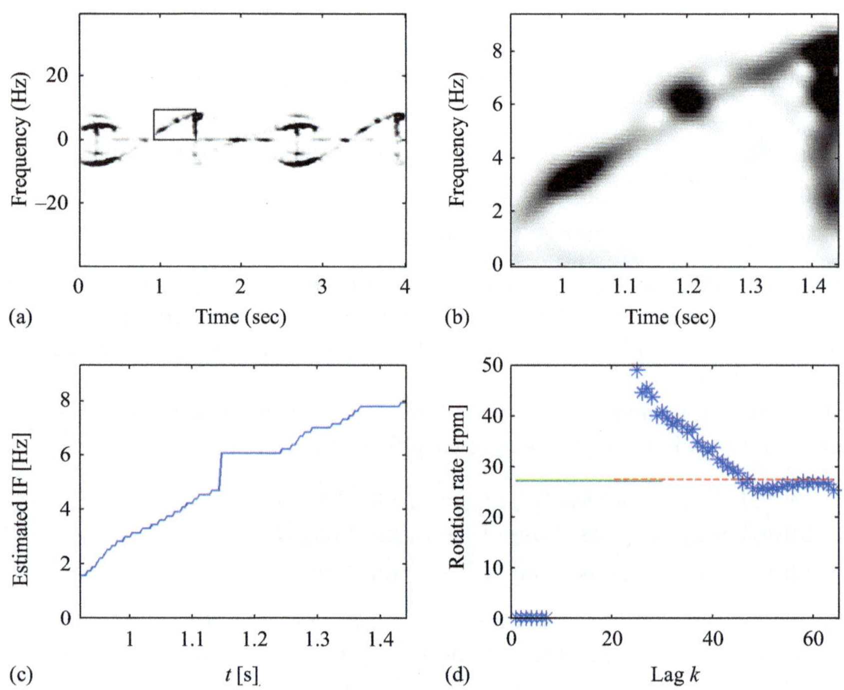

Figure 10.10 (a) TFR of the experimental corner reflector data; (b) Part of the analyzed corner reflector data used for the estimation; (c) The IF of the part of signal shown in (b); (d) Estimation results for various lags k. The solid green line represents the true value (considering whole data set from (a)), while the blue asterisks represent the values estimated by using various lags k

generally found in an indoor environment (for example, a rotating fan) and outdoor environment (for example, a rotating antenna or rotors). Controlled experiments allow us to set the desired rotation rate and to cross check and assess the results.

The blade is set up to simulate real data that might be collected from a similar target such as a rotating antenna or rotating fan or any other rotation of structures on a target. The experiment was conducted with the radar operating at 9.2 GHz. The pulse repetition frequency (PRF) was 1 kHz. The target employed in this experiment was at a range of 300 m from the radar. The S-method is used to depict the micro-Doppler oscillations in TF domain. Figure 10.10 illustrates the results. In order to demonstrate our estimation methods, we have extracted a part of the data signal, which is presented in Figure 10.10(b). The mean estimated frequency over the range of lags $45 \leq k \leq 64$ is $\omega_R = 26.19$ rpm, which is in agreement with the result expected from the whole figure.

In noisy cases, the estimation based on averaging of more observations is recommended, by taking more sets of lag values k or by using intervals like in the case of inverse Radon transform.

10.4 Micro-Doppler Effects Separation Based on the L-Statistics

In this section, the L-statistics-based approach for the rigid body separation is presented. In order to remove the micro-Doppler effect, we perform the TF analysis within the CIT. This approach can be applied to any form of the micro-Doppler effect signature, as far as it is nonstationary in TF plane, during the CIT. This condition is satisfied by very definition of the micro-Doppler as a part of signal with fast varying spectral content. The order statistics is performed based on the spectrogram, while the rigid body signal synthesis is done in the complex TF domain. This approach is very simple to use and produces very good results [11, 12, 20, 21, 23–30]. It can be easily adapted for efficient compensation of a residual, uncompensated, rigid body acceleration in the presence of the micro-Doppler effect as well.

10.4.1 Time-Frequency Analysis and the L-Statistics

The short-time Fourier transform form that will be used here is

$$STFT(t,\Omega) = \int_{-\infty}^{\infty} s(\tau)w(\tau - t)e^{-j\Omega\tau}d\tau \tag{10.37}$$

It has the same spectrogram as the previously defined one (10.14), since they differ in the phase factor only. This STFT form is used here as the one resulting in a simpler reconstruction presentation that follows. Its discrete form is

$$STFT(m,k) = \sum_{i=0}^{M-1} s(i)w(i - m)e^{-j2\pi ik/M} \tag{10.38}$$

where $w(i)$ is a window function used to truncate the considered signal. The window width is M_w, $w(i) \neq 0$ for $-M_w/2 \leq i \leq M_w/2 - 1$. We will use windows zero-padded up to M in order to achieve the same number of samples as in the FT case when all M signal samples are used in the calculation. Due to this zero-padding, we have the same frequency grid in the STFT and in the FT (10.11). Then, we can later easily reconstruct the FT from the STFT, without interpolation, with the concentration close or equal to the concentration of the original FT.

10.4.2 Restoring the High FT Concentration from the STFT

By using a lag window $w(i)$ in the STFT, the concentration in frequency is reduced, as compared to the original FT. For example, if the lag window width is M_w, then the concentration of a sinusoidal signal is reduced M/M_w times, i.e., the STFT-based ISAR/SAR image of a rigid body point (the main lobe of the FT of a sinusoid) would be approximately M/M_w times wider than the original FT-based image of the same reflecting point. However, the concentration could be restored to the original one by summing all the low concentrated STFT (complex) values over discrete time m. Since we calculated $STFT(m, k)$ with the window of the width M_w, usually there are two possibilities for its summation: (a) For all time instants $0 \leq m \leq M - 1$, when the signal $s(i)$ has to be zero-padded for $-M_w/2 \leq i < 0$ and $M \leq i < M + M_w/2 - 1$; (b) For instants $M_w/2 \leq m \leq M - M_w/2$, when zero-padding of $s(i)$ is not used. Reconstruction formula, for the case when the signal is not zero-padded, is

$$\begin{aligned}\sum_{m=M_w/2}^{M-M_w/2} STFT(m,k) &= \sum_{i=0}^{M-1} s(i)\left[\sum_{m=M_w/2}^{M-M_w/2} w(i-m)\right]\exp(-j2\pi ik/M) \\ &= \sum_{i=0}^{M-1} s(i)w_1(i)\exp(-j2\pi ik/M) = S_{w_1}(k)\end{aligned} \tag{10.39}$$

If the signal is not zero-padded $w_1(i)$ is a function of i for $0 \leq i \leq M_w - 2$. Namely, during the summation of the STFT samples (10.39) the first sample of signal $s(0)$ is taken only for $STFT(M_w/2, k)$, the window is then shifted for one sample, and $s(1)$ is taken two times, for $STFT(M_w/2, k)$ and $STFT(M_w/2+1, k)$, and so on. Signal sample $s(i)$ is taken $i + 1$ times, for $i < M_w - 1$. For $M_w - 1 \leq i \leq M - M_w$, each sample is taken exactly M_w times. For $M - M_w + 1 \leq i \leq M - 1$ we have the same situation as for $0 \leq i \leq M_w - 2$, meaning that we will take the last signal sample only ones. To summarize, in the case when the STFT is calculated for each time instant (time step one in the STFT calculation), the resulting window $w_1(i)$ is constant, $w_1(i) = const$, for $M_w - 1 \leq i \leq M - M_w$, for any window. It means that during the most of the CIT interval we have the normalized resulting window $w_1(i)$ being close to the rectangular one, with a small transition at the ending M_w points (Tukay form window). The analysis is not restricted to the step one and rectangular window in the STFT calculation. The same resulting window would be obtained for a step equal to a half of the window width $M_w/2$ and a Hann(ing), Hamming, triangular or rectangular window. The same is valid for steps equal to $M_w/4$, $M_w/8$ .... Resultant window $w_1(i)$ for $w(i)$ being rectangular, Hamming and Hanning, with $M = 256$, $M_w = 16$, and different values of a step used for the STFT calculation, is shown in Figure 10.11, for the case when signal is not zero-padded. Obviously, $w_1(0)$ is close to the rectangular window in all the considered cases.

The FT of the window obtained during the process of reconstruction produces a concentration very close to the full range rectangular window case (i.e., no window). It means that we are able to reconstruct the FT with a concentration close to the one

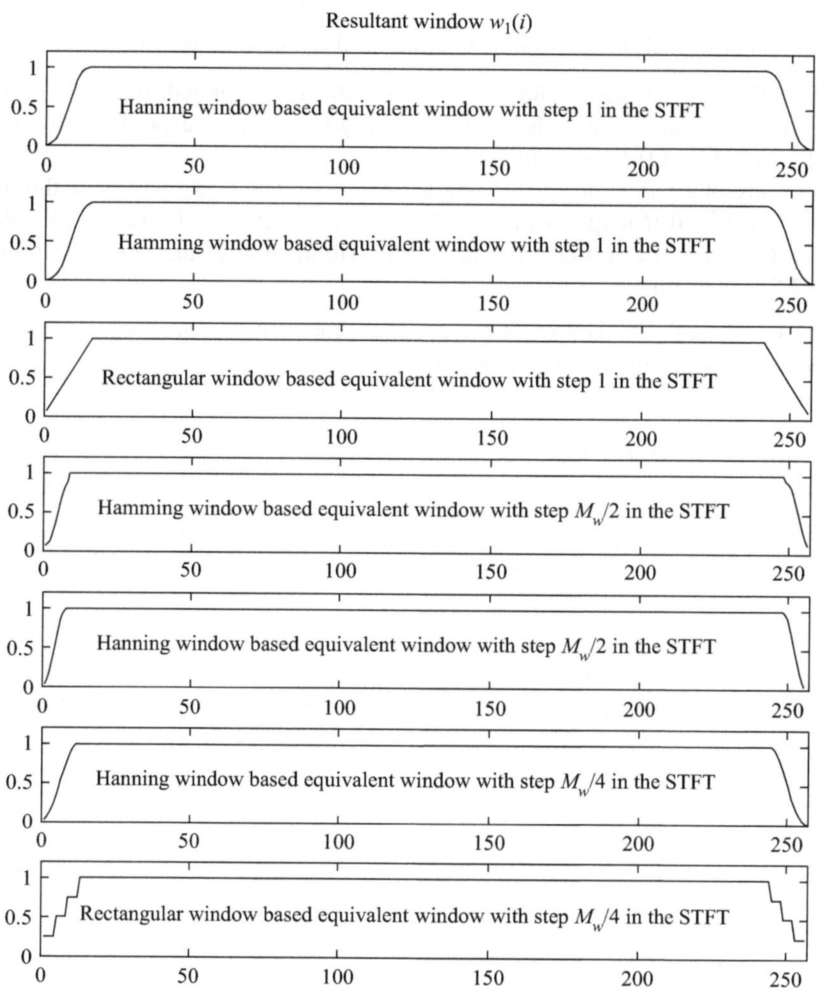

Figure 10.11 The shape of resultant window $w_1(i)$ obtained after summation in (10.39), without signal $s(i)$ zero-padding, for different window types and steps used for the STFT calculation

in the original FT (10.11), by using low concentrated STFTs, calculated with narrow windows. In this way, we will restore the highly concentrated radar image, although we used low concentrated STFTs in the calculation and analysis.

The transition at the ending points of $w_1(i)$ can even be easily overcome by zero-padding the analyzed signal $s(i)$ with M_w samples on both sides. Then, a pure rectangular window $w_1(i)$ would be obtained, for any window $w(i)$.

In order to explain how this mechanism of restoring the original concentration, by summing low-concentrated images, works, consider $s(t) = \exp(j\omega_0 t)$. Its FT is a delta pulse $S(\Omega) = 2\pi\delta(\Omega - \omega_0)$. The STFT of this signal produces

$$STFT(t, \Omega) = W(\Omega - \omega_0)\exp(j(\Omega - \omega_0)t) \tag{10.40}$$

Let us now analyze the result of summing the STFT values over t:

- For $\Omega = \omega_0$ constant values of $W(0)$ will be integrated over an infinite time interval, with the phase $\exp(j(\Omega - \omega_0)t) = \exp(j0)$, producing $\int_{-\infty}^{\infty} STFT(t,\ \Omega)dt \to \infty$ for $\Omega = \omega_0$.
- For any other $\Omega \neq \omega_0$, i.e., when $\Omega = \omega_0 + \theta$, $\theta \neq 0$, we will have the integration $\int_{-\infty}^{\infty} W(\theta)\exp(j\theta t)dt = W(\theta)\int_{-\infty}^{\infty}\exp(j\theta t)dt = 0$. Therefore, all values for $\Omega \neq \omega_0$ are averaged out to zero by summation of the low concentrated STFTs over time.

In the discrete domain the FT reconstruction of the same signal is done by summing values of the discrete form of (10.40)

$$S(k) = \sum_{m=0}^{M-1} W(k - k_0)\exp(j2\pi m(k - k_0)/M) \tag{10.41}$$

with the same conclusions as in the continuous case. Values of $S(k)$, reconstructed when the signal is not zero-padded, are close to (10.41).

10.4.3 Basic Idea for Rigid Body and Micro-Doppler Separation

The presented mechanism of restoring the original concentration of the FT, in the conjunction with the knowledge of the TF patterns behavior of fast-moving and rigid scattering points, lead us to an algorithm for the micro-Doppler free, highly concentrated, radar image. The rigid body and the fast-moving points behave differently in the TF representation of the returned radar signal, within the CIT. The rigid body signal is almost constant in time (stationary), while the fast-varying micro-Doppler part of the signal is highly nonstationary. This part of signal keeps changing its position in the frequency direction.

For the illustration, let us assume that the signal is returned from one point rigid body scatterer and one point fast rotating (micro-Doppler) scatterer. The reflection coefficient of the rigid body is $\sigma_B = 4$, while the reflection coefficient of the fast-moving scatterer is $\sigma_R = 9$ (10.8), (10.22). The STFT representation of the resulting signal is shown in Figure 10.12(a). The rigid body part is at constant frequency for all t within the CIT, while the fast rotating part changes frequency. If we perform

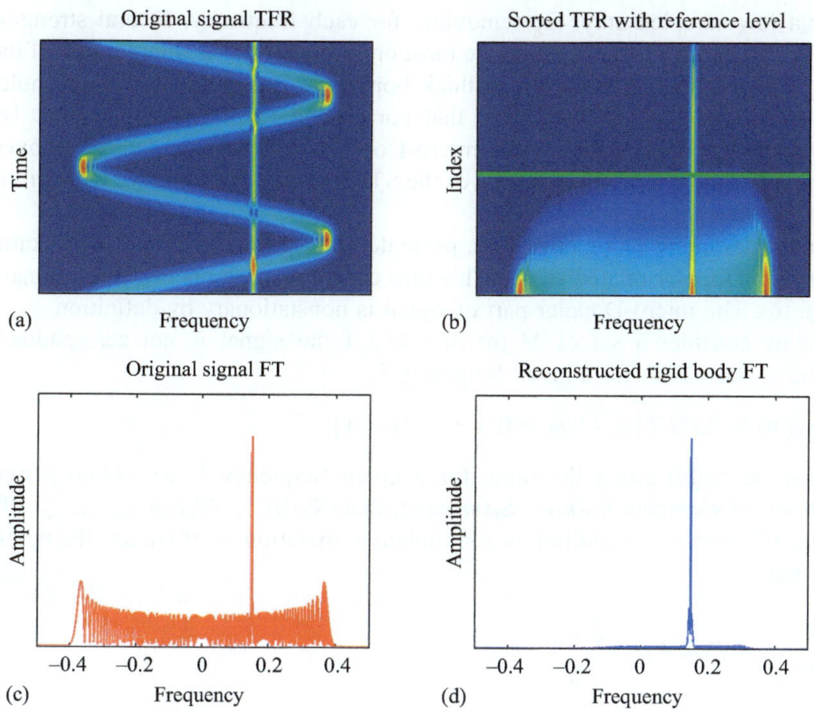

Figure 10.12 Simulated radar signal which correspond to a rigid body reflector with $\sigma_B = 4$ and a rotating reflector with reflection coefficient $\sigma_R = 9$. (a) Absolute value of the STFT, (b) Sorted STFT values, (c) The original FT, (d) The reconstructed FT of the rigid body

sorting over the time axis, as in Figure 10.12(b), we will not change the result of the summation in (10.39) since it is a commutative operation. By summing the STFT values over time, from either of these two plots, presented in Figure 10.12(a) and 10.12(b), we will get the original FT of the corresponding signal, Figure 10.12(c). We analyze here the micro-Doppler which has a moderate strength comparing to the rigid body. However, any value of σ_R from (and including the case without micro-Doppler) $\sigma_R = 0$ up to $\sigma_R \gg \sigma_B$ will not significantly change the pattern.

The basic idea for separating the rigid body and the fast rotating part is in the sorting of STFT values of the returned radar signal along the time axis, within the CIT, Figure 10.12(b). Since the rigid body return is stationary, the sorting procedure will not significantly change the distribution of its values. However, the fast-varying micro-Doppler part of the signal is highly nonstationary, occupying different frequency bins for different time instants (in the case of flashes it exists for some time-instants only). It existence is short in time, for each frequency, over a wide range of frequencies. Thus, after sorting the STFT along the time axis, the micro-Doppler part of the signal have strong values at the wide frequency range, but for a few samples only. By removing, for each frequency, several strongest values of the sorted STFT we eliminate most or all of the micro-Doppler part of the signal. In Figure 10.12(b) the green thick horizontal line presents the threshold, below which are the STFT samples that correspond to the micro-Doppler. By removing these STFT samples, the micro-Doppler will be completely removed from the rigid body. Summing the rest of the STFT values over time we will get the rigid body radar image.

Sinusoidal micro-Doppler pattern, presented in Figure 10.12, is just an example of micro-Doppler related signal. This idea can be applied on any nonstationary signal form. The micro-Doppler part of signal is nonstationary by definition.

Let us consider a set of M (or $M - M_w$ if the signal is not zero-padded) elements of the STFT, for a given frequency k,

$$\mathbf{S}_k(m) = \{STFT(m,k), m = 0, 1, \ldots, M-1\}$$

After sorting $\mathbf{S}_k(m)$ along the time, for a given frequency k, we obtain a new ordered set of elements $\Psi_k(m) \in \mathbf{S}_k(m)$ such that $|\Psi_k(0)| \leq |\Psi_k(1)| \leq \ldots \leq |\Psi_k(M-1)|$. Of course, the addition is commutative operation, so if we use the whole set, we get

$$\sum_{m=0}^{M-1} STFT(m,k) = \sum_{m=0}^{M-1} \Psi_k(m) = S(k)$$

In the L-statistics form of this summation we will, for each k, omit $M_Q = M - M_A$ of the highest values of $\Psi_k(m)$ and produce the L-estimate of $S(k)$, denoted by $S_L(k)$, as

$$S_L(k) = \sum_{m=0}^{M_A-1} \Psi_k(m) \tag{10.42}$$

where $M_A = \text{int}[M(1 - Q/100)]$ is the number of remaining samples and Q is the percent of omitted values.

To illustrate this procedure, we eliminated 40% of the strongest values of the STFT from the previous example, Figure 10.12. In this way, we completely eliminated the micro-Doppler component from the TFR. We are left with the 60% lowest STFT values that contain only the rigid body. The FT reconstruction is performed based on these values only. The reconstructed FT is shown in Figure 10.12(d). The FT of the rigid body is successfully reconstructed by summing 60% of the sorted STFT samples, remained after the micro-Doppler separation.

In the data analysis, this approach, based on elimination of a part of data, before analyzing the rest of the data, is known as the L-statistics [31].

10.4.3.1 Analysis of the Missing Values

Since we have eliminated some of the TFR values, we will analyze the influence of incomplete sum in (10.39). This is the same theory like the L-statistics theory applied to the noisy or non-noisy data [31].

Assume that only points in $m \in \mathbf{D}_k$ are used in summation:

$$S_L(k) = \sum_{m \in \mathbf{D}_k} STFT(m,k) \tag{10.43}$$

where, for each k, $\mathbf{D}_k$ is a subset of $\{0, 1, 2, \ldots, M-1\}$ with M_A elements.

Within the framework of the previous analysis, it means that there will be a highly concentrated component $S(k)$ surrounded by several low-concentrated values $\sum_{m \notin \mathbf{D}_k} STFT(m,k)$. Note that the amplitude of $STFT(m, k)$ is M times lower than the amplitude $S(k)$, since $S(k)$ is obtained as a sum of M values of the STFT. In general, by removing let say $M_Q = (M - M_A)$ values in m, we will get one very highly concentrated pulse, as in $S(k)$, and M_A values of low-concentrated components of the type $STFT(m, k)$, being spread around the peak of $S(k)$ and summed up by different random phases. Only the peak value is summed in phase. Consider:

1. Case for $k = k_0$ corresponding to the position of the rigid body point: At this frequency, all terms in the sum (10.41) are the same and equal to $W(0)$. Thus, the value of $S_L(k)$ does not depend on the positions of the removed samples. Its value is $S_L(k_0) = M_A W(0)$.
2. Case for $k = l + k_0$, where $l \neq 0$: Removed terms in (10.41) are of the form $x_l(m) = W(l)e^{j2\pi ml/M}$. They assume values from the set $\Phi_l = \{W(l)e^{j2\pi ml/M}$, $m = 0, 1, 2, \ldots M-1\}$, with equal probability, for a given l. Statistical mean of these values is $E\{\mathbf{x}_l(m)\} = 0$ for $l \neq 0$, resulting in $E\{S_L(l + k_0)\} = 0$.

The resulting statistical mean for any k is

$$E\{S_L(k)\} = (M - M_Q)W(0)\delta(k - k_0)$$

Variance of a single random variable taking values from set Φ_l is $\text{var}\{\mathbf{x}_l(m)\} = W^2(l)$. We have removed M_Q samples from the summation. The standard deviation of the removed terms is [29]

$$\sigma_r = std\left\{\sum_{m=1}^{M-M_Q} x_l(m)\right\} = W(l)\sqrt{M_Q \frac{M - M_Q}{M - 1}}, \quad \text{for } l \neq 0$$

Therefore, the reconstructed values $S_L(k_0 + l)$ will be zero mean random variables with standard deviation σ_r.

The influence of the number of missing points to the concentration of the reconstructed FT will be illustrated by an example, Figure 10.13. Here we consider a constant frequency signal, without micro-Doppler. Its FT is calculated and presented in Figure 10.13(a). Then the FT is reconstructed based on the 25%, 50%, and 75% of the lowest STFT values for each k. We can see that even by taking a small number of STFT points, we still keep a strong peak, since it is summed in phase, Figure 10.13(b), 10.13(c), and 10.13(d).

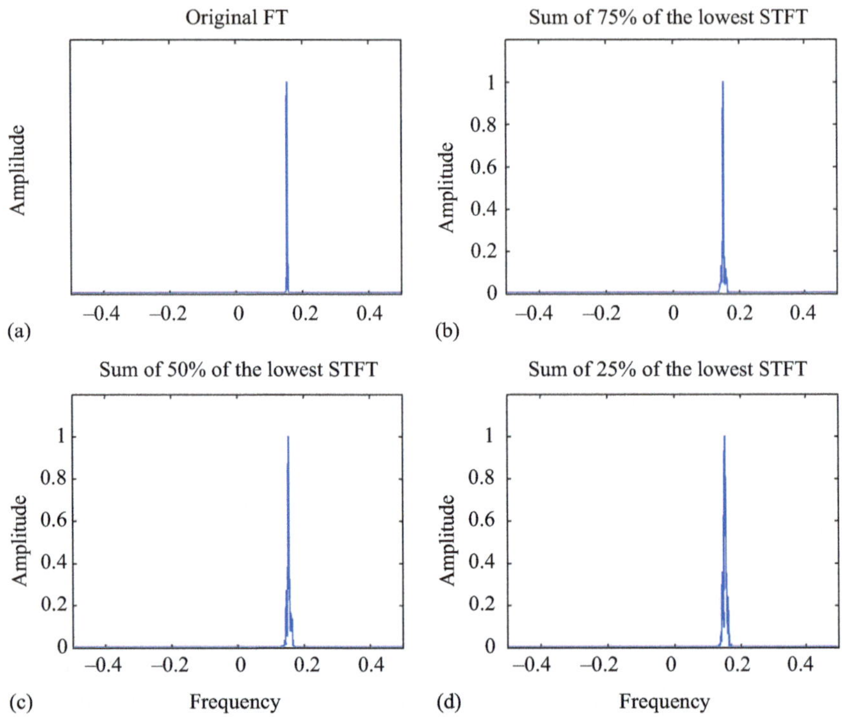

Figure 10.13 The FT of a sinusoidal signal: (a) Original; (b) Reconstructed by summing 75% of the smallest STFT values, for each k; (c) Reconstructed by summing 50% of the smallest STFT values; (d) Reconstructed by summing 25% of the smallest STFT values

---

Example 10 In this example we analyze a signal of the form

$$s(m) = \sigma_B \sum_{i=1}^{K} \exp\{jy_{Bi}m\} + \sigma_R \sum_{i=1}^{P} \exp\{j[y_{R0i}m + A_{Ri}\sin(\omega_{Ri}\, m + \varphi_i)]\} \quad (10.44)$$

with 10 components: $K = 5$ components with constant frequency (used to model rigid body reflectors) and $P = 5$ sinusoidal modulated components (used to model rotating reflectors), with: $\sigma_B = 1$, $\sigma_R = 15$, $y_{Bi} = [1.9\pi, 1.95\pi, 2\pi, 2.05\pi, 2.1\pi]$, $A_{Ri} = [170, 320, 220, 400, 220]$, $y_{R0i} = 0$, $\varphi_i = 0$, and $\omega_{Ri} = [\pi/256, \pi/512, \pi/256, \pi/512, \pi/256]$, for $i = 1, 2, 3, 4, 5$, $M = 1\,024$. The STFT of this signal is shown in Figure 10.14(a). The constant components that correspond to the rigid body are not well separated in the TF plane. Moreover, they are covered by the sinusoidal modulated patterns which represent the micro-Doppler effects of the rotating reflectors. If we sort the STFT values along time axis, then the representation of the rigid body parts does not change significantly, since it is constant during the whole CIT, Figure 10.14(b). On the other hand, the fast rotating parts occupy only small

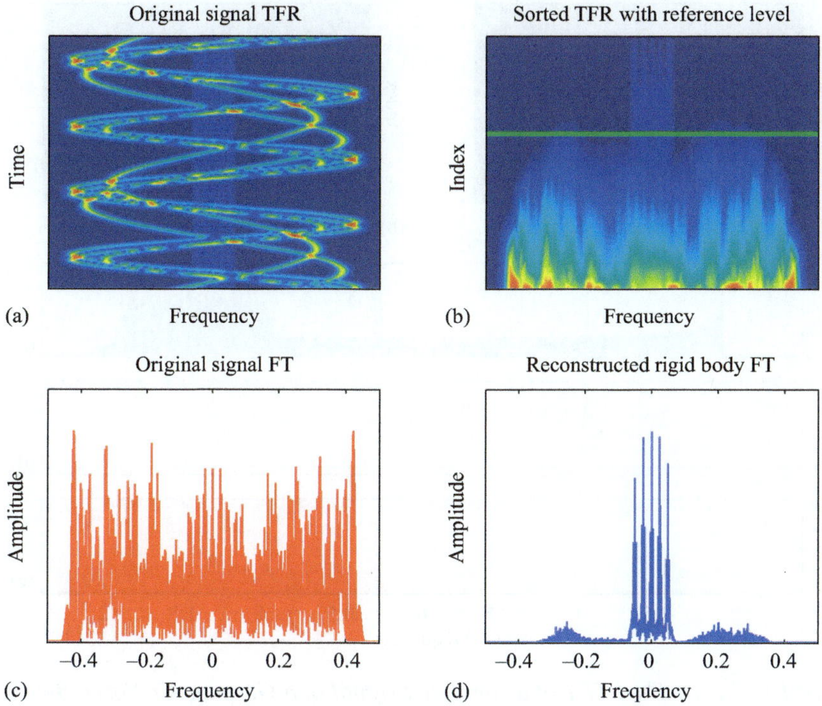

Figure 10.14 *(a) The STFT of a signal consisted of five rigid body components and five sinusoidally modulated m-D components. (b) The sorted STFT of the same signal, with reference level represented by green horizontal line. (c) The original FT of the signal. (d) The FT of rigid body, reconstructed by summing the STFT values remained after sorting and eliminating samples that correspond to the m-D effect*

time intervals over a wide region of frequencies. They lie in high value regions of the sorted transform. Thus, they will be eliminated by removing the 60% of the highest STFT values, for each frequency. The used fixed threshold is depicted by the green horizontal line in Figure 10.14(b). The results are not sensitive to the value of T_{hr}. The reconstructed FT, obtained by summing the rest of the STFT along the time is shown in Figure 10.14(d). We can clearly see five peaks that correspond to the five rigid body reflectors. The original FT is shown in Figure 10.14(c). It cannot be used even to determine the number of components in the analyzed signal.

Example 11 (Helicopter Data Analysis) In this example, we first present a new simulation approach to the data of a German Air Force Bell UH-1D Helicopter known also as "Iroquois" presented in [10]. Here, the simulation is performed according to the variable flashing reflection coefficients, rather than by using a

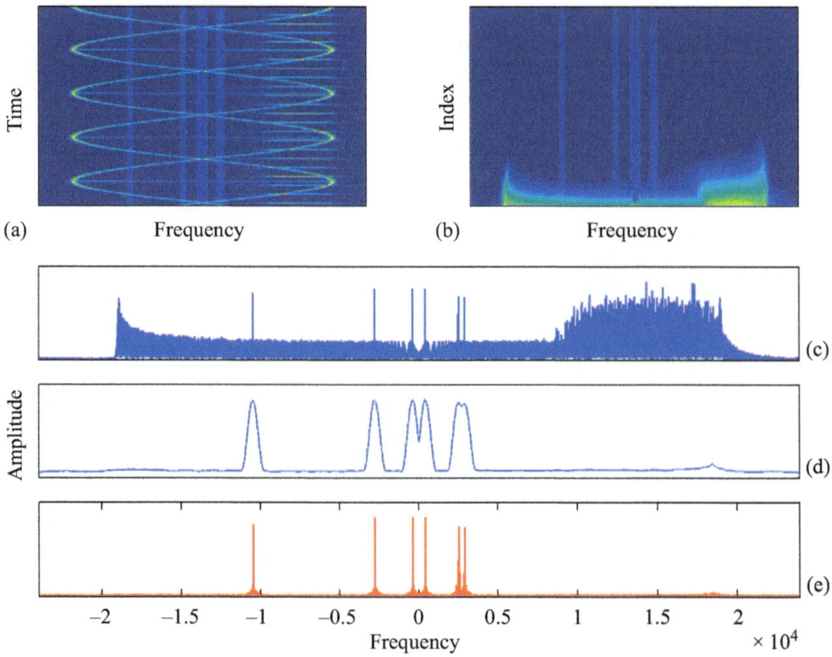

Figure 10.15 (a) The STFT of a simulated signal of a German Air Force Bell UH-1D Helicopter. (b) The sorted STFT of this signal. (c) Original FT of the signal. (d) The FT of the rigid body, reconstructed by summing the lowest absolute STFT values. (e) The FT of the rigid body, reconstructed by the presented method

mathematical form that would produce the data as in [1]. Several effects are emphasized in the TFR, Figure 10.15(a). The stationary patterns along the time-axis correspond to the rigid body reflection. The motion of two main blades is modeled by two rotating reflectors, producing sinusoidal FM signals with a large magnitude in the frequency direction (10.46). The main rotor flashes are simulated by signals producing lines that connect extreme points of the sinusoidal FM signal, along the time axis. The smaller pulses that can be seen on the right-hand side of Figure 10.15(a) correspond to the tail rotor flashes, and they are simulated by taking into account the physical meaning of its appearance. Namely, these flashes correspond to the periodic alignment of the main and tail rotors to maximally reflect the radar signal when they are normal to the line-of-sight. Therefore, we use here an angle dependent reflection coefficient

$$\sigma(t) = \exp(-30|\sin(2\pi t/T_{Rot})|) \tag{10.45}$$

where the reflection takes value 1 when $t = kT_{Rot}/2$ and $|\sin(2\pi t/T_{Rot})| = 0$, while for other t, $30|\sin(2\pi t/T_{Rot})$ assumes high values and the reflection coefficient is small. Note that other effects that can be observed in a radar image, including multipath, are not considered here.

The simplified model of the reflected UH-1D signal can now be written as

$$s(t) = x_{RIG}(t) + x_{ROT}(t) + x_{FL\_M}(t) + x_{FL\_T}(t)$$

where $x_{RIG}(t)$, $x_{ROT}(t)$, $x_{FL\_M}(t)$, and $x_{FL\_T}(t)$ represent signals caused by the rigid body, rotation of the main rotor, and the main and tail rotor flashes, respectively. The signal is considered within the interval of 400 msec, sampled with a rate of $\Delta t = 1/48$ msec. Four sinusoidal components, caused by the rigid body, are at the frequencies -10.5, -2.8, 2.5, and 2.9 kHz. Two components at -0.4 and 0.4 kHz correspond to the modulated time tones commonly added to the data tape [32]. The sinusoidal FM signals, corresponding to the rotation of the main rotor blades, are modeled as

$$x_{ROT}(t) = \sigma_{ROT}[\exp(j2\pi A_{ROT}\sin(2\pi t/T_{ROT})) + \exp(-j2\pi A_{ROT}\sin(2\pi t/T_{ROT}))] \tag{10.46}$$

where $\sigma_{ROT} = 10$, $T_{ROT} = 175$ ms, and $A_{ROT} = 529.19$. The main and tail rotor flashes are modeled as

$$x_{FL\_M}(t) = 2.5\sum_{k=1}^{128} \frac{k+64}{128}\exp(-25|\sin(2\pi t/175)|)\cos(25.98k\sin(2\pi t/175))$$

and

$$x_{FL\_T}(t) = 2.5\sum_{k=64}^{128} \exp(-25|\sin(2\pi t/35.8)|)\exp(j2.66k\sin(4\pi t/35.8))$$

The signal is corrupted by a moderate Gaussian noise. To compare our simulation with the real one (for the micro-Doppler and rigid part values ratio) refer to [1, 32].

The L-statistics based algorithm for the rigid body separation is applied to the simulated helicopter signal. The sorted STFT is shown in Figure 10.15(b). We can see that the STFT values corresponding to the rotating parts are in the high value region. The reconstructed FT is shown in Figure 10.15(e). All five reflectors that correspond to the rigid body are successfully recovered. The original FT is presented in Figure 10.15(c), while the reconstructed FT obtained by summing the absolute values of the remaining STFT samples is presented in Figure 10.15(d).

Example 12 (Noncompensated Rigid Body Acceleration) In this case an accelerating rigid body target is considered and examined. The received radar signal that corresponds to an accelerating target in the ISAR systems is a linear FM. Similarly, in the SAR systems the target motion may induce linear frequency modulation in the received radar signal [13]. Therefore, we simulated three rigid body reflectors as three linear FM components with the chirp-rate α. In order to show that our algorithm will not remove only the micro-Doppler induced by vibrating and rotating targets, here we have also used a more complex form of the micro-Doppler.

The STFT of the analyzed signal is presented in Figure 10.16(a). We can clearly see that, as a result of the acceleration, the TFR of the rigid body part of the signal is not stationary during the time. Consequently, it is difficult to separate it from the micro-Doppler in the sorted STFT, Figure 10.16(b). Namely, if we

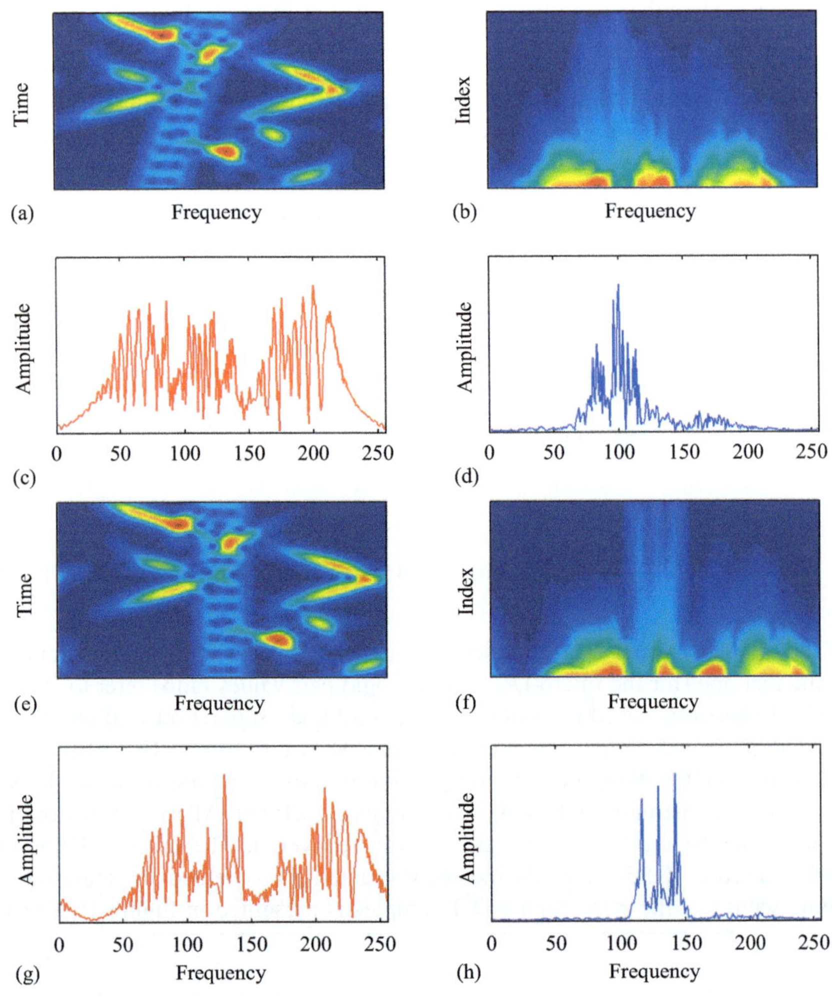

Figure 10.16 Accelerating rigid body with a complex form of the micro-Doppler. (a) TFR of the signal without motion compensation. (b) Sorted TFR of the original signal. (c) Original FT of the analyzed signal. (d) Reconstructed FT of the accelerating rigid body without motion compensation. (e) TFR of the signal after motion compensation. (f) Sorted TFR of acceleration compensated signal. (g) The FT of the original signal with motion compensation. (h) Reconstructed FT of the accelerating rigid body with motion compensation

perform the micro-Doppler separation by removing 50% of the highest STFT samples, as we did it in the examples where there was no need for the motion compensation, we would reconstruct the FT of the rigid body as presented in Figure 10.16(d). Here, we would remove a significant part of the rigid body points, as well.

In the analysis of the rigid body with uncompensated acceleration, we should first compensate the remaining acceleration. This is not possible in the original signal, since the micro-Doppler signatures prevent us from properly compensating the remaining acceleration. However, the application of the presented method for the micro-Doppler removal can solve this problem. We will use the first-order local polynomial Fourier transform (LPFT)

$$LPFT(t,\Omega) = \int_{-\infty}^{\infty} s(\tau)w(\tau - t)e^{-j(\Omega\tau+a\tau^2)}d\tau \tag{10.47}$$

instead of the STFT, where the term $\exp(-ja\tau^2)$ is used to compensate the linear frequency modulation of the rigid body part of the signal, $LPFT(t,\Omega) = FT\{s(\tau)e^{-ja\tau^2}\,w(\tau - t)\}$. The parameter a is not known in advance, but we know that it can take values from a set $\Lambda = [-a_{\max}, a_{\max}]$, where $a_{\max}$ is the chirp rate corresponding to the maximal expected acceleration (positive or negative) [33]. In this example we used $\Lambda = [-2{:}0.25{:}2]$. Now, $\hat{a}$ can be estimated as the value from the set Λ for which we obtain the highest concentration of the reconstructed rigid body (compensated FT) based on the LPFT and the L-statistics, with, for example, $Q = 50\%$. The reconstructed FT, by using 50% of the lower LPFT values, will be denoted by $S_{L,a}(k)$. Its concentration is calculated using the TF concentration measure [18],

$$H(a) = \left(\sum_{k=0}^{M-1} |S_{L,a}(k)|^{1/p}\right)^p \tag{10.48}$$

with $p = 1$. The LPFT, calculated with the estimated optimal value of $\hat{a} = 1.25$, which results from $H(a)$, is shown in Figure 10.16(e). The linear frequency modulation is compensated by $\hat{a}$ in (10.47). Thus, with optimal $\hat{a}$ we have components with almost constant frequency in the TFR representation of the rigid body reflectors. In this way, we have successfully reconstructed the rigid body and removed the micro-Doppler part, as it is presented in Figure 10.16(h). The procedure is not too sensitive to $\hat{a}$. Very good results are obtained with neighboring value $\hat{a} = 1.0$ and $\hat{a} = 1.25$.

Note, that it would be impossible to estimate the chirp-rate $\hat{a}$ from the original signal, without employing the presented algorithm for the micro-Doppler removal.

Example 13 (Real Data) The presented algorithm is illustrated on two sets of real data in this example. The examined data were collected using an X-band radar operating at 9.2 GHz, [12]. The first real data represent three corner reflectors rotating at approximately 60 rpm (all facing radar) and the rigid body observed by the radar with $T_r = 1$ kHz. The STFT of the returned signal, for the given range bin, is shown in Figure 10.17(a). After sorting the STFT over time (Figure 10.17(b)), the constant frequency component corresponding to the rigid body becomes more visible, since the time varying micro-Doppler frequency content is spread over

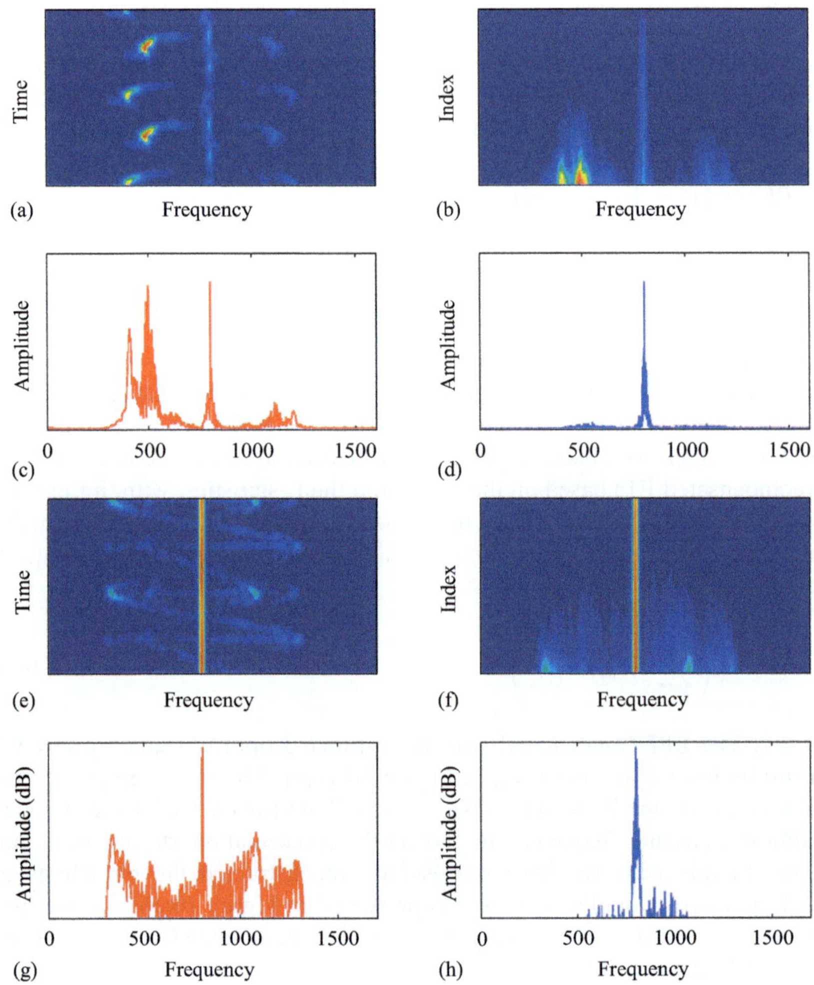

Figure 10.17 Real radar data corresponding to a rigid body and three corner reflectors rotating at ~60 rpm (a–d). (a) The STFT. (b) Sorted STFT. (c) The original FT and (d) the FT reconstructed by summing over 50% of the lowest STFT samples. The same procedure is repeated for real radar data corresponding to a stronger rigid body and two corner reflectors rotating at ~40 rpm (e–h). A logarithmic amplitude scale is used in subplots (g) and (h)

many frequencies, for each frequency bin. Thus, the micro-Doppler frequency content is concentrated at the bottom of the sorted TFR. The rigid body is separated from the micro-Doppler and its FT is successfully reconstructed by using 50% of the lowest STFT values, as shown in Figure 10.17(d). If we compare it to the FT of the original signal, Figure 10.17(c), we can see the improvement in the rigid body presentation.

In the second example, the real radar data corresponding to two outside corner reflectors, rotating at approximately 40 rpm (all facing radar) with rigid body, are analyzed. The same radar as in the previous example is used, while the reflectivity of rigid body is much higher than those of the rotating reflectors. The STFT representation of the observed signal is shown in Figure 10.17(e). The sorted STFT is shown in Figure 10.17(f). The original FT is shown in Figure 10.17(g). The reconstructed FT, obtained by summing 50% of the lowest STFT values is presented in Figure 10.17(h). We have successfully removed most of the micro-Doppler. Moreover, we may use the removed STFT samples in order to estimate features of rotating reflectors. Here, we have used a logarithmic scale to present the reconstructed values, since the micro-Doppler values were very low.

10.4.4 Adaptive Percentage of Missing Values

Now we will discuss how to establish the threshold for the micro-Doppler elimination. There are two possible approaches:

- The first approach is to assume a fixed threshold for the entire ISAR/SAR image: for example, removing Q(%) highest values for each frequency, by knowing that this will not disturb significantly the obtained image, as far as Q is not too high.
- More sophisticated approach is to calculate the adaptive threshold for each range of a radar image. The adaptive threshold can be obtained based on the L-statistics. To use the L-statistics approach, we will sort the STFT values for a given range and a given frequency. If there is a micro-Doppler, then, after sorting the STFT values, there is a region of an increase of the sorted STFT values, Figure 10.18 (upper part). Thus, if we sum the sorted absolute squared values of STFT values over frequency, we will get a function:

$$A(m) = \sum_{k} |\Psi_k(m)|^2 \tag{10.49}$$

Now, it is possible to find the reference level R_L based on the mean of the 10% of the lowest samples, i.e., based on $M/10$ values of the sorted $\Psi_k(m)$, $R_L = T_{hr}\sum_{m=0}^{M/10-1} 10A(m)/M$, with T_{hr} being a parameter, usually from $T_{hr} = 2$ to $T_{hr} = 10$. For example, $T_{hr} = 2$ means that we will use all the values in the sorted $\Psi_k(m)$ whose squared values are up to, for example, two times greater than mean of 10% lowest squared values. Then Q is found as the percent of $A(m)$ below R_L. In the case when there is no micro-Doppler it means that we will use all the values, since

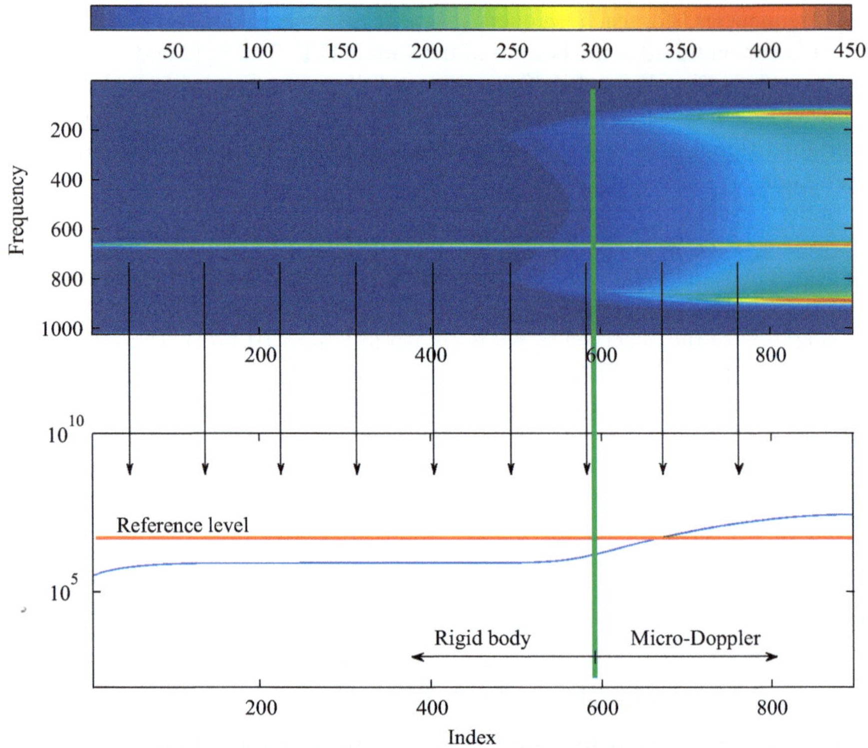

Figure 10.18 Sorted STFT values, $|\boldsymbol{\Psi}_k(m)|$, with a colorbar (upper). The values of $A(m)$, obtained by summing the squared values of the sorted STFT along the frequency, with the adaptive threshold R_L obtained based on the average value of 10% of its smallest values and $T_{hr} = 10$ – thick horizontal line (lower)

the stationary values are close to the mean of the lowest values for all time instants. If there is a micro-Doppler then the value outside of the stationary points will start to increase sharply and the summation will stop. The results are not too sensitive to these values, since $A(m)$ is a fast increasing function when the micro-Doppler starts to appear, Figure 10.18 (lower part).

10.4.5 Algorithm for the Micro-Doppler Effects Removal

The simplest way to use the presented method is in applying the L-statistics approach to all range bins, **with a constant threshold**, for example $Q = 50\%$. In this case, the micro-Doppler will be separated, while the rigid body will not be degraded. However, this way of calculation increases the calculation complexity.

In radar data analysis, only a part of the range bins may contain the micro-Doppler, while most of them will be micro-Doppler free. If there is a micro-Doppler point, then a constant or adaptive threshold for the L-statistics application in the micro-Doppler removal should be used in order to separate the micro-Doppler, while most of the rigid body part of signal would be preserved. If there is no micro-Doppler in the considered range bin, then processing for this range is not necessary. If the FT is already well focused, or if there is no returned target signal in the considered range bin, then the FT can be used as it is. Thus, we can improve the calculation complexity by defining a procedure to avoid processing of the range bins where we can conclude that there is no target return or there is no micro-Doppler part of signal. Of course, **this classification procedure is optional,** and we will not lose anything in the radar image quality if we apply the presented method in some bins with already focused image or where there is no target return. Thus, the thresholds in the next procedure may be chosen in quite conservative way, to be sure not to miss any bin with micro-Doppler, allowing false micro-Doppler detections.

The range bins classification procedure will be presented in a form of an algorithm. In any marginal or indecisive case, we should apply the presented method, as well.

This algorithm consists of the following steps:

Step 1. *Detect whether there is a returned target signal in the considered range bin*, by using

$$\max_k\{|S(k)|\} > \varepsilon \tag{10.50}$$

where $\varepsilon = 0.02 \max\{|S(k,l)\}$ in the noiseless case or $\varepsilon = \max\{0.02\max\{|S(k,l)|\}, 2\sqrt{\sigma}/M\}$ in the presence of noise (here $S(k, l)$ is the 2D FT of the received signal, i.e., full radar image matrix for all ranges and cross-ranges, while $S(k)$ is the radar image for a given range bin). The standard deviation σ of noise in the radar image, for a given range bin, can be estimated as [34]

$$\hat{\sigma}_{re} = \frac{\text{median}\{|\text{Re}\{S(k)\} - \text{Re}\{S(k-1)\}|,\ k = 2,\ldots,M\}}{0.6745\sqrt{2}} \tag{10.51}$$

for the *real* part of $S(k)$. The same procedure is for imaginary part. If there is no target in the returned signal, do not perform the following steps and take $S(k)$ as the radar image for the considered range bin. If there is a target signal, continue.

Step 2. *Detect whether there are the micro-Doppler effects in the considered range bin*:

$$R = \frac{\max\{|S(k)|\}}{\text{mean}\{|S(k)|\}} > 10$$

The previous relation is the simplest concentration measure. In the case of a highly concentrated signal, R is close to M, while for a low concentrated signal, this value is close to 1. We say that the signal is well concentrated (rigid body only) if $R > 10$. This threshold was successfully tested for various scenarios and multicomponent signals.

If the signal is well concentrated take its FT $S(k)$ as the radar image for the considered range bin. If the signal is not well concentrated, continue.

Step 3. *Remove the micro-Doppler effects and reconstruct the FT of the reflectors that correspond to the rigid body in the considered range bin.* Remove the values of the STFT that are higher than the threshold and for each frequency sum the rest along the time. Take the obtained FT $S_L(k)$ (10.43) as the radar image for the considered range bin.

10.5 The Micro-Doppler Signature Tracking by Using the Viterbi Algorithm

The micro-Doppler in the noiseless cases or in the presence of a moderate noise is analyzed in the previous two sections. The micro-Doppler may be efficiently analyzed or removed from the signal by the presented techniques. However, in high noise environments it may be difficult to follow the IF of the micro-Doppler signature. Then, many false TF maxima, far from the true micro-Doppler signature, may be detected. In this section, we will deal with micro-Doppler signatures tracking (IF estimation) in high noise environments. A method based on the Viterbi algorithm [35] is used to reduce the estimation error caused by this kind of disturbance.

Two main assumptions are the basis for the IF estimation, representing a micro-Doppler signature, in a high noise environment. The first one is that there is a high probability that the true IF is at a point having one of the largest TFR values. Namely, we may assume that if the TFR maximum at the considered time instant is not at the IF point, the true IF is at the position of the second, third..., but not as far as, for example, the hundredth highest TFR value for that instant. Thus, a weighting function which assumes greater weights for smaller values of the TFR is adopted. The second assumption is that the IF variation between two consecutive time points is not extremely large. The micro-Doppler signal is an FM modulated, so that we may adopt this assumption.

The algorithm is presented by using the discrete form of an arbitrary TFR of the signal $s(m)$, denoted as $TFR(m,k)$, where $m=1,\dots,M$ represents the time instances and $k=1,\dots,N$ represents the frequency bins. It is assumed that the IF of the signal is continuous function. The used algorithm is an instance of the Viterbi algorithm [34, 36]. It may be presented by the following steps:

Step 1. *Time-frequency representation*: Calculate the discrete form of a TFR of the signal, $TFR(m,k)$. The obtained representation can be considered as an image that has $M{\cdot}N$ points with intensities $TFR(m,k)$.

Step 2. *Labeling*: For each time instant m, we have N TFR values $TFR(m,1)$, $TFR(m,2),\dots,TFR(m,M)$. After sorting these values into non-increasing sequence

$$TFR(m,k_1) \geq TFR(m,k_2) \geq \dots \geq TFR(m,k_j) \geq \dots \geq TFR(m,k_N) \quad (10.52)$$

labeling function $f(\cdot)$ can be defined as

$$f(TFR(m, k_j)) = j - 1 \tag{10.53}$$

that is, the highest TFR value in the considered time instant is labeled by 0, and the lowest value is labeled by $N - 1$. This weighting function realizes the assumption that the points with large TFR values are more important candidates for the IF estimates. The function $f(x)$ is not formed directly by using values of the TFR, since the signal and noise parameters can be time-varying. It means that a particular distribution value at the considered instant may highly probably belong to the signal term, while in other points it can be influenced by disturbances.

Step 3. *Penalty function*: Let us consider two points in the TFR, (m, k_l) and $(m+1, k_j)$ within two consecutive time instants m and $m+1$, where $k_l = 1, \ldots, N$ and $k_j = 1, \ldots, N$. The penalty function can be determined as

$$g(k_l, k_j) = \begin{cases} 0, & |k_l - k_j| \leq \Delta \\ c \cdot (|k_l - k_j| - \Delta), & |k_l - k_j| \geq \Delta \end{cases} \tag{10.54}$$

where Δ is a threshold and c is penalty increase for the penalty function. The threshold Δ represents the penalty for frequency change between the two consecutive time instants. The reasonable choice of Δ would be the maximal expected value of the IF variation between consecutive points. It means that there is no additional penalty due to this function for small IF variation (within Δ points, for two consecutive instants). In the realization we obtained good results by taking Δ which corresponds to a few neighboring points (e.g., $\Delta = 3$). For $\Delta \to \infty$ IF estimation obtained by this algorithm will reduce to the estimation based on the TFR maxima. Note that the presented form of the labeling function $f(x)$ and penalty function $g(x, y)$ is only one possible form.

Step 4. *Path cost*: Now, the best partial path from the first time instant to point $(m+1, k_j)$ should be chosen.

Bearing in mind that for each point (m, k_l) there are N possible paths toward the next time instant $m + 1$ with penalty values

$$g(k_l, 1), g(k_l, 2), \ldots, g(k_l, N) \tag{10.55}$$

total number of paths between two consecutive time instants is N^2 and total numbers of paths in the entire TFR is N^M. This fact makes a direct search for the optimal path impossible. Fortunately, the algorithm can be realized recursively, as an instance of the generalized Viterbi algorithm [36].

The partial best path selection is done using a rule given by [35]

$$\begin{aligned} p(1, k_j) &= f(TFR(1, k_j)) \\ p(m_{i+1}, k_j) &= \arg \min_{1 \leq k_l \leq N} [p(m_i, k_l) + g(k_l, k_j) + f(TFR(m_{i+1}, k_j))] \end{aligned} \tag{10.56}$$

for each point $k_j = 1, \ldots, N$, where $p(m_i, k_l)$ is the partial best path cost value from the initial time instant $m = 1$ (m_1) toward point $p(m_i, k_l)$. Therefore, the partial best paths at the next instant m_{i+1} can be represented as concatenation of partial best path at instant m_i with the point at the new instant $\pi(m_{i+1}) = [\pi(m_i), (k_j)]$, $k_j = 1, \ldots, N$, that produce the minimal value of (10.56). Note that for each path cost value $p(m_i, k_l)$ we should store path between $i-1$ TFR points in order to obtain path cost value $p(m_{i+1}, k_j)$.

Step 5. *Best path*: Minimal path costs $p(M, k_{min})$ for the final time instant $m = M$ is obtained using formula:

$$k_{\min} = \arg \min_{1 \le k_j \le N} p(M, k_j) \tag{10.57}$$

and as we stated in Step 4 for each path cost value $p(m, k_l)$ we stored path $[\pi(1), \pi(2), \ldots, \pi(m-1)]$ needed to obtain path cost value $p(m, k_l)$. Determining the minimal path cost value in the final time instant we also determine the best path from initial toward final time instant.

Example 14 In order to illustrate the algorithm, we will consider a synthetic example. Let us consider a case where in **Step 1**, the *TFR*(*m*, *k*) representation, with $m = 1, \ldots, 5, k = 1, \ldots, 4$, is given by

$$TFR = \begin{bmatrix} 5 & 15 & 1 & 5 & 14 \\ 3 & 6 & 8 & 4 & 6 \\ 12 & 3 & 4 & 12 & 3 \\ 7 & 11 & 14 & 6 & 10 \end{bmatrix} \tag{10.58}$$

In **Step 2**, a label is given to each point, by using (10.53). The labels can be stored in 4×5 matrix. For example, for time instant $m = 1$ labels from the top to bottom are 2, 3, 0, 1. Labeled values for the entire TFR are depicted in Figure 10.19.

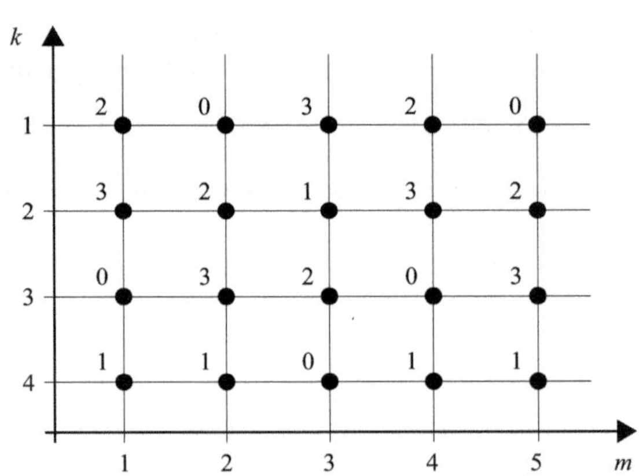

Figure 10.19 Labeled TFR values

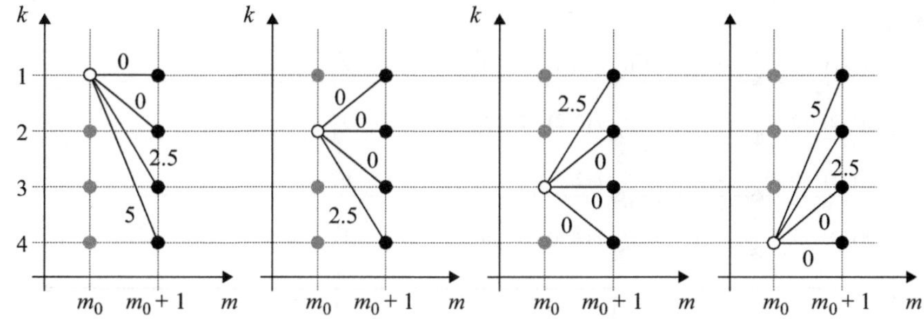

Figure 10.20 Calculation of the partial paths' penalty values

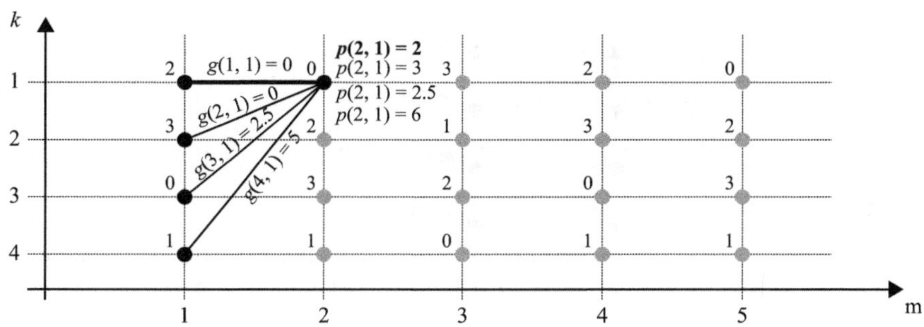

Figure 10.21 Selection of partial best path for the point (2, 1) in time instant $m = 2$ and frequency bin $k = 1$

For the given $TFR(n, k)$ in **Step 3**, penalty value of each partial path for two consecutive time instants m and $m + 1$, where $m = 1, \ldots, M - 1$, can be obtained by using (10.54). Figure 10.20 presents all paths of two consecutive time instants with corresponding penalty values. Here we use threshold $\Delta = 1$ and $c = 2.5$.

In order to illustrate **Step 4**, let us take into consideration only point (2, 1) in time instant $m = 2$ obtained from synthetic representation $TFR(m, k)$ (10.58). Partial paths and their penalty values $g(k_l, 1)$ that lead to that point from points in previous time instant $m = 1$ can be obtained using formula (10.54). In Figure 10.21, for point (2, 1) in time instant $m = 2$ with label value 0 we have four possible paths toward four points (1, 1) to (1, 4) in time instant $m = 1$. The penalty costs of these four paths are 0, 0, 2.5, 5, respectively, and they are depicted in Figure 10.21, as $g(k_l, 1)$, $k_l = 1, \ldots, 4$. Now, using formula (10.56) and labeling function depicted at the same figure, we can conclude that only path leading from point (1, 1) has the minimal path cost value $p(2, 1) = 2$ and hence is the partial best path from all points in time instant $m = 1$, toward the point (2, 1). Thus, we remember ($\pi(1) = [1]$) and the path cost value $p(2, 1) = 2$, see Figure 10.22.

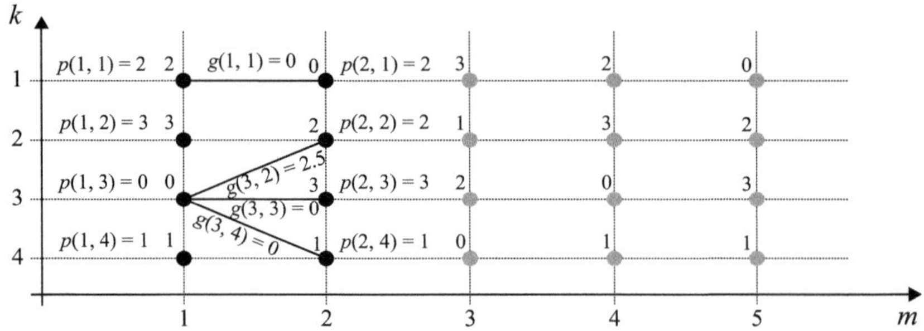

Figure 10.22 Partial best paths' costs p(m, k) for time instants m = 1 and m = 2

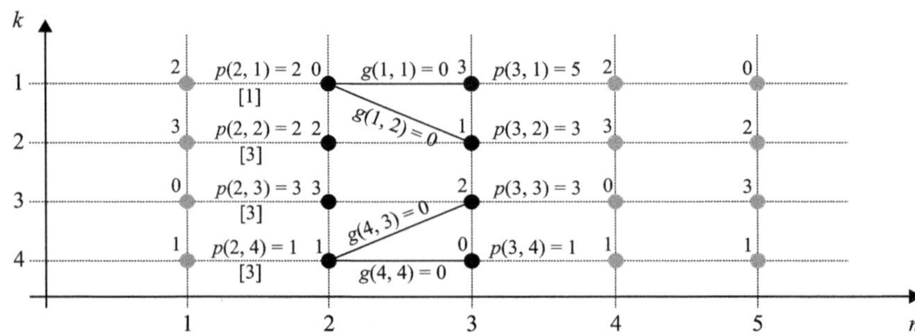

Figure 10.23 Partial best paths' costs p(m, k) for time instants m = 2 and m = 3 and stored path [π(1)] for each k = 1, ..., 4

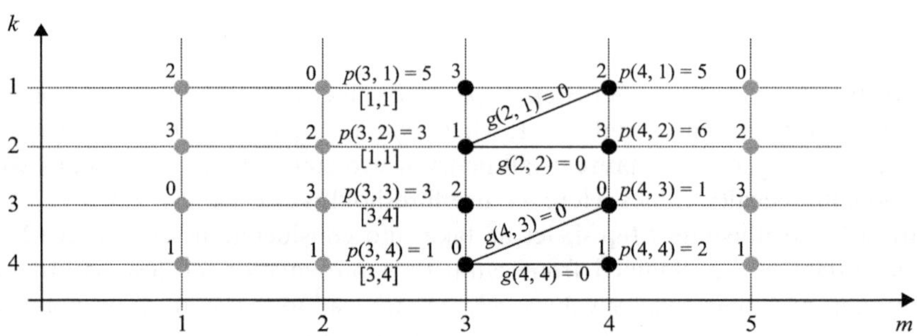

Figure 10.24 Partial best paths' costs p(m, k) for time instants m = 3 and m = 4 and stored path [π(1), π(2)] for each k = 1, ..., 4

For the considered point in this time instant, we have searched over N paths in order to find best partial path. Thus, for the entire time instant we will search over N^2 for paths, and MN^2 for the entire plane.

Repeating the process for each point between two consecutive time instances, we obtain best partial paths for each point as it is depicted in Figures 10.22–10.25.

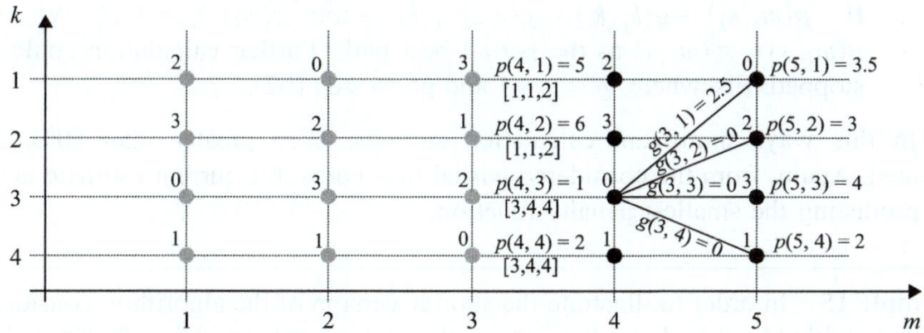

Figure 10.25 Partial best paths' costs p(m, k) for time instants m = 4 and m = 5 and stored path [π(1), π(2), π(3)] for each k = 1, . . . , 4

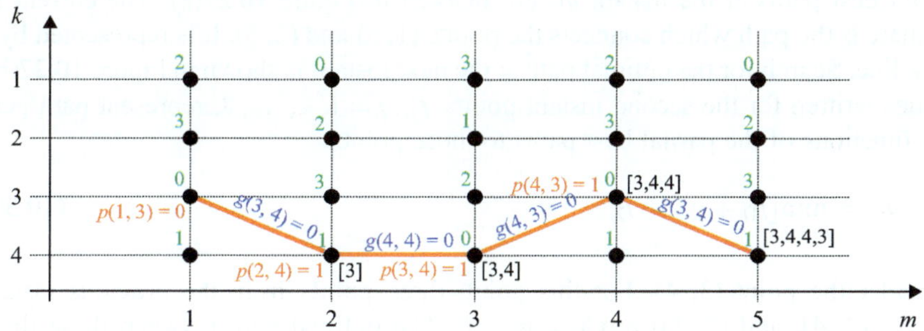

Figure 10.26 Best path estimate

The path costs are depicted in each figure. In Figure 10.25 for two final time instants $m = 4$ and $m = 5$ we have represented calculated path cost values for $p(m, k)$ and stored path values $[\pi(1), \pi(2), \pi(3), \pi(4)]$, for each $k = 1, \ldots, 4$.

From all k paths, arriving at the final time instance, we have to make a decision which one is the most likely to be the best IF estimate (**Step 5**). That decision is done using (10.57), and the path which connects the points (1, 3), (2, 4), (3, 4), (4, 3), and (5, 4) is obtained as the best path to estimate the micro-Doppler signature (IF) of the signal represented by matrix (10.58). It is depicted in Figure 10.26 as red line [3, 4, 4, 3, 4].

In order to further reduce search space for partial best paths (10.56) one may use the fact that $g(k_l, k_j)$ is an increasing function of the distance $|k_l - k_j|$. Then, the procedure could be performed in the following way:

(b1) Set $\varrho = \Delta$;

(b2) $\hat{l} = \arg\min_{l \in [j-\rho, j+\rho]} [p(m_i, k_l) + g(k_l, k_j) + f(TFR(m_{i+1}, k_j))]$, set partial best path from $(m_i, \hat{l})$ in $\pi'(m_{i+1}) = [\pi(m_i), k_j]$

(b3) If $p(m_i, k_{\hat{l}}) + g(k_{\hat{l}}, k_j) < g(k_{j+\varrho+1}, k_j) + \min[\pi(m_i), k_j = 1, \ldots, N]$ set $\pi(m_{i+1}) = \pi'(m_{i+1})$ as the partial best path. Further calculation could be stopped. Elsewhere, $\varrho = \varrho + 1$ and go to step (b2).

In this way, significant calculation savings, even greater than 50%, are obtained. Again, from the considered partial best paths, the current estimate is the one producing the smallest penalty function.

Example 15 In order to illustrate the shorter version of the algorithm, consider a TF plane with $M = 3$ and $N = 8$ points, with assumed values of $f_{ij} = f(WD(m_i, k_j))$, given in Figure 10.27(a). The function $g(k_i, k_j) = g_{ij}$ is calculated by (10.54), where $c = 2.5$ and $\Delta = 1$. Connections from each point, at the considered instant, with 3 nearest points from the previous instant, produces $g_{ij} = 0$, i.e., $g_{ij} = 0$ for $|i - j| \leq 1$. With an increase of $|i - j|$ the function g_{ij} increases for the value of $c = 2.5$. The partial best paths at the instant m_2 are marked in Figure 10.27(a). The current IF estimate is the path which connects the points (1, 5) and (2, 5). It is represented by a thick line. Search for the optimal path at the next instant is shown in Figure 10.27(b). Values written for the second instant points, f'_{2j}, $j = 1, 2, \ldots, 8$, represent path penalty functions of the partial best paths to those points:

$$f'_{2j} = \min_{l=1,8}[f_{1l} + g_{lj}] + f_{2j} \tag{10.59}$$

Consider the point (3, 4). For this point, three points from the previous instant ((2, 3), (2, 4), and (2, 5)) produce $g_{ij} = 0$. The optimal path between these three points and (3, 4) is (2, 5)–(3, 4), since $f'_{25} = 1$. Its value is smaller than f'_{24} and f'_{23}. Furthermore, paths between (3, 4) and any other point from the previous instant

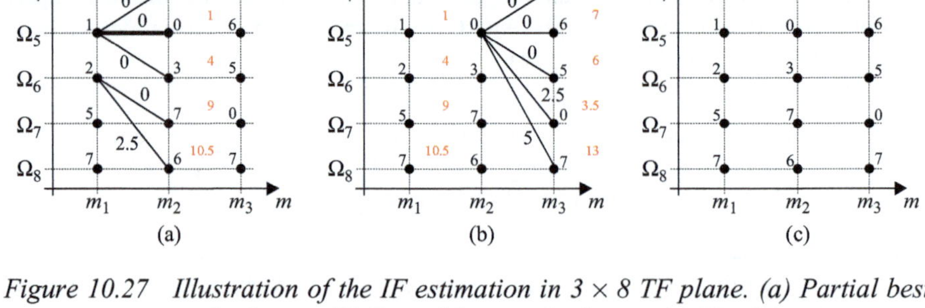

Figure 10.27 Illustration of the IF estimation in 3 × 8 TF plane. (a) Partial best paths for n_2; (b) Partial best paths for n_3; (c) The IF estimate

produce $g_{ij} \geq 2.5$, which is greater than f'_{25}. Thus, for the considered point (3, 4) search for the partial best path could be stopped after only three points. Consider now the point (3, 8). Only two points, (2, 7) and (2, 8), give $g_{ij} = 0$ for their paths to the point (3, 8). Current estimate for the partial best path to the point (3, 8) is (2, 7)–(3, 8), since $f'_{27} = 9 < f'_{28} = 10.5$. However, the connection between (2, 6) and (3, 8) is better than the connection (2, 7)–(3, 8), since $f'_{26} + g_{68} = 4 + 2.5 = 6.5 < 9$. We know that now we should investigate point (2, 6) since $g_{68} = 2.5 < 9$. What is more, we should proceed to the next point (2, 5) since $g_{5,8} = 5 < 6.5$. Then, the path between (2, 5) and (3, 8) is the best, since $f'_{25} + g_{58} = 1 + 5 = 6 < 6.5$. Further search could be stopped since $g_{k,8} \geq 7.5$ for $k \leq 4$. This illustrates that the search procedure can be relatively fast. In the considered case, the optimal path (IF estimate) links the points (1, 1), (2, 1), and (3, 2). Path penalty function for this path assumes the value of 3. It can be easily seen that this path is obtained by linking the partial best path to the point (2, 1) and the point (3, 2). Note that the new estimate could perform an update of the old ones.

Example 16 The presented algorithm is applied on a nonlinear FM signal. Signal with the IF of a sinusoidal shape (as it is in the micro-Doppler part of the radar signal) is used. The Wigner distribution with a Hanning window of the length 256 samples is used. Signal to noise ratio was $A^2/2\sigma^2 = 1/2$, i.e., -3[dB]. The IF estimations based on: the Wigner distribution maxima (dotted green line), the Viterbi algorithm (red line), and the true IF (blue line) are shown in Figure 10.28. The

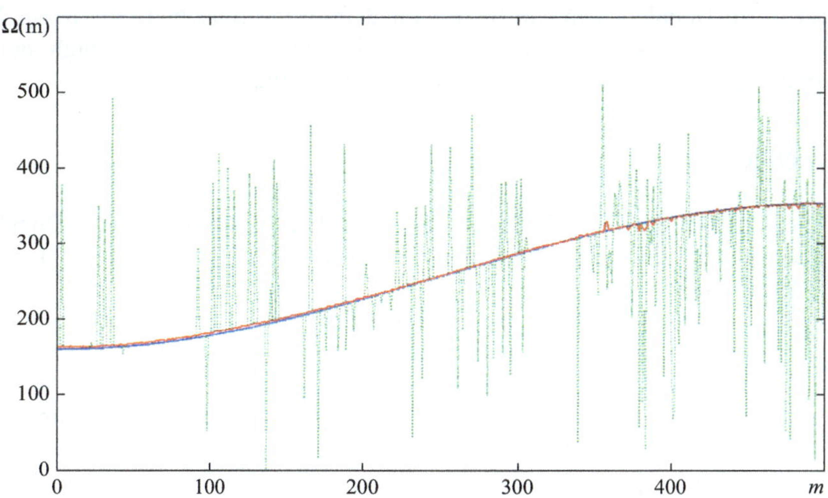

Figure 10.28 IF estimation for the sinusoidal FM signal based on: WD maxima – dotted green line; the Viterbi based algorithm – red line; true IF – blue line

maximal variation between consecutive IF samples was 2. We assumed $\Delta = 3$ in the algorithm. It is obvious that Viterbi based estimation gives the best IF (micro-Doppler) tracking.

In the previously described manner we define a path in a TF plane as an array of m frequency indices $\pi(m)$, while $1 \leq \pi(m) \leq N$ for each and every m. For the obtained path, we calculate the sum of TFR values along the obtained path, that is,

$$J_m = \sum_{m=1}^{M} TFR(m, \pi(m)) \tag{10.60}$$

The quantity defined in this way represents the criteria for determining whether the deterministic component $x(m)$ exists in the $TFR(m, k)$ of the signal $s(m)$. Namely, if J_m has the higher value of the threshold level it can be concluded that the deterministic signal exists in the signal $s(m)$.

10.6 Conclusion

The micro-Doppler effects detection, estimation, and removal are analyzed in this chapter. Since the most common form of the micro-Doppler signal is a sinusoidal modulated one, a simple technique for its analysis, based on the inverse Radon transform is presented. A method for period estimation is presented as well. The presented method is important for the micro-Doppler analysis itself, as well as for the technique based on the inverse Radon transform. In addition, it is shown that a general micro-Doppler form can be separated from the rigid body, by using TF analysis and the L-statistics. Finally, the micro-Doppler effect in high noise environments is analyzed by using the Viterbi algorithm.

References

[1] V. C. Chen, F. Li, S.-S. Ho, and H. Wechsler, "Analysis of micro-Doppler signatures," *IEE Proc. Radar, Sonar, Navigation*, vol. 150, no. 4, pp. 271–276, Aug. 2003.

[2] V. C. Chen, "Micro-Doppler effect in radar: Part I: Phenomenon, physics, mathematics, and simulation study," *IEEE Trans. on Aerosp. Electron. Systems*, vol. 42, no. 1, Jan. 2006.

[3] X. Bai, F. Zhou, M. Xing, and Z. Bao, "High resolution ISAR imaging of targets with rotating parts," *IEEE Trans. on Aerosp. Electron. Systems*, vol. 47, no. 4, pp. 2530–2543, Oct. 2011.

[4] F. Totir and E. Radoi, "Superresolution algorithms for spatial extended scattering centers," *Digital Signal Processing*, vol. 19, no. 5, pp. 780–792, Sep. 2009.

[5] M. Martorella, "Novel approach for ISAR image cross-range scaling," *IEEE Trans. Aerosp. Electron. Systems*, vol. 44, no. 1, pp. 281–294, 2008.

[6] M. Martorella and F. Berizzi, "Time windowing for highly focused ISAR image reconstruction," *IEEE Trans. Aerosp. Electron. Systems*, vol. 41, no. 3, pp. 992–1007, 2005.

[7] Y. Wang and Y.-C. Jiang, "ISAR imaging of ship target with complex motion based on new approach of parameters estimation for polynomial phase signal," *EURASIP Journal on Advances in Signal Processing*, vol. 2011 (2011), Article ID 425203, 9 pages.

[8] T. Sparr and B. Krane, "Micro-Doppler analysis of vibrating targets in SAR," *IEE Proc. Radar Sonar Navigation*, vol. 150, no. 4, pp. 277–283, Aug. 2003.

[9] B. Lyonnet, C. Ioana, and M. G. Amin, "Human gait classification using micro-Doppler time-frequency signal representations," *Radar Conference, 2010 IEEE*, pp. 915–919, May 10–14, 2010.

[10] L. Stanković, T. Thayaparan, and I. Djurović, "Separation of target rigid body and micro-Doppler effects in ISAR imaging," *IEEE Trans. Aerosp. Electron. Systems*, vol. 41, no. 4, pp. 1496–1506, Oct. 2006.

[11] L. Stanković, T. Thayaparan, M. Daković, and V. Popović-Bugarin, "Micro-Doppler removal in the radar imaging analysis," *IEEE Transactions on Aerospace and Electronic and Systems*, vol. 49, no. 2, pp. 1234–1250, Apr. 2013.

[12] T. Thayaparan, L. Stanković, and I. Djurović, "Micro-Doppler based target detection and feature extraction in indoor and outdoor environments," *J. of the Franklin Institute*, vol. 345, no. 6, pp. 700–722, Sep. 2008.

[13] V. C. Chen and H. Ling, *Time-frequency transforms for radar imaging and signal analysis*, Artech House, Boston, USA, 2002.

[14] J. Li and H. Ling, "Application of adaptive chirplet representation for ISAR feature extraction from targets with rotating parts," *IEE Proc. Radar, Sonar, Navigation*, vol. 150, no. 4, pp. 284–291, Aug. 2003.

[15] Y. Wang, H. Ling, and V. C. Chen, "ISAR motion compensation via adaptive joint time-frequency techniques," *IEEE Trans. Aerosp. Electron. Systems*, vol. 38, no. 2, pp. 670–677, 1998.

[16] L. Stanković, M. Daković, and T. Thayaparan, *Time-frequency signal analysis with applications*, Artech House, Boston, March 2013.

[17] L. Stanković, M. Daković, and T. Thayaparan, "Non-stationary signal analysis: Time-frequency approach," *in E-reference book "Signal Processing,"* Elsevier-Oxford Press, 120 pages, Oct. 2013.

[18] L. Stanković, "A measure of some time–frequency distributions concentration," *Signal Processing*, vol. 81, no. 3, pp. 621–631, Mar. 2001.

[19] L. Stanković, T. Thayaparan, V. Popović, I. Djurović, and M. Daković, "Adaptive S-method for SAR/ISAR imaging," *EURASIP Journal on Advances in Signal Processing*, vol. 2008, Article ID 593216, 10 pages, 2008. doi:10.1155/2008/593216.

[20] T. Thayaparan, S. Abrol, and E. Riseborough, "Micro-Doppler feature extraction of experimental helicopter data using wavelet and time-frequency analysis," *RADAR 2004, Proc. of the International Conference on Radar Systems*, 2004.

[21] T. Thayaparan, S. Abrol, E. Riseborough, L. Stanković, D. Lamothe, and G. Duff, "Analysis of radar micro-Doppler signatures from experimental helicopter and human data," *IET Proceedings Radar Sonar Navigation*, vol. 1, no. 4, pp. 288–299, Aug. 2007.

[22] M. Daković and L. Stanković, "Estimation of sinusoidally modulated signal parameters based on the inverse Radon transform," *Proc. of 8th International Symposium on Image and Signal Processing and Analysis* (ISPA), Trieste, Italy, Sep. 2013.

[23] L. Stanković, S. Stanković, and M. G. Amin, "Missing samples analysis in signals for applications to L-estimation and compressive sensing," *Signal Processing*, vol. 94, pp. 401–408, Jan. 2014.

[24] L. Stanković, I. Orović, S. Stanković, and M. G. Amin, "Compressive sensing based separation of non-stationary and stationary signals overlapping in time-frequency," *IEEE Transactions on Signal Processing*, vol. 61, no. 18, pp. 4562–4572, Sep. 2013.

[25] L. Stanković, M. Daković, and S. Vujović, "Concentration measures with an adaptive algorithm for processing sparse signals," *Proc. of 8th International Symposium on Image and Signal Processing and Analysis* (ISPA), Trieste, Italy, Sep. 2013.

[26] T. Thayaparan, L. Stanković, M. Daković, and V. Popović, "Micro-Doppler parameter estimation from a fraction of the period," *IET Signal Processing*, vol. 4, no. 3, pp. 201–212, Jan. 2010.

[27] V. Popović-Bugarin, F. Radenović, and L. Stanković, "Analytical rigid body reconstruction after micro-Doppler removal," *Proc. of 8th International Symposium on Image and Signal Processing and Analysis* (ISPA), Trieste, Italy, Sep. 2013.

[28] L. Stanković, V. Popović-Bugarin, and F. Radenović, "Genetic algorithm for rigid body reconstruction after micro-Doppler removal in the radar imaging analysis," *Elsevier Signal Processing*, vol. 93, no. 7, pp. 1921–1932, Jul. 2013.

[29] L. Stanković, T. Thayaparan, and M. Daković, "Signal decomposition by using the S-method with application to the analysis of HF radar signals in Sea-Clutter," *IEEE Trans. on Signal Processing*, vol. 54, no. 11, pp. 4332–4342, Nov. 2006.

[30] T. Thayaparan, P. Suresh, and S. Qian, "Micro-Doppler analysis of rotating target in SAR," *IET Signal Processing*, vol. 4, no. 3, pp. 245–255, 2010.

[31] I. Djurović, L. Stanković, and J. F. Bohme, "Robust L-estimation based forms of signal transforms and time-frequency representations," *IEEE Trans. Signal Processing*, vol. 51, no. 7, pp. 1753–1761, Jul. 2003.

[32] S. L. Marple, "Special time-frequency analysis of helicopter Doppler radar data," *Time-Frequency Signal Analysis and Processing*, ed. B. Boashash, Elsevier 2004.

[33] I. Djurović, T. Thayaparan, and L. Stanković, "Adaptive local polynomial Fourier transform in ISAR," *Journal of Applied Signal Processing*, vol. 2006, Article ID 36093, 2006.

[34] V. Katkovnik and L. Stanković, "Instantaneous frequency estimation using the Wigner distribution with varying and data driven window length," *IEEE Trans. Signal Processing*, vol. 46, no. 9, pp. 2315–2325, Sep. 1998.

[35] I. Djurović and L. Stanković, "An algorithm for the Wigner distribution based instantaneous frequency estimation in a high noise environment," *Signal Processing*, vol. 84, no. 3, pp. 631–643, Mar. 2004.

[36] G. D. Forney, "The Viterbi algorithm," *Proc. IEEE*, vol. 61, no. 3, pp. 268–278, Mar. 1973.

Chapter 11
Sonar Micro-Doppler Signatures: Principles and Applications

Alessio Balleri and Asif Mehmood

11.1 Introduction

The classification of targets using their sonar micro-Doppler signatures has attracted a growing interest in recent years. Due to a significantly lower propagation speed of acoustic pressure waves, vibrations and rotations of dismounted target components can induce Doppler modulations in acoustic echoes whose variation can reach thousands of hertz even when relatively low-frequency waveforms are transmitted. Similarly to their RF counterpart these modulations contain unique target features that can be used to perform target recognition. Although target classification by micro-Doppler signatures has been widely exploited in the radio frequency regime for radar systems, much less work has been done in acoustics. Acoustic micro-Doppler presents numerous advantages. The hardware can be significantly cheaper and data collection is easier, faster and cost-effective. On the other hand acoustic applications, particularly in air, are limited in range due to the high attenuation of sound waves. In this chapter, we present the principles underlying acoustic micro-Doppler and introduce examples of target classification by acoustic micro-Doppler signatures. We introduce recent applications and show and discuss the results of a number of recent trials that resulted in a collection of micro-Doppler signatures of a variety of targets.

In the last decade there has been a growing interest in applying micro-Doppler techniques to air acoustic sensors operating at ultrasound frequencies. In particular the literature shows a number of investigations where ultrasound micro-Doppler target signatures have been used to detect and classify humans and animals with operating frequencies ranging from 20 kHz up to around 80 kHz. There are a number of reasons for this; first the electromagnetic spectrum is becoming over-congested and hence the use of acoustics can become a valid substitute in situations where long-range detection is not a tight requirement. Second the hardware, including the transducers and the electronics, is on the average much cheaper than its radar counterpart although, recently, a range of low-power and cost-effective super-heterodyne radars have become available off-the-shelf. In particular, for a much lower price, it is possible to operate at frequencies that correspond to small wavelengths and hence high Doppler shifts; for example the Doppler shift due to a target that moves with a speed of 10 m/sec at 40 kHz in the ultrasound regime is the same as that achieved by a radar operating at 34 GHz. On the other hand the use of acoustic sensors has a significant drawback compared to RF systems. The attenuation of acoustic pressure waves in air increases dramatically with frequency and the maximum operating range is often limited to a few tens of metres.

In [1], Zhang *et al.* present a portable acoustic micro-Doppler radar system with an operational frequency of 40 kHz and report the results of a set of measurements in which human gait signatures were acquired indoor and outdoor. The system was composed of two transducers, one transmitting loudspeaker and one receiving microphone, followed by an amplification chain and a data acquisition (DAQ) card that was used to sample data at a rate of 1 MHz. During the experiments, the system was placed at human knees height. The experiment demonstrated that collection of micro-Doppler signatures of humans at this range of frequencies is feasible. This experiment was repeated in [2] to gather micro-Doppler signatures of (a) a 9.5 lb Miniature Pinscher in a 30 ft long indoor corridor, (b) an American Pitbull and a horse in a farm and (c) two personnel targets walking together in a parking lot. The experimental setup was the same as that of [1] but an accelerometer was placed on the humans' ankle to verify the micro-Doppler signatures. The authors showed that the signatures of these four-legged animals are significantly different to those of humans.

A low-cost (about 20 USD) Acoustic Doppler Sensor (ADS) made of off-the-shelf components was developed by the authors of [3] to collect micro-Doppler signatures of personnel targets at 40 kHz. Micro-Doppler signatures of humans walking towards and away from the sensor were collected with the ADS positioned at about waist height. A small loudspeaker, with a diameter as small as 1 wavelength (8.6 mm), was used to achieve a sonar beamwidth of about 60 degrees. The Doppler signatures were then used to extract Cepstral features to feed a Bayesian classifier modelled with Gaussian mixtures. Experiments were repeated for different locations and setups with about 30 targets (males and females). Classification performance was assessed between targets walking away versus target walking towards the sensor, male versus female and identification performance was also assessed.

Following this work, Zhang *et al.* [4] significantly reduced their 1 MHz sampling rate of [1] by deploying a simple and cost-effective bandpass sampling approach. They repeated their experiments with the modified system and gathered micro-Doppler signatures of eight humans, five males and three females, with the new system and the old system simultaneously. The signatures were used to extract Cepstral features from the data and to perform classification with a Bayesian classifier based on Gaussian mixture models. Classification performance is compared to that obtained in [3].

Another 40 kHz ultrasound sensor for collection of micro-Doppler signatures of personnel targets is presented in [5]. As in previous work, this is a low-cost, compact system characterised by a low processing power. This system was deployed in a variety of experiments aiming to demonstrate that micro-Doppler signatures can be successfully used to classify between different human actions. The experiments were carried out under various geometries and with targets moving on and off a treadmill. The authors showed that classification performance of a *k*-Nearest-Neighbour (*k*-NN) classifier reached values over 95% average correct classification when 50% of the signatures were randomly selected to form the training set. The same sensor was used in [6] to investigate classification performance of a *k*-NN classifier as a function of what the authors named 'clusters', for recognition of gender and for biometrics. An experimental trial aiming at collecting multimodal acoustic data including micro-Doppler signatures is described in [7].

In this chapter we introduce the theory behind sonar micro-Doppler signatures and highlight advantages and disadvantages with respect to their radar counterpart. We present the results of two experimental trials carried out at (1) the U.S. Army Research Laboratory and (2) University College London in which sonar micro-Doppler signatures of personnel targets were collected at 40 and 80 kHz, respectively.

11.2 Micro-Doppler Theory

The analytic signal of a pure tone $s(t)$ is defined as the signal $\hat{s}(t)$, such that $s(t) = Real\{\hat{s}(t)\}$, and is generally expressed in polar format as

$$\hat{s}(t) = e^{j2\pi f_0 t} \tag{11.1}$$

The Doppler shift induced by a target moving with a constant radial velocity v with respect to a sonar (or radar) system is

$$f_D = 2f_0 \frac{v}{c_s} \tag{11.2}$$

where f_0 is the carrier frequency of the active sensor and C_s is the speed of propagation of sound in air (or water). In the presence of a complex target, which is composed of a number of parts M each one moving with a velocity component $v_i(t)$, the resulting Doppler shift is given by the sum of each single Doppler shift as

$$f_D(t) = \sum_{i=1}^{M} 2f_0 \frac{v_i(t)}{C_s} \tag{11.3}$$

The analytic signal of the echo return from such a target is given by

$$\hat{s}_R(t) = e^{j2\pi f_0 t} \times e^{j2\pi f_D(t)t} \tag{11.4}$$

Mixing the received signal $\hat{s}_R(t)$ with the conjugate of the transmitted signal $\hat{s}(t)$ as follows

$$\hat{s}_R(t) \times \hat{s}(t)^* = e^{j2\pi f_D(t)t} \tag{11.5}$$

allows extraction of the Doppler signature from the data. This is the component of the signal that contains the micro-Doppler information on the target and that can be used for target recognition and classification. The bandwidth of the resulting signal is normally much smaller than the carrier frequency with the information contained in the lower frequencies. It becomes apparent, looking at (11.2) that because the Doppler shift is inversely proportional to the transmitted wavelength $\lambda = \frac{C_s}{f_0}$, and because the speed of sound is much lower than the speed of light, it is much easier to obtain high Doppler shifts using acoustics than RF devices.

The micro-Doppler signature can be represented in the two-dimensional time–frequency space by a short-time Fourier transform (STFT). This can be computed by calculating the Fourier transform of a series of 50% overlapping sliding windows $x_{i(n)}$ of a given length N as

$$\mathrm{STFT}(i, K) = \sum_{n=0}^{N-1} x_i(n) e^{-j2\pi \frac{nK}{N}} \quad K = 0, \ldots, N-1 \tag{11.6}$$

where $x_{i(n)} = \hat{s}_R(k) w(n)$ is the ith window, $w(n)$ is a suitable weighting function and $k = n + i\frac{N}{2}$.

The frequency resolution can be approximated as the inverse of the duration of a window $T_w = \frac{N}{f_s}$, where f_s is the sampling rate, and therefore only Doppler shifts which are greater than $\frac{1}{T_w}$ and that correspond to velocities

$$v > \frac{C_s}{2 f_0 T_w} \tag{11.7}$$

will be clearly visible. Equation (11.7) shows that the use of higher frequency has the additional advantage to induce a wider micro-Doppler bandwidth and that, given a certain frequency resolution, small movements are easier to detect when the carrier frequency is higher.

11.3 Applications

11.3.1 *Micro-Doppler Signature Collection at 40 kHz*

In this section we present the results of an experimental trial carried out at the U.S. Army Research Laboratory to collect and characterise the contribution of human body components to sonar micro-Doppler signatures [8]. An ultrasonic Doppler sonar (UDS) was designed using two ultrasonic ceramic sensors (MATSU/PAN EFR-RCB40K 54). These sensors have a resonance frequency of 40 kHz, a typical bandwidth (at 6 dB) of 2 kHz, and a directivity (at –6 dB) of 55°. Both sensors were placed in a compact plastic enclosure. One of the sensors emitted an ultrasonic wave and the other acted as a receiver as shown in Figure 11.1. An HP 3314A signal generator applied a 3 V continuous wave electrical signal at 40 kHz to the transmitter. Data acquisition and recording were conducted using a two-channel, 24-bit DAQ (Echo Indigo IO) and a laptop computer with the software (Spectra Lab).

11.3.1.1 Description of the Experiment

The measurements of the ultrasound from a person walking on the treadmill in the laboratory environment were conducted using the UDS. The speed of the treadmill was set to 1.34 m/sec. The UDS was placed on a tripod at variable heights depending upon which part of the human body was to be illuminated. The UDS on

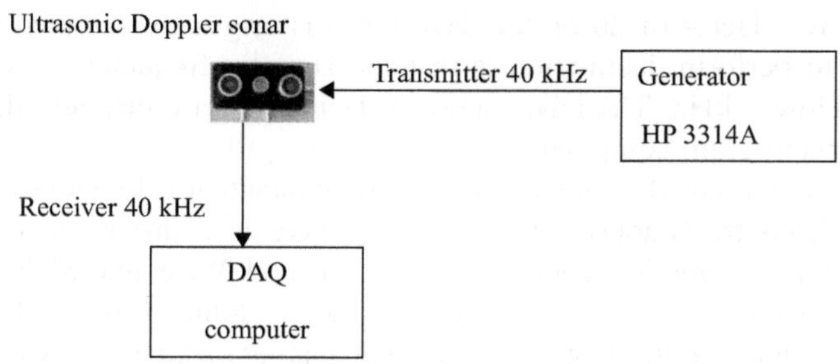

Figure 11.1 Block diagram of the measuring system

the tripod was placed close and aligned with the intended target (left foot, left lower leg, left thigh or the chest) to capture the maximum reflected signal from that target. We were interested in isolating different human body segments acoustically and then capturing the reflected signal from each of these segments individually. This was done by using two acoustic screens made of cardboard. The height of the bigger screen is greater or equal to the height of the walker on the treadmill. These screens are placed between the transducers and the walker, and are perpendicular to the transducers. Apertures were made in the bigger screen by cutting openings at different heights corresponding to walker's foot, lower leg and thigh heights. The dimensions (height × width) of these apertures for the foot, lower leg and the thigh were 0.077 m × 0.14 m, 0.33 m × 0.14 m and 0.38 m × 0.14 m, respectively. First data was taken from the left foot of a walking person on the treadmill. The aperture in the screen for the left foot was left open and all the other apertures in the cardboard were closed. The transmitted acoustic wave passed through the aperture and struck the left foot and a part of the right foot. The reflection from the right foot was blocked by the smaller screen and only the reflection from the left foot could pass through the aperture in the bigger screen back to the co-located receiving transducer. The captured signal from the left foot was then digitised using a DAQ card with a sampling rate of 96 and 48 kHz antialiasing filter [9]. The digitised data were stored on the laptop for the subsequent analysis in MATLAB. This process was repeated for all the remaining body segments and data was stored for further analysis.

11.3.1.2 Results

The UDS is capable of resolving motions of objects whose dimensions are equal to or larger than the wavelength of the acoustic waves. The resolution of the 40 kHz system employed is 8.6 mm and the attenuation is about 1.3 dB/m. The velocity resolution is determined by the sampling rate of the DAQ and the window size of the STFT as in (11.6). Given a sampling frequency of 4.8 kHz (after decimation) and a window size of 256 samples, the frequency resolution is 18.75 Hz, corresponding to a velocity resolution of 0.081 m/sec when 40 kHz transducers are used. The results presented in this chapter are obtained by performing STFT as given in (11.6). Window size of 256 samples is found to be optimal for a sampling frequency of 4.8 kHz. This sampling frequency is obtained after performing the decimation by a factor of 20 on the data that was originally sampled at 96 kHz. The reason to perform decimation is that the Doppler frequency of our interest falls well below 2 kHz. The advantages of decimation are efficient data storage and faster spectrogram computation.

The results presented in this chapter are semi-quantitatively compared with the results calculated by Bradley [10]. However, there is a difference between our approach and the assumption made in Bradley's method. We extracted different body segments' velocities from a person walking on the treadmill focusing UDS such that it is perfectly aligned to the centre of the intended target making an angle of 0° while Bradley computed body segments' velocities assuming that a person is walking on the ground and receiver is placed 20 m from a 1 m high target, thus making an angle of 2.86°. Bradley's method considers the bulk velocity that is the average velocity of the walking human in addition to the segments's velocity. In our method there is only a negligible bulk velocity component, if any, since the walker is on the treadmill and not on the ground. In Bradley's method the average velocity of a walking person is assumed 1.34 m/sec. Hence the velocity calculations made in Bradley method are 1.34 m/sec more than the velocity values extracted in our experiment.

The results presented here are from segments of the left leg (foot, lower leg and thigh) and the torso (chest) of a walking human. For conciseness, the prefix 'left' is omitted from different body segments in the discussion to follow.

First data is obtained by illuminating the foot of a person walking on the treadmill at 1.34 m/sec towards the UDS, and its spectrogram is shown in Figure 11.2.

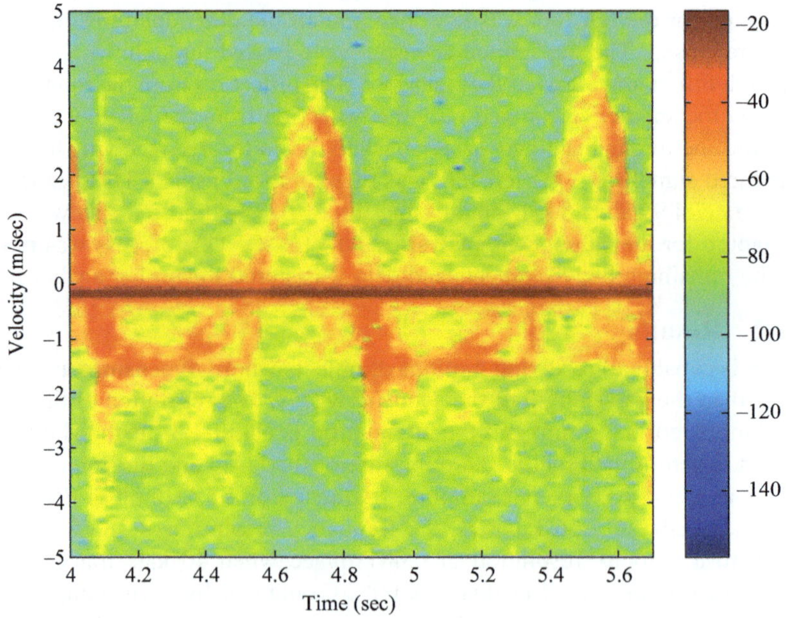

Figure 11.2 Foot spectrogram [8]

The spectrogram is plotted from 4.00 to 5.70 sec. Heel strike of the foot on the treadmill occurs at 4.10 sec and at 4.15 sec the foot (toe-down) is on the treadmill belt. The foot stays on the treadmill from 4.10 to 4.50 sec and then takes off and swings towards the sensors. In Figure 11.2, during the foot contact with the treadmill belt, a backward foot drag is observed. The Doppler velocity for the foot drag is from -1.65 to -1.15 m/sec. The mean velocity of the foot drag is approximately equal to the treadmill speed. At 4.50 sec the foot swings towards the UDS and produces a wide range of velocities seen between 4.6 sec and about 4.8 sec and also shows a sudden rise in the velocity. The velocity width of the spectrogram is due to the reflected signal from the proximal and distal segments of the foot. In Figure 11.2, the area between these two ends is filled in and represents the Doppler velocity bandwidth and is variable at different times depending upon which part of the foot is exposed to the UDS. At the beginning of the foot swing at 4.35 sec there is a wide Doppler velocity bandwidth that becomes narrow at 4.60 sec and widens again at 4.65 sec. However, during the remaining phase of the swing, we see a constant Doppler velocity bandwidth until the foot hits the treadmill belt at 4.9 sec. At 4.85 sec, the apparent increase in Doppler velocity bandwidth is because of some leakage from the right foot acoustic scattering. The acoustic scattering from the right foot is also seen in the spectrogram at 4.25 sec with Doppler velocity of 0.88 m/sec and at 5.05 sec with Doppler velocity of 0.86 m/sec. However, during most of the time of this measurement, the scattering from the right foot is blocked with a screen and has a little effect on the spectrogram. The maximum velocity observed during the first foot swing at the distal end is 3.50 m/sec at 4.75 sec. At the same time, the maximum velocity for the proximal end is 2.60 m/sec. However, the foot velocities at 5.54 sec in the second foot swing at the distal end and at the proximal end are 4.20 and 3.40 m/sec, respectively. It is the range in the velocity that we call the Doppler velocity bandwidth. It can also be observed that the slopes in the rising and falling of the foot swing are asymmetrical. The slope for the rising part of the foot swing computed from 4.37 to 4.72 sec is 12.38 m/sec^2. On the other hand the slope computed for the falling part of the foot swings from 4.77 to 4.93 sec is -27.63 m/sec^2, and is 2.23 times steeper than the rising slope. The amplitude of the foot velocity is about –30 dB (re 1 V) that corresponds to the strength of the acoustic returns.

It is our interest to compare these temporal observations with that predicted by Bradley [10] and provide a connection to the biomechanics of the foot swing. When these two methods are compared two significant differences are observed: first as already mentioned there is no bulk velocity and second there is a backward drag when a person walks on the treadmill. In the case when a person walks on the ground a bulk velocity is observed but no backward drag. Therefore, the velocity results shown in Bradley's model-based spectrogram is about 1.34 m/sec more than the experimentally extracted velocity of a walking human foot. The distal and proximal velocities are observed in both Bradley's model and our experimental approaches. The Doppler velocity bandwidth observed in our method and the one calculated in Bradley's method is almost same. The changes in Doppler velocity bandwidth at 4.35–4.88 sec shown in Figure 11.2 compare well with

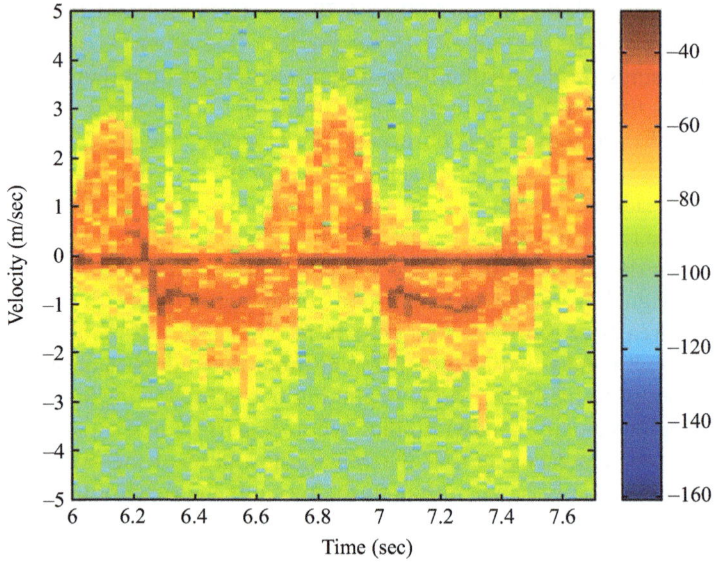

Figure 11.3 Lower leg spectrogram [8]

Bradley's foot results shown in Figure 11.2(a) in [10] and are understood as follows.

Referring to the time when the foot is on the treadmill and at the greatest distance from the sensor, the proximal end of the foot or heel initially lifts to rotate about the ankle and then the distal end of the foot or toe lifts off the treadmill (4.35–4.55 sec). During this time period, the Doppler velocity has the largest bandwidth. As the foot swings forward, the distal and proximal foot ends become aligned with the transducer and the bandwidth narrows (4.55–4.65 sec) and then again broadens as the foot continues to rotate about the ankle causing the bandwidth to increase (4.65–4.75 sec). As the foot approaches contact the rotation reverses, the bandwidth narrows and a heel contact occurs (4.88 sec). During the later time period, the velocity decreases more rapidly compared to the lift off phase. This observed asymmetry indicates greater acceleration and therefore a larger force on the foot during heel strike compared to toe lift off. Authors have used force plates on treadmills to measure these forces and this same asymmetry is observed in these measurements [11]. See, for example, Figure 11.5 in [11].

The returns from the lower leg are shown in Figure 11.3, and constitute the reflections from the proximal end of the foot and from the distal end of the thigh. Its amplitude is stronger than the foot and is most pronounced when the lower leg is perpendicular to the treadmill surface, and can be seen in Figure 11.3 from 6.1 to 6.5 sec. This is due to the fact that the maximum cross section of the lower leg is exposed to the transducer. Similar to the foot spectrogram, the lower leg spectrogram also exhibits the backward drag on the treadmill from 6.3 to 6.7 sec and at this time the leg is not swinging. The amplitude of the leg spectrogram is about –25 dB, and is more than the foot spectrogram because of its larger cross section. The maximum velocity of the lower leg is 3.45 m/sec at 7.65 sec that occurs at its distal end and the velocity at the proximal end at 7.65 sec is 0. The leg's velocity is smaller than the foot velocity and this is in agreement with the results presented by Bradley [10]. The area between the proximal end and the distal end is completely filled in and can be verified by the Bradley's method. It is shown in Figure 11.3 that the proximal end of the leg generates a very small velocity compared to its distal end, and therefore has a greater Doppler velocity bandwidth. The Doppler velocity bandwidth of the leg is greater than the velocity bandwidth of the foot because of its greater length. This is in agreement with the results presented in Bradley's method [10].

The third segment extracted experimentally is the upper leg or the thigh of a walking person on the treadmill. The Doppler velocity versus time of the thigh is shown in Figure 11.4, and the maximum Doppler velocity observed is 0.64 m/sec at 5.35 sec. The Doppler velocity of the thigh is smaller but its amplitude (5 dB) is greater than the foot and the lower leg. During human walk, thigh spectrogram exhibits smaller swing compared to both the foot and the lower leg but stronger amplitude due to its larger cross section area. Like the lower leg, the area between the proximal end and the distal end of the lower leg is also completely filled in. The Doppler velocity at the proximal end is almost zero because it occurs at the axis of rotation. The reflections from the right thigh exhibiting maximum velocity can also be observed at 5.75 sec. The small screen used to block the right thigh does not work effectively when the right thigh moves backward. However, during right leg forward motion a significant backscattering is blocked. This is because the gap between

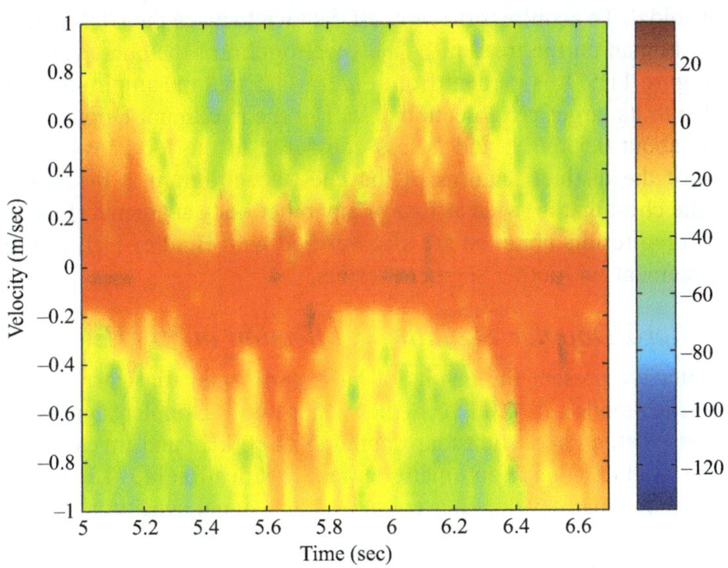

Figure 11.4 Thigh spectrogram [8]

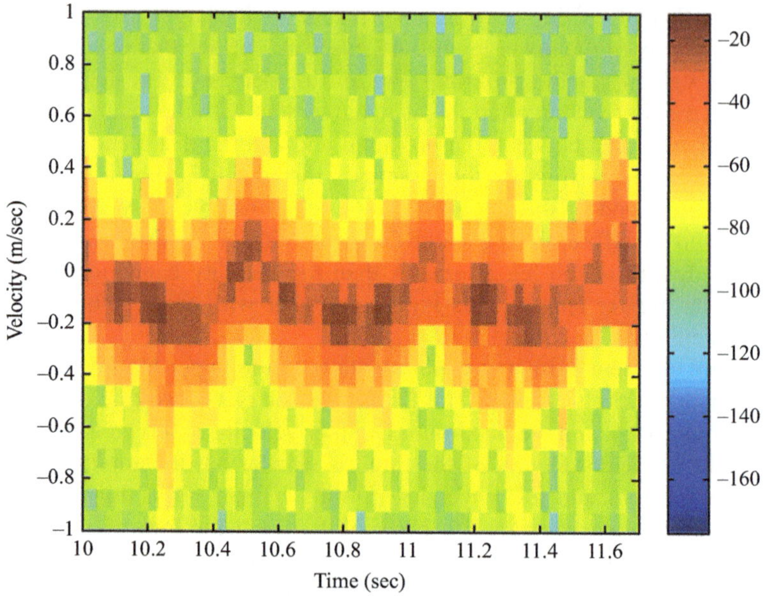

Figure 11.5 Chest spectrogram [8]

two thighs is smaller than the legs or the feet and is harder to isolate acoustically. The thigh results shown in Figure 11.4 are also verified by Bradley's method.

Another human body segment that is a major contributor in human motion Doppler signature is the torso. Torso exhibits a unique Doppler signature that is almost sinusoidal. To acquire torso or chest data we focused the UDS to the chest of the walking human on the treadmill. The spectrogram of the chest velocity data is shown in Figure 11.5. It is evident from Figure 11.5 that the amplitude of the torso is strong due to its larger cross section area. But the maximum Doppler velocity of the chest is about 0.5 m/sec and is less compared to the foot, lower leg and the thigh. One of the distinct features of the chest spectrogram is the vibration frequency of the chest that is twice as that of any other body segments (foot, lower leg etc.) [12]. Torso results obtained are also verified with Bradley [10] and are found to be in agreement with other segment results.

11.3.2 Micro-Doppler Signature Collection at 80 kHz

In this chapter we present the results of an experimental trial carried out at University College London in 2011 to gather sonar micro-Doppler signatures of personnel targets with an acoustic radar operating at 80 kHz and investigate whether these can be used to achieve automatic target classification [13–15].

The acoustic radar developed is composed of a signal generator, a loudspeaker, a microphone, two pre-amplifiers that amplify the signals before transmission and after recording and a DAQ card that digitises the recorded signal. The waveforms are generated by a National Instrument NI PXI-6733 card capable of generating up to 500 MSamples/sec (32 bit resolution) per channel. Each card can transmit on eight channels and therefore eight different waveforms could potentially be transmitted simultaneously. The generated waveform is given as an input to the pre-amplifier (Ultra Sound Advice S55A) capable of generating a maximum output of 140 V peak to peak. The loudspeaker provides a monitor output which is 1/100 of the actual output of the amplifier. This allows monitoring of the output and of the gain of the amplifier. The signal is transmitted by a loudspeaker (Ultra Sound Advice) which can nominally operate between 20 kHz and over 200 kHz. It can generate an output level greater than 105 dB S.P.L. (sound pressure level) between 20 and 50 kHz (measured at 0.25 m) and greater than 85 dB S.P.L. up to 150 kHz. The round-shaped active area of the loudspeaker has got a 50 mm diameter [16]. In the receive section, echoes are captured by a microphone capsule (Ultra Sound Advice UM3 capsule). The sensitive element of the capsule is a 1.8 cm diameter disk made of a very thin aluminised polyester film. Nominally, its sensitivity is better than −57 dB from 20 to 120 kHz and better than −70 dB at 180 kHz. The maximum gain obtainable by the amplifier is nominally ×220 at 100 kHz [17]. The amplified received signal is digitised by a DAQ card (NI PXI-6133) which is capable of receiving eight channels simultaneously with a sampling frequency up to 500 kHz per channel. The digitised data is stored in a file and the signal processing is performed off-line with MATLAB.

11.3.2.1 Description of the Experiment

The data containing micro-Doppler signatures were gathered at UCL between 2010 and 2011. The microphone and the loudspeaker were arranged in a pseudo-monostatic configuration next to each other as shown in Figure 11.6. Their active areas were contained on the same plane perpendicularly to the ground floor with their phase centres positioned 12 cm far from each other and both at a height of 105 cm with respect to the ground floor. During the experiments the personnel targets were facing the microphone and the loudspeaker at a distance of 2 m. A 10 sec long continuous wave tone at 80 kHz with 0.3 V amplitude was generated

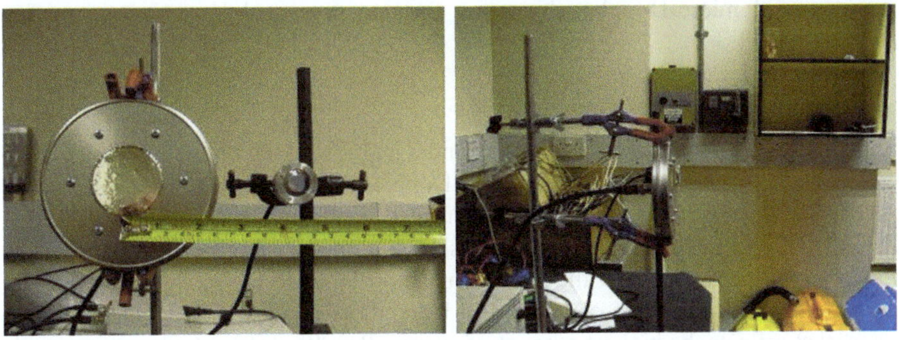

Figure 11.6 Frontal and lateral views of the space arrangements of the microphone and the loudspeaker during the experiment [14, 15]

with the NI PXI-6733 card to ensonify the targets. The gain of the loudspeaker pre-amplifier, measured by using the monitor output, was about 20 dB. Echoes were recorded with the microphone and then sampled at a rate of 500 kHz. Micro-Doppler signatures were gathered for three different personnel targets undertaking various actions:

- walking,
- running,
- walking while carrying an object in one hand,
- walking while carrying an object with both hands,
- walking with a heavy backpack on the shoulders.

When walking or running the targets were moving on a Pro Fitness manual treadmill and were ensonified from the back to avoid any masking effects. This was deployed to remove the main Doppler shift contribution from the Doppler signatures, and to keep the signal-to-noise ratio (SNR) of the received signal as constant as possible during the recording. Also the treadmill removed any movement constraint on the targets and therefore allowed longer recordings.

A recording of the background clutter was removed from the raw data for each target recording. The analytic signal of the difference signal was mixed with the complex conjugate of the analytic signal of the transmitted tone to obtain the baseband signal containing the micro-Doppler signatures. The analytic signals were obtained by applying the MATLAB *Hilbert.m* function to the original real signals. The mixer output was filtered with an ideal lowpass filter cutting off all the frequencies over 2.5 kHz. It was experimentally observed, in fact, that the micro-Doppler signatures were contained within the frequencies below 2.5 kHz. The resulting signal was down-sampled of a factor 50 leading to a final sampling frequency of 10 kHz. This was done to reduce the load of the data without altering the information contained in the frequencies below 2.5 kHz. The STFT was applied to all the recordings to visualise the recorded micro-Doppler signatures. This was calculated by using 30 msec long 50% overlapping windows weighted with a Hanning function and by computing their respective Fourier transform as in (11.6). The measured micro-Doppler signatures for one personnel target are given in Figures 11.7–11.9.

11.3.2.2 Results

Figure 11.7 shows the signature of a personnel target walking on the treadmill. The swinging arms contribute with the highest Doppler shifts which show peaks around 1 kHz. The Doppler shift due to the feet instead, because of the geometry, is attenuated such that the feet contribution is self-contained inside the Doppler spread induced by the movement of the torso. This contributes to the micro-Doppler signatures within a bandwidth of about 100 Hz and, as expected, is the target component that reflects the highest amount of energy. Results clearly show a gait period T_{sig} of about 1 sec. Because two arms are moving the repetition rate of Doppler peaks is 0.5 sec corresponding to a repetition frequency of 2 Hz.

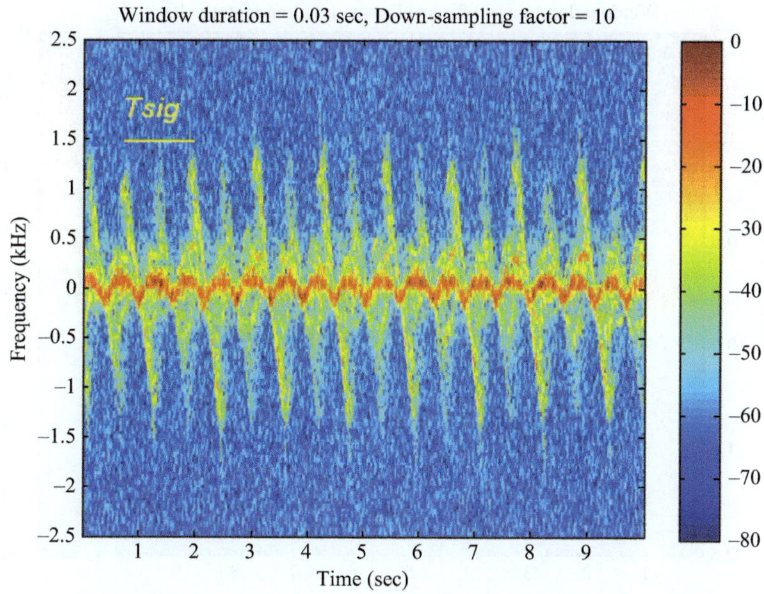

Figure 11.7 Micro-Doppler signature of a personnel target walking on a treadmill. The signature period T_{sig} indicates the time between two one-arm swings [14, 15]

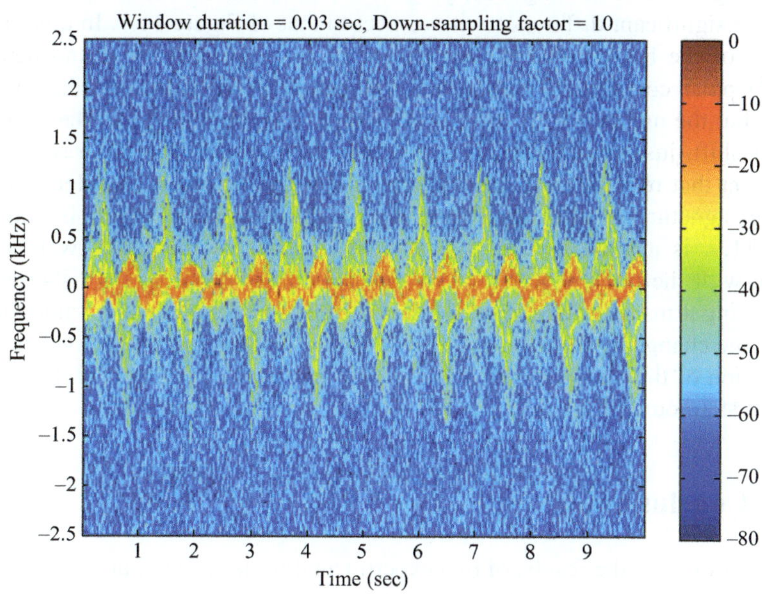

Figure 11.8 Micro-Doppler signature of a personnel target walking on a treadmill with a book in a hand [14, 15]

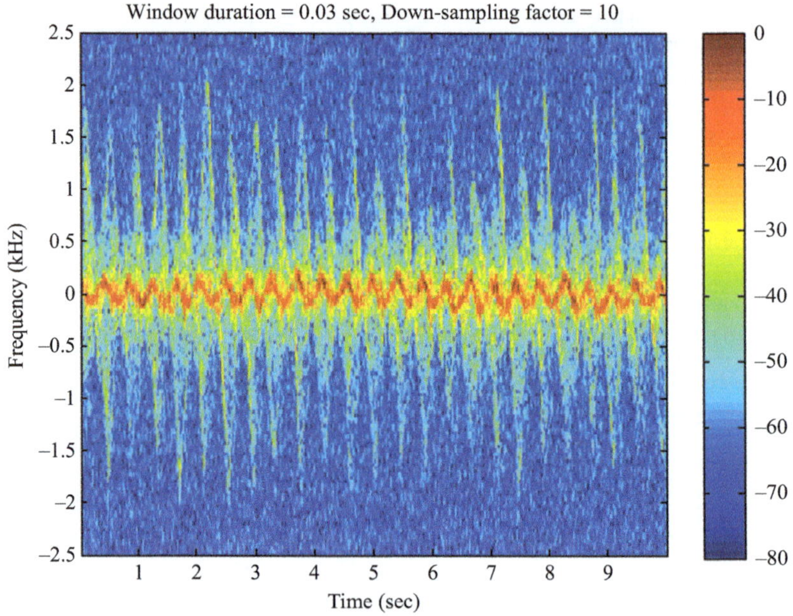

Figure 11.9 Micro-Doppler signature of a personnel target running on a treadmill [14, 15]

The signature of the target carrying a book in one hand, shown in Figure 11.8, presents a significant difference with respect to that of Figure 11.7. In this case the presence of the book limits the movement of the arm carrying it and hence its micro-Doppler contribution disappears. The period of the signature remains close to 1 Hz but the number of Doppler peaks inside one period halves. The maximum Doppler shift, just over 1 kHz, is also lower in this case. The torso remains the component that reflects the highest energy and with the lowest Doppler shift.

The signature of the target running on the treadmill, which is shown in Figure 11.9, is again significantly different to those of the target walking and walking with the book. The period of the signatures becomes much shorter with both moving arms contributing with higher Doppler peaks that reach values around 2 kHz. The change of velocity, and hence of the signature period, is evident in the contribution of the torso as well, which remains the target component that reflects the highest amount of energy.

11.4 Conclusions

We have presented the results of two experimental trials carried out to collect sonar micro-Doppler signatures of personnel targets at 40 and 80 kHz.

The Doppler signatures of different segments of human body such as foot, lower leg, thigh and torso are extracted experimentally. The Doppler velocities of each of these segments are then computed from their Doppler signatures. The measurements are made using a 40 kHz UDS by illuminating one segment of a walking human body at a time while acoustically screening the other human body segments. The experimental results thus obtained are verified with Bradley's method. As only the foot, leg and torso motions are the major features in whole human body motion, the experimental extraction of these features may be used as a basis to distinguish walking human from walking quadrupeds.

Measurements of micro-Doppler signatures at 80 kHz have shown there are features that change with the type of human action undertaken and that therefore micro-Doppler signatures might be a strong candidate for human action classification.

This chapter shows that ultrasound sensors could represent a cost-effective alternative to radar systems for applications that do not require long-range detection. They certainly demonstrate that target detection and possibly classification by ultrasound micro-Doppler signatures are achievable and they might potentially open up a new research avenue looking at the exploitation of these techniques in future systems.

The method presented in this chapter is an alternative to video analysis popular to display joints movements and rotations of the body segments of the quadrupeds.

References

[1] Z. Zhang, P. O. Pouliquen, A. Waxman, and A. G. Andreou, 'Acoustic micro-doppler radar for human gait imaging', *Journal of the Acoustical Society of America*, vol. 121, no. 3, pp. 110–113, 2007.

[2] Z. Zhang, P. Pouliquen, A. Waxman, and A. Andreou, 'Acoustic micro-Doppler gait signatures of humans and animals', in *CISS '07. 41st Annual Conference on Information Sciences and Systems*, 2007, pp. 627–630, March 2007.

[3] K. Kalgaonkar and B. Raj, 'Acoustic Doppler sonar for gait recognition', in *IEEE Conference on Advanced Video and Signal Based Surveillance*, 2007. AVSS 2007, pp. 27–32, September 2007.

[4] Z. Zhang and A. Andreou, 'Human identification experiments using acoustic micro-Doppler signatures', in *Argentine School of Micro-Nanoelectronics, Technology and Applications*, 2008. EAMTA 2008, pp. 81–86, September 2008.

[5] S. Dura-Bernal, G. Garreau, C. Andreou, A. Andreou, J. Georgiou, T. Wennekers, *et al.*, 'Human action categorization using ultrasound micro-Doppler signatures', in *Proceedings of the Second International Conference on Human Behavior Understanding, HBU'11* (Berlin, Heidelberg), pp. 18–28, Springer-Verlag, 2011.

[6] G. Garreau, C. Andreou, A. Andreou, J. Georgiou, S. Dura-Bernal, T. Wennekers, *et al.*, 'Gait-based person and gender recognition using micro-Doppler signatures', in *IEEE 2011 Biomedical Circuits and Systems Conference (BioCAS)*, pp. 444–447, November 2011.

[7] J. Georgiou, P. Pouliquen, A. Cassidy, G. Garreau, C. Andreou, G. Stuarts, *et al.*, 'A multimodal-corpus data collection system for cognitive acoustic scene analysis', in *45th Annual Conference on Information Sciences and Systems (CISS)*, 2011, pp. 1–6, March 2011.

[8] A. Mehmood, J. M. Sabatier, M. Bradley, and A. Ekimov, 'Extraction of the velocity of walking human's body segments using ultrasonic Doppler', *Journal of the Acoustical Society of America*, vol. 128, no. 5, pp. 316–322, 2010.

[9] A. Ekimov and J. M. Sabatier, 'Human motion analyses using footstep ultrasound and Doppler ultrasound', *Journal of the Acoustical Society of America*, vol. 123, no. 6, pp. 149–154, 2008.

[10] M. Bradley and J. M. Sabatier, 'Applications of Fresnel–Kirchhoff diffraction theory in the analysis of human-motion Doppler sonar grams', *Journal of the Acoustical Society of America*, vol. 128, no. 5, pp. 248–253, 2010.

[11] R. Kram, T. M. Griffin, J. M. Donelan, and Y. H. Chang, 'Force treadmill for measuring vertical and horizontal ground reaction forces', *Journal of Applied Physiology*, vol. 85, no. 2, pp. 764–769, 1998.

[12] P. V. Dorp and F. Groen, 'Feature-based human motion parameter estimation with radar', *IET Proceedings on Radar, Sonar and Navigation*, vol. 2, no. 2, pp. 135–145, 2008.

[13] A. Balleri, K. Woodbridge, and K. Chetty, 'Frequency-agile non-coherent ultrasound radar for collection of micro-Doppler signatures', in *2011 IEEE Radar Conference*, pp. 45–48, May 2011.

[14] A. Balleri, K. Chetty, and K. Woodbridge, 'Classification of personnel targets by acoustic micro-Doppler signatures', *IET Proceedings on Radar, Sonar and Navigation*, vol. 5, pp. 943–951, December 2011.

[15] H. Griffiths, A. Balleri, and C. Baker, 'Biologically inspired and multi-perspective target recognition', in D. Blacknell and H. Griffiths (Eds.), *Radar Automatic Target Recognition (ATR) and Non-Cooperative Target Recognition (NCTR)*, Institution of Engineering and Technology, pp. 177–211, London, 2013.

[16] 'Instruction manual for ultrasound loudspeaker system', Ultra Sound Advice, www.ultrasoundadvice.co.uk.

[17] 'SP3 PSU Operating Instruction', Ultra Sound Advice, www.ultrasoundadvice.co.uk.

Chapter 12

Radar Micro-Doppler Signature of Wind Turbines

Fanxing Kong, Yan (Rockee) Zhang, and Robert Palmer

12.1 Introduction

Wind power industry has developed rapidly around the world in the past decade. The world total installed wind power capacity reached 239 000 MW by the end of year 2011 [1], which is almost 10 times of 10 years ago. The development of this green and renewable energy solution helps diversify the energy portfolio. It also relieves the energy dependency on fossil fuels and reduces greenhouse gas emissions. Countries that demand large electricity consumption such as the United States, China, Germany, India are all on the top of the list of annual wind power installation in recent years. As in the United States, the total wind power capacity has reached 51 630 MW by the end of the third quarter 2012 [2]. A report released by the Department of Energy [3] has discussed the scenario of providing 20% of the nation's electricity need using wind power by year 2030.

As the wind power industry thrives, an increasing number of wind farms are being constructed around the world, where many utility-scale wind turbine generators have been installed to efficiently harvest wind energy. These extremely large man-made structures have raised concerns on their Electromagnetic Interference (EMI) to current radar networks. Wind turbines built too close to radars can potentially jam radar receivers or create partial beam blockage due to the significant amount of backscattered Electromagnetic (EM) energy. Diffractions caused by closely spaced wind turbines may result in shadowing effect in the wind farm proximity [4]. Even at ranges up to several tens of kilometers, these structures can still cause clutter effects that may degrade nearby radar performance. Unlike conventional ground clutter, which is essentially stationary from pulse to pulse, wind turbines feature complicated micro-Doppler characteristics primarily due to the blade rotation. Thus, conventional ground clutter filtering techniques have failed to mitigate this type of interference.

Early investigations by researchers at the University of Michigan in the late 1970s revealed that large windmills can have EMI on television signals at very high frequency (VHF) and ultra high frequency (UHF) bands [5]. But the impact of large wind turbines on radars was not systematically studied until early 2000s by British scientists [6]. As both the number and the size of wind turbines have increased dramatically in recent years, the potential impacts of wind turbines on radar performance have been gradually realized and Wind Turbine Clutter (WTC), as a recently recognized new type of radar clutter, starts to draw more attention from the radar community. In the United States, the issue was first discussed in the Department of Defense (DOD) report [4] to the congress regarding the effect of wind farms on military readiness, where the potential degradation of radar performance was analyzed on long range defense radar, Air Surveillance Radar (ASR), and weather radar. The two-part paper [7, 8] summarized the DOD sponsored wind turbine radar measurement campaign carried out by the AFRL (Air Force Research Laboratory), where spurious radar detection and tracking of aircraft caused by wind turbines were simulated to address potential impacts of wind turbines built in radar vicinity.

Weather radar is the type of radar used for detecting, locating precipitation, and estimating its motions. It operates the same way as other radar does by transmitting EM wave and receiving backscattered signal, only that the targets of interest for weather radar are in different precipitation forms, such as rain, snow, and hail. WSR-88D (Weather Surveillance Radar – 88 Doppler) or NEXRAD (Next Generation Weather Radar) is the current weather radar network in the United States operated by the National Weather Service, including 160 high-resolution S-band Doppler weather radars built in the late 1980s, which are being upgraded with polarimetric capabilities. Also, 48 active C-band Terminal Doppler Weather Radars (TDWR) are being primarily used for wind shear observation in or near major US airports. The negative impact of wind turbines built in the vicinity of weather radar on its capability of storm detection and tracking was discovered in early 2000s [9]. The first experimental campaign to study wind turbine weather radar signal characteristics was launched by the Radar Operations Center (ROC) [10, 11]. It has been found that weather radar observations of wind turbines have similar micro-Doppler signatures as early AFRL experiments suggested [6–8]. Several cases of wind farms built close to WSR-88D affecting radar operations have been reported [9, 10, 12–16].

12.2 Interactions between Radar and Wind Farms

Large commercial-scale or utility-scale wind turbines are becoming major concerns of radar operations. An example of such wind turbine is shown in Figure 12.1, the state-of-the-art 1.6 MW wind turbine generator manufactured by the General Electric. Such horizontal axis wind turbine typically consists of the tower, nacelle, hub, and rotor blades, as shown in Figure 12.2. The tubular tower is made of thick steel and is usually divided into two or three cylinder or frustum segments for transportation convenience. The nacelle hosts the gearbox, generator, and mainframe and is placed on top of the tower. The hub, covered by the nose cone, is usually made of cast iron and has built-in mechanism to actively pitch the rotor blade connected to it. Though these three parts may contribute the most to the radar returns because they are made of metal, their radar returns concentrate at low frequencies in the Doppler domain, thus making it easy to suppress by conventional

Figure 12.1 The GE 1.6 MW wind turbine in Minco wind farm, north Grady County, Oklahoma

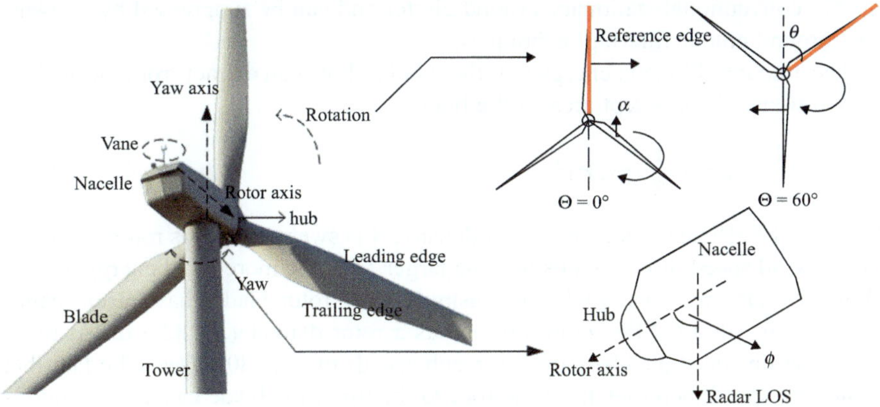

Figure 12.2 The horizontal axis wind turbine structure: the blade typically rotates clockwise viewed from the nose direction and the leading edge is the thicker edge of the blade, while the trailing edge is the thinner one. The blade position θ ($0^\circ \leq \theta < 360^\circ$) is the angle swept by the reference leading edge from the upper vertical position. The aspect angle ϕ ($0^\circ \leq \phi < 360^\circ$) is the angle from RLOS to the rotor axis. α is the blade angle formed by the leading and trailing edge from the same blade

notch filters. However, the rotor blades made of composite material that is far less reflective to EM waves compared with metal create complex Doppler spectrum due to blade rotation. This spectrum varies with both time and frequency, presenting a serious problem for Doppler radar operations.

Other than the blade rotation, the wind turbine has two other major degrees of freedom: yaw and pitch. The yaw mechanism is realized by several motor-driven gears at the bottom of the nacelle, which are controlled according to the wind direction sensed by a wind vane on top of the nacelle to actively yaw the blade rotation plane into the wind, to achieve maximum efficiency. The pitch system feathers the blade to change the angle of attack, which adapts the blade rotation rate to wind speed. The blade rotation rate varies in a small range, usually from 10 to 20 rpm for modern wind turbines. The yaw and pitch are slow motions to radar; thus, they can be considered as stationary during each radar observation. Therefore, the blade rotation is the "micro motion" of the wind turbine, resulting in unique micro-Doppler radar signatures [17]. However, it is worth noting that different yaw and pitch will lead to different micro-Doppler spectrum patterns, as will be shown later in this chapter.

The general specification of the GE 1.6 MW wind turbine is listed in Table 12.1. Rated speed is the minimum wind speed for the wind turbine to output the name capacity. Cut in speed is the wind speed at which the wind turbine starts operating and generating electricity. The cut out speed is the wind speed to shut down the wind turbine for safety reasons. The wind turbine stalls if the wind speed is less than cut in speed or higher than cut out speed. In both cases, WTC degenerates to conventional stationary ground clutter and can be suppressed by conventional ground clutter filtering techniques.

The amount of kinetic energy that the wind turbine can extract from the airflow is proportional to the swept area of the blades:

$$E = \frac{1}{2}\gamma\rho A v^3 = \frac{1}{8}\gamma\rho\pi D^2 v^3 \tag{12.1}$$

where γ is the efficiency factor, ρ is air density, A is swept area, D is rotor diameter, and v is wind speed. It is obvious that the larger D is, the more wind energy a wind turbine can capture. This is why increasingly large rotor blades have been manufactured. The GE 1.6 MW wind turbine has a rotor diameter of 82.5 m. At maximum rotation speed, its blade tip can reach speeds of over 80 m/sec. The fact that the moving speed increases from the root to the tip of the blade causes continuous

Table 12.1 Specifications of GE 1.6 MW wind turbine

Parameter	Value
Capacity	1.6 MW
Hub height	80/100 m
Blade length	40.3 m
Blade diameter	82.5 m
Cut in speed	3.5 m/sec
Rated speed	11.5 m/sec
Cut out speed	25 m/sec
Rotor speed	Variable
Blade material	Glass-fiber reinforced plastic

Doppler response, which occupies the entire spectrum and possibly aliases depending on radar Nyquist velocity. This indicates that all radar targets are subjected to the WTC contamination regardless of their motions, if they fall into the same resolution volume with an operating wind turbine.

From (12.1), E is also proportional to the third power of wind speed. Thus, sufficient wind speed is critical to meet the rated capacity. For this reason, wind farms are only located at places with ample wind resource based on past meteorological surveys. In addition, wind farms are often sited close to the city to lower the electricity transmission loss, while many weather radars are also located at population-dense area to provide forecast and severe weather surveillance for that region. Therefore, sometimes, it is unavoidable to have wind farms constructed within the Radar Line of Sight (RLOS). Figure 12.3 gives an example of radar locations on top of wind resource distribution over the contiguous United States. It is a known fact that the wind speed increases almost exponentially with height in the lower boundary layer due to less surface friction at higher altitudes. The hub height is therefore high above the ground. Wind turbines also prefer to be built on higher ground, up on the ridge if possible, where the wind is steady with less turbulence. However, higher hub height means that wind farms need to be located further away from the radar to be out of the RLOS, which is not always possible.

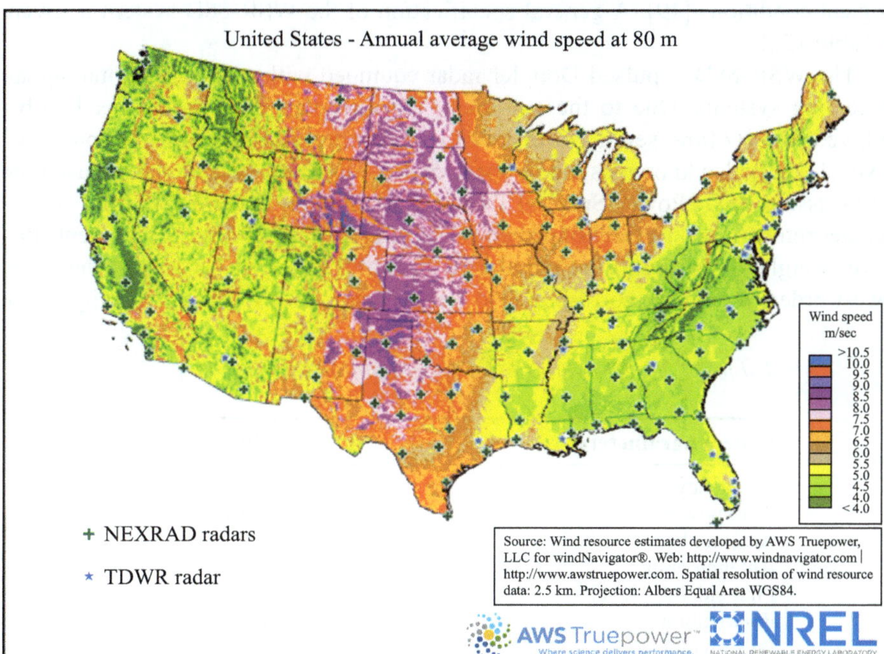

Figure 12.3 Weather radar locations on top of wind resource of the contiguous United States. (Picture from: http://www.roc.noaa.gov/WSR88D/WindFarm/WindFarm_Index_GreatFalls.aspx?wid=)*

12.3 Impacts of Wind Turbines on Existing Weather Radar Operation

This section provides some case studies for wind turbine impact on existing weather radar operations. For ASR, there is normally a significant difference in altitude between aircrafts and wind turbines, which can be used to distinguish one from another. In fact, the concurrent beam processing [18] is based on this assumption. However, the hydrometeors which weather radar is interested in are spatially inseparable from the wind turbines residing in the same resolution volume on the scale of radar observations. This has made it even more difficult for weather radar to mitigate WTC effects. Therefore, in this chapter, we will mainly focus on wind turbine observations from weather radar and its impact on the WSR-88D radar network.

The WSR-88D radar's parabolic dish antenna forms a pencil beam in contrast to the fan beam of most ASRs. The antenna pedestal rotates in azimuth for 360° at progressively higher elevation angle for a complete scan of Volume Coverage Pattern (VCP), which is a series of predefined antenna elevation angle sets, pedestal rotation rate, pulse width, and Pulse Repetition Frequency (PRF) for optimum meteorological applications. There are seven VCPs for the precipitation mode and two VCPs for clear air mode. Different VCPs are activated in different meteorological conditions [19]. A general specification of the WSR-88D system is given in Table 12.2.

The WSR-88D is pulsed-Doppler radar equipped with real-time digital signal processing systems. Due to the extremely large amount of data that need to be archived, the *IQ* time series (level I data) are further processed to base products (level II data), also known as the moment estimates, which are calculated based on the Gaussian PSD (Power Spectral Density) assumption that models PSD of the weather return signal as Gaussian shape. The PSD of weather returns represents the power-weighted distribution of radial velocities of the scatters [20]; thus, the three lowest-order moments correspond to reflectivity, mean radial velocity, and

Table 12.2 WSR-88D brief specifications

System parameters	Value
Frequency	2 700 MHz
Peak power	750 kW
Antenna beam width	1°
Antenna gain	44.5 dB
First sidelobe level	26 dB
Polarization	Linear*
Pulse width	1.57~4.57 μsec
PRF	320~1 300 Hz
Receiver dynamic range	93 dB

*The legacy system is horizontal polarization, but the dual-polarized upgrade has been implemented at many sites and is still ongoing.

spectrum width. The covariance calculation has been widely used to estimate the moments for its simplicity by estimating the Auto Correlation Function (ACF):

$$\hat{R}(\ell) = \frac{1}{M}\sum_{m=0}^{M-1} V^*(m-\ell)V(m) \tag{12.2}$$

where $V(m)$ is the IQ time series and ℓ is the lag. Obviously, $\hat{R}(0)$ corresponds to power estimate, which is proportional to reflectivity:

$$\hat{P} = \hat{R}(0) - N = \frac{1}{M}\sum_{m=0}^{M-1} V^*(m)V(m) - N \tag{12.3}$$

where N is the noise power. The phase of $\hat{R}(1)$ is the average phase difference between adjacent pulses, which is proportional to the mean radial velocity:

$$\hat{v} = -\left(\frac{\lambda}{4\pi t_p}\right)\angle\hat{R}(1) \tag{12.4}$$

where λ is the radar wavelength, t_p is the Pulse Repetition Time (PRT), and $\angle\hat{R}(1)$ is the phase of lag 1 estimate of the ACF. The spectrum width can be estimated by

$$\hat{\sigma}_v = \frac{\lambda}{2\pi T_s\sqrt{2}}\left|\ln\left(\frac{\hat{P}}{|\hat{R}(1)|}\right)\right|^{\frac{1}{2}} \tag{12.5}$$

Figure 12.4 shows a typical weather PSD and the reconstructed Gaussian PSD based on moment estimates from (12.2) to (12.5).

Ground clutter such as buildings, terrain, and trees causing strong radar returns can severely affect moment estimate accuracy. An example of such ground clutter contamination is shown in Figure 12.5. The most significant spectral component is found to be around 0 m/sec, indicating it corresponds to ground clutter, since most ground clutter remains stationary within a short dwell time. The weather return is approximately 20 m/sec, and it is worth noting that part of the weather spectrum aliases in this example. The weather return is almost 20 dB lower than the ground clutter return. This high CSR (Clutter-to-Signal Ratio) will definitely result in biased moment estimates. However, with conventional notch filter, the moving weather returns can be easily distinguished from stationary ground clutter in the Doppler domain as shown in Figure 12.5(b).

The GMAP (Gaussian Model Adaptive Processing) technique has been adopted in the WSR-88D system for stationary and near stationary ground clutter filtering [19]. But in the case of WTC, GMAP is only able to suppress clutter caused by stationary parts of the wind turbine, i.e., the tower, nacelle, and hub. Clutter introduced by blade rotation can extend throughout the entire Doppler spectrum as shown in Figure 12.6. The rotor blades have lower radar return compared to the stationary parts, yet if comparable with weather return, it can significantly bias moment estimations.

Earth curvature makes the radar beam progressively higher above the ground as the wave propagates in normal atmospheric conditions. Figure 12.7 shows

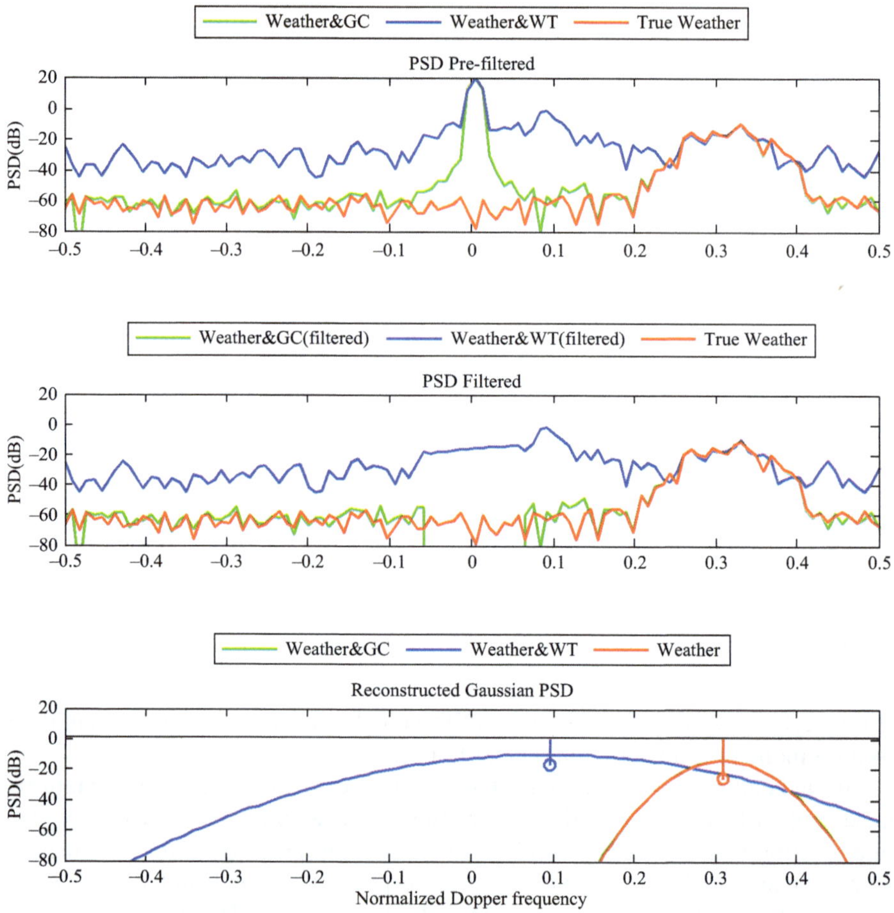

Figure 12.4 The PSD estimate of weather signal and the reconstructed Gaussian PSD based on moment estimates

heights of the lower bond of radar beam at different elevation angles of WSR-88D radar. The GE 1.6 MW wind turbine has a hub-and-blade combined height of 121.25 m, so its clutter effect exists up to 80 km away from the WSR-88D radar. The second lowest beam, however, will skip the top of the wind turbine further than 12 km. Thus, WTC impact on WSR-88D has mostly been observed at the lowest elevation scan.

An example of the wind farm impact on WSR-88D radar at the lowest elevation angle is shown in Figure 12.8 at Dodge City, Kansas (radar ID: KDDC). The two wind farms are located at the city of Montezuma (Southwest) and Spearville (Northeast), respectively. The former is approximately 40 km from the radar and the latter is about 22 km. They all fall into the yellow RLOS zone as the upper left image indicates. It is worth noting that this area is not symmetric around the radar site due to the fact that the terrain is higher in the west and lowest in the east. In the

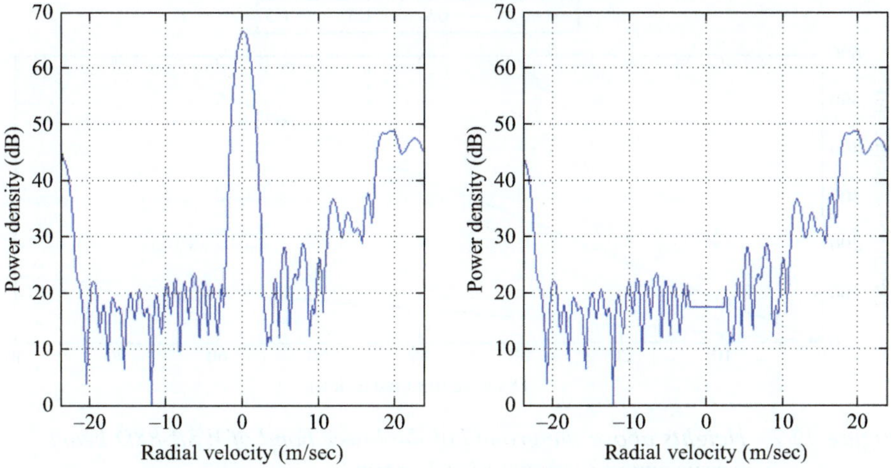

Figure 12.5 Example of the PSD of moving weather with stationary ground clutter returns and the ground clutter filtered PSD using GMAP

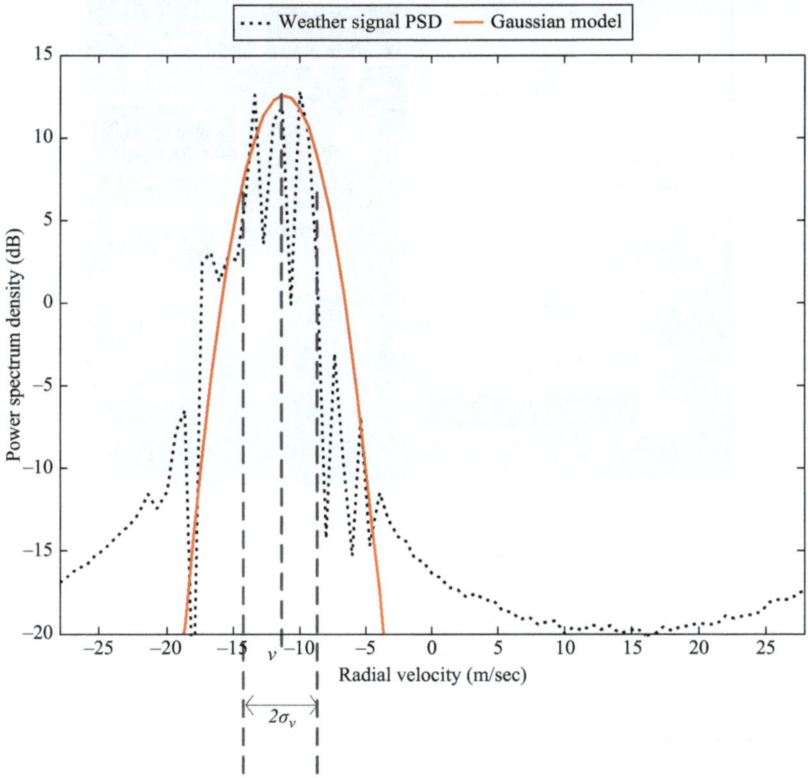

Figure 12.6 The PSD comparison of weather return, weather with stationary ground clutter, and weather with WTC (top); the PSD after applying GMAP (middle) and reconstructed PSD using moment estimates

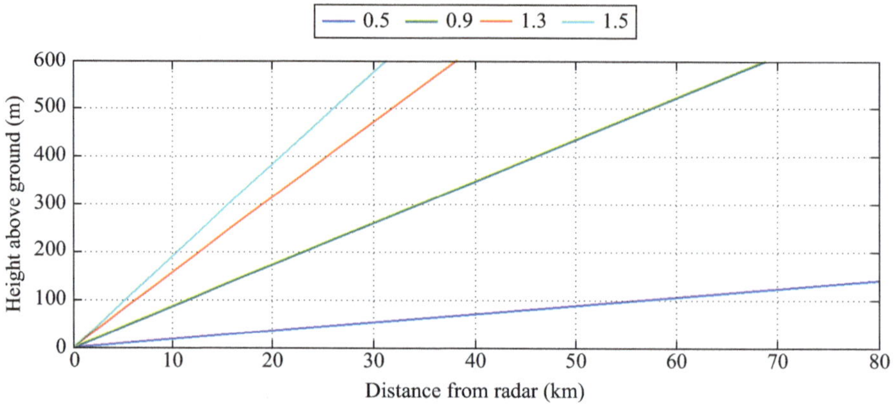

Figure 12.7 Heights above the ground of the lower bond of WSR-88D beam versus range (assuming level ground)

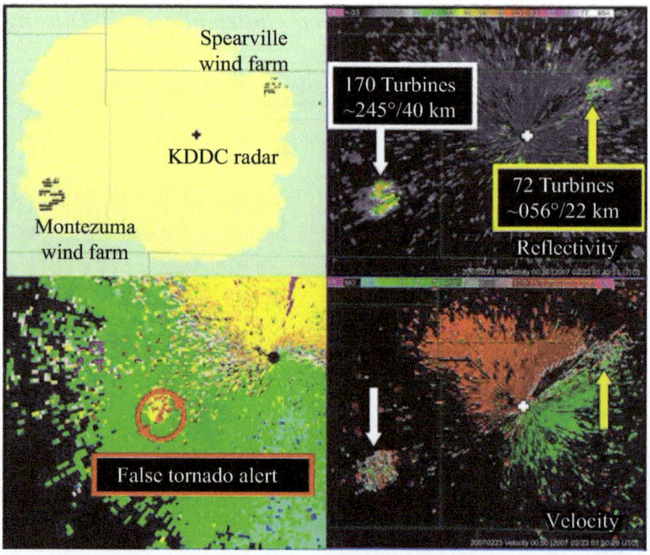

Figure 12.8 Wind farm impacts on KDDC radar at Dodge City, Kansas: upper right, locations of KDDC radar and the two wind farms in the RLOS; upper right: the reflectivity estimates; lower right: the radial velocity estimates; lower left: a false tornado alert issued by the tornado detection algorithm. (Picture from: http://www.roc.noaa.gov/WSR88D/WindFarm/TurbinesImpactOn.aspx?wid=dev)

upper right image, even after applying GMAP ground clutter filter, the two wind farms are still clearly visible due to the high reflectivity from rotating blades. The mean radial velocity estimates render meaningless results at the wind farms in the lower right image. The opposite radial velocity directions detected at the

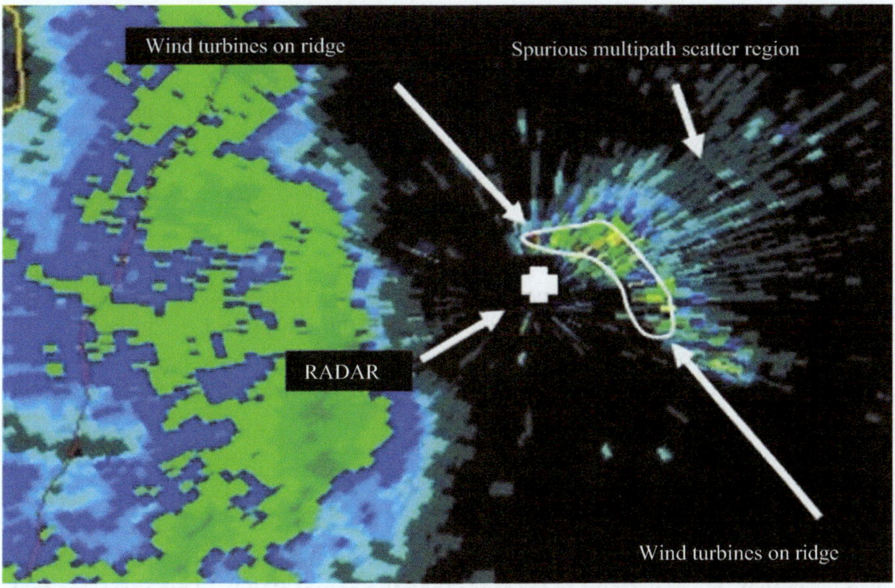

Figure 12.9 The multi-path effects observed by the WSR-88D radar near Ft Drum, NY. (Picture from: http://www.roc.noaa.gov/WSR88D/WindFarm/TurbinesImpactOn.aspx?wid=dev)

Montezuma wind farm resembled tornado signatures and triggered the detection algorithm to issue a false tornado alert at that time.

The impact increases dramatically as the wind turbines are placed closer to the radar, not only because the wind turbines protrude to higher elevation scans, but more importantly, the multipath effect may occur as they become closer to the radar. Multiple bounces of waves between wind turbines and the ground and among wind turbines will expand WTC affected area, in which case multiple images of wind turbines appear behind the wind farm as in Figure 12.9. This may also happen when the wind turbines are far from the radar, but the higher-order backscattered EM energy caused by multiple bounces will be much lower and can be ignored in that case. The wind farm in Figure 12.9 is less than 10 km from the radar site, but the contaminated area extends much further than the actual wind farm. This effect can severely degrade radar data quality and cause large bias in moment estimation.

12.4 Radar Cross Section (RCS) of Wind Turbines

The RCS of the wind turbine is important in evaluating its impact on nearby radars, i.e., the higher RCS the wind turbine has, the more clutter contamination is expected on radar data. Some efforts have been made to calculate wind turbine RCS by numerical methods [8, 21], but the extremely large electric scale and multiple degrees of freedom of wind turbines can make the computational load unmanageable. Hence, high frequency approximations [22] and simplified models [23] have also been exploited in wind turbine RCS predictions. Both approaches calculate the RCS from scattered field based on the definition of RCS as the overall target scattered power (assuming uniform scatter) over incident power density:

$$\sigma = \lim_{r\to\infty} 4\pi r^2 \left|\frac{E_s}{E_i}\right|^2 \tag{12.6}$$

where E_s and E_i are the power density of scattered and incident wave, respectively. The definition assumes that the distance between the target and the radar goes to infinity, i.e., the incoming wave for both the target and the radar need to be plane wave, which is commonly known as the far field constraint, and is enforced in practice by

$$r > \frac{2D^2}{\lambda} \tag{12.7}$$

where D is the maximum dimension of radar antenna and target. The wind turbines are so large that this constraint is far more than the actual RLOS. This indicates that for the interest of most ground based radars, the definition of RCS is not suitable for evaluating wind turbine impact. In these real cases, the RCS loses some of the merits in characterizing radar target signatures, for example, it becomes dependent on the distance between the wind turbine and the radar. Similar situation occurs for the scaled model measurements in the following section. However, the concept of RCS can still be used to quantify how much EM energy is scattered back to the radar; therefore, we would refer to more general terms such as "return power" and "return power ratio" in the following discussions.

The RCS of wind turbine assuming far field may give a rough estimate on the potential clutter level, but will most likely overestimate the clutter effect. Under the plane wave assumption, the incident wave onto the wind turbine comes from the same direction, but the actual incident angle across the wind turbine varies due to the large dimension. For example, if the distance is 10 km, hub height is 100 m, and the radar tower is 10 m tall, then the incident angle variation across the tower alone is over 0.5°. This variation can make significant difference in near-specular reflection, i.e., the tower is illuminated from the broadside, which is the case for most ground based radars. The scattered wave from different parts of the wind turbine will also coherently add with each other, resulting in range dependent radar returns. Basic techniques to estimate near-field radar return of wind turbines and how it varies with range are discussed in [24]. Another important fact that is directly related to the return power is the number of wind turbines within each radar resolution volume, which is determined by wind turbine spacing, as well as radar resolution and distance between wind turbines and the radar. If there are multiple wind turbines in one resolution volume, the scattered field from each wind turbine may constructively or destructively combine with each other and the actual return can only be analyzed on a case-by-case basis.

12.5 Micro-Doppler Signatures of Scaled Wind Turbine Model

It is desirable to study the micro-Doppler signatures of wind turbine in a "controlled" environment, so we can have a better understanding of the connections between the physical parameters and the spectrum characteristics. This is extremely difficult for a full-sized wind turbine, but can be realized if the wind turbine has a manageable size. A scaled Wind Turbine Model (WTM) [22, 25] is customized for such a purpose, as shown in Figure 12.10. The model has hub height of 1.2 m and a rotor diameter of 0.7 m. The model preserves the geometry of an actual GE wind turbine. A steel rod is embedded within each blade to mimic the lightning rods in actual wind turbine blades. A servomotor with shaft is installed inside the nacelle to drive the blade rotation with the desired speed. The entire model is mounted on top of a base rotor for yaw control. The motors mimic the rotation and yaw of the actual wind turbine, and provide real-time position feedback, which can be synchronized with radar measurements.

The model can easily fit into an anechoic chamber, which provides a "quiet" EM environment and isolates the wind turbine radar signature emulation from environmental interference. Both frequency and time domain measurements are performed. The results are analyzed and compared in the following sections.

12.5.1 Frequency Domain Measurement

Frequency domain measurements are carried out using a Vector Network Analyzer (VNA) [22]. The measurement is set up as shown in Figure 12.11. Two horn antennas (transmit and receive) are the interface between DUT (Device Under Test) and the VNA in this OTA (Over The Air) measurement. The DUT, which is the WTM in this case, is about 4 m away from the antennas. An automated software program remotely controls both the rotors and the VNA. The S-parameter S_{21} is measured while the rotor blade holds stationary before it increments by $\Delta\theta$ for the next measurement.

Single-tone measurement is subject to possible reflections from the wall and ceiling of the chamber despite being covered by absorbing materials. These reflections can be separated from WTM returns by gating in the time domain. The frequency domain measurements are thus transformed into the time domain by IFT (Inverse Fourier Transform). The transformed return power measurement is shown in Figure 12.12(a), where the brightest line corresponds to the WTM. The averaged power return across all blade positions is shown in Figure 12.12(b). As is shown, reflections from the back wall and ceiling are right behind the WTM, after which is the multipath reflection. Other small peaks in the time domain return power profile are due to impedance mismatch at the ports of antennas, cables, adapters, etc.

The micro-Doppler spectrum or the spectrogram of the frequency measurement can be synthesized from stationary measurements as shown in [22]. An example is given in Figure 12.13. This stationary measurement in which the blades

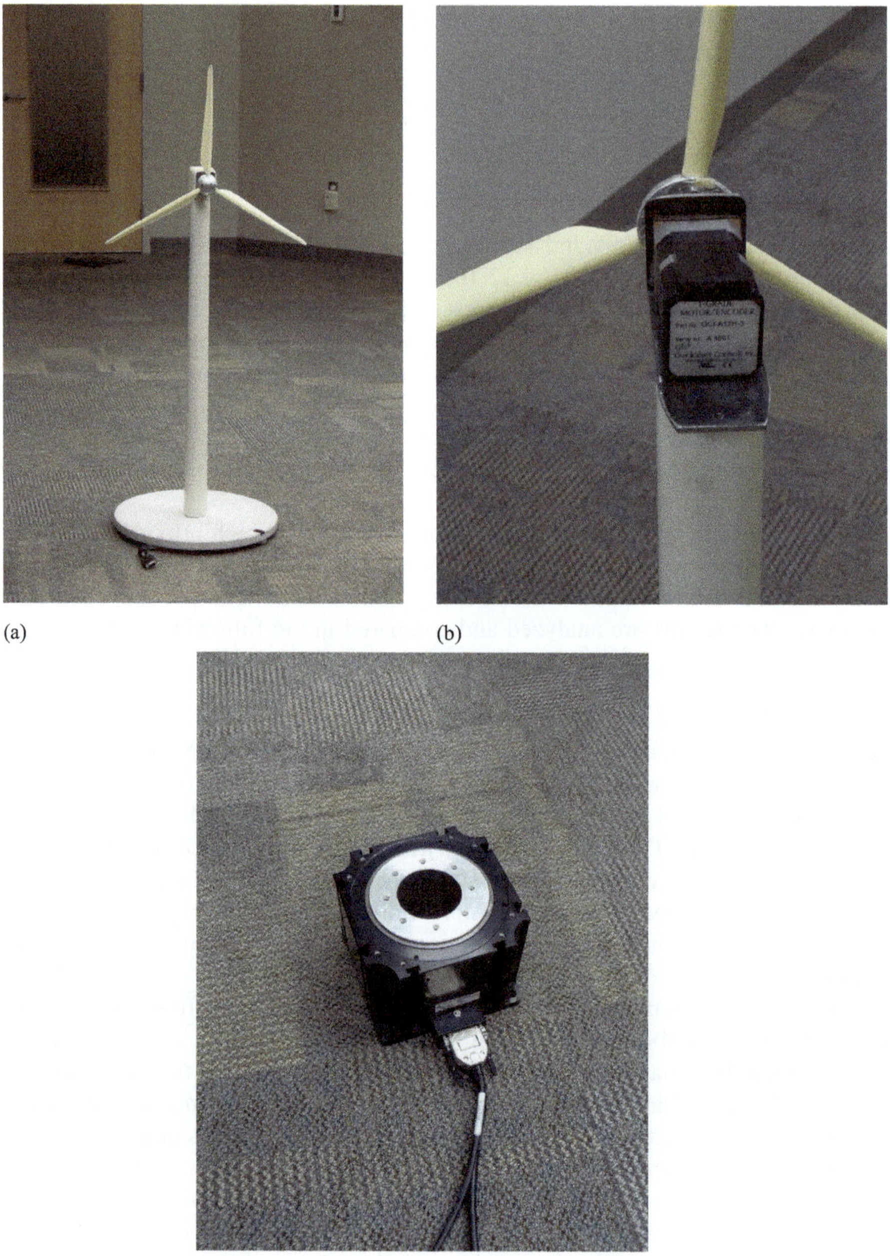

(a) (b) (c)

Figure 12.10 The scaled wind turbine model: (a) WTM placed in the middle of the anechoic chamber; (b) the rotor in the opened nacelle; (c) the base rotor used to mimic the yaw motion

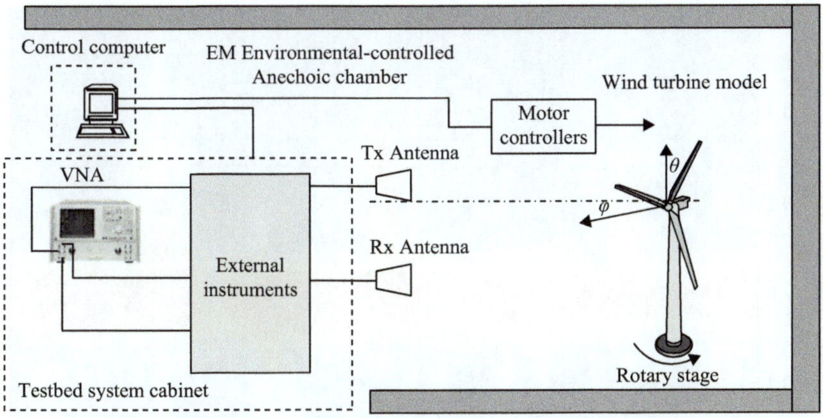

Figure 12.11 The frequency domain measurement setup: both the VNA and motor controllers are remotely controlled

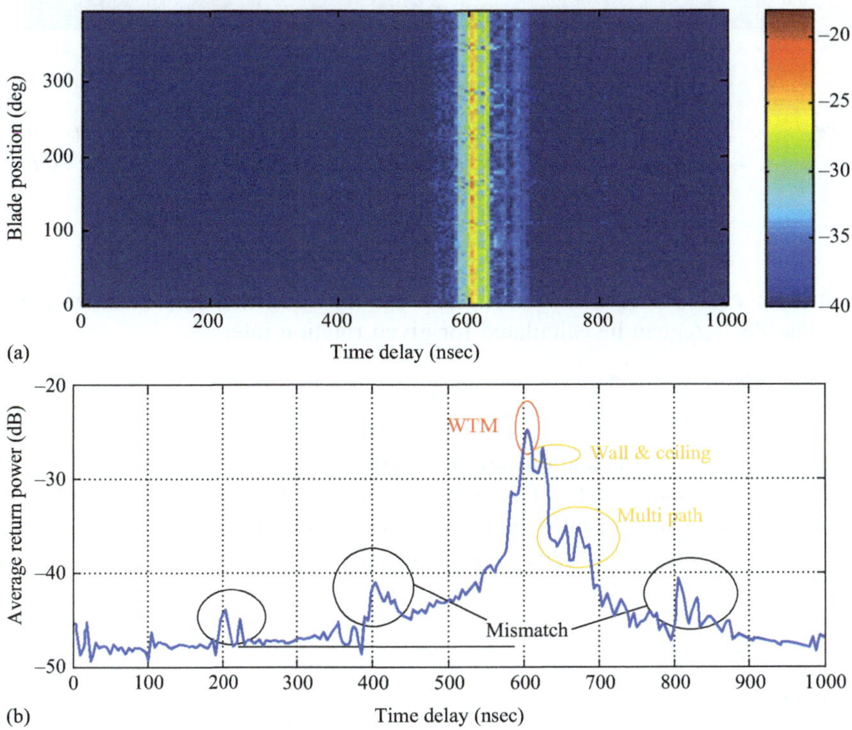

Figure 12.12 The return power versus time delay derived from frequency domain measurement using IFFT: (a) the return power at different time delay and blade position; (b) the return power averaged across all blade positions showing where the WTM is in the time domain

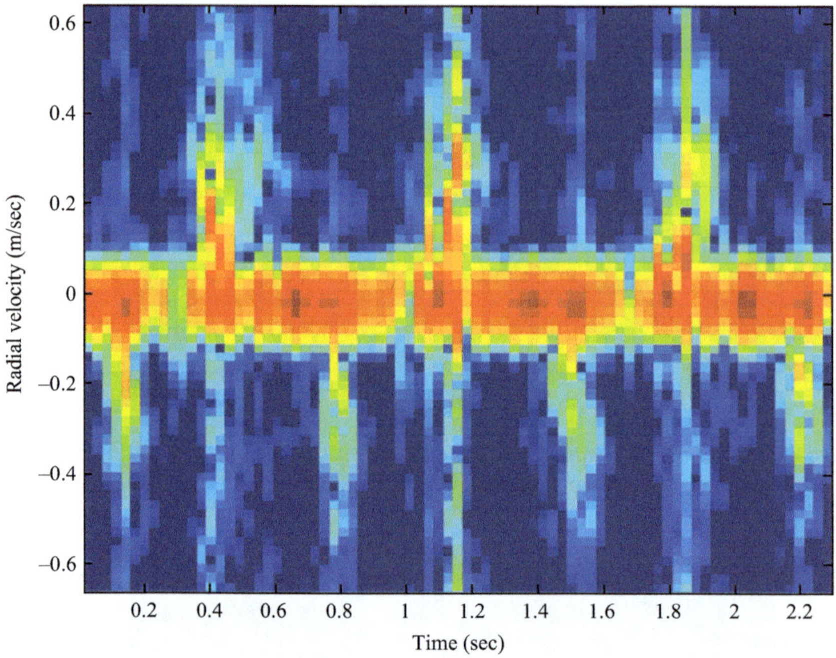

Figure 12.13 Spectrogram of WTM from frequency domain measurements: the frequency is 40 GHz with frequency span of 200 MHz. The rotor blade increments 0.5° from one measurement to the next

stay still for each measurement is a "slow motion" of the actual radar measurement. The PRT (t_p) can be calculated for given rotation rate:

$$t_p = \frac{\Delta\theta}{\omega} \tag{12.8}$$

where ω is the emulated rotation rate in radians per second. Figure 12.13 is generated based on rotation rate of 28.5 rpm to match the realistic spectrogram in section 12.6. So in about a little over one period (2.3 sec), there are seven flashes. From (12.8), the aliasing velocity can be calculated as

$$v_a = \frac{\lambda}{4t_p} = \frac{\lambda\omega}{4\Delta\theta} \tag{12.9}$$

And the tip velocity normalized to the aliasing velocity is

$$\overline{v_{tip}} = \frac{\frac{D}{2}\omega}{v_a} = \frac{2D\Delta\theta}{\lambda} \tag{12.10}$$

Table 12.3 System specifications for the Laboratory Radar

System parameters	Value
Operating frequency	10.5 GHz
Peak power	126 mW
Antenna gain	12 dBi
Transmit pulse width	6~200 nsec
PRF	95~24 400 Hz
System bandwidth	100 MHz

where D is the rotor diameter. To avoid aliasing, i.e., $\overline{v_{tip}} < 1$, the angle increment between each measurement needs to satisfy:

$$\Delta\theta < \frac{\lambda}{2D} \tag{12.11}$$

D is fixed, hence, the smaller the wavelength, or the higher the frequency, the finer increment is required to avoid spectrum aliasing. The spectrogram emulation in Figure 12.13 is apparently aliased due to limited rotor precision.

12.5.2 Time Domain Measurement

The frequency domain approach afore mentioned is limited to stationary measurement due to relatively long cycle time of the VNA, i.e., the minimum time interval between consecutive measurements. Due to this limit, the time needed for measurements of all possible blade and yaw positions is tremendous. For this reason, the Laboratory Radar (LR) is developed for direct time domain measurements [26].

LR is a high-resolution, dual-polarized pulsed scatterometer designed for indoor measurement. It transmits a narrow pulse at X-band to achieve fine range resolution, and is designed with waveform reconfigurable capability to adjust pulse width and PRF. The general specifications of the system are listed in Table 12.3. Note that the nominal peak power is in simultaneous dual-polarized mode, when horizontally (H) and vertically (V) polarized waves are transmitted within the same pulse. It will be halved from the nominal value in the alternative dual-polarized mode, when H and V polarized waves interleave from pulse to pulse.

Figure 12.14 shows the system block diagram of the LR. The same coherent oscillator is used for both transmit and receive. Direct conversion is adopted to make the system small and compact. The data acquisition system is triggered by the same clock that modulates the transmit signal to avoid pulse-to-pulse jitters. The baseband *I/Q* signal is digitized, calibrated, and then processed for storage and display. The measurement configuration is similar to the frequency domain measurement except that VNA is replaced with LR. The photo of configuration is shown in

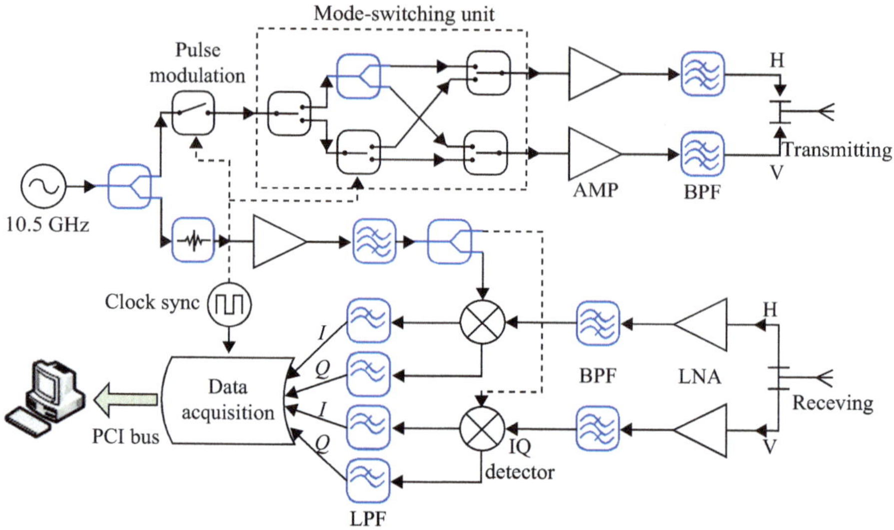

Figure 12.14 The system block diagram of LR: direct conversion is adopted for system size reduction

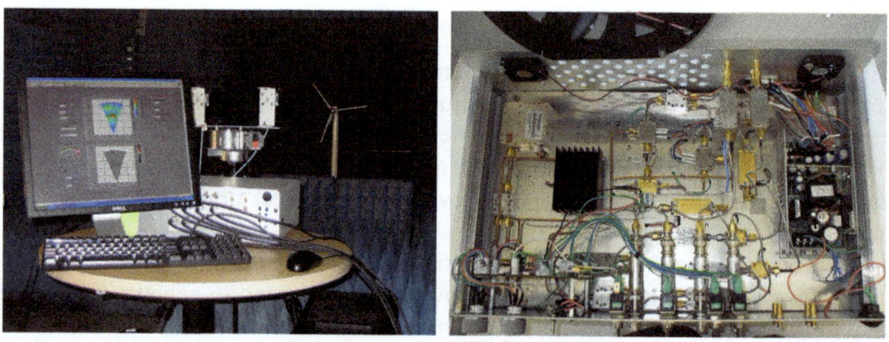

(a) (b)

Figure 12.15 The laboratory radar setup: (a) ongoing measurement; (b) inside the LR transceiver (top view)

Figure 12.15(a). The top view of the inside of LR transceiver is shown in Figure 12.15(b). All parts used are commercial off-the-shelf (COTS) components except for the customized control and synchronization board. In addition, a software program is developed to synchronize WTM motions with data acquisitions. As a result of the faster cycle time or PRT, the LR is able to continuously sample the WTM returns while the blade is rotating (dynamic measurement). The measurement is carried out in the sequence of "stop-spin-sample-go": the WTM rotates for integer number of revolutions at each aspect angle ϕ, as defined in Figure 12.2, while the LR keeps sampling, and then yaws to $\phi + \Delta\phi$ until it turns back to the start.

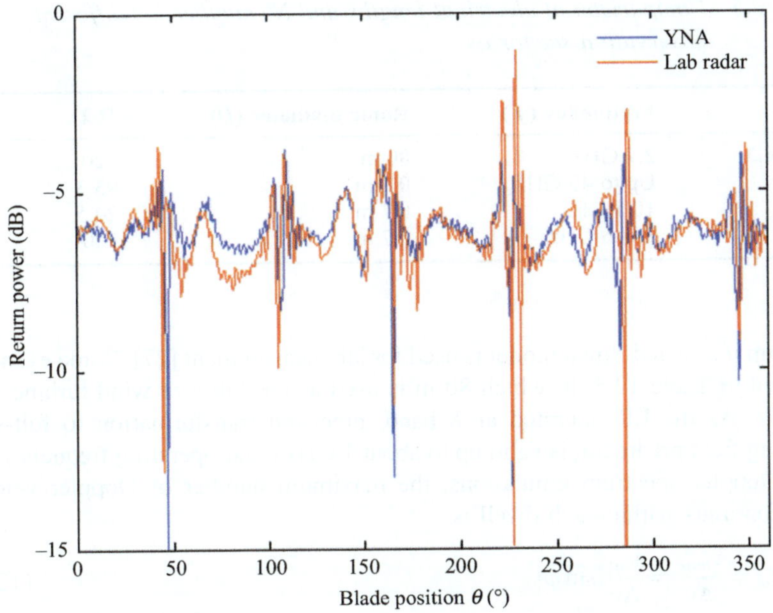

Figure 12.16 Comparison of WTM return-power varying with blade positions using both measurement methods. For the purpose of comparing the two approaches, the measured results are not necessarily aligned accurately with blade position

Table 12.4 Comparison of measurement time requirements for two approaches

	Cycle time	Single revolution	All rotation and all yaw
Frequency domain	2.24 (sec)	56 (min)	14 (days)
Time domain	0.66 (msec)	3 (sec)	18 (min)

The return power variation in terms of blade position obtained from both stationary/frequency domain and dynamic/time domain measurements (with tower and blade) are compared in Figure 12.16, where the two closely match with each other. However, the time needed to perform a measurement is significantly shortened in the dynamic measurement. A direct comparison of the time required to perform each measurement is listed in Table 12.4. The pulse repetition interval (PRI) of LR is set up to be 1 515 Hz and the blade rotation rate is 20 rpm. The yaw increment $\Delta\phi$ is 1° for both cases, while $\Delta\theta$ is much finer for the time domain measurement.

The LR is also an important instrument for *real-time micro-Doppler* radar signature emulation and validation. There are, however, limitations on the attainable precisions. Commercial wind turbines manufactured in the recent decade have rotor diameter ranging from 60 to 100 m. Measurement frequency is supposed to be

Table 12.5 Comparison of electrical lengths and N_d number for different observation scenarios

Radar	Frequency (f)	Rotor diameter (D)	D/λ	N_d
WSR-88D	2.7 GHz	80 m	720	126
VNA	Up to 40 GHz	0.7 m	93.3	16
LR	10.5 GHz	0.7 m	24.5	4
MMLR	308.6 GHz	0.7 m	720	126

scaled-up if a scaled-down model is used for lab measurement [27]. Some examples are listed in Table 12.5, in which 80 m is used as the full-size wind turbine rotor diameter. As the LR operated at X-band, precision transformation to full-scale scattering field prediction is valid up to about 1 GHz radar operating frequency. For micro-Doppler spectrum emulations, the maximum number of Doppler velocity resolution units within each dwell is

$$N_d = \frac{v_{max}}{\Delta v} = \frac{v_{tip}}{\Delta v}|\sin\phi| \tag{12.12}$$

where v_{max} is the maximum radial velocity, which is v_{tip} projected onto the direction of RLOS; Δv is radial velocity resolution and is proportional to the Doppler resolution:

$$\Delta v = \frac{1}{2}\lambda \cdot \Delta f_d = \frac{\lambda}{2T} \tag{12.13}$$

where Δf_d is the Doppler resolution, T is the dwell time or the temporal resolution of the spectrogram and is given by

$$T = \frac{\Theta}{\omega} \tag{12.14}$$

where Θ is the rotor blade swept angle in T. Substituting (12.14) into (12.13) and the tip velocity $v_{tip} = D\omega/2$ into (12.12), we have

$$N_d = \frac{D\Theta}{\lambda}|\sin\phi| \tag{12.15}$$

N_d characterizes the "fineness" of the micro-Doppler spectrum. The larger N_d is, the more details can be observed within the "spectrum flashes." The values of N_d for full-size wind turbine radar measurement and scale measurements are shown in Table 12.5. In order to match the same N_d of observing full-size wind turbine using WSR-88D radar with the WTM, a Millimeter LR (MMLR) over 300 GHz is needed.

This relative spectral resolution loss is the cause of difference between the micro-Doppler signatures of scaled measurements and the actual wind turbine radar measurements, which will be shown later in section 12.6. From Table 12.4, one can see that the Doppler spectrum of the scaled model measurement is expected to be approximately 30 times coarser than that of the full-size wind turbine measurements using WSR-88D radar.

Examples of emulated micro-Doppler spectrum of wind turbines using the X-band LR and time domain measurements are shown in Figure 12.17. Note that the spectrogram is unlike the line structures we are going to see in some actual radar measurements; the flashes here are single sided, because they are not aliased. The flashes occur when one blade is perpendicular to RLOS, i.e., θ becomes multiple times of 60°. In fact, from the time series of return power back in Figure 12.16, the six major peaks correspond to these flashes, with rapid oscillations in a very short angular increment. When the blade is perpendicular to RLOS, or vertical when radar elevation angle is small, which is the case for ground radar operating low elevation scans, the blade is aligned with the tower. It is believed that this alignment leads to constructive in-phase coherent integration of backscattered EM field and causes high return power. Because the blade rotation quickly breaks the in-phase lineup and individual scattering centers having range shifting in scales comparable to radar wavelength cause significant oscillations due to coherent field integration, the peak drops rapidly and starts oscillating.

In addition, scattering centers along the blade have the maximum velocity projection onto RLOS when the blade is vertical. The trajectory in Figure 12.17(a) shows a sinusoidal variation of the blade tip, but finer Doppler resolution is needed for better interpretation. It is worth noting that the peaks corresponding to the leading edge occur almost exactly at θ becoming multiple times of 60°, but earlier for peaks caused by the trailing edge. This phase difference is found to be about the same as the blade angle α defined in Figure 12.2 and exists because the trailing edges turn to vertical earlier as a result of the blade shape. Figure 12.17(b) shows the spectrogram of the WTM as it yaws to $\phi = 0°$, i.e., the radar beam illuminates the WTM from nose direction. At this position, theoretically, if the incident field is plane wave, the return is constant as the blade rotates, but there will be some slow variations in practice due to near-field effect. Hence, the spectrogram only has a centerpiece but no flashes. In this case, the wind turbine is equivalent to stationary ground clutter and can be mitigated by conventional ground clutter filtering.

The return power of two linear polarizations (Horizontal and Vertical) in relation to the rotation and yaw is measured and shown in Figure 12.18. Both HH and VV polarization have similar features, with the most prominent being the periodicity due to blade rotation. As the blade rotates for one revolution, six significant peaks are observed at most aspect angles except for angles around $\phi = 0°$ (360°) and $\phi = 180°$. The WTM rotation plane is perpendicular to RLOS for these cases. And as has been discussed in the spectrogram, there are only slow variations at these positions. In fact, as φ approaches 0°(360°) or 180°, the specular reflection from the front/back of blades starts to dominate and the two peaks resulting from both edges of the blade merge. As a result of the asymmetric blade structure, the return power is not symmetric about $\varphi = 180°$. In fact, an arbitrary point on the blade passing through vertical position above the nacelle at (θ, φ) will pass vertically below the nacelle at $(\theta - 60°, 360° - \varphi)$.

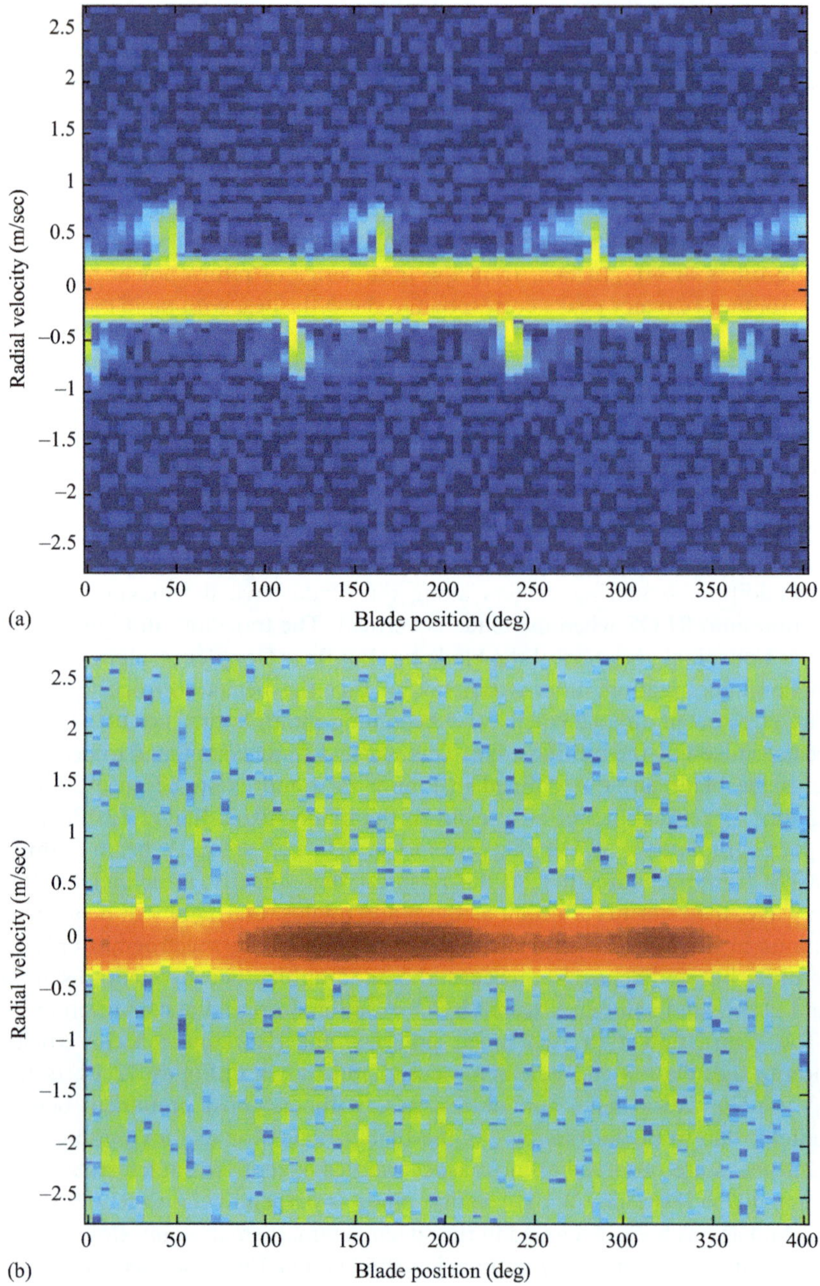

Figure 12.17 Spectrograms of time domain measurements of WTM: (a) The aspect angle $\phi = 90°$; (b) When $\phi = 0°$

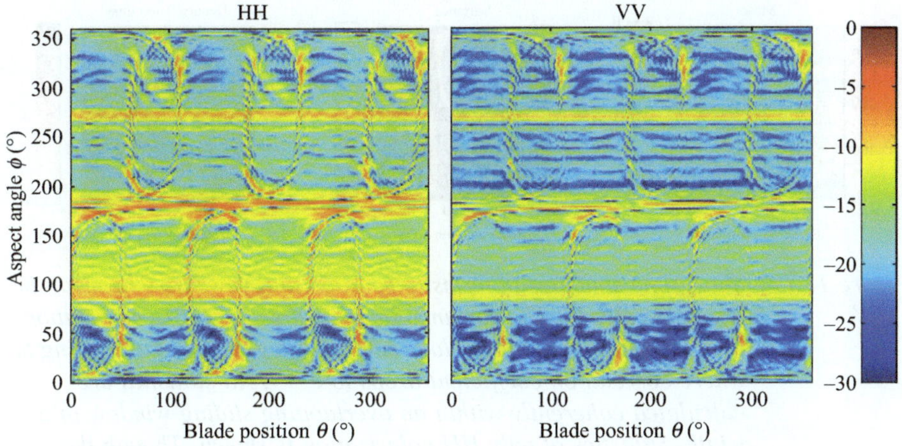

Figure 12.18 Measured WTM radar return power variation (decibel scale) in terms of blade position and aspect (yaw) angle

The periodicity of rotation fluctuation can be used to distinguish wind turbines from the desired radar targets such as aircrafts and weather. However, the dwell time during which the wind turbine is continuously illuminated needs to be long enough to observe this periodicity, which is not quite practical for some current radar operations. Assuming the wind turbine is far from the radar and the azimuth dimension of the wind turbine is small compared to the beam width of the radar antenna, the swept angle of the rotor blade within the dwell time is

$$\Theta = \omega \frac{\Psi}{\Omega} \tag{12.16}$$

where ψ is antenna beam width and Ω is radar scan rate. For example, if the wind turbine spins at 15 rpm and the radar scans at 18°/sec with 1° beam width, then $\Theta = 5°$. With such a short observation window, the WTM returns are found to have diverse statistics from scan to scan, as shown in Figure 12.19. The short-time mean varies as the blade rotates, indicating the statistical nonstationarity. The variance is exceptionally higher due to the fast oscillation where the return power peaks occur. This large variance is one of the radar signatures that can help identify wind turbines. With large variance, the decorrrelation time at such positions are expected to be lower, as shown in Figure 12.19(c). In fact, less than 0.5° decorrelation angle is observed at these positions.

Many large modern wind turbines have variable rotation rate, so it is reasonable to assume uniformly distributed random blade position for every radar visit from scan to scan. The statistics of the rotation fluctuation thus characterizes the yaw fluctuation since the latter is the slow variation over a relatively long time. The mean and standard deviation of the return power in terms of aspect angle are shown in Figure 12.20. The mean power clearly have three major peaks at $\varphi = 90°$, 180°, and 270°, which correspond to the three aspect positions shown in Figure 12.20.

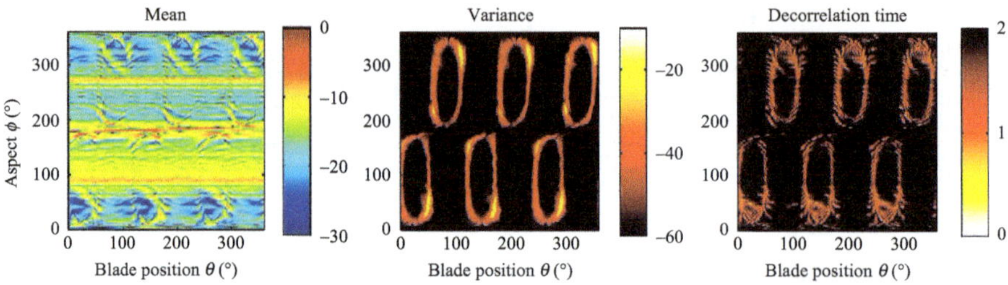

Figure 12.19 Short-time statistics in terms of rotation and yaw: (a) The mean in decibel scale. (b) The variance in decibel scale. (c) Decorrelation time (degrees) in terms of blade rotation (degrees) corresponding to where correlation coefficient drops to $e^{-1/2}$. All statistics are calculated coherently within an overlapping sliding window of 5° width. Statistics of only HH polarization is shown. Though the statistics are still shown in sequence of θ, it can be of any sequence for actual radar operations

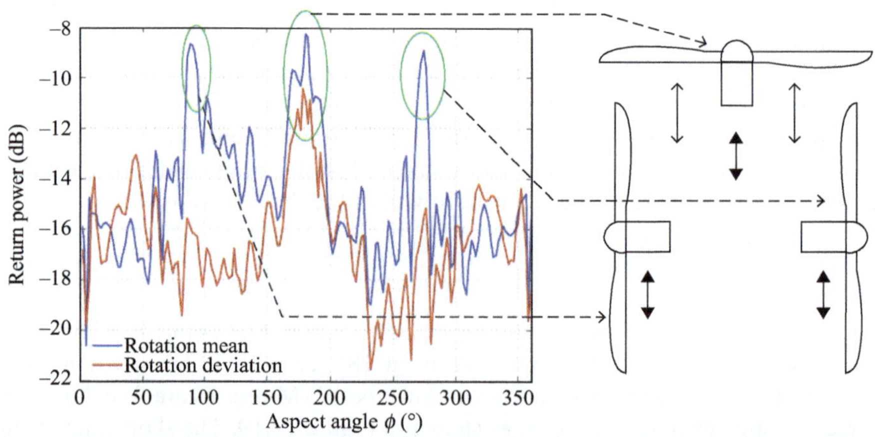

Figure 12.20 Mean and standard deviation of return power in terms of aspect angle: both the mean and deviation at each aspect angle are estimated across one complete revolution. The result is HH polarization

As the wind turbine yaws to 90° and 270°, one side of the nacelle is perpendicular to the RLOS, resulting in specular reflections, which causes the mean return power to rise. The peak at $\varphi = 180°$, however, is caused by the reflection from the back of the nacelle. The standard deviation is also maximized at $\varphi = 180°$, which can be attributed to the reflection from the backside of the blades. Therefore, the dominant scattering mechanism is specular reflection and proper design of nacelle shape that reduces normal incidence may suppress the interference to radars.

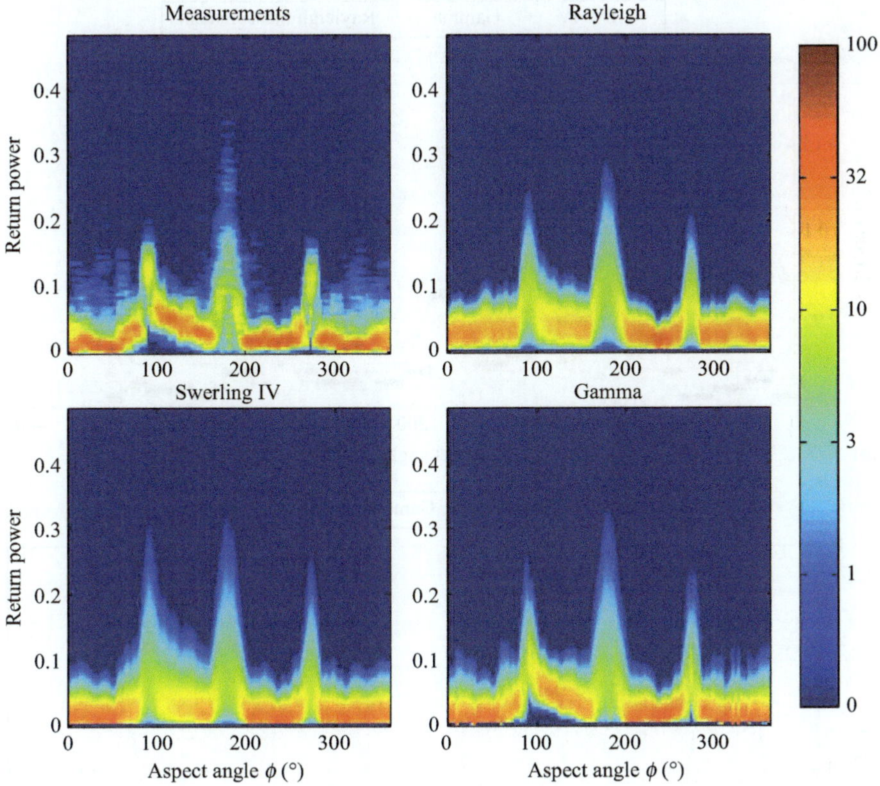

Figure 12.21 PDF fittings: (a) Comparison of PDF fittings, where the probability density is mapped to different colors. The aspect deviation $\varphi_\Delta = 6°$, and the rotation fluctuation samples are decimated by 10 to decorrelate measurements. Maximum likelihood estimation has been used for all fittings. (b) Comparison of mode. (c) Comparison of RMSE for different distribution fittings

Due to blade rotation, the return power of a wind turbine fluctuates from pulse to pulse. It is therefore necessary to model the Probability Density Function (PDF) of the return power fluctuations. Since the aspect angle is determined by wind direction, which normally varies quickly, it is assumed $\varphi = \varphi_0 \pm \varphi_\Delta$, where φ_Δ describes the variation. The comparison of PDF estimates, along with Rayleigh, Swerling IV, and Gamma fittings, are shown in Figure 12.21. The mode corresponding to the peak of the PDF, along with the fitting Root Mean Square Error (RMSE), is shown in Figure 12.21(b) and 12.21(c), respectively, for comparison. The RMSE is defined as

$$e = \sqrt{\frac{1}{\Sigma}\int_0^{\Sigma} |p_m(\sigma) - p_f(\sigma)|^2 d\sigma} \tag{12.17}$$

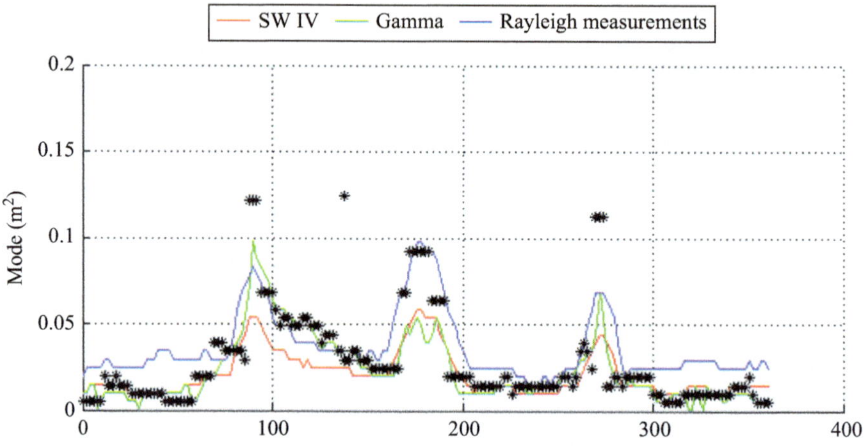

(b)

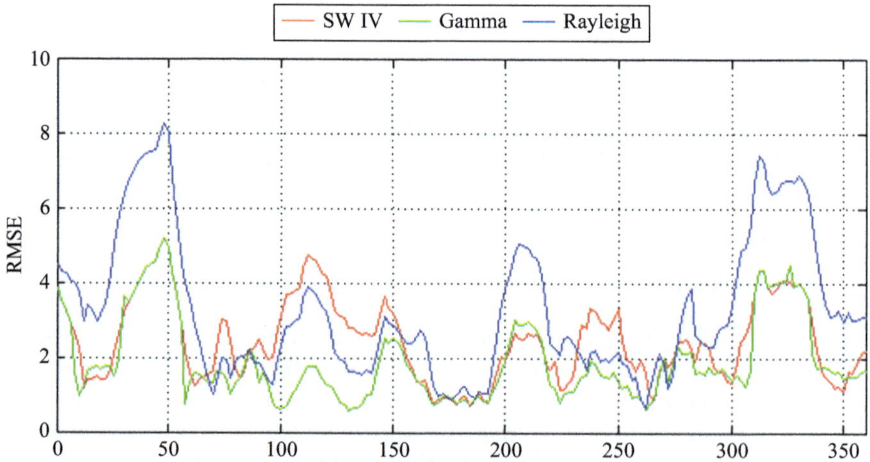

(c)

Figure 12.21 (*Continues*)

where Σ is the dynamic range of measured return power, $p_m(\sigma)$ and $p_f(\sigma)$ are the measurement and fitted distribution, respectively. Both Rayleigh and Swerling IV are single-parameter distributions and are given in (12.6) and (12.7), respectively, where $\overline{\sigma}$ is the mean.

$$p_S(\sigma) = \frac{2\sigma}{\overline{\sigma}} e^{-\frac{\sigma^2}{\overline{\sigma}}} \tag{12.18}$$

$$p_R(\sigma) = \frac{4\sigma}{\overline{\sigma}^2} e^{-\frac{2\sigma}{\overline{\sigma}}} \tag{12.19}$$

$$p_G(\sigma) = \frac{\sigma^{k-1}}{\Gamma(k)\theta^k} e^{-\frac{\sigma}{\theta}} \tag{12.20}$$

The Gamma distribution is determined by the shape parameter k and scale parameter θ, where the former is constrained to $k > 1$ during the fitting because of nonzero modes in the measurement distribution. Swerling IV is equivalent to Gamma with $k = 2$. The comparison shows that Gamma fitting has the overall best fitting performance of all three. However, when the actual mode of measurements is close to zero, all models have difficulties approximating the actual distribution.

Lastly, the dual-polarized (HH and VV) radar signatures of the WTM have been measured and derived [28]. Stokes parameters are used to describe the polarization state of the wind turbine:

$$Q_{HV} = E_H E_H^* - E_V E_V^* \tag{12.21}$$

$$U_{HV} = E_H E_V^* + E_V E_H^* \tag{12.22}$$

$$V_{HV} = -i(E_H E_V^* - E_v E_H^*) \tag{12.23}$$

where E_H and E_V are scattered field in horizontal and vertical polarization, respectively, Q_{HV} is the degree of polarization, U_{HV} and V_{HV} are the shape parameters of the polarization ellipse, from which the cross-correlation coefficient is derived:

$$\rho_{HV} = \frac{\sqrt{U_{HV}^2 + V_{HV}^2}}{2|E_H E_V^*|} \tag{12.24}$$

The cross-correlation coefficient ρ_{HV} is a normalized measure of the correlation between HH and VV polarization states.

Q_{HV} and ρ_{HV} derived from time domain WTM measurements are shown in Figure 12.22. It has been shown in Figure 12.18 that HH measurements are much higher than VV. As a result, Q_{HV} in Figure 12.22(a) is found to be greater than zero for most blade and yaw positions, indicating that a vertically polarized radar may

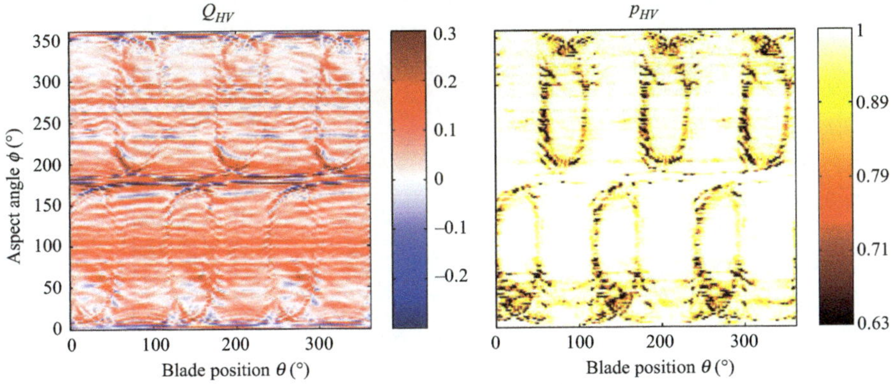

Figure 12.22 Dual-polarized signatures in terms of rotation and yaw: (a) Degree of polarization Q_{HV}. (b) Cross correlation coefficient ρ_{HV}. Both have been averaged on an overlapping sliding window of 5°

be less susceptible to wind turbine interference. The cross-correlation coefficient in Figure 12.22(b) shows similar periodicity as the return power variation. It is clear that ρ_{HV} has relatively low values at positions where the return power peaks occur (the flash positions), which could be a result of scattering phase difference at these positions.

12.6 Time-Frequency Analysis of Wind Turbine Micro-Doppler Signatures from Operational Radars

12.6.1 Micro-Doppler Signatures from WSR-88D Weather Radar

The impacts of wind turbine on WSR-88D radar operations have been discussed in section 12.3. Here, more detailed and realistic micro-Doppler phenomenology from actual radars is analyzed. The blade rotation is oscillatory micro motion with respect to the entire wind turbine structure. It causes extremely complicated Doppler modulations, which is best perceived by continuous observation, i.e., the spotlight mode of radar. Such continuous temporal evolution of Doppler spectrum is manifested by the example in Figure 12.23 by implementing short-time Fourier transform. The strong centerpiece is attributed to the stationary parts of the wind turbine, i.e., the tower and nacelle. Since the signal is not clutter-filtered, the hub rotation may contribute to some of the low frequency variations near the centerpiece. Similar to the scaled-model measurement, the periodic flashes caused by blade rotation are the key feature of the spectrogram. The periodicity is due to the

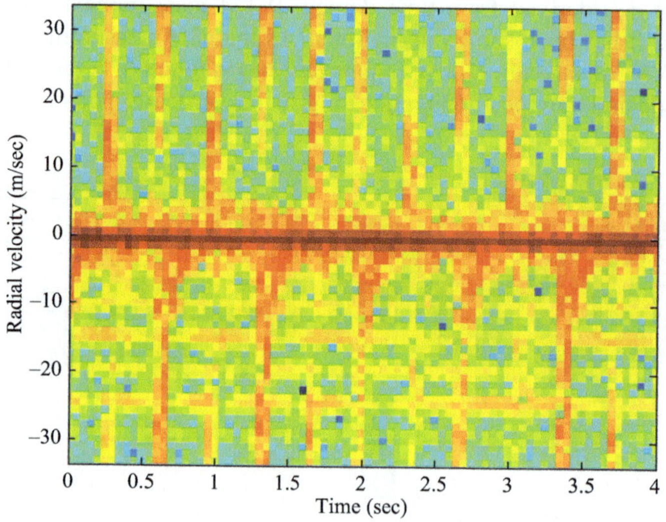

Figure 12.23 Spectrogram of the wind turbine radar return in spotlight mode from a WSR-88D radar. The Doppler frequency is converted to radial velocity to map spectral components to different parts of the rotor blades

symmetric rotor blade structure, which repeats itself for every 120°. The rotation rate for this particular wind turbine (Vestas V47) is 28.5 rpm, so each revolution takes about 2.1 sec. Within such period, six near uniformly distributed flashes are found in Figure 12.23, meaning the flashes must be a result of the blades turning to certain positions every 60°, i.e., half of the periodicity. Our intuitive estimate of these positions is when a blade turns to be perpendicular to RLOS, either above the nacelle ($\theta = 0°$, $\theta = 120°$, $\theta = 240°$) or below it ($\theta = 60°$, $\theta = 180°$, $\theta = 300°$), as shown in Figure 12.2, where the blade position θ is defined. Because it is extremely difficult to accurately synchronize the actual blade positions with the radar observations, this estimate needs to be further verified with other experiments. The centerpiece has much high power density than the flashes as the tower and nacelle are made of metal while the blades are less reflective composite material.

The spectral component corresponding to the highest radial velocity is supposed to be attributed to blade tip, which is moving at extremely high speed due to large rotor diameter. For the example shown, the blade tip moves at 70 m/sec, yet the highest PRF of the WSR88-D radar is 1 300 Hz (Table 12.2), which gives a maximum unambiguous velocity of 36 m/sec. This is adequate for observation of storm motions, but the aliasing effect for the wind turbine observation is obvious in Figure 12.23. Thus, each flash extends almost the entire Doppler domain, forming a vertical line. These vertical lines and the horizontal centerpiece result the unique 2-D spatial spectrum in Figure 12.24. The spatial spectrum is the 2-D Fourier transform of spectrogram:

$$F(u, v) = \int\int f(x, y) e^{-j2\pi(ux+vy)} dxdy \tag{12.25}$$

where u and v are the horizontal and vertical spatial frequency, respectively. First, the v axis ($u = 0$) corresponds to flat horizontal structures in the spectrogram, whose horizontal frequency is near zero, which corresponds to the centerpiece of the spectrogram. And then the flashes are projected along the u axis, where the harmonics are the result of periodicity. Ideally, if the flashes can be modeled as white noise, i.e., flat across the Doppler domain, they will project to evenly spaced dots on v axis. The actual Doppler spectrum, however, will fluctuate and lead to the spatial frequency expansion along v.

12.6.2 Micro-Doppler Signatures of Wind Turbines from a Mobile X-band Radar

While the full-size wind turbine is not controllable as in lab measurements, a mobile radar can change its relative positions and aspect with respect to the wind turbine. And most importantly, non-aliased measurements of full-size wind turbines can be performed, which can reveal more details in the spectrogram. Subsequently, a field study was launched using the RaXPol (Rapid X-band Polarized) mobile radar at the Advanced Radar Research Center (ARRC) of the University of Oklahoma. The RaXPol is an experimental mobile weather radar, featuring rapid sampling rate and dual-polarization capability [29]. A picture of the radar is shown

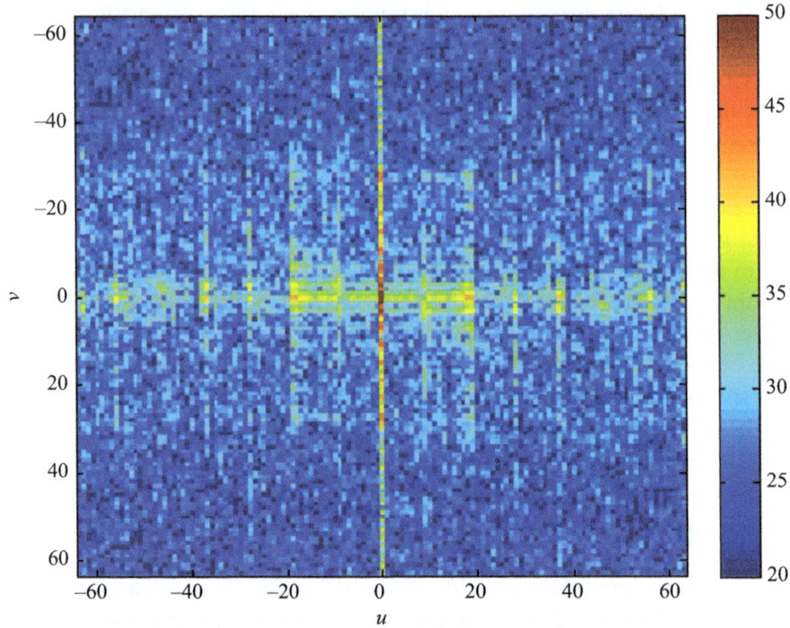

Figure 12.24 The 2-D spatial spectrum of the spectrogram of wind turbine radar return

in Figure 12.25(a). The wind farm chosen for this study is the one at Minco, OK, where over a hundred wind turbines have been installed in the past two years. These wind turbines are state-of-the-art GE 1.6 MW as has been shown in Figure 12.2.

The radar truck was parked at the red pin (circled in yellow) in Figure 12.25(b), and the locations of all the wind turbines are also pinned in the figure. The radar beam is pointed to 292° from the north. The wind turbine T4 at about 7 350 m falls in RLOS, and the wind direction at that time was north-northeast, so the rotor blades are close to parallel to RLOS, i.e., the aspect angle ϕ is close to 270°. The PRF is first set to 2 kHz, and the spectrogram is shown in Figure 12.26(a). The relative spectral resolution is much better compared to scaled measurements, but it is still aliased. The PRF is then raised to 10 kHz, and the spectrogram is again plotted in Figure 12.26(b).

The GE 1.6 MW has variable rotation rate. From Figure 12.26(b), the tip speed is found to be close to 80 m/sec. The period (estimated by counting flashes) is about 3.4 sec, so the rotation rate is estimated to be 17.6 rpm. The rotor diameter is 82.5 m as given in Table 12.1; thus, the tip speed is about 76 m/sec, which is about what is observed from Figure 12.26(b). Through the comparison, we found that SNR (Signal-to-Noise Ratio) in Figure 12.26(b) seems much higher than (a); this is because the noise floor seen in the former case is not really due to noise. If we zoom in to the same radial velocity range as in (a), we can see that many spectrum

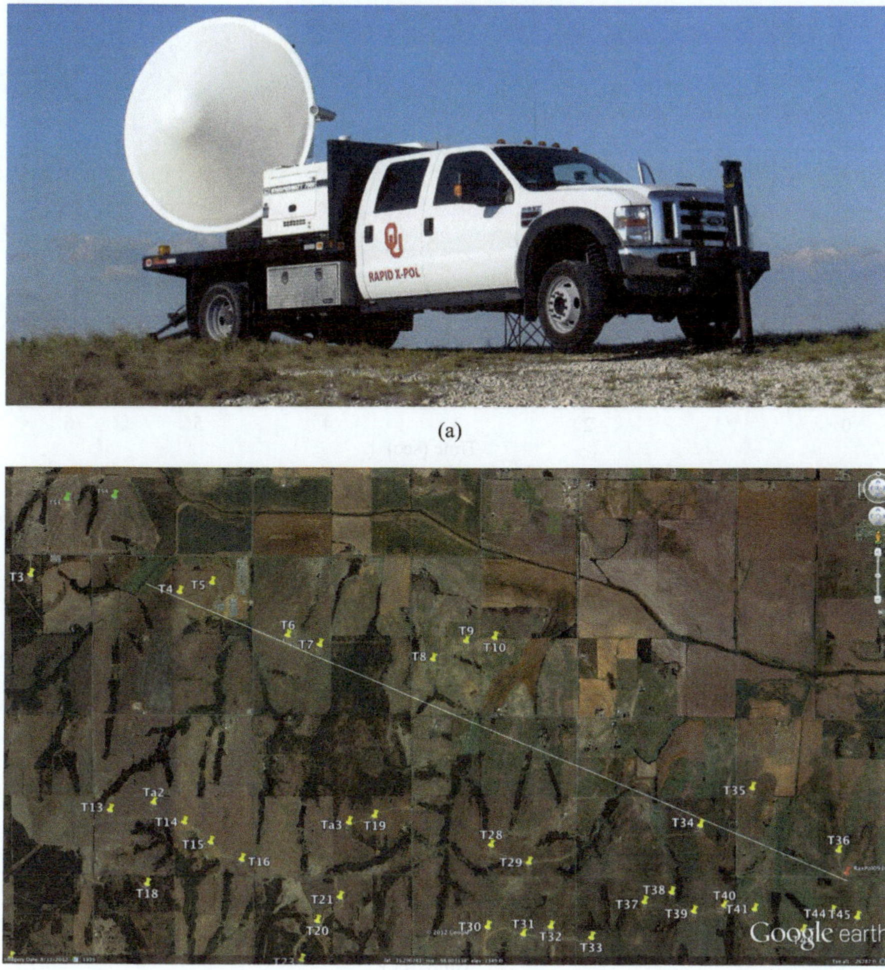

(a)

(b)

Figure 12.25 A mobile radar was deployed at Minco, OK for wind turbine Doppler signature study: (a) The RaXPol radar truck (picture from: http://arrc.ou.edu/radars.html). (b) RaXPol and wind turbine locations on Google map

components corresponding to higher velocities fold in this narrow range and coherently add on top of each other, resulting in a cluttered background alike noise floor.

The non-aliasing full-size wind turbine spectrogram possesses some interesting features. The two sides of the spectrogram are extremely different from each other. The negative side, representing the blade coming toward the radar, where the flash is supposed to be caused by the blade sweeping down the nacelle and perpendicular

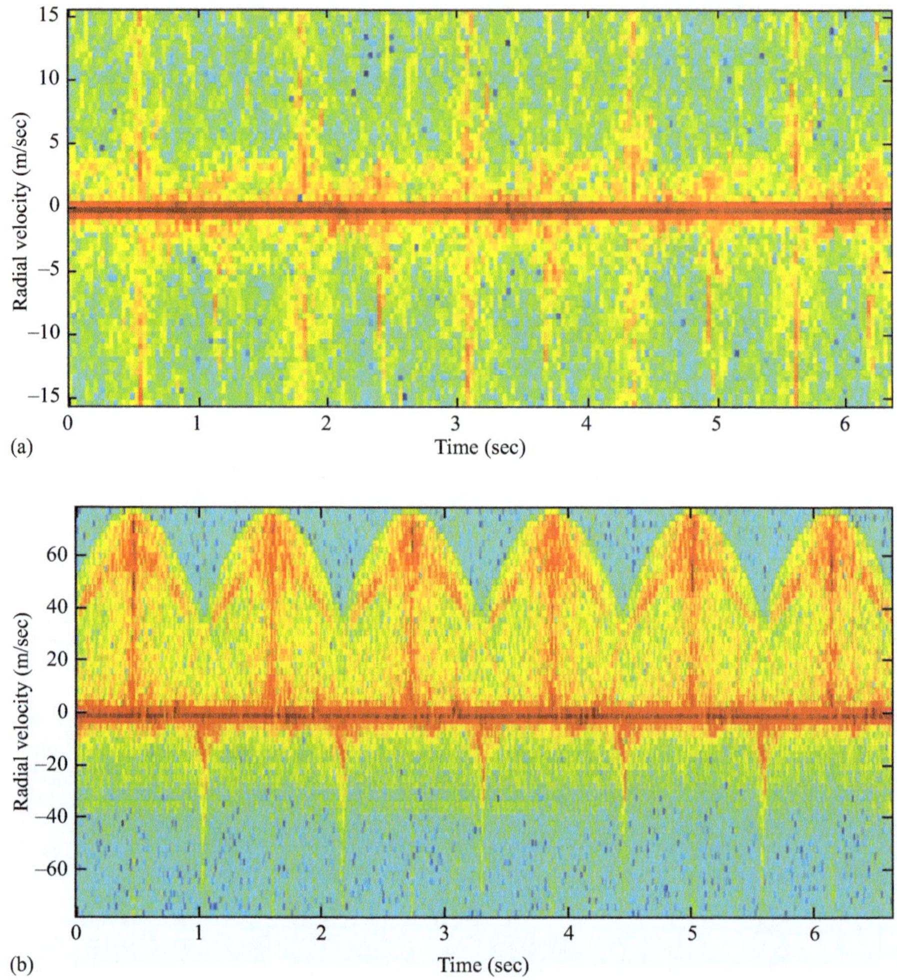

Figure 12.26 Spectrogram of GE 1.6 MW wind turbine from field measurements: (a) PRF = 2 kHz; (b) PRF = 10 kHz

to the RLOS, has significantly lower return power than the positive side. At $\phi = 270°$, the negative flashes correspond to the leading edge sweeping downward and coming toward radar, while the positive flashes are the trailing edge sweeping upward and going away from the radar. The two types of flashes have comparable return power levels at low radial velocities as in Figure 12.26(b), but as the radial velocity increases, the return power from the leading edge almost vanishes, while the flashes corresponding to the trailing edge remain high. This observation challenges the assumption that flashes from leading edge is stronger than those from trailing edge because the leading edge is thicker. One possible reason is that as it goes to the tip of the blade, the edge diffraction caused by the thin trailing edge remains the same, but the specular reflection area of the thick leading edge is becoming smaller, which significantly reduces its power returns. Besides the trajectory corresponding to the blade tip, another sinusoid trajectory is found corresponding to about ¾ of the blade length, which may be caused by some structural discontinuity, but further measurements and studies need to be performed to confirm. There are many other interesting features about the spectrogram, like the complex modulation within each sector on the positive side, etc., and more analysis will be reported in subsequent publications.

12.7 Mitigation of Wind Turbine Clutter

The most straightforward, and perhaps also the most efficient way to mitigate this clutter effect, is to study the LOS avoidance in the planning phase and carefully site the wind farms far away from the radar or take advantage of terrain blocking [30]. However, siting is a very comprehensive procedure and it is often unavoidable to have wind farms within the radar LOS.

For wind turbines sited within radar LOS, there are two general approaches to mitigate the clutter effect: (1) reducing the RCS of wind turbines [31, 32] and (2) reducing the impacts of clutter by improving radar system (antenna, processing, etc.). The former is constrained by both the aerodynamic requirements of the wind turbine and cost factors. Although Vestas claimed the "stealth" wind turbine prototype has been first tested and the RCS can be significantly reduced, no such wind turbines is planned for practical use in the near future. Furthermore, as the size of wind turbines becomes larger, it will be more and more difficult to manage their RCS.

For primary air traffic control and surveillance radars, airplanes are, in most cases, at higher altitude than the wind turbine top. This spatial separation has been used to mitigate the WTC by concurrent beam processing. Enhanced CFAR (Constant False Alarm Rate) detection method [18] has also been proposed for future system upgrade. The gap-filling concept [33] has been investigated to make up radar coverage loss due to wind turbines, but the implementation relies on terrain proofing and it also has significant cost constraints.

Unlike ASR, the targets of interest for weather radar are various forms of precipitation, such as rain, snow, hail, etc., which are often spatially inseparable from wind turbines for low elevation scans. The difficulty in separating wind turbines and weather radar returns in both the spatial and Doppler domains has prompted ideas of combining information from both domains. Range-Doppler processing [34] exploits the difference between wind turbine and weather radar return signals in the range-Doppler domain, alleviates Doppler spectral components identified as WTI, and reconstructs weather signal based on the spatial-continuity assumption. The Adaptive Spectrum Processing (ASP) technique [35] makes use of the clustering nature of weather signal and adaptively constructs band pass filters for mitigation of WTI.

Due to its complicated nonstationary micro-Doppler signature, conventional clutter filter techniques have failed in suppressing the WTCs. Many methods have been proposed and are effective in certain applications, but only some are introduced briefly in this section. The readers are encouraged to search for more recent publications. The authors believe that as we understand better the micro-Doppler radar signatures of the wind turbine, more effective solutions will emerge.

12.8 Summary

The radar micro-Doppler signatures of wind turbines have been discussed in this chapter. Due to the extremely large dimension and blade rotation, an increasingly large number of wind turbines have become potential interference to radar operations. The interactions between wind turbines and radar especially weather radars are analyzed, and some examples have been given to showcase the clutter characteristics. As a result of the blade rotation, wind turbines have unique micro-Doppler radar signatures that are best perceived in the time-frequency analysis. The Doppler flashes extend through the entire spectrum, making it difficult to mitigate in the Doppler domain. The scaled measurements of the wind turbine model have been implemented to better understand the complex micro-Doppler radar signatures and its dependency on wind turbine motions in a controlled laboratory environment. Both frequency and time domain measurement approaches have been used. Some interesting EM phenomenon and statistical variables are studied from the scaled measurement results. However, the scaling factor limits our ability to interpret the spectrum by compromising the relative Doppler spectrum resolution. Hence, a mobile radar has been deployed to better perceive the details of wind turbine micro-Doppler signatures.

The wind turbine is a new form of radar interference that features extremely complicated micro-Doppler signatures. The radar return variations of wind turbines with respect to their position, shape, material, and radar parameters, etc. are still being studied and there are many aspects that are not covered in this chapter. More extensive studies from physical models, radar signature analysis to mitigation approaches are needed.

References

[1] S. Gsanger and J.-D. Pitteloud, "World Wind Energy Report 2011," World Wind Energy Association, Bonn, 2012.

[2] AWEA, "AWEA U.S. Wind Industry Third Quarter 2012 Market Report," 2012.

[3] U.S. Department of Energy, "20% Wind Energy by 2030, Increasing Wind Energy's Contribution to U.S. Electricity Supply," 2008.

[4] Office of the Director of Defense Research and Engineering, "Report to the congressional defense committees: the effect of windmill farms on military readiness," Department of Defense, USA, 2006.

[5] T. B. A. Senior, D. L. Sengupta, and J. E. Ferris, "TV and FM Interference by Windmills," University of Michigan Radiation Laboratory Report 014438-1-F, Ann Arbor, MI, 1977.

[6] G. J. Poupart, "Wind farms impact on radar aviation interests," QinetiQ Ltd, Tech. Rep. W/14/00614/00/REP, 2003.

[7] A. Buterbaugh, B. M. Kent, K. C. Hill, G. Zelinski, R. Hawley, L. Cravens, *et al.*, "Dynamic Radar Cross Section and Radar Doppler Measurements of Commercial General Electric Windmill Power Turbines Part 2: Predicted and Measured Doppler Signatures," presented at the AMTA symposium, St Louis, MO, 2007.

[8] B. M. Kent, K. C. Hill, A. Buterbaugh, G. Zelinski, R. Hawley, L. Cravens, *et al.*, "Dynamic Radar Cross Section and Radar Doppler Measurements of Commercial General Electric Windmill Power Turbines Part 1: Predicted and Measured Radar Signatures." *Antennas and Propagation Magazine, IEEE*, pp. 211–219, April 2008

[9] R. J. Vogt, T. D. Crum, J. T. Snow, R. Palmer, B. Isom, D. W. Burgess, *et al.*, "An Update on Policy Considerations of Wind Farm Impacts on WSR-88D Operations," presented at the 24th International Conference on Interactive Information and Processing Systems (IIPS) for Meteorology, Oceanography, and Hydrology, New Orleans, LA, 2008.

[10] R. J. Vogt, J. R. Reed, T. D. Crum, J. T. Snow, R. Palmer, B. Isom, *et al.*, "Impacts of Wind Farms on WSR-88D Operations and Policy Considerations," presented at the 23rd International Conference on Interactive Information and Processing Systems (IIPS) for Meteorology, Oceanography, and Hydrology, San Antonio, TX, 2007.

[11] B. M. Isom, R. D. Palmer, G. S. Secrest, R. D. Rhoton, D. Saxion, T. L. Allmon, *et al.*, "Detailed Observations of Wind Turbine Clutter with Scanning Weather Radars," *Journal of Atmospheric and Oceanic Technology*, vol. 26, pp. 894–910, 2009.

[12] R. J. Vogt, T. D. Crum, M. J. B. Sandifer, E. J. Ciardi, and R. Guenther, "A Way Forward Wind Farm – Weather Radar Coexistance," presented at the Wind Power Conference, Chicago, IL, 2009.

[13] R. J. Vogt, T. D. Crum, W. Greenwood, E. J. Ciardi, and R. Guenther, "Recent Efforts to Improve Estimates of and Mitigate Potential Wind Turbine Clutter Impacts on WSR-88Ds," presented at the 27th Conference Interactive Information Processing Systems (IIPS), Seattle, WA, 2011.

[14] R. J. Vogt, T. D. Crum, and E. J. Ciardi, "How to Successfully Work with NOAA on Wind Turbine – NEXRAD Radar Interference Issues, an Update," presented at the Wind Power Conference, Atlanta, GA, 2012.

[15] R. J. Vogt, T. Crum, J. Sandifer, R. Steadham, T. L. Allmon, G. Secrest, *et al.*, "Continued Progress in Assessing and Mitigating Wind Farm Impacts on WSR-88Ds," presented at the 25th Conference on International Interactive Information and Processing Systems (IIPS) for Meteorology, Oceanography and Hydrology, Phoenix, AZ, 2009.

[16] D. W. Burgess, T. D. Crum, and R. J. Vogt, "Impacts of Wind Farms on WSR-88D Radars," presented at the 24th International Conference on Interactive Information Processing Systems (IIPS) for Meteorology, Oceanography and Hydrology, New Orleans, LA, 2008.

[17] V. C. Chen, *The Micro-Doppler Effect in Radar*. Norwood, MA: Artech House, 2011.

[18] P. Drake, "Overview of Raytheon Wind Farm Mitigation Techniques and Test Results," presented at the CNS/ATM, Orlando, FL, 2011.

[19] Office of the Federal Coordinator for Meteorological Service and Supporting Research, "Dopper Radar Meteorological Observations – Part C WSR-88D Products and Algorithms," Washington, DC, 2006.

[20] R. J. Doviak and D. S. Zrnić, *Doppler Radar and Weather Observations*, 2nd ed. Mineola, New York: Dover Publications, INC., 2006.

[21] A. Naqvi, S. T. Yang, and H. Ling, "Investigation of Doppler Features from Wind Turbine Scattering," *IEEE Antennas and Wireless Propagation Letters*, vol. 9, pp. 485–488, 2010.

[22] Y. Zhang, A. Huston, R. D. Palmer, R. Albertson, F. Kong, and S. Wang, "Using Scaled Models for Wind Turbine EM Scattering Characterization: Techniques and Experiments," *Instrumentation and Measurement, IEEE Transactions*, pp. 1–9, 2010.

[23] Y. Bai, F. Kong, Y. Zhang, and R. Palmer, "Development of Simple Radar Signature Models of Wind Turbine for Knowledge-Aided Wind Turbine Radar Interference Analysis," presented at the 3rd Conference in Weather, Climate and the New Energy Economy, New Orleans, LA, 2012.

[24] F. Kong, Y. Zhang, and R. Palmer, "Impacts of Wind Turbine Siting on Radar Clutter Mitigation," in *IET International Radar Conference*, Xian, China, 2013.

[25] F. Kong, Y. Zhang, and R. Palmer, "Wind Turbine Radar Interference Studies by Polarimetric Measurements of a Scaled Model," *IEEE Transactions on Aerospace Electronic Systems*, Vol. 49, no. 3, pp. 1589-1600, July 2013.

[26] F. Kong, Y. Zhang, R. Palmer, and Y. Bai, "Wind Turbine Radar Signature Characterization by Laboratory Measurements," in *Radar Conference (RADAR), 2011 IEEE*, pp. 162–166, 2011.

[27] D. C. Jenn, *Radar and Laser Cross Section Engineering*, 2nd ed. AIAA, Reston, VA, 2005.

[28] F. Kong, Y. Zhang, and R. D. Palmer, "Wind Turbine Radar Interference Studies by Polarimetric Measurements of a Scaled Model," *Aerospace and Electronic Systems, IEEE Transactions*, vol. 49, pp. 1589–1600, 2013.

[29] A. L. Pazmany and H. B. Bluestein, "A Mobile, Rapid–Scanning, X-Band, Polarimetric (RaXPol) Doppler Radar System," presented at the 35th Conference on Radar Meteorology, Pittsburg, PA, 2011.

[30] R. Palmer, F. Kong, and Y. Zhang, "The Near-Field Radar Cross Section Estimation of Wind Turbine: Mitigating Radar Clutter Caused by Wind Turbines by Proper Siting," presented at the Fourth Conference on Weather, Climate, and the New Energy Economy, Austin, TX, 2013.

[31] J. Pinto, J. C. G. Matthews, and C. Sarno, "Radar Signature Reduction of Wind Turbines Through the Application of Stealth Technology," in *3rd European Conference on Antennas and Propagation, 2009. EuCAP 2009. 3rd European Conference*, pp. 3886–3890, 2009.

[32] L. S. Rashid and A. K. Brown, "Radar Cross-Section Analysis of Wind Turbine Blades with Radar Absorbing Materials," in *Radar Conference (EuRAD), 2011 European*, pp. 97–100, 2011.

[33] E. Aarholt and C. A. Jackson, "Wind Farm Gapfiller Concept Solution," in *Radar Conference (EuRAD), 2010 European*, pp. 236–239, 2010.

[34] F. Nai, R. D. Palmer, and S. M. Torres, "Range-Doppler domain signal processing to mitigate wind turbine clutter," in *Radar Conference (RADAR), 2011 IEEE*, Kansas City, MI, pp. 841–845, 2011.

[35] F. Kong, Y. Zhang, and R. Palmer, "Wind Turbine Clutter Mitigation for Weather Radar by Adaptive Spectrum Processing," in *Radar Conference (RADAR), 2012 IEEE*, Atlanta, GA, 2012.

Index